Computational Methods
for
Linear Control Systems

Prentice Hall International Series

in Systems and Control Engineering

M.J. Grimble, Series Editor

Computational Methods for
Linear Control Systems

**P.Hr. Petkov, N.D. Christov, and
M.M. Konstantinov**

Prentice Hall

New York · London · Toronto · Sydney · Tokyo · Singapore

First published 1991 by
Prentice Hall International (UK) Ltd
66 Wood Lane End, Hemel Hempstead
Hertfordshire HP2 4RG
A division of
Simon & Schuster International Group

Typeset in 10/12 pt Times by
MCS, Salisbury

Printed and bound in the United States

Library of Congress Cataloging-in-Publication Data

Petkov, P. Hr. (Petko Hr.), 1948–
    Computational methods for linear control systems/P. Hr. Petkov,
N.D. Christov, and M.M. Konstantinov.
        p. cm.
    Includes bibliographical references and index.
    ISBN 0-13-161803-2
    1. Automatic control. 2. Linear systems. I. Christov, N. D.
(Nikolai D.), 1948– . II. Konstantinov, M. M. (Mihail M.), 1948–
. III. Title.
    TJ213.P4278 1991
    629.8--dc20   91-6903
                                                                CIP

British Library Cataloguing in Publication Data

Petkov, P. Hr. (Petko Hr) *1948 Feb. 18–*
    Computational methods for linear control systems.
    1. Control systems. Design. Use of computers
    I. Title II. Christov, N. D. (Nikolai D.) *1948 Dec. 16–*
    III. Konstantinov, M. M. (Mihail M.) *1948 Mar. 5–*
    629.8320285

    ISBN 0-13-161803-2

---

1 2 3 4 5   95 94 93 92 91

To our parents

Contents

List of Algorithms

Preface

Although the computer-aided design of control systems has been an area of increasing activity in recent years, the computational aspects of control systems analysis and design have largely been avoided in textbooks on modern control theory. The description of methods suitable for computer implementation is spread over a great number of journal articles, symposium papers and technical reports, which makes it difficult to choose an appropriate method for solving a certain design problem.

The purpose of this book is to give a systematic presentation of methods oriented towards the computer analysis and design of linear control systems. From the large variety of existing techniques for linear systems, we consider the methods intended to solve some of the basic problems arising in the state–space analysis and design of continuous and discrete time-invariant multivariable systems. The final goal is to present numerical algorithms which solve reliably the analysis and design problems using finite precision computer arithmetic. The theoretical proofs are usually omitted, and we are concerned mainly with the numerical behaviour of the different methods in the presence of rounding errors. The description of best current methods finishes with algorithms written in an informal language, which may be used to develop working programs. Several algorithms presented in the book are implemented as Fortran programs, contained in the package LISPACK and the interactive system SYSLAB, which may be obtained from the authors on request.

The volume is intended for use as a textbook for graduate and postgraduate students in control engineering, but it is hoped that it will also serve as a reference book for control engineers and applied mathematicians. The reader is assumed to have an acquaintance with the fundamentals of linear control theory and matrix analysis, as well as some experience in the computer solution of mathematical problems.

We want to express our gratitude to D. Boley, R. Byers, J. Demmel, S. Hammarling, B. Kågström, A. Laub, G. Miminis, C. Moler, S. Nash, C. Paige, G. Stewart, P. Van Dooren, C. Van Loan and R. Ward whose works greatly influenced the contents of this book. We are very obliged to Leslie Fletcher who made several valuable suggestions. We are also grateful to the staff of Prentice Hall for their efforts in improving the quality of the manuscript.

Introduction

The use of computers in the analysis and design of linear control systems requires implementation of computational methods which work reliably in the presence of rounding errors. Such methods should be developed on the basis of the results of linear control theory and numerical analysis. It is now widely recognized that many analysis and design methods, proposed earlier in linear system theory, which do not respect the properties of the computer arithmetic, break down even for low-order problems.

Since the state–space analysis and design of linear control systems is related to the use of matrices, the development of computational methods for linear systems is strongly influenced by the progress in numerical linear algebra. The availability of high-quality algorithms and software for solving linear algebra problems has stimulated the appearance of sophisticated methods and programs for the computer-aided design of linear systems. However, it is the authors' opinion that the development of reliable and efficient computational methods for analysis and design of linear control systems is still in its early stages.

The implementation of a numerical method on a computer may produce an inaccurate result for several reasons. First of all, the problem solved may be ill conditioned, so that small perturbations in the data may lead to large changes in the solution. This phenomenon requires investigation of the sensitivity of the problem, which is done by perturbation analysis. Another source of errors may be the numerical instability of the method used. The detection of the numerical instability requires study of the effect of rounding errors on the result, which is done by the rounding error analysis of the algorithm. This is usually the most difficult part of the numerical analysis of a given method. It should be stressed, however, that the calculated error bound on the result is usually of little significance and the purpose of the rounding error analysis is actually to reveal the causes for introduction of large errors in the result.

The preceding considerations predetermine the structure of the presentation in this book. Before the description of particular computational methods for solving a given analysis or design problem, we usually investigate the sensitivity of this problem to perturbations in the data. The presentation of a method is followed by a discussion of its numerical properties and efficiency. To illustrate the numerical behaviour of the method we give an example and compare the computed result with the exact one. The examples are solved by the package LISPACK and the interactive system SYSLAB, developed by the authors, on VAX 11/750† using double precision. Lower precision is simulated by shortening the machine word, as done in the 'matrix laboratory' MATLAB, developed by Professor C.B. Moler.

The following note concerns the software for analysis and design of linear control systems. Although the algorithms given in the book may be used to write efficient programs, it should be borne in mind that the development of quality computer programs is not a trivial task. It requires attention to many details, some of which are outwith the scope of this text. In the Appendix we refer to several sources which may be used to gather information about the available software for the computer-aided design of linear control systems.

Here is a brief description of the contents of the book.

In Chapter 1 we present the fundamentals of the numerical computations involving matrices. After a short exposition of the basic facts concerning vectors and matrices, we consider the rounding errors arising in floating-point arithmetic and their effect on the basic matrix operations. Next we discuss the important subject of conditioning and numerical stability. The rest of the chapter is devoted to the main problems arising in matrix computations – the solution of linear equations and the eigenvalue computations. In this connection we consider in detail the orthogonal transformations, which ensure lowest error bounds, and their implementation in the computation of various useful matrix decompositions, such as QR, Schur and singular value decomposition.

In Chapter 2 we have a concise review of the basic results in the state–space analysis and design of continuous and discrete-time linear control systems. Along with the presentation of the main analysis and design problems, we give some rudimentary methods for their solution. (In the succeeding chapters we show that most of these methods are not suitable for computer implementation.) This chapter may be skipped by the reader who has the corresponding background in linear system theory.

Chapter 3 is devoted to one of the important problems in the analysis of control systems – the solution of state equations. For continuous-time linear systems this problem reduces to the computation

† VAX is trademark of Digital Equipment Corporation.

of the exponential of the state matrix. We present several methods for evaluating the matrix exponential and compare their numerical properties. The chapter ends with the exposition of methods for discretization of continuous-time systems.

Chapter 4 begins with the discussion of different methods for stability analysis of continuous and discrete-time systems. A closely related problem is the numerical solution of Lyapunov and Sylvester matrix equations, which also arise in several other analysis and design problems. Next we consider the controllability/observability analysis of linear control systems and the reduction into canonical forms by numerically stable methods. This chapter ends with a discussion of methods for system balancing and model reduction.

Computational methods for state–space design are presented in Chapters 5 and 6.

Chapter 5 is devoted to the solution of the pole assignment problem. At the beginning of this chapter we discuss the numerical difficulties related to the pole assignment problem. We consider next several computational methods, most of them relying on orthogonal transformations. We give also some algorithms for stabilization of linear control systems, which do not involve explicit assignment of the closed-loop poles. At the end of this chapter we give methods for design of state observers.

Chapter 6 is entirely devoted to the computational solution of continuous and discrete matrix Riccati equations arising in the quadratic optimization of linear systems. After studying the sensitivity of these equations we present, in detail, several methods for their solution.

In the last chapter we describe computations related to the geometric theory of linear control systems. After considering the computation of basic subspaces, distances and angles between subspaces we present methods for determining controllable/unobservable subspaces, controlled invariant and controllability subspaces.

1

Numerical Matrix Computations

In this chapter we describe in brief numerical methods for matrix computations, which are used in the subsequent presentation.

It is assumed that the reader is familiar with the basic matrix operations but has no knowledge of the effect of finite precision arithmetic on the computations.

We shall be concerned mainly with the solution of linear algebraic equations and eigenvalue problems. Along with this we shall present some matrix decompositions such as QR decomposition and singular value decomposition, which play an important role in all computations involving matrices. One of our aims is to show that the systematic use of orthogonal transformations in solving matrix problems not only makes it possible to keep rounding errors small, but also allows development of methods which are suitable for computer implementation.

1.1 VECTORS, MATRICES AND LINEAR SPACES

In this section we give notations and definitions of vectors, matrices and linear spaces used throughout the book.

The vectors are denoted by lower-case letters and the same letter with a subscript refers to the corresponding vector component, e.g.

$$x = [x_i] = \begin{bmatrix} x_1 \\ x_2 \\ \vdots \\ x_n \end{bmatrix} \tag{1.1}$$

is a (column) n-vector.

The matrices are denoted by upper-case letters. The same lower-case letter with a double subscript indicates the corresponding element, e.g.

$$A = [a_{ij}] = \begin{bmatrix} a_{11} & a_{12} & \cdots & a_{1m} \\ a_{21} & a_{22} & \cdots & a_{2m} \\ \vdots & \vdots & & \vdots \\ a_{n1} & a_{n2} & \cdots & a_{nm} \end{bmatrix} \tag{1.2}$$

is an $n \times m$ matrix. An $n \times m$ matrix A is also represented as

$$A = [a_1, a_2, \ldots, a_m] \tag{1.3}$$

where the n-vector a_j is the jth column of A.

All vectors and matrices are defined over the set $\mathbb{R}$ of real numbers (i.e. their elements are real numbers) unless otherwise stated.

If all components of a vector are zero then it is called a *zero vector* and is denoted by 0. Similarly, a matrix having all its elements being equal to zero, is said to be a *zero matrix* and is denoted also by 0.

The $n \times m$ matrices with $n \neq m$ are referred to as *rectangular*. The $n \times n$ matrices are said to be *square* and n is the *order* of such matrices.

The matrix

$$A^T = [a_{ji}] = \begin{bmatrix} a_{11} & a_{21} & \cdots & a_{n1} \\ a_{12} & a_{22} & \cdots & a_{n2} \\ \vdots & \vdots & & \vdots \\ a_{1m} & a_{2m} & \cdots & a_{nm} \end{bmatrix}$$

is the *transpose* of the matrix (1.2). If $A = [a_{ij}]$ is a complex matrix (i.e. matrix with complex elements) then the *conjugate transpose* of A is denoted by $A^H = [\bar{a}_{ji}]$, where $\bar{a}$ is the complex conjugate of a. The transpositions are defined for vectors as well, e.g. the row n-vector

$$x^T = [x_1, x_2, \ldots, x_n]$$

is the transpose of the vector (1.1).

The $n \times n$ *identity* (or *unit*) *matrix* is denoted by I_n,

$$I_n = \begin{bmatrix} 1 & & & \mathbf{0} \\ & 1 & & \\ & & \ddots & \\ \mathbf{0} & & & 1 \end{bmatrix}$$

If the order of I_n is clear from the context, the subscript n will be omitted.

For a given square matrix A the matrix B which satisfies $AB = I$ is said to be the *inverse* of A and is denoted by A^{-1}. Whenever A^{-1} exists, the matrix A is referred to as *nonsingular*, otherwise A is said to be *singular*. The notation A^{-T} is used for $(A^{-1})^T = (A^T)^{-1}$.

The *determinant* of a square matrix A is denoted by $\det(A)$.

A square matrix with a special pattern of zero entries is

diagonal	if $a_{ij} = 0$ for $i \neq j$
upper bidiagonal	if $a_{ij} = 0$ for $i > j$ or $j > i+1$
tridiagonal	if $a_{ij} = 0$ for $|i - j| > 1$
upper triangular	if $a_{ij} = 0$ for $i > j$
strictly upper triangular	if $a_{ij} = 0$ for $i \geq j$
upper Hessenberg	if $a_{ij} = 0$ for $i > j+1$

Similar definitions hold replacing 'upper' by 'lower' and inversing the corresponding inequalities.

A diagonal matrix A is denoted by

$$A = \text{diag}(a_{11}, a_{22}, \ldots, a_{nn})$$

If the diagonal elements of a triangular matrix are equal to 1 it is said to be *unit triangular*.

A square matrix is

symmetric	if $A^T = A$
positive definite	if $x^T A x > 0$ for $x \neq 0$
positive semidefinite (or *non-negative definite*)	if $x^T A x \geq 0$ for $x \neq 0$
nilpotent	if $A^k = 0$ for some integer $k \geq 0$

If A is positive semidefinite then there exists a unique semidefinite B such that $B^2 = A$. The matrix B is referred to as the *square root* of A. If A is positive definite, so is B.

A complex matrix A is *Hermitian* if $A^H = A$.

If A_{ij} are $n_i \times m_j$ matrices then

$$A = [A_{ij}] = \begin{bmatrix} A_{11} & A_{12} & \ldots & A_{1p} \\ A_{21} & A_{22} & \ldots & A_{2p} \\ \vdots & \vdots & & \vdots \\ A_{q1} & A_{q2} & \ldots & A_{qp} \end{bmatrix}$$

is a $q \times p$ *block matrix* with A_{ij} being its (i, j)th block. If $A_{ij} = 0$ for $i \neq j$ then A is a *block diagonal*, and if $A_{ij} = 0$ for $i > j$ then A is a *block upper triangular matrix*.

The *Kronecker product* of the $n \times m$ matrix A and the matrix B is defined as

$$A \otimes B = \begin{bmatrix} a_{11}B & a_{12}B & \ldots & a_{1m}B \\ a_{21}B & a_{22}B & \ldots & a_{2m}B \\ \vdots & \vdots & & \vdots \\ a_{n1}B & a_{n2}B & \ldots & a_{nm}B \end{bmatrix}$$

If S and T are nonsingular matrices of orders n and m then the transformation of the $n \times m$ matrix A to SAT is referred to as *equivalent transformation*. If A, T are square matrices of the same order, with T being nonsingular, then $T^{-1}AT$ is a *similarity transformation* of the matrix A.

If x, y are n-vectors and α is a scalar then $z = x + y$ is an n-vector with components $z_i = x_i + y_i$ and αx is an n-vector with components αx_i. A set of vectors with these properties is said to be a *linear space*. The linear space of real column n-vectors is further denoted by $\mathbb{R}^n$.

A system of $k + 1$ vectors $x, y, ..., z, u \in \mathbb{R}^n$ is said to be *linearly dependent* if either $u = 0$ or there exist k numbers $\alpha, \beta, ..., \gamma$ such that

$$u = \alpha x + \beta y + \cdots + \gamma z \tag{1.4}$$

otherwise the vectors $x, y, ..., z, u$ are *linearly independent*.

Any system of n linearly independent vectors in $\mathbb{R}^n$ forms a *basis* for $\mathbb{R}^n$. The number n is the *dimension* of $\mathbb{R}^n$. The n columns of I_n constitute the *canonical basis* of $\mathbb{R}^n$.

Let $x, y, ..., z$ be a system of k linearly independent vectors in $\mathbb{R}^n$. The set of all n-vectors which may be represented in the form (1.4) is a *k-dimensional subspace* of $\mathbb{R}^n$ which is *spanned* by $x, y, ..., z$.

A subspace $X \subset \mathbb{R}^n$ is *invariant* for the $n \times n$ matrix A (or X is *A-invariant*) if $x \in X$ implies $Ax \in X$.

Two vectors $x, y \in \mathbb{R}^n$ are said to be *orthogonal* if $x^T y = 0$. Two subspaces $X, Y \subset \mathbb{R}^n$ are orthogonal if $x^T y = 0$ for all $x \in X$, $y \in Y$. A system of nonzero pairwise orthogonal vectors is linearly independent.

Let A be an $n \times m$ matrix represented in the form (1.3). The number of linear independent columns a_j of A is said to be the *rank* of A and is denoted by rank(A). It is also equal to the number of linearly independent rows of A. The matrix A is of *full column (row) rank* if rank(A) = m (rank(A) = n). The matrix A is of *full rank* if rank(A) = min$\{m, n\}$.

The set of all n-vectors Ax, $x \in \mathbb{R}^m$, is the *range* (or *image*) of the $n \times m$ matrix A and is denoted by Rg(A). It is a subspace of $\mathbb{R}^n$ of dimension rank(A) and is spanned by the linearly independent columns of A.

The set of all vectors $x \in \mathbb{R}^m$ satisfying the condition $Ax = 0$ is said to be the *kernel* (or *null space*) of the $n \times m$ matrix A. It is a subspace of $\mathbb{R}^m$ and is denoted by Ker(A). Its dimension is equal to m-rank(A).

If A is an $n \times n$ matrix then both Rg(A) and Ker(A) are invariant subspaces of A.

An $n \times n$ real matrix $A = [a_1, a_2, ..., a_n]$ is *orthogonal* if $A^T A = I_n$. The columns of the orthogonal matrix A are pairwise orthogonal $(a_i^T a_j = 0, i \neq j)$ and have the property $a_i^T a_i = 1$.

An $n \times n$ complex matrix A is *unitary* if $A^H A = I_n$. The columns a_i of a unitary matrix satisfy $a_i^H a_j = 0$ $(i \neq j)$ and $a_i^H a_i = 1$.

EXERCISES

1.1 Let A, B and C be upper triangular matrices with A and B being of the same order and C being nonsingular. Show that AB and C^{-1} are also upper triangular. What about the case when A, B and C are upper Hessenberg matrices?

1.2 Prove that any system of $n + 1$ vectors in $\mathbb{R}^n$ is linearly dependent.

1.3 Let $x, y, ..., z$ be a system of n orthogonal vectors in $\mathbb{R}^n$ such that $x^T x = y^T y = \cdots = z^T z = 1$, and let $u, v, ..., w$ be any system of n-vectors satisfying

$$(u^T u)^{1/2} + (v^T v)^{1/2} + \cdots + (w^T w)^{1/2} < 1$$

Prove that the system $x + u, y + v, ..., z + w$ is linearly independent.

1.4 Show that $\text{rank}(AB) \leq \min\{\text{rank}(A), \text{rank}(B)\}$.

1.5 Show that a 2×2 orthogonal matrix has one of the forms

$$\begin{bmatrix} c & s \\ -s & c \end{bmatrix} \quad \text{or} \quad \begin{bmatrix} c & s \\ s & -c \end{bmatrix}, \text{ where } c = \cos \alpha, \ s = \sin \alpha \text{ for some } \alpha.$$

1.6 Find the general form of a 3×3 orthogonal matrix as a function of three parameters.

1.2 VECTOR AND MATRIX NORMS

Working with vectors or matrices it is convenient to have single numbers which somehow characterize their size. For this reason certain functions of vector or matrix elements are used which are called *norms*.

The norm of a real n-vector x is denoted by $\| x \|$, and all vector norms satisfy the relations

$\| x \| \geq 0$ the equality taking place if and only if $x = 0$
$\| cx \| = | c | \ \| x \|$ for an arbitrary scalar c
$\| x + y \| \leq \| x \| + \| y \|$

The following vector norms are commonly used:

$\| x \|_1 = | x_1 | + | x_2 | + \cdots + | x_n |$
$\| x \|_2 = \sqrt{(x_1^2 + x_2^2 + \cdots + x_n^2)} = \sqrt{(x^T x)}$
$\| x \|_\infty = \max\{| x_i | : i = 1, 2, ..., n\}$

The norm $\|x\|_2$ is the Euclidean length of the vector x. This norm is invariant under orthogonal transformations, for if Q is an orthogonal matrix $(Q^TQ = I_n)$ then

$$\|Qx\|_2^2 = x^TQ^TQx = x^Tx = \|x\|_2^2$$

Example 1.1

If $x = [-6, 4, 2, -5]^T$ then

$$\|x\|_1 = 17 \qquad \|x\|_2 = 9 \qquad \|x\|_\infty = 6$$

Let $\bar{x}$ be an approximation to x. Then the *absolute error* in $\bar{x}$ is defined as

$$\Delta = \|\bar{x} - x\|$$

where $\|.\|$ is some of the vector norms, and the *relative error* in $\bar{x}$ is

$$\delta = \frac{\|\bar{x} - x\|}{\|x\|} \qquad x \neq 0$$

It may be shown that if the components of $\bar{x}$ and x coincide up to t decimal digits then

$$\frac{\|\bar{x} - x\|}{\|x\|} \approx 10^{-t}$$

Similarly, the norm of a real $m \times n$ matrix A is denoted $\|A\|$ and all matrix norms satisfy the relations

$$\|A\| \geq 0 \qquad \text{with equality taking place if and only if } A = 0$$
$$\|cA\| = |c| \, \|A\| \qquad \text{for an arbitrary scalar } c$$
$$\|A + B\| \leq \|A\| + \|B\|$$

The most frequently used matrix norms are the *Frobenius norm*

$$\|A\|_F = \sqrt{\sum_{i=1}^{m} \sum_{j=1}^{n} a_{ij}^2}$$

and the *p*-norms $(p = 1, 2, ..., \infty)$

$$\|A\|_p = \max\left\{\frac{\|Ax\|_p}{\|x\|_p} : x \neq 0\right\} \tag{1.5}$$

All these norms satisfy the inequality

$$\|AB\| \leq \|A\| \, \|B\|$$

It may be shown that

$$\| A \|_1 = \max\left\{ \sum_{i=1}^{m} | a_{ij}| : j = 1, 2, ..., n \right\}$$

$$\| A \|_\infty = \max\left\{ \sum_{j=1}^{n} | a_{ij}| : i = 1, 2, ..., m \right\}$$

The norm $\| A \|_2$ is sometimes called the *spectral norm*.

A matrix norm which satisfies (1.5) is said to be *subordinate* to the corresponding vector norm. In contrast, the Frobenius norm is not subordinate to any vector norm.

In accordance with (1.5) the vector norm and the subordinate matrix norm satisfy

$$\| Ax \| \leq \| A \| \, \| x \| \tag{1.6}$$

Any matrix and vector norms for which (1.6) holds are called *consistent*. The Frobenius norm is consistent with the vector 2-norm, i.e.

$$\| Ax \|_2 \leq \| A \|_F \, \| x \|_2$$

It is obvious that if an *n*-vector x is considered as an $n \times 1$ matrix, then $\| x \|_2 = \| x \|_F$.

Example 1.2

If

$$A = \begin{bmatrix} -5 & 1 & -3 \\ 6 & -2 & 4 \\ -2 & -9 & -3 \\ 7 & 3 & 9 \end{bmatrix}$$

then

$$\| A \|_F = 18 \qquad \| A \|_1 = 20 \qquad \| A \|_2 = 15.4095 \qquad \| A \|_\infty = 19$$

Both $\| A \|_F$ and $\| A \|_2$ have the important property that they are invariant under orthogonal transformations, i.e. if Q and Z are orthogonal matrices, then

$$\| QAZ \|_F = \| A \|_F \qquad \| QAZ \|_2 = \| A \|_2$$

The use of one or another matrix norm depends on the application. The norm $\| A \|_2$, for example, usually gives tighter bounds on the result of matrix computations than the norm $\| A \|_F$. On the other hand, $\| A \|_F$ is easier to compute in comparison with $\| A \|_2$. Also, the Frobenius norm of $|A|$ (here $|A|$ denotes the matrix with elements $|a_{ij}|$) is the same as that of A which is useful in some cases.

The vector and matrix norms should be used with some care. For instance, if A, B are matrices of equal dimensions and

$$|a_{ij}| < |b_{ij}| \tag{1.7}$$

then

$$\|A\|_F < \|B\|_F \tag{1.8}$$

(in general, this is not valid for the 2-norm). However, if (1.8) holds then (1.7) may not be fulfilled.

Example 1.3

Let

$$A = \begin{bmatrix} 3 & -1 \\ 1 & 5 \end{bmatrix} \qquad B = \begin{bmatrix} 2 & 4 \\ 8 & -4 \end{bmatrix}$$

Then

$$\|A\|_F = 6 \quad \text{and} \quad \|B\|_F = 10$$

but

$$|a_{11}| > |b_{11}| \quad \text{and} \quad |a_{22}| > |b_{22}|$$

EXERCISES

1.7 Show that for any n-vector x the following is satisfied:

$$\|x\|_2 \leq \|x\|_1 \leq \sqrt{n}\,\|x\|_2$$
$$\|x\|_\infty \leq \|x\|_2 \leq \sqrt{n}\,\|x\|_\infty$$
$$\|x\|_\infty \leq \|x\|_1 \leq n\,\|x\|_\infty$$

1.8 Prove the Cauchy–Schwartz inequality

$$|x^T y| \leq \|x\|_2 \|y\|_2$$

1.9 Show that for an $m \times n$ matrix A the following relations between the various norms exist:

$$\frac{1}{\sqrt{m}}\,\|A\|_1 \leq \|A\|_2 \leq \sqrt{n}\,\|A\|_1$$

$$\frac{1}{\sqrt{n}}\,\|A\|_\infty \leq \|A\|_2 \leq \sqrt{m}\,\|A\|_\infty$$

$$\| A \|_2 \le \| A \|_F \le \sqrt{n} \| A \|_2$$
$$\| A \|_2^2 \le \| A \|_1 \| A \|_\infty$$

1.10 Prove that

$$\| A^k \| \le \| A \|^k$$

1.3 MATRIX ALGORITHMS

The implementation of numerical methods for computer solution of matrix problems is done in the form of programs, written in some algorithmic language like Fortran or Pascal. The coding of an algorithm as a program, however, can lead to difficulties in reading the program and, in addition, the program may deviate from the original description of the algorithm for several reasons. That is why we shall later describe the algorithms for matrix computations in an informal language which permits easy reading of the algorithms as well as development of working programs. It should be pointed out, however, that the same algorithm may be realized with several computer programs which may differ in their details. These details may have a significant effect on the computational results.

The informal language subsequently used resembles the one proposed by Stewart (1973b). It will be introduced through examples.

Consider first the summation of two $m \times n$ matrices A and B,

$$C = A + B$$

which may be done by the following algorithm.

Algorithm 1.1 Summation of two matrices

For $i = 1, 2, ..., m$
  For $j = 1, 2, ..., n$
    $c_{ij} = a_{ij} + b_{ij}$

This algorithm requires mn additions.

When implemented as a Fortran program, Algorithm 1.1 has the drawback that the elements of A and B are not accessed in sequential mode since the inner loop goes across a row. This is due to a particularity of Fortran which stores the matrices by columns in the one-dimensional computer memory. Sequential access is preferable on operating systems with some kind of *paging* since it results in significant improvement in performance. That is why the Fortran program will be more efficient if the algorithm is *column oriented*, i.e. the matrix elements are accessed in columnwise instead of rowwise. Thus Algorithm 1.1 should be rearranged for efficiency in the following way.

Algorithm 1.2 Summation of two matrices with column access

For $j = 1, 2, ..., n$
  For $i = 1, 2, ..., m$
    $c_{ij} = a_{ij} + b_{ij}$

Another important aspect of the algorithm efficiency is the memory requirement since the matrices may occupy a large amount of storage. If one of the matrices A or B is not necessary for the next computation it may be overwritten by the sum to save storage. In the algorithm that follows the result overwrites the first matrix.

Algorithm 1.3 Economic summation of two matrices

For $j = 1, 2, ..., n$
  For $i = 1, 2, ..., m$
    $a_{ij} \leftarrow a_{ij} + b_{ij}$

Algorithm 1.3 requires only $2mn$ places for storing the input data and the result.
    The multiplication of two square matrices A and B may be done by the following row-oriented algorithm.

Algorithm 1.4 Multiplication of two matrices

For $i = 1, 2, ..., n$
  For $j = 1, 2, ..., n$

$$c_{ij} = \sum_{k=1}^{n} a_{ik}b_{kj}$$

This algorithm may be written also in a more detailed form to show the inner loop calculation.

For $i = 1, 2, ..., n$
  For $j = 1, 2, ..., n$
    Set $s = 0$
    For $k = 1, 2, ..., n$
      $s \leftarrow s + a_{ik}b_{kj}$
    $c_{ij} = s$

Algorithm 1.4 requires n^3 multiplications and the same number of additions. A situation like this is frequently encountered in matrix algorithms. That is why the estimation of the time required for a matrix computation is usually done by the number of basic floating-point operations – *flops*. A flop is the amount of work necessary to perform the operation

$$s \leftarrow s + a_{ik}b_{kj}$$

or in Fortran

```
S = S + A(I,K) * B(K,J)
```

Each flop involves one floating-point multiplication, one floating-point addition, a few subscript and index calculations and a few storage references. Thus the matrix multiplication requires n^3 flops.

In computing the number of flops the following identities are useful

$$\sum_{k=1}^{n} k = \tfrac{1}{2}n^2 + \tfrac{1}{2}n \qquad \sum_{k=1}^{n} k^2 = \tfrac{1}{3}n^3 + \tfrac{1}{2}n^2 + \tfrac{1}{6}n$$

The rearrangement of Algorithm 1.4 for column access to the elements of A and B is a little tricky and may be done as follows.

Algorithm 1.5 Multiplication of two matrices with column access

For $j = 1, 2, ..., n$
Set $c_{ij} = 0 \qquad i = 1, 2, ..., n$
  For $k = 1, 2, ..., n$
    $c_{ij} \leftarrow c_{ij} + a_{ik}b_{kj} \qquad i = 1, 2, ..., n$

To reduce the storage for matrix multiplication, it is possible to use the following representation. Let the matrix A be partitioned by rows, i.e.

$$A = \begin{bmatrix} a_1 \\ a_2 \\ \vdots \\ a_n \end{bmatrix}$$

Then

$$AB = \begin{bmatrix} a_1 B \\ a_2 B \\ \vdots \\ a_n B \end{bmatrix}$$

This means that after $a_i B$ is computed, the row a_i is no longer necessary and may be overwritten by $a_i B$. For computation of $a_i B$ an additional array with n elements is required. This leads to the Algorithm 1.6.

Algorithm 1.6 Economic multiplication of two matrices

For $i = 1, 2, ..., n$

$$s_j = \sum_{k=1}^{n} a_{ik} b_{kj} \qquad j = 1, 2, ..., n$$

$$a_{ij} \leftarrow s_j \qquad j = 1, 2, ..., n$$

This algorithm is obviously not suitable for ensuring column access to the elements of A.

EXERCISES

1.11 Write an algorithm for transposing a square matrix in the same array.

1.12 Write an algorithm for multiplication of rectangular matrices.

1.13 Write an algorithm for multiplication of a matrix with the transpose of another matrix.

1.14 Develop an algorithm for multiplication of two matrices which stores the result in the array for the second matrix.

1.15 Write an algorithm for computation of 1-norm of a matrix.

1.16 Develop an efficient algorithm for computing the product $A^T A$.

1.17 Develop an efficient algorithm for computing the product $U^T T U$, where T is an upper triangular matrix.

1.4 ROUNDING ERRORS

When performed in finite precision arithmetic, the computations with matrices may be greatly affected by *rounding errors*. This is because the real number system is approximated in the computer by a *floating-point number system*, which contains a finite rather than an infinite set of real numbers. The floating-point system is characterized by the *base* β, the *precision t* and the *exponent range* $[m, M]$. Here β and t are positive integers, m is a negative integer and M is a positive integer. In this system each *t-digit base β floating-point number* is represented in the *normalized form*

$$\pm 0.d_1 d_2 \dots d_t \times \beta^e \tag{1.9}$$

where the base β integers $d_1, \dots, d_t$ (the *significant digits*) satisfy the inequalities

$$1 \leqq d_1 \leqq \beta - 1$$
$$0 \leqq d_i \leqq \beta - 1 \qquad i = 2, 3, \dots, t$$

and $m \leqq e \leqq M$. The integer e is said to be the *exponent* and the number

$$f = \frac{d_1}{\beta} + \frac{d_2}{\beta^2} + \dots + \frac{d_t}{\beta^t} \qquad \frac{1}{\beta} \leqq f \leqq 1$$

to be the *fractional part* (*mantissa*) of the number (1.9).

The most frequently used number bases are 2, 8, 10 and 16. The parameters t, m and M vary greatly from one machine to another.

The floating-point number, represented by (1.9), has the value

$$x = \pm f\beta^e$$

The smallest positive number, represented in the floating-point system, is

$$r = \beta^{m-1}$$

and the largest one is

$$R = \beta^M (1 - \beta^{-t})$$

On most computers two floating-point systems are used which are called *single precision* and *double precision*. Later we shall use the notion of *working precision* for the precision actually used in particular computations.

Let x be any real number which satisfies

$$r \leqq |x| \leqq R$$

The *rounded value* of x is defined as the floating-point number x_R which is nearest to x under the condition that if x lies exactly between two floating-point numbers, then x_R will be that of the larger modulus. The *chopped value* x_C of x is defined as the nearest floating-point number x_C whose modulus is smaller than $|x|$.

To analyze the influence of the rounding or chopping errors (we shall use the generic name rounding errors for both) on the computations, it is necessary to have an estimation of the size of these errors. It may be proved that

$$|x_R - x| \leqq \tfrac{1}{2}\beta^{e-t} \tag{1.10}$$

and

$$|x_C - x| \leqq \beta^{e-t} \tag{1.11}$$

Using (1.10) and (1.11) and bearing in mind that $\beta^{e-1} \leqq |x|$, it is easy to show that

$$x_R = x(1 + \delta_R) \qquad |\delta_R| \leqq \tfrac{1}{2}\beta^{1-t}$$

and

$$x_C = x(1 + \delta_C) \qquad |\delta_C| \leqq \beta^{1-t}$$

where δ_R is the relative rounding error and δ_C is the relative chopping error.

The inequalities

$$|\delta_R| = \frac{|x_R - x|}{|x|} \leqq \tfrac{1}{2}\beta^{1-t}$$

$$|\delta_C| = \frac{|x_C - x|}{|x|} \leqq \beta^{1-t} \qquad x \neq 0$$

are important since their right-hand sides depend only on the floating-point system and are independent of the size of the number, which is approximated.

Let x and y be floating-point numbers. Later we shall denote by

$$\mathrm{fl}(x + y) \qquad \mathrm{fl}(x - y) \qquad \mathrm{fl}(x \times y) \qquad \mathrm{fl}(x/y)$$

the result of an addition, subtraction, multiplication and division respectively in floating-point computations.

Let $\square$ be any of the operations $+$, $-$, $\times$, $/$. In analyzing the effect of rounding errors we shall assume that the floating-point operation is performed in such a way that

$$\mathrm{fl}(x \,\square\, y) = (x \,\square\, y)(1 + \delta) \qquad |\delta| \leqq \varepsilon \tag{1.12}$$

where

$$\varepsilon = \begin{cases} \frac{1}{2}\beta^{1-t} & \text{if rounded arithmetic is used} \\ \beta^{1-t} & \text{if chopped arithmetic is used} \end{cases}$$

The quantity ε is said to be the *relative machine precision* of the computer used and represents the smallest positive floating-point number, which being added to 1, gives a result larger than 1, i.e.

$$\mathrm{fl}(1 + \varepsilon) > 1$$

Usually $\varepsilon \gg r$.

The assumption (1.12) means that the result of a floating-point operation is obtained so that the exact result of this operation is represented as a floating-point number using rounding or chopping. For many computers this idealization is not valid but Equation (1.12) still holds for some larger ε, provided that a *guard digit* is used in the computations. We refer the reader to Vandergraft (1978) for details in the work of the floating-point arithmetic.

Equation (1.12) may be used only when

$$r \le |x \,\square\, y| \le R$$

If $|x \,\square\, y| > R$ then an *overflow* occurs and the computations terminate since there is no reasonable approximation for the result. If $|x \,\square\, y| < r$ then an *underflow* occurs and on most machines the result is set to zero. It should be noted that in contrast to the overflow, the underflow may not be destructive to the computations.

Equation (1.12) shows that the elementary arithmetic operations are accomplished with small relative errors.

The relative machine precision ε has an important role in analyzing the rounding errors in numerical computations. Its exact determination, however, is complicated and may differ for the individual operations on some machines. Sophisticated algorithms for determining ε and the other parameters of the floating-point system may be found in Gentleman and Marovich (1974), and in Cody (1988).

It should be stressed that the systematic use of (1.12) in the error analysis of computational algorithms leads to estimates which are usually pessimistic. This is a consequence of the probabilistic character of the relative rounding errors which are randomly distributed in the interval $[-\varepsilon, \varepsilon]$.

For convenience, instead of (1.12) we shall sometimes also use the representation

$$\mathrm{fl}(x \,\square\, y) = \frac{(x \,\square\, y)}{1 + \delta} \qquad |\delta| \le \varepsilon$$

There exists a special case when the multiplication in floating-point arithmetic gives an exact result. If x is a floating-point number and

$$r \le |x\beta^k| \le R$$

then

$$\mathrm{fl}(x\beta^k) = x\beta^k$$

This shows that a floating-point number may be *scaled* with a power of the base without any error.

Another basic arithmetic operation used in computational algorithms is the extraction of a square root. This operation is usually performed in a such way that

$$\mathrm{fl}(\sqrt{x}) = \sqrt{x}(1 + \delta) \qquad |\delta| \le \varepsilon$$

The *rounding error analysis* of computational algorithms is done by using the expressions for the errors arising in the elementary arithmetic operations. In this analysis we assume that there are no over-flows or underflows in the corresponding algorithm.

Consider first the sum of n floating-point numbers $x_1, x_2, ..., x_n$. It is assumed that the operations are performed in the following way

$$s_2 = \mathrm{fl}(x_1 + x_2)$$
$$s_k = \mathrm{fl}(s_{k-1} + x_k) \qquad k = 3, ..., n$$

Using (1.12) we obtain

$$s_2 = (x_1 + x_2)(1 + e_2) \qquad |e_2| \le \varepsilon$$
$$s_k = (s_{k-1} + x_k)(1 + e_k) \qquad |e_k| \le \varepsilon \qquad k = 3, ..., n$$

In this way

$$\mathrm{fl}(x_1 + \cdots + x_n) = x_1(1 + f_1) + \cdots + x_n(1 + f_n) \tag{1.13}$$

where

$$1 + f_1 = (1 + e_2)\ldots(1 + e_n)$$
$$1 + f_k = (1 + e_k)\ldots(1 + e_n) \qquad k = 2, ..., n$$

and

$$(1 - \varepsilon)^{n-1} \le 1 + f_1 \le (1 + \varepsilon)^{n-1} \tag{1.14}$$
$$(1 - \varepsilon)^{n-k+1} \le 1 + f_k \le (1 + \varepsilon)^{n-k+1} \qquad k = 2, ..., n$$

Since in the problems considered later n is less than a few hundred and $\varepsilon < 10^{-6}$ for existing computers, the bounds in (1.14) may be approximated with sufficient accuracy by

$$(1 + \varepsilon)^k = 1 + k\varepsilon \tag{1.15}$$

$$(1 - \varepsilon)^k = 1 - k\varepsilon \tag{1.16}$$

taking into account only the linear terms in ε.

Later we shall assume everywhere that the approximative equalities (1.15), (1.16) hold. The accounting of higher order terms in ε makes it possible to obtain more accurate bounds on the errors, but complicates the analysis greatly.

From (1.14), (1.15) and (1.16) it follows that

$$\begin{aligned} |f_1| &\le (n - 1)\varepsilon \\ |f_k| &\le (n - k + 1)\varepsilon \qquad k = 2, \ldots, n \end{aligned} \tag{1.17}$$

The relations (1.13) and (1.17) show that the computed sum of $n \ge 3$ numbers is not necessarily characterized by a small relative error, although this is true for the sum of two numbers. This is a consequence of the fact that there may be a *subtractive cancellation* between the components of the sum, and if these components have large moduli, then large errors may be introduced in the final result.

Example 1.4

Let $\beta = 10$, $t = 4$ and it is necessary to find the sum of the numbers

$$x_1 = 0.8134 \times 10^3 \qquad x_2 = 0.3547 \times 10^3 \qquad x_3 = -0.1168 \times 10^4$$

performing correct rounding.

The summation in the order mentioned above gives

$$s_2 = 0.1168 \times 10^4 \qquad \text{and} \qquad s_3 = 0.0$$

while the exact answer is 0.1.

This phenomenon is known as *catastrophic cancellation* in floating-point arithmetic and it occurs when small numbers are obtained by subtraction from large numbers with the same sign, or by addition of large numbers of opposite signs. In this case, the cancellation of high-order (left-most) digits causes the right-most digits to become important and thus the relative error in the result becomes very large. The subtractive cancellation is a major source of errors in numerical computations.

One of the frequently used operations in matrix algorithms is the computation of the inner product $x^T y$ of n-vectors x and y,

$$s_n = \text{fl}(x^T y) = \text{fl}(x_1 y_1 + \cdots + x_n y_n)$$

where the components x_i, y_i are floating-point numbers. It is assumed that the computations are performed in the following order

$$t_k = \mathrm{fl}(x_k y_k)$$
$$s_1 = t_1 \qquad s_k = \mathrm{fl}(s_{k-1} + t_k) \qquad k = 2, \ldots, n$$

Using (1.12) we have

$$t_k = x_k y_k (1 + f_k) \qquad |f_k| \leqq \varepsilon$$
$$s_k = (s_{k-1} + t_k)(1 + g_k) \qquad |g_k| \leqq \varepsilon$$

and

$$s_n = x_1 y_1 (1 + e_1) + \cdots + x_n y_n (1 + e_n) \tag{1.18}$$

$$|e_1| \leqq n\varepsilon$$
$$|e_k| \leqq (n - k + 2)\varepsilon \qquad k = 2, \ldots, n \tag{1.19}$$

Equations (1.18) and (1.19) show that $\mathrm{fl}(x^T y)$ may have a large relative error because of the cancellation that may take place between the terms of $x^T y$.

If we substitute y with x we find that

$$s_n = x_1^2 (1 + e_1) + x_2^2 (1 + e_2) + \cdots + x_n^2 (1 + e_n)$$
$$|e_k| \leqq n\varepsilon \tag{1.20}$$

Since $x_1^2, \ldots, x_n^2$ are non-negative numbers, (1.20) may be represented as

$$s_n = (x_1^2 + x_2^2 + \cdots + x_n^2)(1 + \delta) \qquad |\delta| \leqq n\varepsilon$$

This result shows that the scalar product $\mathrm{fl}(x^T x)$ always has a small relative error.

Consider now the errors arising in some matrix computations. Since each matrix element is stored in the computer as a floating-point number, instead of a given matrix $A = [a_{ij}]$ we shall work with a matrix $\bar{A}$ whose elements are

$$\bar{a}_{ij} = a_{ij}(1 + e_{ij}) \qquad |e_{ij}| \leqq \varepsilon$$

Hence

$$\bar{A} = A + E \qquad \|E\|_F \leqq \varepsilon \|A\|_F$$

This equation shows that before any computations are done, the initial data is contaminated by error of size $\varepsilon \|A\|$. That is why we cannot expect to obtain the result of any computation with A with an error much less than $\varepsilon \|A\|$.

The sum of two matrices A and B, whose elements are floating-point numbers, satisfies

$$C = \mathrm{fl}(A + B) = A + B + F$$

where

$$c_{ij} = (a_{ij} + b_{ij})(1 + e_{ij}) \qquad |e_{ij}| \leq \varepsilon$$

and

$$\| F \|_F \leq \varepsilon \| A + B \|_F$$

In this way the addition of two matrices is always performed with a small relative error.

For the computed product of two $n \times n$ matrices it may be found that

$$\mathrm{fl}(AB) = AB + F \tag{1.21}$$

where, in accordance with (1.18) and (1.19),

$$\| F \|_F \leq n\varepsilon \| A \|_F \| B \|_F \tag{1.22}$$

Since $\| A \| \| B \|$ may be very large in comparison to $\| AB \|$ because of subtractive cancellations in computing AB, we see from (1.21) and (1.22) that the product of two matrices may be obtained with a large relative error. In this connection, it should be pointed out that it is especially dangerous to compute the powers of a matrix using subsequent multiplications.

Example 1.5

Let the third power of

$$A = \begin{bmatrix} -20.857\,1 & -9.064\,67 & 12.995\,5 \\ -43.192\,4 & -18.752\,8 & 26.899\,4 \\ -63.586\,0 & -27.621\,9 & 39.610\,0 \end{bmatrix}$$

be obtained from

$$\bar{A}^2 = \mathrm{fl}(A \times A)$$
$$\bar{A}^3 = \mathrm{fl}(\bar{A}^2 \times A)$$

using floating-point arithmetic with $2^{-15} \approx 3 \times 10^{-5}$. The final result is

$$\bar{A}^3 = \begin{bmatrix} -0.057\,004\,0 & -0.024\,739\,7 & 0.035\,445\,2 \\ -0.785\,385 & -0.341\,064 & 0.489\,075 \\ -0.154\,087 & -0.067\,112\,0 & 0.096\,242\,9 \end{bmatrix}$$

while the exact answer (up to six significant digits) is

$$A^3 = \begin{bmatrix} 0.009\,061\,77 & 0.003\,936\,13 & -0.005\,643\,15 \\ 0.018\,754\,5 & 0.008\,146\,94 & -0.001\,167\,97 \\ 0.027\,610\,9 & 0.011\,993\,8 & -0.017\,194\,9 \end{bmatrix}$$

We see that even the signs of the elements of the computed result are wrong. To explain this unfortunate computation it must be taken into account that $\| A \|_F \approx 100$ but $\| A^3 \|_F \approx 0.043$. Hence A^3 should have small elements and the only way to obtain them is through subtractive cancellations. This leads to a large relative error in the final result.

Equation (1.22) shows the advantages of using orthogonal transformations in matrix computations. If either A or B is orthogonal then the corresponding Frobenius norm is equal to $\sqrt{n}$ and the computed product is obtained with a small relative error.

EXERCISES

1.18 Verify (1.10) and (1.11).

1.19 Verify (1.14).

1.20 Verify (1.22).

1.21 Show that the following algorithm computes a reasonable approximation of the relative machine precision.

> Set $\varepsilon = 1$
> For $k = 1, 2, \ldots$
> $\quad \varepsilon \leftarrow \varepsilon/2$
> Set $\varepsilon 1 = \mathrm{fl}(1 + \varepsilon)$
> If $\varepsilon 1 = 1$, terminate the loop
> $\quad \varepsilon \leftarrow 2\varepsilon$

1.22 Analyze the following algorithm for determining the relative machine precision.

> Set $a = \mathrm{fl}(4./3.)$
> Set $b = \mathrm{fl}(a - 1.)$
> Set $c = \mathrm{fl}(b + b + b)$
> $\varepsilon = |c - 1.|$

1.23 Show that if

$$\mathrm{fl}(\mathrm{tol} + |x|) = \mathrm{tol}$$

for some positive tol, then

$$|x| < \varepsilon \, \mathrm{tol}$$

1.24 Show that if δ is the relative error in the norm of A then the number of the exact decimal digits in the elements of A is approximately equal to

$$p = \log_{10}\left(\frac{1}{\delta}\right) + 1$$

1.25 Derive a bound on the relative error in computing the power of a matrix.

1.5 CONDITIONING AND NUMERICAL STABILITY

There are two reasons which may lead to unsatisfactory results when performing numerical computations on a computer. These are the *ill conditioning* of some problems and the *numerical instability* of some computational methods.

Example 1.6

Consider the fifth-order equation

$$s^5 + 7s^4 + 19.4s^3 + 26.6s^2 + 18.0384s + 4.8384 = 0$$

The roots of this equation are

$$s_1 = -1, \; s_2 = -1.2, \; s_3 = -1.4, \; s_4 = -1.6, \; s_5 = -1.8$$

Note that the roots are well separated from one another.

Suppose now that the coefficient of s^4 is changed from 7 to 7.001. The roots of the new equation, rounded correctly to six decimal digits, are

$$-1.056\,32 \pm 0.024\,2279i$$
$$-1.483\,19 \pm 0.234\,691i$$
$$-1.921\,98$$

Thus a relative change of order 10^{-3} in the coefficient 7 causes four roots to become complex with imaginary parts which are about 100 times larger than the perturbation. It must be stressed that these larger changes in the roots are not results of rounding errors and are independent of the computational algorithm which is used for finding the roots. The real cause is the sensitivity of the roots to changes in the coefficients.

Example 1.7

Consider as a second example the solution of the quadratic equation

$$ax^2 + 2bx + c = 0$$

where

$$a = 0.001 \qquad b = 300 \qquad c = -60.000\,01$$

One of the solutions is given by

$$x_1 = \frac{-b + \sqrt{(b^2 - ac)}}{a} \tag{1.23}$$

and its exact value is 0.1.

When computing the solution through (1.23) it is necessary to find first the square root of $300^2 + 0.001 \times 60.000\,01$. If this root is computed correctly to eight digits by floating-point arithmetic with $\varepsilon = 2^{-23} \approx 1.2 \times 10^{-7}$ then the value of x_1, given by (1.23), is

$$\bar{x}_1 = 0.061\,035\,156$$

Thus there is no correct digit in the result (the relative error $|\bar{x}_1 - x_1|/|x_1|$ is about 0.39).

The reason for this inaccurate answer is seen from the intermediate result

$$(-b + \sqrt{(b^2 - ac)})/a = (-300.000\,00 + 300.000\,06\ldots)/0.001$$
$$= 0.06\,103\,515\,6\ldots$$

In this way the first seven digits of the eight-digit square root are cancelled by the value of $-b$, so that no correct significant digit remains in the final result. This is an extreme example of a subtractive cancellation which makes the rounding errors, performed in the previous steps, significant. This cancellation is absent in the alternative to (1.23)

$$x_1 = \frac{c}{-b - \sqrt{(b^2 - ac)}}$$

whose eight-digit realization gives an approximation of x_1, which is exact to seven digits.

The difference between the above examples is that the first one presents a problem whose solution is extremely sensitive to variations in the data, while the second one presents a problem whose solution is not computed in a reasonable way. Following Stewart (1973b) these situations may be made mathematically more precise in the following way.

Let us assume a function $y = f(x)$, where x is an m-vector and y is an n-vector, defined in the domain D. The m components of the argument x are the data which determine the problem, and the n components of y are the answer. The computational problem is to find an approximation to $f(x)$ for a given x.

In practice, only an approximation x^* of the data is known and in the best case it is possible to compute $f(x^*)$. That is why the accuracy attainable in the computations is limited by the nature of the function f. For some classes of problems $f(x^*)$ and $f(x)$ may differ very much. Such problems are called *ill conditioned*. Every attempt to solve an ill-conditioned problem with inaccurate data will lead to an inexact solution. Example 1.6 illustrates a problem which is ill conditioned. It is necessary to bear in mind that the ill conditioning depends neither on the floating-point system which is used in the computations nor on the algorithm implemented.

The implementation of a numerical method for solving the computational problem, corresponding to f, leads to the definition of a new function f^* which for a given data x gives an approximate solution $f^*(x)$. Clearly, it cannot be expected that f^* will solve an ill-conditioned problem more exactly than the data warrants. However, it is not desirable that f^* introduces large errors itself. This means that the following is required from the function f^*. Near every x from the domain D there must exist some x^* from D, such that $f(x^*)$ is near to $f^*(x)$. In other words, it is required that the method gives a solution, which is near the solution of a problem with slightly changed data. Methods with such a property are called *numerically stable*. Equation (1.23) presents an unstable way for computing the root x_1. It should be pointed out, however, that an unstable method may sometimes give an exact (as the data warrants) result.

The case when a numerically stable method is used for solving a well-conditioned problem is illustrated in Figure 1.1. Owing to the stability of the method, the computed $f^*(x)$ should be near to some exact $f(x^*)$, where x^* is near to x. Since the problem is well conditioned and x^* is near to x, $f(x^*)$ is near to $f(x)$. In this way if a stable method is implemented to a well-conditioned problem then the computed solution is near to the exact one.

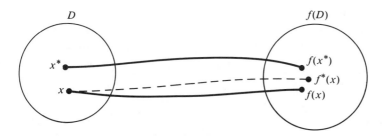

Figure 1.1 Solution of a well-conditioned problem with a stable method.

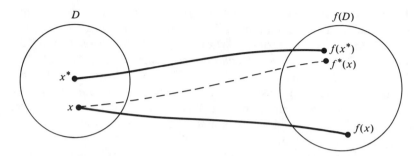

Figure 1.2 Solution of an ill-conditioned problem with a stable
method.

When a stable method is used to solve an ill-conditioned problem
(Figure 1.2), the computed solution and the exact solution may differ
greatly, since $f(x^*)$ may be very far from $f(x)$. However, because $f(x^*)$
and $f^*(x)$ are close, the difference between $f(x)$ and $f^*(x)$ should be
of the same size as the difference between $f(x)$ and $f(x^*)$. That is why,
although the method may be stable, the difference between the computed solution and the exact answer may be very large. The same error,
however, may be obtained if the initial data is slightly perturbed. If this
equivalent perturbation is smaller than the uncertainty which already
exists in the data, then the errors introduced by the method are not
larger than the data warrants. In this way the stable algorithm does not
increase the sensitivity of the problem with respect to variations in the
input data. Finally, Figure 1.3 illustrates an unstable method
implemented in some problem. Here the instability of the method is
reflected by a larger perturbation in the problem.

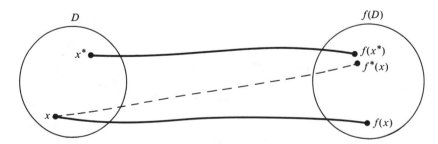

Figure 1.3 Solution of a problem with an unstable method.

Example 1.8

Consider the function f, defined by

$$f(x) = x_1 + x_2 + x_3$$

which corresponds to the problem of computing the sum of three numbers, and which may or may not be ill conditioned depending on the data.

As shown by Example 1.4, the function $f(x)$, relevant to the problem, may be computed in floating-point arithmetic with a large relative error due to the catastrophic cancellation that may take place for some $x = [x_1, x_2, x_3]^T$. Using the properties of floating-point arithmetic, the function f^* associated with the computational procedure is

$$f^*(x) = x_1(1 + f_1) + x_2(1 + f_2) + x_3(1 + f_3) \qquad |f_1|, |f_2| \leq 2\varepsilon$$
$$|f_3| \leq \varepsilon$$

In this way for a given vector x there exists a near vector

$$x^* = [x_1^*, x_2^*, x_3^*]^T = [x_1(1 + f_1), x_2(1 + f_2), x_3(1 + f_3)]^T$$

such that $f^*(x) = f(x^*)$. This shows that floating-point arithmetic is a stable method for computing a sum, independently of the fact that the computations may be accompanied by large relative errors.

The above example illustrates the technique of *backward error analysis*. In this analysis, the error bounds for the basic arithmetic operations are used to show that the computed solution of some problem is close to the exact solution of a slightly perturbed problem. In this way the errors introduced by the method are represented as a small perturbation in the original data. This is sufficient to show that the method used in the computation is stable.

The proof that a given method is numerically stable does not mean that this method will give an exact answer. To find the error in the solution it is necessary to find the bound on $f(x^*) - f(x)$ as a function of $x^* - x$ (i.e. it is necessary to estimate the conditioning of the problem) which is done by so-called *perturbation analysis*. Knowing the size of $x^* - x$ and an estimation of the conditioning, it is possible to determine the size of the error in the result. In this way the error is estimated in a two-step procedure involving analysis of the problem conditioning and analysis of the numerical properties of the method.

Besides the backward error analysis there is also the so-called *forward error analysis*, when the exact solution of a given problem is compared with the computed solution. In this analysis one attempts to obtain a bound on the error in the final result accounting, step by step, for both the effect of rounding errors and the original data uncertainty. The results obtained by this technique frequently depend on the conditioning of the problem, so that it cannot be used independently in estimating the stability of the method. Forward error analysis is sometimes used in conjunction with backward error analysis.

The above concepts may be stated quantitatively as follows (see also Demmel (1987b)). Let Δx be a perturbation in the data such that $x^* = x + \Delta x \in D$. Then the perturbation in the solution $y = f(x)$ of the computational problem $x \mapsto f(x)$ is $\Delta y = f(x^*) - f(x)$. Denote by $\delta_x = \|\Delta x\|/\|x\|$ and $\delta_y = \|\Delta y\|/\|y\|$ the relative perturbations in the data and the solution respectively (we assume that x and y are not zero vectors since this can always be achieved by suitable shifts).

Consider the number

$$c(f, x) = \lim_{d \to 0} \max\{\delta_y/\delta_x : \|\Delta x\| \le d\}$$

The computational problem is *well posed* if $c(f, x) < \infty$, and *well posed in D* if $c(f, x) < \infty$ for all $x \in D$. The number $c(f, x)$ is said to be the *condition number* (or conditioning) of the problem.

If the perturbation Δx is small then

$$\delta_y \le c(f, x)\delta_x + g(\|\Delta x\|)$$

where $g(z)/z \to 0$ as $z \to 0$. Moreover, there exists a perturbation Δx such that

$$\delta_y = c(f, x)\delta_x$$

within first-order terms in $\|\Delta x\|$.

If the *Jacobi matrix* $f'(x) = [\partial f_i(x)/\partial x_j]$ exists then

$$c(f, x) = \|f'(x)\| \, \|x\|/\|y\|$$

The problem of estimating $c(f, x)$ as shown above is that $f'(x)$ is difficult to compute in most cases.

The computational problem is referred to as well conditioned if the quantity $c(f, x)$ is 'small', and ill conditioned if $c(f, x)$ is 'large'. However, these concepts are usually interpreted in the framework of a particular floating-point computing environment: the problem is well conditioned if $\varepsilon c(f, x) \ll 1$ and very ill conditioned if $\varepsilon c(f, x) \approx 1$.

There is another important characteristic of a well-posed computational problem – the *distance between x and the set $B \subset D$ of all $b \in D$* with $c(f, b) = \infty$,

$$d(f, x) = \min\{\|b - x\| : c(f, b) = \infty\}$$

The problems $b \mapsto f(b)$, $b \in B$, are said to be *infinitely ill-conditioned*, or *ill posed*. The computational solution of such problems usually leads to large errors.

It is clear from the above definitions that $d(f, x)$ and $c(f, x)$ have somewhat opposite behaviour: when $d(f, x)$ is small then $c(f, x)$ is large and vice versa. For a wide class of computational problems one has $d(f, x)c(f, x) \approx$ constant.

Consider finally the problem $x \mapsto f(x)$ in a floating-point computing environment with relative machine precision ε. If a backwardly stable numerical method is used then the computed solution y^* is usually the exact solution of a problem with slightly perturbed data:

$$y^* = f(x^*) \qquad x^* = x + \Delta x$$

where $\| \Delta x \| \leq \varepsilon a \| x \|$ and a is a moderate constant. If the problem is well posed we get

$$\| y^* - y \| = \| f(x + \Delta x) - f(x) \| \leq c \| \Delta x \| \leq \varepsilon a c \| x \|$$

where $c = c(f, x)$ is the condition number of the problem. The main difficulty in applying the above elegant bound for the error in the computed solution is the estimation of the constant a (and eventually c).

Now we are in a position to formulate the requirements for the computational algorithms intended to solve numerical problems:

1. The algorithm must be numerically stable. If numerical instability is unavoidable, it should be made obvious, i.e. the algorithm should give warning in cases when it introduces large errors. These algorithms are called *reliable*.
2. The algorithm must be efficient, i.e. it must use computational operations as little as possible.
3. The algorithm must need as little storage as possible.
4. The algorithm should be organized so that overflows and underflows are avoided.
5. The algorithm should lead to a short and easy-to-understand program.

It should be noted that these requirements are frequently conflicting. Reducing the necessary storage, for instance, usually increases the number of operations and vice versa. However, the number of operations and the amount of storage are less important than the requirement for numerical stability.

EXERCISES

1.26 Try to develop an algorithm for solving the quadratic equation $ax^2 + 2bx + c = 0$ which avoids under- and overflows as well as subtractive cancellations.

1.27 Evaluate the sensitivity of the roots of $ax^2 + bx + c = 0$ to perturbations in a, b and c.

1.28 Show by an example that the computation of e^{-x} by the infinite series

$$e^{-x} = 1 - x + \frac{x^2}{2!} - \frac{x^3}{3!} + \cdots$$

is numerically unstable for large x.

1.29 Prove that the multiplication of a matrix by a sequence of orthogonal matrices is stable from a backward error analysis point of view.

1.6 SOLUTION OF LINEAR ALGEBRAIC EQUATIONS

1.6.1 Conditioning of a System of Linear Equations

In this section we consider the numerical solution of systems of linear algebraic equations, written in the form

$$Ax = b$$

where A is a given nonsingular $n \times n$ matrix, b is a given n-vector and x is the unknown n-vector.

In practice, the elements of A and b are usually not known exactly, but even if they are known and are exactly represented in the computer, rounding errors are introduced during the solution. The effect of rounding errors may in turn be expressed by equivalent perturbations in the original data. That is why it is important to know what is the sensitivity of the solution x to changes in the elements of A and b.

Assume first that the matrix A is known exactly and the vector b has some uncertainty Δb, which leads to an uncertainty Δx in the solution. Hence

$$A(x + \Delta x) = b + \Delta b$$

so that

$$\Delta x = A^{-1}\Delta b$$

In this way

$$\|\Delta x\| \leq \|A^{-1}\|\,\|\Delta b\| \qquad (1.24)$$

where $\|\,.\,\|$ denotes any vector norm and consistent matrix norm. From

$b = Ax$ it follows that

$$\| b \| \leq \| A \| \ \| x \| \tag{1.25}$$

The multiplication of (1.24) and (1.25) yields

$$\| \Delta x \| \ \| b \| \leq \| A \| \ \| A^{-1} \| \ \| x \| \ \| \Delta b \| \tag{1.26}$$

Assuming $b \neq 0$ it is found from (1.26) that

$$\frac{\| \Delta x \|}{\| x \|} \leq \| A \| \ \| A^{-1} \| \frac{\| \Delta b \|}{\| b \|} \tag{1.27}$$

For any nonsingular matrix A the product $\| A \| \ \| A^{-1} \|$ is said to be the *condition number* of A, cond(A). From the property of the matrix norms,

$$\| AA^{-1} \| \leq \| A \| \ \| A^{-1} \|$$

it follows that

$$\text{cond}(A) = \| A \| \ \| A^{-1} \| \geq 1$$

The value of the condition number depends on the underlying norm. Frequently the 1-norm is used since it is computed very easily. We shall also use the notation $\text{cond}_p(A) = \| A \|_p \| A^{-1} \|_p$ for the corresponding p-norm.

If V is an orthogonal matrix and $B = VA$ then

$$\| B \|_2 \| B^{-1} \|_2 = \| VA \|_2 \| A^{-1} V^{\mathrm{T}} \|_2 = \| A \|_2 \| A^{-1} \|_2$$

This equality shows that if the 2-norm is used, the condition number is invariant with respect to equivalent orthogonal transformations. Also, this condition number is equal to 1 if A is a multiple of an orthogonal matrix.

From (1.27) we obtain that

$$\frac{\| \Delta x \|}{\| x \|} \leq \text{cond}(A) \frac{\| \Delta b \|}{\| b \|}$$

The inequality reveals that the condition number limits from above the ratio of the relative uncertainty in the solution to the relative uncertainty in the right-hand-side vector b.

If cond(A) = 1 then the relative uncertainty in x is not greater than the relative uncertainty in b. However, if cond(A) = 10^6, then $\| \Delta x \| / \| x \|$ may be 10^6 times $\| \Delta b \| / \| b \|$. If cond(A) is small then the matrix A is said to be *well conditioned* (with respect to the problem of solving linear equations). If cond(A) is large, then the matrix A is *ill conditioned* (with respect to the same problem).

Example 1.9

Let

$$A = \begin{bmatrix} 2 & -7 & 1.0001 \\ -3 & 4 & -8 \\ 1 & -6 & -2 \end{bmatrix} \qquad b = \begin{bmatrix} -3.9999 \\ -7 \\ -7 \end{bmatrix}$$

The exact solution of the system $Ax = b$ is $x = [1, 1, 1]^T$. The matrix A is ill conditioned since $\text{cond}_2(A) \approx 682\ 358$. If the vector b is changed to

$$\tilde{b} = [-4, \ -7, \ -7]^T$$

then the solution of the system becomes

$$\tilde{x} = [5, \ 2, \ 0]^T$$

Thus a relative perturbation in the right-hand side

$$\frac{\|\tilde{b} - b\|_2}{\|b\|_2} \approx 9.3659 \times 10^{-6}$$

leads to a relative perturbation in the solution

$$\frac{\|\tilde{x} - x\|_2}{\|x\|_2} \approx 2.4495$$

In this way

$$\frac{\|\tilde{x} - x\|_2}{\|x\|_2} \approx 261\ 533 \ \frac{\|\tilde{b} - b\|_2}{\|b\|_2}$$

which is near to the worst possible case.

The condition number plays the same role with respect to perturbations in A as it does to perturbations in b. Let there be an uncertainty ΔA in the matrix A and let the vector b be known exactly. If $x = A^{-1}b$ and

$$x + \Delta x = (A + \Delta A)^{-1}b \tag{1.28}$$

then

$$\Delta x = [(A + \Delta A)^{-1} - A^{-1}]b \tag{1.29}$$

The matrix $A + \Delta A$ may be singular if there are no restrictions posed on ΔA. Writing

$$A + \Delta A = A(I + A^{-1}\Delta A)$$

it may be proved that $A + \Delta A$ is nonsingular if

$$\| A^{-1}\Delta A \| < 1$$

It is assumed in (1.28) and (1.29) that this inequality is fulfilled.
Setting $B = A + \Delta A$ in the identity

$$B^{-1} - A^{-1} = A^{-1}(A - B)B^{-1}$$

it is found from (1.29) that

$$\Delta x = -A^{-1}\Delta A (A + \Delta A)^{-1}b = -A^{-1}\Delta A (x + \Delta x)$$

and hence

$$\| \Delta x \| \leq \| A^{-1} \| \, \| \Delta A \| \, \| x + \Delta x \|$$

Finally

$$\frac{\| \Delta x \|}{\| x + \Delta x \|} \leq \mathrm{cond}(A) \, \frac{\| \Delta A \|}{\| A \|}$$

In this way the uncertainty in x related to $x + \Delta x$ is limited by the relative uncertainty in the matrix A, multiplied by $\mathrm{cond}(A)$.

It may be shown in a similar way that if both A and b are perturbed then the relative perturbation in the solution x satisfies

$$\frac{\| \Delta x \|}{\| x \|} \leq \frac{\mathrm{cond}(A)}{1 - \| A^{-1}\Delta A \|} \left(\frac{\| \Delta A \|}{\| A \|} + \frac{\| \Delta b \|}{\| b \|} \right)$$

It must be stressed that the uncertainty in x because of the uncertainty in the data does not depend on the method used to find the solution, nor does it depend on the rounding errors.

The norm $\| \Delta A \|$ of the smallest perturbation ΔA such that $A + \Delta A$ is singular, is said to be the *distance of A to the set of singular matrices* and is denoted by $d(A)$. It may be proved that $d(A) = 1/\| A^{-1} \|$ if A is nonsingular, and $d(A) = 0$ if A is singular. Hence the relative perturbation which makes the (nonsingular) matrix A singular is

$$\frac{d(A)}{\| A \|} = 1/\mathrm{cond}(A)$$

1.6.2 Gaussian Elimination

The system of linear equations $Ax = b$ is easily solved if the matrix A has some special form, in particular diagonal or triangular. Using the method of *Gaussian elimination* every nonsingular matrix A may be reduced to upper triangular form

$$U = M_{n-1} \ldots M_2 M_1 A$$

using a sequence of lower triangular matrices $M_1, M_2, \ldots, M_{n-1}$. The original problem is then equivalent to the upper triangular system

$$Ux = M_{n-1} \ldots M_2 M_1 b$$

which gives a simple way of finding x.

The Gaussian elimination for an $n \times n$ matrix A may be described in matrix terms as follows. Suppose that $a_{11} \neq 0$ and let $m_{i1} = a_{i1}/a_{11}$ where $i = 2, \ldots, n$. For $i = 2, \ldots, n$ the first row of A is multiplied by m_{i1} and the result is subtracted from the ith row. This yields

$$A_2 = M_1 A = \begin{bmatrix} a_{11} & a_{12} & a_{13} & \ldots & a_{1n} \\ 0 & a_{22}^{(2)} & a_{23}^{(2)} & \ldots & a_{2n}^{(2)} \\ \vdots & \vdots & \vdots & & \vdots \\ 0 & a_{n2}^{(2)} & a_{n3}^{(2)} & \ldots & a_{nn}^{(2)} \end{bmatrix}$$

where

$$M_1 = \begin{bmatrix} 1 & & & & \\ -m_{21} & 1 & & \text{\Large 0} & \\ -m_{31} & 0 & 1 & & \\ \vdots & \vdots & \vdots & \ddots & \\ -m_{n1} & 0 & 0 & \ldots & 1 \end{bmatrix}$$

Suppose next that $a_{22}^{(2)} \neq 0$ and let

$$m_{i2} = a_{i2}^{(2)}/a_{22}^{(2)} \qquad i = 3, \ldots, n$$

Multiplying then A_2 by

$$M_2 = \begin{bmatrix} 1 & & & & \\ 0 & 1 & & \text{\Large 0} & \\ 0 & -m_{32} & 1 & & \\ \vdots & \vdots & \vdots & \ddots & \\ 0 & -m_{n2} & 0 & \ldots & 1 \end{bmatrix}$$

we obtain

$$A_3 = M_2 A_2 = \begin{bmatrix} a_{11} & a_{12} & a_{13} & \ldots & a_{1n} \\ 0 & a_{22}^{(2)} & a_{23}^{(2)} & \ldots & a_{2n}^{(2)} \\ 0 & 0 & a_{33}^{(3)} & \ldots & a_{3n}^{(3)} \\ \vdots & \vdots & \vdots & & \vdots \\ 0 & 0 & a_{n3}^{(3)} & & a_{nn}^{(3)} \end{bmatrix}$$

Continuing in this way we find at the $(n-1)$th stage the upper triangular matrix

$$U = A_n = M_{n-1} \dots M_2 M_1 A = \begin{bmatrix} a_{11} & a_{12} & \cdots & a_{1n} \\ & a_{22}^{(2)} & \cdots & a_{2n}^{(2)} \\ & & \ddots & \vdots \\ \text{\huge 0} & & & a_{nn}^{(n)} \end{bmatrix} \qquad (1.30)$$

where

$$M_k = \begin{bmatrix} 1 \\ 0 & 1 & & & & \text{\huge 0} \\ \vdots & \vdots & \ddots \\ 0 & 0 & & 1 \\ 0 & 0 & & -m_{k+1,k} & 1 \\ \vdots & \vdots & & \vdots & \vdots & \ddots \\ 0 & 0 & & -m_{nk} & 0 & \cdots & 1 \\ & & & \underbrace{\quad}_{k} \end{bmatrix} \qquad (1.31)$$

and

$$m_{ik} = \frac{a_{ik}^{(k)}}{a_{kk}^{(k)}} \qquad a_{kk}^{(k)} \neq 0 \qquad i = k+1, \dots, n$$

The quantities $a_{kk}^{(k)}$ where $k = 1, \dots, n-1$, called *pivots*, play an important role in Gaussian elimination.

Defining the matrix

$$L = M_1^{-1} M_2^{-1} \dots M_{n-1}^{-1}$$

it may be shown that

$$L = \begin{bmatrix} 1 \\ m_{21} & 1 & & & \text{\huge 0} \\ m_{31} & m_{32} & 1 \\ \vdots & \vdots & \vdots & \ddots \\ m_{n1} & m_{n2} & m_{n3} & \cdots & 1 \end{bmatrix}$$

Now the matrix A is represented as a product of the unit lower triangular matrix L and the upper triangular matrix U, i.e.

$$A = LU \qquad (1.32)$$

The decomposition (1.32) is called *LU decomposition*.

A useful property of Gaussian elimination is that the elements of the matrix A can be overwritten with the corresponding elements of the matrices L and U as they are produced. For example, after two steps of the decomposition of a 4×4 matrix, the array A would have the form

$$
\begin{bmatrix}
u_{11} & u_{12} & u_{13} & u_{14} \\
m_{21} & u_{22} & u_{23} & u_{24} \\
m_{31} & m_{32} & a_{33}^{(3)} & a_{34}^{(3)} \\
m_{41} & m_{42} & a_{43}^{(3)} & a_{44}^{(3)}
\end{bmatrix}
$$

The Gaussian elimination requires $\frac{1}{3}n^3 - \frac{1}{3}n$ flops.

Example 1.10

If

$$
A = \begin{bmatrix}
2 & 4 & 1 & -3 \\
-4 & -2 & 3 & 5 \\
8 & 10 & 2 & -13 \\
-6 & 0 & 4 & 5
\end{bmatrix}
$$

then

$$
M_1 = \begin{bmatrix}
1 & 0 & 0 & 0 \\
2 & 1 & 0 & 0 \\
-4 & 0 & 1 & 0 \\
3 & 0 & 0 & 1
\end{bmatrix}
\qquad
A_2 = M_1 A = \begin{bmatrix}
2 & 4 & 1 & -3 \\
0 & 6 & 5 & -1 \\
0 & -6 & -2 & -1 \\
0 & 12 & 7 & -4
\end{bmatrix}
$$

$$
M_2 = \begin{bmatrix}
1 & 0 & 0 & 0 \\
0 & 1 & 0 & 0 \\
0 & 1 & 1 & 0 \\
0 & -2 & 0 & 1
\end{bmatrix}
\qquad
A_3 = M_2 A_2 = \begin{bmatrix}
2 & 4 & 1 & -3 \\
0 & 6 & 5 & -1 \\
0 & 0 & 3 & -2 \\
0 & 0 & -3 & -2
\end{bmatrix}
$$

$$
M_3 = \begin{bmatrix}
1 & 0 & 0 & 0 \\
0 & 1 & 0 & 0 \\
0 & 0 & 1 & 0 \\
0 & 0 & 1 & 1
\end{bmatrix}
\qquad
A_4 = M_3 A_3 = \begin{bmatrix}
2 & 4 & 1 & -3 \\
0 & 6 & 5 & -1 \\
0 & 0 & 3 & -2 \\
0 & 0 & 0 & -4
\end{bmatrix}
$$

In this way $A = LU$, where

$$L = \begin{bmatrix} 1 & 0 & 0 & 0 \\ -2 & 1 & 0 & 0 \\ 4 & -1 & 1 & 0 \\ -3 & 2 & -1 & 1 \end{bmatrix} \qquad U = \begin{bmatrix} 2 & 4 & 1 & -3 \\ 0 & 6 & 5 & -1 \\ 0 & 0 & 3 & -2 \\ 0 & 0 & 0 & -4 \end{bmatrix}$$

1.6.3 Gaussian Elimination with Pivoting

Gaussian elimination fails if the pivot at some stage of the reduction is equal to zero. In such a case it is necessary to interchange rows and columns to find a new pivot which is not zero. This is always possible if the matrix to be decomposed is nonsingular. Small pivots must also be avoided since they lead to large errors in the solution.

Example 1.11

It is necessary to solve the system

$$\begin{bmatrix} 0.0002 & 1.00 \\ 2.00 & 1.00 \end{bmatrix} \begin{bmatrix} x_1 \\ x_2 \end{bmatrix} = \begin{bmatrix} 1.00 \\ 3.00 \end{bmatrix}$$

using chopped arithmetic with $\beta = 10$ and $t = 3$. The exact solution of this system (to ten digits) is

$$x = \begin{bmatrix} 1.000\ 100\ 010 \\ 0.999\ 799\ 980 \end{bmatrix}$$

Applying Gaussian elimination one obtains

$$\begin{bmatrix} 0.0002 & 1.00 \\ 0.00 & -10\ 000 \end{bmatrix} \begin{bmatrix} x_1 \\ x_2 \end{bmatrix} = \begin{bmatrix} 1.00 \\ -10\ 000 \end{bmatrix} \qquad x_2 = 1.00 \qquad x_1 = 0.00$$

The use of Gaussian elimination with interchanging of the first and second row yields

$$\begin{bmatrix} 2.00 & 1.00 \\ 0.00 & 1.00 \end{bmatrix} \begin{bmatrix} x_1 \\ x_2 \end{bmatrix} = \begin{bmatrix} 3.00 \\ 1.00 \end{bmatrix} \qquad x_2 = 1.00 \qquad x_1 = 1.00$$

It must be stressed that the small pivot is not necessarily an indication of ill conditioning. In the above example $\text{cond}_2(A) \approx 2.618$ so that the matrix A is very well conditioned.

In this way, if at the kth stage the element $a_{kk}^{(k)}$ is very small, it is necessary to choose another element, say $a_{p_k,q_k}^{(k)}$, as a pivot. If we interchange the rows k and p_k and the columns k and q_k, the element $a_{p_k,q_k}^{(k)}$

will emerge in the (k, k)th position. Now the elimination may continue with the new pivot. Clearly, it is obligatory that $p_k \geq k$ and $q_k \geq k$ since in the opposite case the interchange will destroy the zeros introduced earlier. The resulting algorithm, together with the rule for choosing pivots, is known as *Gaussian elimination with complete pivoting*. This algorithm is reliable, but it is much more time-consuming, since the search of the pivot is realized between $(n - k + 1)^2$ elements of the matrix A_k. That is why the search is usually done for the element of maximum modulus only in the kth column of A_k. The rows are then interchanged to place this element on the pivot position. If the pivot is zero, the zeros necessary are already present in the kth column and it is possible to take $M_k = I_n$. The corresponding algorithm is called *Gaussian elimination with partial pivoting*. If we define an *elementary permutation matrix P_k* as a unit matrix with kth and p_kth columns interchanged, this elimination may be expressed as

$$U = A_n = M_{n-1}P_{n-1}M_{n-2}P_{n-2} \dots M_1 P_1 A \qquad (1.33)$$

Denoting $P = P_{n-1} \dots P_1$ and

$$L = P(M_{n-1}P_{n-1} \dots M_1 P_1)^{-1}$$

one obtains

$$PA = LU$$

Similarly to the case of Gaussian elimination without pivoting, it is possible to show that L is a unit lower triangular matrix.

Let A be an $n \times n$ matrix, $n > 1$. The following algorithm overwrites A with the result of Gaussian elimination with permuted rows. The multipliers m_{ij} overwrite a_{ij}. The indices p_k $(k = 1, 2, \dots, n - 1)$ of the interchanged rows are stored in the vector p.

Algorithm 1.7 Gaussian elimination with partial pivoting

For $k = 1, 2, \dots, n - 1$
    Find $p_k \geq k$ such that $| a_{p_k,k}| = \max | a_{i,k}|$ $k \leq i \leq n$
    If $a_{p_k,k} = 0$, step k
    $a_{kj} \leftrightarrow a_{p_k,j}$ $(j = k, k + 1, \dots, n)$
    $a_{ik} \leftarrow m_{ik} = a_{ik}/a_{kk}$ $(i = k + 1, k + 2, \dots, n)$
    For $j = k + 1, k + 2, \dots, n$
        $a_{ij} \leftarrow a_{ij} - m_{ik}a_{kj}$ $(i = k + 1, k + 2, \dots, n)$

This algorithm requires approximately $\frac{1}{3}n^3$ flops and $\frac{1}{2}n^2$ comparisons.

1.6.4 Cholesky Decomposition

If A is a symmetric positive definite matrix, then it may be represented as

$$A = U^T U \tag{1.34}$$

where U is an upper triangular matrix. This representation is known as the *Cholesky decomposition* of a positive definite matrix. It is determined uniquely if we pose the condition for positiveness of the diagonal elements of U.

Example 1.12

The positive definite matrix

$$A = \begin{bmatrix} 16 & -8 & 12 & -4 \\ -8 & 8 & -20 & 12 \\ 12 & -20 & 94 & -50 \\ -4 & 12 & -50 & 31 \end{bmatrix}$$

is decomposed as $A = U^T U$, where

$$U = \begin{bmatrix} 4 & -2 & 3 & -1 \\ 0 & 2 & -7 & 5 \\ 0 & 0 & 6 & -2 \\ 0 & 0 & 0 & 1 \end{bmatrix}$$

　　　The Cholesky decomposition may be obtained by a variant of the Gaussian elimination. However, in accordance with (1.34) the elements of U may be determined directly from

$$\sum_{k=1}^{i} u_{ki} u_{kj} = a_{ij} \qquad j = 1, \ldots, n; \; i = 1, \ldots, j$$

in the sequence $u_{11}, u_{12}, u_{22}, \ldots, u_{1n}, \ldots, u_{nn}$. Since A is symmetric, it is necessary to work only with its upper half. The elements of U can overwrite the corresponding elements of A. This gives the following algorithm.

Algorithm 1.8 Cholesky decomposition

For $j = 1, 2, ..., n$
　For $i = 1, 2, ..., j - 1$

$$a_{ij} \leftarrow u_{ij} = \left(a_{ij} - \sum_{k=1}^{i-1} u_{ki} u_{kj} \right) \bigg/ u_{ii}$$

$$a_{jj} \leftarrow u_{jj} = \sqrt{ \left(a_{jj} - \sum_{k=1}^{j-1} u_{kj}^2 \right) }$$

The computation of the diagonal element u_{jj} requires the square root of a quantity which is positive if and only if the matrix is positive definite.

The number of the necessary flops to carry out Algorithm 1.8 is

$$\sum_{j=1}^{n} \sum_{i=1}^{j} (i - 1) + \sum_{j=1}^{n} (j - 1) = \tfrac{1}{6} n^3 - \tfrac{1}{6} n$$

Thus the Cholesky decomposition requires half of the operations necessary for Gaussian elimination. This is to be expected since the symmetry of A is taken into account in the algorithm.

1.6.5 Solution of Linear Systems

If the LU decomposition of a matrix A is known, the solution of the system

$$Ax = b$$

reduces to the solution of two systems of equations with triangular matrices

$$Ly = b \tag{1.35}$$

and

$$Ux = y \tag{1.36}$$

Consider the system (1.35). The first equation of this system is

$$l_{11} y_1 = b_1$$

so that

$$y_1 = b_1 / l_{11}$$

In general, if $y_1, y_2, ..., y_{i-1}$ are known, the ith equation of (1.35) may be solved for y_i by the Algorithm 1.9.

Algorithm 1.9 Solution of a linear system with lower triangular matrix

For $i = 1, 2, ..., n$

$$y_i = \left(b_i - \sum_{j=1}^{i-1} l_{ij} y_j \right) \bigg/ l_{ii}$$

This algorithm is known as *forward elimination*. It may produce overflows if some of the diagonal elements l_{ii} are very small. The number of the necessary flops is

$$\sum_{i=1}^{n} (i-1) = \tfrac{1}{2} n^2 - \tfrac{1}{2} n$$

The solution of the system (1.36) may be done in the sequence $x_n, x_{n-1}, ..., x_1$ by Algorithm 1.10.

Algorithm 1.10 Solution of a linear system with upper triangular matrix

For $i = n, n-1, ..., 1$

$$x_i = \left(y_i - \sum_{j=i+1}^{n} u_{ij} x_j \right) \bigg/ u_{ii}$$

This algorithm is called *back substitution*. It also requires $\tfrac{1}{2} n^2 - \tfrac{1}{2} n$ flops.

When Gaussian elimination with partial pivoting is used, the system

$$Ax = b$$

is reduced to

$$A_n x = M_{n-1} P_{n-1} \ldots M_2 P_2 M_1 P_1 b$$

where A_n is an upper triangular matrix defined from (1.33), the matrices M_k $(k = 1, 2, ..., n-1)$ have the form (1.31) and $P_1, P_2, ..., P_{n-1}$ are elementary permutations coded in the vector $p = [p_1, p_2, ..., p_{n-1}]^T$. In this case the solution for x may be found by Algorithm 1.11.

Algorithm 1.11 Solution of a linear system with matrix decomposed by Gaussian elimination with partial pivoting

For $k = 1, 2, \ldots, n - 1$
   Interchange b_k and b_{p_k}
   $b_i \leftarrow b_i - m_{ik} b_k \qquad i = k + 1, k + 2, \ldots, n$
Solve $A_n x = b$ by Algorithm 1.10

The solution may be obtained in the place of b which saves memory. This algorithm requires $n^2 - n$ flops.

Algorithms 1.7 and 1.11 together constitute a full path for finding the solution of a linear system of equations. This path requires approximately $\frac{1}{3} n^3 + n^2$ flops.

If A is symmetric and positive definite the solution of $Ax = b$ may be computed by the following algorithm.

Algorithm 1.12 Solution of a linear system with positive definite matrix decomposed by the algorithm of Cholesky

Solve $U^T y = b$ by Algorithm 1.9
Solve $Ux = y$ by Algorithm 1.10

This algorithm requires about n^2 flops. Adding $\frac{1}{6} n^3$ flops for finding the Cholesky decomposition by Algorithm 1.8, the full path for solving $Ax = b$ with positive definite A will require $\frac{1}{6} n^3 + n^2$ flops.

The above algorithms may be used for solving linear equations in the form

$$AX = B \tag{1.37}$$

where B and X are $n \times m$ matrices. Representing B and X as

$$B = [b_1, b_2, \ldots, b_m]$$
$$X = [x_1, x_2, \ldots, x_m]$$

Equation (1.37) is reduced to the equations

$$Ax_j = b_j \qquad j = 1, 2, \ldots, m$$

which are solved separately for each x_j. The decomposition of A need only be done once.

When the LU decomposition of a matrix A is known, the inverse A^{-1} may be found efficiently from

$$A^{-1} = U^{-1} L^{-1}$$

Since U^{-1} and L^{-1} may be obtained in place of U and L, respectively, the computation of the inverse may be performed in the same array used to store A. Obtaining A^{-1} in this way requires approximately $\frac{2}{3}n^3$ flops ($\frac{1}{3}n^3$ flops for positive definite matrices). This shows that the solution of a linear system $Ax = b$ by computing first A^{-1} and then multiplying A^{-1} by b is not efficient, since the solution may be found directly by $\frac{1}{3}n^3$ flops. Also, the errors in the solution are larger when using A^{-1} explicitly.

1.6.6 Estimation of the Conditioning

After the linear system $Ax = b$ is solved, it is important to estimate the accuracy of the computed solution. This may be done by the condition number, which characterizes the sensitivity of x to changes in A and b. It should be pointed out that the exact value of cond(A) is not necessary, only its order of magnitude being of interest.

In computing cond(A) the 1-norm of a vector is frequently used since it is determined easily and also since the subordinate matrix norm is computed directly from the columns a_j of A by

$$\| A \|_1 = \max_j \| a_j \|_1$$

If $\| A \|$ is determined, the problem of finding cond(A) reduces to the determination of A^{-1} and then of $\| A^{-1} \|$. This, however, requires many more computations than obtaining x. That is why several methods are proposed for computing a sufficiently precise estimation of cond(A), which do not require many more operations than the solution of $Ax = b$. Such a method is used in the program package LINPACK (Dongarra *et al.*, 1979) and may be described in general terms as follows.

The basic problem in computing the condition number is to find a satisfactory approximation for $\| A^{-1} \|$ without computing the columns of A^{-1}.

By definition

$$\| A^{-1} \| = \max_{y \neq 0} \frac{\| A^{-1}y \|}{\| y \|}$$

The idea for estimating cond(A) is to generate some vector y, solve $Az = y$ and then estimate

$$\| A^{-1} \| \approx \frac{\| z \|}{\| y \|}$$

The system $Az = y$ may be solved using the decomposition of A, computed in the solution of $Ax = b$. To avoid overflows when A is close to a singular matrix, instead of cond(A) one may estimate $1/\text{cond}(A)$, which is given by

$$\text{rcond}(A) = \frac{1}{\text{cond}(A)} = \frac{\|y\|}{\|A\|\,\|z\|}$$

Since

$$\|A^{-1}\| \geqq \frac{\|z\|}{\|y\|}$$

the estimation in fact satisfies

$$\frac{1}{\text{rcond}(A)} \leqq \text{cond}(A)$$

This estimation will be sufficiently precise if y is chosen so that $\|z\|/\|y\|$ is as large as possible. In LINPACK, y is obtained by solving the system $A^Ty = e$, where, up to a scalar multiplier, e is a vector with components ± 1. Using the decomposition $A = LU$, this includes the solution of $U^Tw = e$ and then the solution of $L^Ty = w$. The components of e are determined during the computation of w so that $\|w\|$ will be large. The details of this procedure are complicated and depend on the actual decomposition of A, which is used in the solution of $Ax = b$. A singular matrix A will give a small, eventually zero value, of rcond(A). If rcond(A) is so small that

$$\text{fl}(1.0 + \text{rcond}(A)) = 1.0$$

then A is referred to as *singular to working precision*.
    The estimation of the condition number is done by Algorithm 1.13.

Algorithm 1.13 Estimation of the condition number of a matrix

Compute $\|A\|$
Decompose A as $A = LU$
Solve $U^Tw = e$ choosing e_i so as to obtain the maximum of $\|w\|$
Solve $L^Ty = w$
Solve $Lv = y$
Solve $Uz = v$

Obtain $\text{rcond}(A) = \dfrac{\|y\|}{\|A\|\,\|z\|}$

For positive definite matrix instead of L one must use U^T.

In this way the estimation of the norm of the inverse includes the solution of four systems of equations with triangular matrices. This requires approximately $4n^2$ flops, i.e. much less than the necessary $\frac{2}{3}n^3$ flops for computing A^{-1} ($\frac{1}{3}n^3$ for positive definite matrices).

Experiments show that with this estimation of rcond(A) the quantity $1/\text{rcond}(A)$ is usually of the same order of magnitude as the actual condition number.

1.6.7 Error Analysis and Accuracy Estimation

Let A, b and x be the given matrix, the right-hand-side vector and the solution, obtained by Gaussian elimination with pivoting. We shall assume that the elements of A and b are floating-point numbers. It is possible to show that the computed solution satisfies exactly the perturbed equation

$$(A + \Delta A)x = b \tag{1.38}$$

where ΔA is a matrix whose elements have small moduli. The aim of the error analysis is to find the dependence of $\| \Delta A \|$ on $\| A \|$ and on ε and then to estimate the accuracy of the solution.

Equation (1.38) is an example of backward error analysis when the effect of the rounding errors is expressed as an equivalent perturbation in the data.

The first stage in the solution is the decomposition of A into the product LU. Define the *growth factor* ρ by

$$\rho = \max_{i,j,k} | a_{ij}^{(k)} | / \| A \|_1$$

The value of ρ is a measure of the growth of matrix elements during Gaussian elimination.

Using an analysis similar to those presented in Forsythe and Moler (1967, ch. 21) it may be proved that the matrices L and U, computed by Gaussian elimination with pivoting, satisfy

$$LU = A + E \tag{1.39}$$

where

$$\| E \|_1 \leq n^2 \rho \varepsilon \| A \|_1$$

Equation (1.39) shows that the computed L and U form the exact decomposition of a slightly perturbed matrix. If ρ is not very large, the perturbations will have the size of the rounding errors of the elements of A. In the case of complete pivoting, no matrix was found for which

ρ was greater than n. That is why Gaussian elimination with complete pivoting is a numerically stable method. In the case of partial pivoting, there is a possibility that the elements of the matrices $A_2, ..., A_n$ will grow so that $\rho \leq 2^n$. That is why Gaussian elimination with partial pivoting may be considered as a mildly unstable method.

Having L and U the solution of $Ax = b$ is found by successive solution of the triangular systems $Ly = b$ and $Ux = y$. The error analysis made in Forsythe and Moler (1967, ch. 21) shows that the computed solution for y fulfils

$$(L + \Delta L)y = b$$

where ΔL is a lower triangular matrix whose elements satisfy

$$|\Delta l_{ij}| \leq n\varepsilon |l_{ij}|$$

Hence

$$\|\Delta L\|_1 \leq n^2 \varepsilon \|L\|_1$$

so that the perturbations in L have the size of the machine precision. This means that Algorithm 1.9 is numerically stable and the same holds for Algorithm 1.10.

In this way the computed solution x satisfies the equation

$$(L + \Delta L)(U + \Delta U)x = b$$

or, since $LU = A + E$,

$$(A + E + \Delta LU + L\Delta U + \Delta L\Delta U)x = b$$

It may be shown that the elements of L and U fulfil the inequalities

$$|l_{ij}| \leq 1$$
$$|u_{ij}| \leq \rho \|A\|_1$$

so that

$$\|L\|_1 \leq n \qquad \|U\|_1 \leq n\rho \|A\|_1$$
$$\|\Delta L\|_1 \leq n^2 \varepsilon \quad \text{and} \quad \|\Delta U\|_1 \leq n^2 \rho \varepsilon \|A\|_1$$

Since $n^2\varepsilon \ll 1$ the product $\|\Delta L\|_1 \|\Delta U\|_1$ may be neglected.

If we denote

$$\Delta A = E + \Delta LU + L\Delta U$$

then

$$\|\Delta A\|_1 \leq \|E\|_1 + \|\Delta L\|_1 \|U\|_1 + \|L\|_1 \|\Delta U\|_1$$

Hence the solution, computed by Gaussian elimination with pivoting, satisfies the equation

$$(A + \Delta A)x = b \tag{1.40}$$

where

$$\| \Delta A \|_1 \le (2n^3 + n^2)\rho\varepsilon \| A \|_1 \tag{1.41}$$

The dominant term in (1.41) is $2n^3\rho\varepsilon \| A \|_1$ which comes from the solution of triangular systems. In practice, however, the error bounds in solving triangular systems may be attained only in exceptional cases. In fact, the factor $2n^3 + n^2$ in (1.41) almost always may be replaced by n.

From (1.40) and (1.41) it follows that

$$\| Ax - b \|_1 \le (2n^3 + n^2)\rho\varepsilon \| A \|_1 \| x \|_1$$

This inequality shows that the size of the *residual* $Ax - b$ is of the order of the rounding errors in A even if cond(A) is large. That is why the size of the residual is not a reliable measure of the solution accuracy.

Let A be a nonsingular matrix and $x^* = A^{-1}b$ be the exact solution of $Ax = b$. Based on the consideration of the conditioning of a linear system it may be written that

$$\| x - x^* \|_1 \le (2n^3 + n^2)\rho\varepsilon \; \text{cond}_1(A)\| x \|_1$$

This result shows that if cond(A) is large, then the relative error in x may be very large. That is why, independently of the fact that the solution is obtained in a numerically stable way, it may be very inaccurate if the matrix is ill conditioned. To estimate the accuracy one may use the following rough rule of thumb. If we work on a machine whose arithmetic has a precision equivalent to p decimal digits, then the computed solution of a system with condition number cond(A) $\approx 10^q$ will have at worse only $p - q$ exact significant digits. The rule is rough, since cond(A) gives the upper bound on the error. In general, if $q \ge p$, i.e. cond(A) $\ge 1/\varepsilon$ then it is possible that there is no correct significant digit in the solution.

The error analysis of the Cholesky algorithm for positive definite matrices shows that the computed factor U satisfies

$$U^T U = A + E$$

where

$$\| E \| \le \rho\varepsilon \| A \|$$

The main difference from the general case is that the growth factor ρ has now an order of at most n^2. Moreover, in almost all cases ρ is of order 1. That is why the Cholesky algorithm is unconditionally stable.

Analogically to the general case, the computed solution of $Ax = b$, for A positive definite, satisfies

$$\| Ax - b \|_1 \le \sigma\varepsilon \| A \|_1 \| x \|_1$$
$$\| x - x^* \|_1 \le \sigma\varepsilon \, \mathrm{cond}_1(A) \| x \|_1$$

where σ is of order n^2 and x^* is the exact solution.

1.6.8 Scaling

The error bound obtained in (1.41) shows tnat the error ΔA will be small in comparison to the norm of A, i.e. in comparison to the largest element of A. If all elements of A have nearly equal moduli this result is completely satisfactory. If, however, there is a very large difference between the elements of A, then there is a danger that the small elements will be entirely changed by the error. If these elements are important for the reduction, the result will be inaccurate.

Example 1.13

Let the first equation of the system

$$\begin{bmatrix} 0.0002 & 1.00 \\ 2.00 & 1.00 \end{bmatrix} \begin{bmatrix} x_1 \\ x_2 \end{bmatrix} = \begin{bmatrix} 1.00 \\ 3.00 \end{bmatrix}$$

be multiplied by 10^5. In a result we obtain the system

$$\begin{bmatrix} 20.0 & 100\,000 \\ 2.0 & 1.0 \end{bmatrix} \begin{bmatrix} x_1 \\ x_2 \end{bmatrix} = \begin{bmatrix} 100\,000 \\ 3.00 \end{bmatrix}$$

Since $| a_{11} | > | a_{21} |$ the partial pivoting does not require interchanging of the equations. That is why the elimination done in chopped arithmetic with $\beta = 10$, $t = 3$, yields

$$\begin{bmatrix} 20.0 & 100\,000 \\ 0 & -10\,000 \end{bmatrix} \begin{bmatrix} x_1 \\ x_2 \end{bmatrix} = \begin{bmatrix} 100\,000 \\ -10\,000 \end{bmatrix}$$

The solution of this system is

$$x_2 = 1.00$$
$$x_1 = 0.00$$

(the true solution of the system up to three digits is $x_1 = 1.00$ and $x_2 = 1.00$). Hence the computed solution is the same as the solution obtained without pivoting (see Example 1.11). In this way the scaling of the first equation leads to the wrong choice of pivot and to a large error in the final result.

The error analysis presented above guarantees that the computed solution is exact for a matrix $A + \Delta A$, where the elements of ΔA are of order $10^{-3}\| A \|_1$. However, since $\| A \|_1 = 100\,001$, the elements of ΔA may be, and they are, of order 100. A perturbation of this size in the large coefficient (100 000) is insignificant, but it changes the second equation entirely and hence the result.

Since the cause of large errors lies in the large differences between the elements of A, it is natural to make an attempt to scale the rows and the columns of A, so that the elements have nearly equal moduli. This process is called *equilibration of a matrix*. A frequently recommended method is to scale A so that the element of maximum size in each row and column is between 1 and 10. The original matrix from Example 1.13 satisfies this requirement so that the matrix is reduced safely by Gaussian elimination with partial pivoting. In some cases, however, this scaling may not lead to satisfactory results.

EXERCISES

1.30 Show that for an arbitrary non-zero scalar c, the following is fulfilled

$$\text{cond}(cA) = \text{cond}(A)$$

1.31 Show that

$$\text{cond}(A^{-1}) = \text{cond}(A)$$
$$\text{cond}(AB) \leq \text{cond}(A)\text{cond}(B)$$

1.32 Write an algorithm for the Gaussian elimination of a rectangular matrix.
1.33 Write column-oriented versions of Algorithms 1.9 and 1.10.
1.34 Write an algorithm for finding the inverse of A by using LU decomposition.
1.35 Perform detailed error analysis of Algorithms 1.9 and 1.10.
1.36 Show that if $\bar{X}$ is the inverse of A, computed with Gaussian elimination with pivoting, then

$$\| A\bar{X} - I \|_1 \leq (2n^3 + n^2)\rho\varepsilon \| A \|_1 \| \bar{X} \|_1$$

1.7 ELEMENTARY ORTHOGONAL TRANSFORMATIONS

Most of the computational problems involving matrices are solved by reduction of these matrices to some special form (diagonal, triangular, and so on), using a sequence of elementary orthogonal transformations, implemented to ensure the numerical stability of the corresponding method. Two kinds of elementary orthogonal transformations are usually exploited, namely *plane rotations* and *elementary reflections.*

A plane (or *Givens*) rotation in the (i, j)-plane is a matrix of the form

$$
R = \begin{array}{c} \\ \\ i\{ \\ \\ j\{ \\ \\ \end{array}
\begin{bmatrix}
1 & & & & & & & \\
& \ddots & & & & & 0 & \\
& & 1 & & & & & \\
& & & p_1 & & q & & \\
& & & & \ddots & & & \\
& & & -q & & p_1 & & \\
& 0 & & & & & \ddots & \\
& & & & & & & 1
\end{bmatrix}
$$

where

$$p^2 + q^2 = 1$$

The plane rotation is easily applied on a vector. If $y = Rx$, then y differs from x only in the ith and jth components which are given by

$$y_i = px_i + qx_j$$
$$y_j = px_j - qx_i$$

The main purpose of the plane rotation is to introduce zero in a single element of a vector or matrix. If a is a given vector, it is possible to construct a plane rotation which annuls the jth element of a. In fact, denoting $b = Ra$ and setting

$$p = \frac{a_i}{\sqrt{(a_i^2 + a_j^2)}} \qquad q = \frac{a_j}{\sqrt{(a_i^2 + a_j^2)}} \tag{1.42}$$

then $p^2 + q^2 = 1$ and

$$b_i = \sqrt{(a_i^2 + a_j^2)}$$
$$b_j = 0$$

Example 1.14

Let $a = [6, 3, -8, 2]^T$. It is required to zero the third element of a. This may be done by

$$R = \begin{bmatrix} p & 0 & q & 0 \\ 0 & 1 & 0 & 0 \\ -q & 0 & p & 0 \\ 0 & 0 & 0 & 1 \end{bmatrix}$$

where

$$p = \frac{6}{\sqrt{(6^2 + 8^2)}} = 0.6$$

$$q = \frac{-8}{\sqrt{(6^2 + 8^2)}} = -0.8$$

In a result $Ra = [10, 3, 0, 2]^T$.

A sequence of plane rotations may be used to introduce several zeros in specified positions in vectors or matrices.

Given scalars a_i and a_j the following algorithm constructs a plane rotation so that

$$\begin{bmatrix} p & q \\ -q & p \end{bmatrix} \begin{bmatrix} a_i \\ a_j \end{bmatrix} = \begin{bmatrix} g \\ 0 \end{bmatrix}$$

$p^2 + q^2 = 1$ and $g = \sqrt{(a_i^2 + a_j^2)}$. Here the elements a_i and a_j are scaled which allows overflows and destructive underflows to be avoided.

Algorithm 1.14 Construction of a plane rotation

Set $s = |a_i| + |a_j|$
If $s = 0$
   Set $p = 1$, $q = 0$, $g = 0$
   Quit
Set $p = a_i/s$, $q = a_j/s$
Set $r = \sqrt{(p^2 + q^2)}$
$p \leftarrow p/r$, $q \leftarrow q/r$
$g = sr$

This algorithm requires about 8 flops.

Let a_i and a_j be floating-point numbers. The exact rotation

$$R = \begin{bmatrix} p & q \\ -q & p \end{bmatrix}$$

determined by Algorithm 1.14, fulfils (1.42). Owing to rounding errors, one obtains a matrix $\bar{R}$ with elements $\bar{p}$ and $\bar{q}$ instead of p and q. Performing an analysis similar to those presented in Wilkinson (1965, ch. 3), it may be shown that

$$\bar{p} = p(1 + \delta p) \qquad \bar{q} = q(1 + \delta q) \tag{1.43}$$

$$|\delta p| \leq 5\varepsilon \qquad |\delta q| \leq 5\varepsilon \tag{1.44}$$

It follows from (1.43) and (1.44) that the computed $\bar{p}$ and $\bar{q}$ have small relative errors. It may be written that

$$\bar{p} = p + \Delta p \qquad \bar{q} = q + \Delta q \qquad \bar{R} = R + \Delta R$$

where

$$\Delta p = p\delta p \qquad \Delta q = q\delta q$$

$$\Delta R = \begin{bmatrix} \Delta p & \Delta q \\ -\Delta q & \Delta p \end{bmatrix}$$

Hence

$$\| \Delta R \|_2 = \sqrt{(\Delta p^2 + \Delta q^2)} \leq 5\varepsilon$$
$$\| \Delta R \|_F = \sqrt{(2(\Delta p^2 + \Delta q^2))} \leq 5\sqrt{2}\,\varepsilon$$

which shows that the computed rotation $\bar{R}$ is near to the exact R corresponding to a_i and a_j.

The computed plane rotation is usually applied on a matrix in order to perform a similarity transformation RAR^T. The left multiplication of an $n \times n$ matrix A by a plane rotation†

$$RA = \begin{bmatrix} 1 & & & & & & \\ & \ddots & & & & 0 & \\ & & 1 & & & & \\ & & & p & q & & \\ & & & -q & p & & \\ & & & & & 1 & \\ & 0 & & & & & \ddots \\ & & & & & & & 1 \end{bmatrix} \begin{bmatrix} & & & & & \\ x \ldots x & a_{ij} & x \ldots x \\ x \ldots x & a_{i+1,j} & x \ldots x \\ & & & & & \end{bmatrix}$$

changes only the ith and $(i + 1)$th rows of A. The resulting product may overwrite A in the same array as shown in Algorithm 1.15.

† The symbol x here denotes unspecified elements of A.

Algorithm 1.15 Left multiplication by plane rotation

For $j = 1, 2, \ldots, n$
  Set $z = a_{ij}$
  $a_{ij} \leftarrow pz + qa_{i+1,j}$
  $a_{i+1,j} \leftarrow pa_{i+1,j} - qz$

The algorithm requires $4n$ flops.
  Denoting

$$b = \mathrm{fl}(\bar{R}a) \qquad a = [a_1, a_2, \ldots, a_n]^{\mathrm{T}} \qquad a_i = a_{ij}$$

for some $1 \leq j \leq n$, we have that

$b_k = a_k \qquad k \neq i, i+1$
$b_i = \mathrm{fl}(\bar{p}a_i + \bar{q}a_{i+1}) = \bar{p}a_i(1 + f_1) + \bar{q}a_{i+1}(1 + f_2)$
$b_{i+1} = \mathrm{fl}(-\bar{q}a_i + \bar{p}a_{i+1}) = -\bar{q}a_i(1 + f_3) + \bar{p}a_{i+1}(1 + f_4)$
$\qquad |f_k| \leq 2\varepsilon \qquad k = 1, 2, 3, 4$

Hence

$$\begin{bmatrix} b_i \\ b_{i+1} \end{bmatrix} = \begin{bmatrix} p & q \\ -q & p \end{bmatrix} \begin{bmatrix} a_i \\ a_{i+1} \end{bmatrix} + \begin{bmatrix} \Delta p & \Delta q \\ -\Delta q & \Delta p \end{bmatrix} \begin{bmatrix} a_i \\ a_{i+1} \end{bmatrix} + \begin{bmatrix} \bar{p}f_1 & \bar{q}f_2 \\ -\bar{q}f_3 & \bar{p}f_4 \end{bmatrix} \begin{bmatrix} a_i \\ a_{i+1} \end{bmatrix}$$

Using the bounds on Δp, Δq and f_k and setting

$$\mathrm{fl}(\bar{R}a) - Ra \equiv e$$

we obtain

$$\begin{aligned} \| e \|_2 &\leq \sqrt{(\Delta p^2 + \Delta q^2)} \sqrt{(a_i^2 + a_{i+1}^2)} + 2\varepsilon \sqrt{[2(\bar{p}^2 + \bar{q}^2)]} \sqrt{(a_i^2 + a_{i+1}^2)} \\ &\leq (5 + 2\sqrt{2})\varepsilon \| a \|_2 \\ &< 8\varepsilon \| a \|_2 \end{aligned}$$

Combining these results for $j = 1, 2, \ldots, n$ we get

$$\| \mathrm{fl}(\bar{R}A) - RA \|_{\mathrm{F}} < 8\varepsilon \| A \|_{\mathrm{F}} \tag{1.45}$$

The result obtained is an example of forward error analysis. From a backward error analysis point of view this result can also be represented as

$$\mathrm{fl}(\bar{R}A) = R(A + \Delta A_1) \tag{1.46}$$

Taking into account that R is orthogonal, it follows from (1.45) that

$$\| \Delta A_1 \|_{\mathrm{F}} < 8\varepsilon \| A \|_{\mathrm{F}}$$

This shows that the computed product is exact for a slightly perturbed matrix A.

The right multiplication of an $n \times n$ matrix A by a plane rotation R is done similarly. It may be shown that

$$\mathrm{fl}(A\bar{R}^\mathrm{T}) = (A + \Delta A_2)R^\mathrm{T}$$

where, again

$$\| \Delta A_2 \|_\mathrm{F} < 8\varepsilon \| A \|_\mathrm{F}$$

Thus, a similarity transformation of a matrix A with a plane rotation R yields

$$\mathrm{fl}(\bar{R}A\bar{R}^\mathrm{T}) = R(A + \Delta A)R^\mathrm{T}$$

where

$$\| \Delta A \|_\mathrm{F} < 16\varepsilon \| A \|_\mathrm{F}$$

An *elementary* (or *Householder*) *reflection* is a matrix of the form

$$U = I - \frac{uu^\mathrm{T}}{h}$$

where u is a vector and

$$h = \tfrac{1}{2} \| u \|_2^2 \tag{1.47}$$

It follows from this definition that U is symmetric and orthogonal.

The elementary reflection is used to introduce several zeros in a vector or a matrix simultaneously. For a given vector a it is possible to find a reflection U, such that

$$Ua = -ge_1 \qquad e_1 = [1, \, 0, \, \ldots, \, 0]^\mathrm{T} \tag{1.48}$$

(the sign $-$ is chosen for convenience). The constant g is determined up to the sign from the fact that U is orthogonal,

$$g = \pm \| a \|_2 \tag{1.49}$$

It may be shown that if the vector u is chosen as

$$u = a + ge_1 \tag{1.50}$$

and if h fulfils (1.47), then (1.48) holds.

Example 1.15

Let $a = [2, \, -6, \, 5, \, 4]^\mathrm{T}$. If the sign in (1.49) is taken as positive, then

$$g = 9 \qquad u = [11, \, -6, \, 5, \, 4]^\mathrm{T} \qquad h = 99$$

$$U = \frac{1}{99} \begin{bmatrix} -22 & 66 & -55 & -44 \\ 66 & 63 & 30 & 24 \\ -55 & 30 & 74 & -20 \\ -44 & 24 & -20 & 83 \end{bmatrix}$$

and $Ua = [-9, 0, 0, 0]^T$.

The analysis of (1.50) shows that if g and a_1 have different signs, then a subtractive cancellation may take place in the computation of u_1 by (1.50). That is why g is taken with the same sign as a_1, i.e.

$$g = \text{sign}(a_1) \| a \|_2$$

From (1.47) and (1.50) it also follows that

$$h = g^2 + ga_1 = gu_1 \tag{1.51}$$

It should be pointed out that the determination of the elementary reflection by (1.50) and (1.51) is not unique. In particular, u and h may be scaled by g as proposed in Dongarra *et al.* (1979, ch. 9), which is advantageous in some applications (in this case $h = u_1$).

The following algorithm finds u, h and g such that for a given n-vector a the following is fulfilled

$$\left(I - \frac{uu^T}{h} \right) a = -ge_1$$

The components of the vector u overwrite those of a. In computing $\| a \|_2$ the elements of a are scaled to avoid overflows and destructive underflows.

Algorithm 1.16 Construction of an elementary reflection

Set $s = \displaystyle\sum_{i=1}^{n} | a_i |$

If $s = 0$
    Set $u = 0$, $g = 0$
    Quit
$a_i \leftarrow u_i = a_i / s$ $(i = 1, 2, ..., n)$

Set $t = \displaystyle\sum_{i=1}^{n} u_i^2$

Set $g = \text{sign}(u_1)\sqrt{t}$
$h = t + u_1 g$
$u_1 \leftarrow u_1 + g$
$g \leftarrow sg$

This algorithm requires approximately $2n$ flops.

Let the elements of a be floating-point numbers. The detailed error analysis, done as in Wilkinson (1965, ch. 3), using the results from Section 1.4, shows that instead of the exact elementary reflection U, one is computing in practice a matrix

$$\bar{U} = I - \frac{\bar{u}\bar{u}^{\mathrm{T}}}{\bar{h}}$$

where

$$\bar{u} + u + \Delta u \qquad \bar{h} = h(1 + \delta h) \tag{1.52}$$

$$\|\Delta u\|_2 \leq \frac{n+6}{2} \varepsilon \|u\|_2 \qquad |\delta h| \leq (n+5)\varepsilon$$

It may be proved that

$$\|\bar{U} - U\|_2 \leq (4n + 22)\varepsilon$$

This inequality shows that the computed reflection $\bar{U}$ is near to the exact reflection U corresponding to the given a.

It is instructive to see the effect of the right sign choice in Equation (1.50).

Example 1.16

Let

$$a = [0.8316 \times 10^0, \ 0.3917 \times 10^{-3}, \ 0.2651 \times 10^{-3}]^{\mathrm{T}}$$

Using floating-point arithmetic with $\varepsilon = 0.5 \times 10^{-3}$, which corresponds to four-digit decimal arithmetic with correct rounding, we have

$$\bar{t} = \mathrm{fl}(a_1^2 + a_2^2 + a_3^2) = 0.6916 \times 10^0$$
$$\bar{g} = \mathrm{fl}(\sqrt{\bar{t}}) = 0.8316 \times 10^0$$

If we use a stable choice of the sign, then

$$\bar{h} = \mathrm{fl}(\bar{t} + a_1\bar{g}) = 0.1383 \times 10^1$$
$$\bar{u}_1 = \mathrm{fl}(a_1 + \bar{g}) = 0.1663 \times 10^1$$
$$\bar{u} = [0.1663 \times 10^1, \ 0.3917 \times 10^{-3}, \ 0.2651 \times 10^{-3}]^{\mathrm{T}}$$

The exact values under this choice of sign are (up to eight digits)

$$g = 0.831\,600\,13 \times 10^0 \qquad h = 0.138\,311\,75 \times 10^1$$
$$u = [0.166\,320\,01 \times 10^1, 0.391\,700\,00 \times 10^{-3}, 0.265\,100\,00 \times 10^{-3}]^{\mathrm{T}}.$$

It may be verified that the computed quantities satisfy the inequalities (1.52).

Using the unstable choice of sign we obtain

$$\bar{g} = -0.8316 \times 10^0$$
$$\bar{h} = \text{fl}(\bar{t} + a_1\bar{g}) = 0.4144 \times 10^{-4} \qquad \bar{u}_1 = \text{fl}(a_1 + \bar{g}) = 0.0000 \times 10^0$$
$$\bar{u} = [0.0000 \times 10^0, \ 0.3917 \times 10^{-3}, \ 0.2651 \times 10^{-3}]^\text{T}$$

The exact values corresponding to this choice of sign are (again up to eight digits)

$$h = 0.111\ 853\ 46 \times 10^{-6}$$
$$u_1 = -0.134\ 503\ 90 \times 10^{-6}$$
$$u = [-0.134\ 503\ 90 \times 10^{-6}, \ 0.391\ 700\ 00 \times 10^{-3},$$
$$0.265\ 100\ 00 \times 10^{-3}]^\text{T}$$

It may be shown that in this case $\bar{U} = I - \bar{u}\bar{u}^\text{T}/\bar{h}$ is not even approximately orthogonal and thus it is not close to U. This is a consequence of the fact that $\bar{h}$ has a very large relative error due to subtractive cancellation.

After determination of U it is frequently required to compute the product UA where A is a given $n \times m$ matrix. If A is written as

$$A = [a_1, a_2, ..., a_m]$$

then

$$UA = [Ua_1, Ua_2, ..., Ua_m]$$

To reduce the computations the matrix U is not formed explicitly using the fact that

$$Ua_j = a_j - h^{-1}(u^\text{T}a_j)u$$

This gives the following algorithm which overwrites the matrix A by UA. Note that it uses column access to the elements of A.

Algorithm 1.17   Left multiplication by an elementary reflection

For $j = 1, 2, ..., m$

Set $t = h^{-1} \sum_{i=1}^{n} u_i a_{ij}$

$a_{ij} \leftarrow a_{ij} - t u_i \qquad (i = 1, 2, ..., n)$

This algorithm requires $2nm + m$ flops.

A straightforward but cumbersome error analysis shows that

$$\| \mathrm{fl}(\bar{U}A) - UA \|_F \leq (6n + 27)\varepsilon \| A \|_F$$

From the backward error analysis point of view it may be written that

$$
\begin{aligned}
\mathrm{fl}(\bar{U}A) &= U(A + \Delta A_1) \\
\| \Delta A_1 \|_F &\leq (6n + 27)\varepsilon \| A \|_F
\end{aligned}
\tag{1.53}
$$

In this way the computed product $\mathrm{fl}(\bar{U}A)$ may be considered as a result of the exact multiplication of U with a slightly perturbed matrix A.

It should be noted that the bound (1.53) does not depend explicitly on the number of the columns of A. This bound is also valid if the vector a, which determines $\bar{U}$, is a column of A.

The computation of the product AU, where A is a given $m \times n$ matrix, may be performed by the following column-oriented algorithm which requires an additional array with m elements necessary for computation of the product Au.

Algorithm 1.18 Right multiplication by an elementary reflection

Set $t_i = 0 \qquad (i = 1, 2, \ldots, m)$
For $j = 1, 2, \ldots, n$
$\qquad t_i \leftarrow t_i + a_{ij}u_j \qquad (i = 1, 2, \ldots, m)$
For $j = 1, 2, \ldots, n$
$\qquad a_{ij} \leftarrow a_{ij} - h^{-1}t_iu_j \qquad (i = 1, 2, \ldots, m)$

The product computed by Algorithm 1.18 fulfils

$$
\begin{aligned}
\mathrm{fl}(A\bar{U}) &= (A + \Delta A_2)U \\
\| \Delta A_2 \|_F &\leq (6n + 27)\varepsilon \| A \|_F
\end{aligned}
$$

EXERCISES

1.37 Verify (1.43) and (1.44).
1.38 Write an algorithm for right multiplication of a matrix by a plane rotation.
1.39 Show that the elementary reflection U is *involutory*, i.e. $U^2 = I$.
1.40 Verify (1.52).
1.41 Give an efficient algorithm for the construction of 2×2 and 3×3 elementary reflections.
1.42 Compare the efficiency and accuracy in implementing plane rotations and 2×2 elementary reflections.

1.8 QR DECOMPOSITION AND LEAST SQUARES SOLUTIONS

1.8.1 Orthogonal Triangularization

Using a sequence of elementary orthogonal transformations, every $n \times m$ matrix may be reduced to upper triangular (if $n \geq m$) or trapezoidal (if $n < m$) form. This reduction is known as *orthogonal triangularization* or *Householder reduction*. The orthogonal triangularization has several important applications in the computational solution of matrix problems.

The algorithm for orthogonal triangularization produces elementary reflections $U_1, U_2, ..., U_l$ such that the product $A_{l+1} = U_l U_{l-1} ... U_1 A$ has the desired upper triangular (trapezoidal) form. This algorithm is an orthogonal analogue of the Gaussian elimination without pivoting.

Assume for definiteness that $n > m$. First an $n \times n$ elementary reflection U_1 is determined so that

$$U_1 a_1 = [r_{11}, 0, ..., 0]^T$$

where a_1 is the first column of A. When this reflection is applied to A one obtains

$$A_2 = U_1 A = \begin{bmatrix} r_{11} & r_{12} & \cdots & r_{1m} \\ 0 & & & \\ \vdots & & A_2' & \\ 0 & & & \end{bmatrix}$$

A second reflection U_2' is chosen to transform the first column a_2' of A_2' as

$$U_2' a_2' = [r_{22}, 0, ..., 0]^T$$

The left multiplication of A_2 by

$$U_2 = \begin{bmatrix} 1 & 0 \\ 0 & U_2' \end{bmatrix}$$

yields

$$A_3 = U_2 A_2 = \begin{bmatrix} r_{11} & r_{12} & r_{13} & \cdots & r_{1m} \\ 0 & r_{22} & r_{23} & \cdots & r_{2m} \\ 0 & 0 & & & \\ \vdots & \vdots & & A_3' & \\ 0 & 0 & & & \end{bmatrix}$$

This transformation does not affect the first column of A_2.

Continuing in this way, at the mth stage of the reduction we obtain the matrix

$$A_{m+1} = U_m A_m = U_m U_{m-1} \dots U_1 A = \begin{bmatrix} r_{11} & r_{12} & \dots & r_{1m} \\ & r_{22} & \dots & r_{2m} \\ & & \ddots & \vdots \\ 0 & & & r_{mm} \end{bmatrix} = \begin{bmatrix} R \\ 0 \end{bmatrix}$$

where R is upper triangular (note that some of the diagonal elements r_{ii} may be equal to zero).

If $n < m$ the reduction of A into upper trapezoidal form requires $n - 1$ elementary reflections.

Example 1.17

If

$$A = \begin{bmatrix} 3 & -1 & 6 \\ 5 & 2 & -7 \\ -4 & -8 & -3 \\ 6 & -5 & -9 \end{bmatrix}$$

then up to three significant digits

$$U_1 = \begin{bmatrix} -0.323 & -0.539 & 0.431 & -0.647 \\ -0.539 & 0.780 & 0.176 & -0.264 \\ 0.431 & 0.176 & 0.859 & 0.211 \\ -0.647 & -0.264 & 0.211 & 0.684 \end{bmatrix}$$

$$A_2 = U_1 A = \begin{bmatrix} -9.27 & -0.970 & 6.36 \\ 0 & 2.01 & -6.85 \\ 0 & -8.01 & -3.12 \\ 0 & -4.99 & -8.82 \end{bmatrix}$$

$$U_2 = \begin{bmatrix} 1 & 0 & 0 & 0 \\ 0 & -0.209 & 0.830 & 0.517 \\ 0 & 0.830 & 0.430 & -0.355 \\ 0 & 0.517 & -0.355 & 0.779 \end{bmatrix}$$

$$A_3 = U_2 A_2 = \begin{bmatrix} -9.27 & -0.970 & 6.36 \\ 0 & -9.65 & -5.72 \\ 0 & 0 & -3.90 \\ 0 & 0 & -9.31 \end{bmatrix}$$

$$U_3 = \begin{bmatrix} 1 & 0 & 0 & 0 \\ 0 & 1 & 0 & 0 \\ 0 & 0 & -0.386 & -0.922 \\ 0 & 0 & -0.922 & 0.386 \end{bmatrix}$$

$$A_4 = U_3 A_3 = \begin{bmatrix} -9.27 & -0.970 & 6.36 \\ 0 & -9.65 & -5.72 \\ 0 & 0 & 10.1 \\ 0 & 0 & 0 \end{bmatrix}$$

Suppose that A has linearly independent columns (this means that by necessity $n \geq m$). Defining the orthogonal matrix $Q^T = U_m U_{m-1} \ldots U_1$, the matrix A_{m+1} may be represented as

$$A_{m+1} = Q^T A = \begin{bmatrix} R \\ 0 \end{bmatrix}$$

where R is an $m \times m$ upper triangular matrix. If Q is partitioned as

$$Q = [Q_1, \ Q_2]$$

where Q_1 has m columns, then

$$A = Q A_{m+1} = [Q_1, \ Q_2] \begin{bmatrix} R \\ 0 \end{bmatrix} = Q_1 R \qquad (1.54)$$

The representation (1.54), in which Q_1 has orthonormal columns and R is upper triangular, is called *QR decomposition* of A.

The matrix R in (1.54) is obtained explicitly by the orthogonal triangularization. The matrix Q_1 may be computed as the first m columns of the product

$$Q = U_1 U_2 \ldots U_m \qquad (1.55)$$

It should be noted, however, that in most applications the matrix Q or Q_1 is not required in explicit form and it is sufficient to know the factorized form (1.55).

1.8.2 QR Decomposition with Column Pivoting

If there is a linear dependence between the columns of A (i.e. $\operatorname{rank}(A) = r < m$) then some of the diagonal elements of the matrix R in the QR decomposition of A are zero. The result of the reduction in this case may look as follows

$$Q^T A = \begin{bmatrix} x & x & x & x \\ 0 & x & 0 & x \\ 0 & 0 & 0 & x \\ 0 & 0 & 0 & 0 \\ 0 & 0 & 0 & 0 \end{bmatrix}$$

where x denotes a nonzero element. Here the first and the third columns of A are linearly dependent and rank$(A) = 3$.

In practice it is more convenient to obtain the first $r = $ rank(A) diagonal elements of R as nonzero since this facilitates the determination of the rank of the transformed matrix. For this purpose it is necessary to interchange the columns of A during the reduction so that the linearly independent columns of A appear first. This yields *QR decomposition with column pivoting* which allows presentation of every $n \times m$ matrix $A, n > m$, as

$$AP = Q \begin{bmatrix} R \\ 0 \end{bmatrix}$$

where the first $r = $ rank(A) diagonal elements of R are nonzero and P is a permutation matrix which reflects the moving of the columns during the reduction.

The QR decomposition with column pivoting may be obtained as follows. At the kth stage of the reduction one is searching for the column of the $(n - k + 1) \times (m - k + 1)$ matrix A_k' situated in the lower right corner of A_k, which is of maximal norm. This column is interchanged with the kth column of A_k and after that the reduction is performed. To reduce the computations, the norms of the columns of A may be determined in advance and their values updated at each stage in order to take into account the removing of rows. This strategy results in a matrix R whose elements satisfy

$$r_{kk}^2 \geq r_{kj}^2 + \cdots + r_{jj}^2 \qquad j = k + 1, \ldots, m$$

This means that if $r_{kk} = 0$ then the rows from k to m are also zero and hence rank$(A) = k - 1$. In this way the QR decomposition with column pivoting may be used to determine the rank of a matrix. However, since the condition $r_{kk} = 0$ is not exactly fulfilled in practice, a suitable threshold should be used (see for details Dongarra *et al.* (1979, ch. 9)).

The updating of the norms may be done as follows. Let the matrix A_k' be represented as

$$A_k' = [a_k, \ a_{k+1}, \ \ldots, \ a_m]$$

After the reduction of a_k one obtains

$$U_k' A_k' = \begin{bmatrix} r_{kk} & r_{k,k+1} & \cdots & r_{km} \\ 0 & a_{k+1}' & \cdots & a_m' \end{bmatrix}$$

Since U_k' is orthogonal it follows that

$$\| a_j \|_2^2 = \left\| \begin{matrix} r_{kj} \\ a_j' \end{matrix} \right\|_2^2 = r_{kj}^2 + \| a_j' \|_2^2 \qquad j = k + 1, \ldots, m$$

and

$$\| a_j' \|_2 = \| a_j \|_2 \sqrt{\left(1 - \left(\frac{r_{kj}}{\| a_j \|_2}\right)^2\right)} \tag{1.56}$$

This allows us to find easily the norm of a_j' from that of a_j. It should be noted, however, that with $r_{kj}/\| a_j \|_2$ approaching 1 there may be a significant loss of accuracy in using (1.56) because of subtractive cancellation. In such cases a recomputation of $\| a_j' \|_2$ from the elements of a_j' is required.

Algorithm 1.19 determines elementary reflections $U_1, ..., U_l$, $l = \min\{n - 1, m\}$ and a permutation matrix P such that

$$A_{l+1} = U_l ... U_1 A P = Q^T A P \qquad Q = U_1 U_2 ... U_l$$

is upper triangular (trapezoidal) and its first columns are linearly independent. The permutations are coded in the vector p so that

$$P = [e_{p_1}, e_{p_2}, ..., e_{p_m}]$$

where e_j $(j = 1, 2, ..., m)$ is the jth column of the unit matrix.

Algorithm 1.19 QR decomposition with column pivoting

Set $p_j = j$ $\qquad (j = 1, 2, ..., m)$

Set $q_j = \| a_j \|_2 = \sqrt{\sum_{i=1}^{n} a_{ij}^2}$ $\qquad (j = 1, 2, ..., m)$

For $k = 1, 2, ..., \min\{n, m\}$
    Find s so that
        $q_s = \max_{k \leq j \leq m} q_j$

    If $q_s = 0$, quit
    If $s \neq k$
        $a_{ik} \leftrightarrow a_{is}$ $\qquad (i = 1, 2, ..., n)$
        $q_s \leftarrow q_k$
        $p_k \leftrightarrow p_s$
    $q_k \leftarrow 0$
    If $k = n$, quit
    Determine an elementary reflection U_k' such that

$$U_k' \begin{bmatrix} a_{kk} \\ a_{k+1,k} \\ \vdots \\ a_{nk} \end{bmatrix} = \begin{bmatrix} r_{kk} \\ 0 \\ \vdots \\ 0 \end{bmatrix}$$

$a_{kk} \leftarrow r_{kk}$

If $k < m$

   For $j = k + 1, k + 2, \ldots, m$

$$\begin{bmatrix} a_{kj} \\ a_{k+1,j} \\ \vdots \\ a_{nj} \end{bmatrix} \leftarrow U_k' \begin{bmatrix} a_{kj} \\ a_{k+1,j} \\ \vdots \\ a_{nj} \end{bmatrix}$$

   If $q_j = 0$, step j

   $q_j \leftarrow q_j \sqrt{(1 - (a_{kj}/q_j)^2)}$

The construction of the reflection

$$U_k' = I - \frac{u_k u_k^{\mathrm{T}}}{h_k} \qquad u_k = [u_{kk}, \ldots, u_{nk}]^{\mathrm{T}}$$

is done by Algorithm 1.16. The last $n - k$ elements of the vector u_k may overwrite the corresponding elements of A. The quantity u_{kk} should be stored in the kth position of an auxiliary one-dimensional array.

     The implementation of U_k' on A_k' is accomplished by Algorithm 1.17. This algorithm requires approximately $2mnr - r^2(m + n) + \frac{2}{3}r^3$ flops, where $r = \mathrm{rank}(A)$.

Example 1.18

If Algorithm 1.19 is applied to

$$A = \begin{bmatrix} 1 & 5 & 9 \\ 2 & 6 & 10 \\ 3 & 7 & 11 \\ 4 & 8 & 12 \end{bmatrix}$$

then $P = [e_3, \ e_1, \ e_2]$ and (up to three decimal digits) one obtains

$$Q = \begin{bmatrix} -0.426 & -0.720 & 0.544 & -0.0626 \\ -0.474 & -0.275 & -0.679 & 0.489 \\ -0.521 & 0.169 & -0.275 & -0.790 \\ -0.568 & 0.614 & 0.409 & 0.363 \end{bmatrix}$$

$$R = \begin{bmatrix} -21.1 & -5.21 & -13.2 \\ 0 & 1.69 & 0.847 \\ 0 & 0 & 0 \\ 0 & 0 & 0 \end{bmatrix}$$

1.8.3 Error Analysis of the Orthogonal Triangularization

Let A be an $n \times m$ matrix whose elements are floating-point numbers. Based on the results from Sections 1.4 and 1.7 we can find the error arising in the implementation of $l = \min\{n - 1, m\}$ by successive elementary reflections beginning with the matrix A. The desired mathematical operation is

$$A_1 = A$$
$$A_{k+1} = \hat{U}_k A_k \qquad k = 1, \ldots, l$$

where the elementary reflection $\hat{U}_k$ is constructed so that zeros are introduced in the positions from $k + 1$ to n of the kth column of $\hat{U}_k A_k$.

The computed quantities are

$$\bar{A}_1 = A$$
$$\bar{A}_{k+1} = \text{fl}(\bar{U}_k \bar{A}_k) \qquad k = 1, \ldots, l$$

where $\bar{U}_k$ is the computed approximation of the true matrix U_k which would introduce zeros in the elements $k + 1$ to n of the kth column of the matrix $U_k \bar{A}_k$.

An error matrix is defined as

$$F_k = \text{fl}(\bar{U}_k \bar{A}_k) - U_k \bar{A}_k = \bar{A}_{k+1} - U_k \bar{A}_k \qquad k = 1, \ldots, l \qquad (1.57)$$

Using the results from Section 1.7 it may be written that

$$\| F_k \|_F \leq f_{n-k+1} \| \bar{A}_k \|_F \qquad k = 1, \ldots, l$$

where

$$f_j = (6_j + 27)\varepsilon$$

Combining Equations (1.57) for $k = 1, \ldots, l$ we obtain

$$\bar{A}_{l+1} = U_l \ldots U_1 A + U_l \ldots U_2 F_1 + U_l \ldots U_3 F_2 + \cdots + F_1$$

From the orthogonality of the matrices U_k it follows that

$$\| A_{l+1} - U_l \ldots U_1 A \|_F \leq \sum_{k=1}^{l} f_{n-k+1} \| \bar{A}_k \|_F$$

$$\leq \| A \|_F \sum_{k=1}^{l} f_{n-k+1}$$

$$\leq (6n - 3l + 30)l\varepsilon \| A \|_F \qquad (1.58)$$

It is convenient to define the error matrix

$$\Delta A = U_l \ldots U_2 \bar{A}_{l-1} - A$$

Then

$$\bar{A}_{l+1} = U_l \ldots U_1 (A + \Delta A) \tag{1.59}$$

and from Equation (1.58)

$$\| \Delta A \|_F \leq (6n - 3l + 30)l\varepsilon \| A \|_F \tag{1.60}$$

Equation (1.59) and the bound (1.60) show that the computed matrix $\bar{A}_{l+1}$ is the exact result of an orthogonal transformation of a matrix $A + \Delta A$, where $\| \Delta A \|_F$ is small compared to $\| A \|_F$. This proves that Algorithm 1.19 is numerically stable.

If in (1.58) we take $A = I$, it is possible to obtain the error in the computed product of l approximated elementary reflections. This gives

$$\| \text{fl}(\bar{U}_l \ldots \bar{U}_1) - U_l \ldots U_1 \|_F \leq \sqrt{n}\,(6n - 3l + 30)l\varepsilon$$

The bound obtained shows that the computed product differs slightly from an orthogonal matrix.

1.8.4 Solution of the Least Squares Problem

The *linear least squares problem* is to find a solution of the (overdetermined) linear system $Ax \approx b$ in the sense that the m-vector x must satisfy

$$\rho(x) = \| b - Ax \|_2^2 = \min \tag{1.61}$$

where A is a given $n \times m$ matrix and b is an n-vector. This problem always has a solution which is unique, if and only if A has linearly independent columns, i.e. if $\text{rank}(A) = m \leq n$. Later we shall assume that this condition is fulfilled.

Let x be the solution of the least squares problem (1.61). Then it may be proved that the *residual vector* $r = b - Ax$ satisfies

$$A^T r = 0$$

so that

$$A^T Ax = A^T b \tag{1.62}$$

The square system of Equations (1.62) is called the *system of normal equations*. Since A has linearly independent columns, $A^T A$ is nonsingular and the solution x may be obtained as

$$x = (A^T A)^{-1} A^T b \tag{1.63}$$

If we define

$$A^+ = (A^T A)^{-1} A^T$$

then (1.63) is written as

$$x = A^+ b$$

The matrix A^+ is called the *pseudo-inverse* or *generalized inverse of Moore–Penrose*.

Analogically to the case when $m = n$, the condition number of A is defined as

$$\text{cond}_2(A) = \| A \|_2 \| A^+ \|_2$$

The use of the condition number gives the possibility of estimating the accuracy of the solution of the least squares problem. More precisely, let E be an error matrix such that rank $(A + E)$ remains equal to m. Let x be the solution of (1.61) and let $\bar{x}$ be the solution of the same problem with A replaced by $A + E$. Then it may be proved (see Stewart (1977)) that

$$\frac{\| \bar{x} - x \|_2}{\| x \|_2} \leqq 1.6 \left[\text{cond}_2(A) + \text{cond}_2^2(A) \, \frac{\| r \|_2}{\| A \|_2 \| x \|_2} \right] \frac{\| E \|_2}{\| A \|_2}$$

Let rank$(A) = m$. Then

$$\text{cond}_2^2(A) = \text{cond}_2(A^T A)$$

In fact, since

$$\| A \|_2^2 = \| A^T A \|_2$$

from the definition of A^+ we obtain that

$$\| A^+ \|_2^2 = \| A^+ A^{+T} \|_2 = \|(A^T A)^{-1}\|_2$$

Hence

$$\text{cond}_2^2(A) = (\| A \|_2 \| A^+ \|_2)^2 = \text{cond}_2(A^T A)$$

Equation (1.63) gives one of the possible ways for solving the least squares problem. Since rank$(A) = m$, the matrix $A^T A$ is positive definite and may be decomposed by the Cholesky algorithm 1.8 (see Section 1.6). This leads to the following example algorithm.

Set $C = A^T A$
Set $d = A^T b$
Solve $Cx = d$ using Algorithms 1.8 and 1.12

This algorithm has some disadvantages which make it inappropriate for the computational solution of the least squares problem. The first difficulty arises from the fact that $\text{cond}_2(A^T A) = \text{cond}_2^2(A)$ so that the conditioning of the system of normal equations may be much worse than that of the original problem. The second difficulty is related to the fact that the computed matrix A^TA may be not positive definite so that the Cholesky algorithm cannot be used.

Example 1.19

Let

$$A = \begin{bmatrix} 1.000 & 1.020 \\ 1.000 & 1.000 \\ 1.000 & 1.000 \end{bmatrix}$$

The matrix A^TA, rounded to four significant digits, is

$$C = \begin{bmatrix} 3.000 & 3.020 \\ 3.020 & 3.040 \end{bmatrix}$$

and it is not positive definite within this accuracy.

The solution of the least squares problem may be done in a numerically stable way using the QR decomposition of the matrix A,

$$A = Q\begin{bmatrix} R \\ 0 \end{bmatrix} = Q_1 R$$

Since A has linearly independent columns, R is a nonsingular $m \times m$ upper triangular matrix.
Setting

$$Q^Tb = \begin{bmatrix} c \\ d \end{bmatrix}$$

where c is an m-vector, and multiplying $r = b - Ax$ with Q^T, one obtains

$$Q^Tr = \begin{bmatrix} c \\ d \end{bmatrix} - \begin{bmatrix} R \\ 0 \end{bmatrix}x = \begin{bmatrix} c - Rx \\ d \end{bmatrix} \tag{1.64}$$

From the orthogonality of Q it follows that

$$\| r \|_2^2 = \| Q^Tr \|_2^2 = \| c - Rx \|_2^2 + \| d \|_2^2 \tag{1.65}$$

Since d does not depend on x, $\rho(x) = \| r \|_2^2$ will be minimized if x is chosen so that

$$c - Rx = 0 \tag{1.66}$$

Then it follows from (1.65) that $\rho(x) = \| d \|_2^2$. Moreover, from (1.64) we obtain

$$Q^\mathrm{T} r = \begin{bmatrix} 0 \\ d \end{bmatrix}$$

so that

$$r = Q \begin{bmatrix} 0 \\ d \end{bmatrix} = Q_2 d \tag{1.67}$$

Example 1.20

Let

$$A = \begin{bmatrix} 3 & -1 & 6 \\ 5 & 2 & -7 \\ -4 & -8 & -3 \\ 6 & -5 & -9 \end{bmatrix} \quad \text{and} \quad b = \begin{bmatrix} 3 \\ -1 \\ -6 \\ 2 \end{bmatrix}$$

Using the orthogonal triangularization of A (see Example 1.17) we obtain (up to three decimal digits)

$$c = \begin{bmatrix} -4.31 \\ -2.99 \\ 3.50 \end{bmatrix} \quad d = 3.19$$

$$x = \begin{bmatrix} 0.693 \\ 0.104 \\ 0.347 \end{bmatrix} \quad r = \begin{bmatrix} 1.06 \\ 2.24 \\ 1.36 \\ -1.49 \end{bmatrix}$$

Equations (1.64), (1.66) and (1.67) give the solution of the linear least squares problem $\| Ax - b \|_2^2 = \min$ using the QR decomposition of the matrix A. The product $Q^\mathrm{T} b$ may be computed in the form $U_m U_{m-1} \ldots U_1 b$ without forming Q. It is possible to compute r in (1.67) in the same way using the factorized form of Q.

If the QR decomposition of the $n \times m$ matrix A, rank$(A) = m$, is obtained by Algorithm 1.19, then the least squares problem may be solved by Algorithm 1.20. The residual vector $r = Ax - b$, corresponding to the minimum, overwrites b.

Algorithm 1.20 Solution of the linear least squares problem by QR decomposition

For $k = 1, 2, ..., m$
  $b \leftarrow U_k b$
Set $c_i = b_i$ $(i = 1, 2, ..., m)$
Solve $Rx = c$ by Algorithm 1.10
  $x \leftarrow Px$
Set $b_i = 0$ $(i = 1, 2, ..., m)$
For $k = m, m - 1, ..., 1$
  $b \leftarrow U_k b$

This algorithm requires approximately n^2 flops to form $Q^T b$ and $\frac{1}{2} n^2$ flops to solve $Rx = c$.

Consider the numerical properties of Algorithm 1.20. For simplicity we shall neglect the column pivoting during the QR decomposition of A. Using the error analysis of the orthogonal triangularization, it is possible to write

$$Q^T(A + E) = \begin{bmatrix} R \\ 0 \end{bmatrix} \tag{1.68}$$

and

$$Q^T(b + e) = \begin{bmatrix} c \\ d \end{bmatrix}$$

where

$$\| E \|_F \leq (6n - 3m + 30)m\varepsilon \| E \|_F$$
$$\| e \|_2 \leq (6n - 3m + 30)m\varepsilon \| b \|_2$$

The solution of $Rx = c$ satisfies

$$(R + F)x = c \tag{1.69}$$

where the elements of F fulfil $|f_{ij}| < m\varepsilon |r_{ij}|$ so that

$$\| F \|_F < m\varepsilon \| R \|_F = m\varepsilon \| A \|_F$$

If we set

$$G = E + Q \begin{bmatrix} F \\ 0 \end{bmatrix}$$

then from the orthogonality of Q it follows that

$$\| G \|_F = \| E \|_F + \| F \|_F \leq (6n - 3m + 31)m\varepsilon \| A \|_F$$

Also

$$Q(A + G) = \begin{bmatrix} R + F \\ 0 \end{bmatrix} \qquad (1.70)$$

From (1.70), (1.68) and (1.69) we obtain that the computed solution x is the exact solution of the least squares problem

$$\|(A + G)x - (b + e)\|_2^2 = \min$$

with slightly perturbed matrix A and vector b. This proves the numerical stability of Algorithm 1.20.

To estimate the accuracy of the solution it is necessary to know the condition number of A. If the QR decomposition with column pivoting of A is performed, then the conditioning may be estimated by

$$\text{cond}_2(A) \approx r_{11}/r_{mm}$$

Algorithms 1.19 and 1.20 may be applied to solve square systems $Ax = b$ but this approach is less efficient (although more reliable) than Gaussian elimination with pivoting.

The QR decomposition may be used also to solve undetermined systems of equations $Ax = b$, where A is an $n \times m$ matrix with $n < m$. In this case the QR decomposition with column pivoting of A has the form

$$Q^T A P = [R, S]$$

where P is a permutation matrix and R is an $n \times n$ upper triangular matrix. If A has n linearly independent columns ($\text{rank}(A) = n$), then the column pivoting will lead to a nonsingular matrix R, and a solution of $Ax = b$ is given by

$$x = P \begin{bmatrix} R^{-1} Q^T b \\ 0 \end{bmatrix}$$

Consider finally the case $\text{rank}(A) = r < \min\{n, m\}$ when

$$Q^T A P = \begin{bmatrix} R_1 & R_2 \\ 0 & 0 \end{bmatrix}$$

In this decomposition the $r \times r$ matrix R_1 is nonsingular. Hence the solution of the least squares problem is

$$x = P \begin{bmatrix} R_1^{-1}(c - R_2 y) \\ y \end{bmatrix} \qquad c = [I_r, 0] Q^T b$$

where the $(m - r)$-vector y is arbitrary. For the minimum norm solution, y is obtained from the linear equation

$$(I_{m-r} + R_2^T R_1^{-T} R_1^{-1} R_2)y = R_2^T R_1^{-T} R_1^{-1} c$$

EXERCISES

1.43 Write a computer program implementing Algorithm 1.20.

1.44 Write an algorithm for orthogonal triangularization utilizing plane rotations instead of elementary reflections.

1.45 Write an algorithm for the decomposition of an $n \times m$ matrix A, $n < m$, as

$$A = [L, 0]Q$$

where L is a lower triangular and Q is an orthogonal matrix.

1.46 Prove that the pseudo-inverse A^+ of A satisfies

$$AA^+A = A$$
$$A^+AA^+ = A^+$$
$$A^+A = (A^+A)^T$$
$$AA^+ = (AA^+)^T$$

1.47 Derive an error bound for the solution of the linear least squares problem.

1.48 Give an algorithm for least squares solution of the overdetermined system of equations $Ax = b$, where A is an $n \times m$ matrix with $n > m$, $\text{rank}(A) < m$, by using QR decomposition with column pivoting.

1.9 THE EIGENVALUE PROBLEM

1.9.1 Eigenvalues and Eigenvectors

One of the most important and difficult problems in the area of matrix computations is the determination of the eigenvalues and eigenvectors of a square matrix. This problem has a simple theoretical formulation but its computational solution may have associated considerable difficulties.

The *algebraic eigenvalue problem* consists in the determination of those values of λ for which the system of linear equations

$$Ax = \lambda x \tag{1.71}$$

where A is a given $n \times n$ matrix, has a nontrivial solution.

The equation (1.71) may be written in the form

$$(\lambda I - A)x = 0 \tag{1.72}$$

and a nontrivial solution exists if and only if the matrix $\lambda I - A$ is singular, i.e.

$$\det(\lambda I - A) = 0 \tag{1.73}$$

Expanding the determinant on the left-hand side of (1.73) one obtains

$$\lambda^n + a_n\lambda^{n-1} + \cdots + a_2\lambda + a_1 = 0 \tag{1.74}$$

Equation (1.74) is called the *characteristic equation* of the matrix A and the polynomial on the left-hand side of this equation is referred to as the *characteristic polynomial*. Equation (1.74) always has n roots and they may be complex even if the matrix A is real. These roots are called *eigenvalues* or *characteristic values* of the matrix A. If the eigenvalue λ is repeated k times we say that λ is a *multiple eigenvalue* with multiplicity k. If $k = 1$ the eigenvalue is called *simple*. To any eigenvalue λ corresponds at least one solution for x. This solution is referred to as an *eigenvector* or a *characteristic vector*, corresponding to λ. If x is a solution of (1.72) then cx is also a solution for any $c \neq 0$, so that the eigenvectors are determined up to an arbitrary nonzero scalar multiple.

Since the characteristic polynomial of a matrix with real elements has real coefficients, the complex eigenvalues of a real matrix occur in conjugate pairs. If the matrix is symmetric, it has only real eigenvalues.

The set of the eigenvalues of A is called the *spectrum* of A and is denoted by $\lambda(A)$.

Example 1.21

The matrix

$$A = \begin{bmatrix} 8 & -1 & -5 \\ -4 & 4 & -2 \\ 18 & -5 & -7 \end{bmatrix}$$

has a characteristic equation

$$\det(\lambda I - A) = \lambda^3 - 5\lambda^2 + 24\lambda - 20 = 0$$

The roots of this equation, i.e. the eigenvalues of A, are

$$\lambda_1 = 1 \qquad \lambda_2 = 2 + 4i \qquad \lambda_3 = 2 - 4i$$

The eigenvectors, corresponding to λ_1, λ_2 and λ_3, are respectively

$$x_1 = \begin{bmatrix} 1 \\ 2 \\ 1 \end{bmatrix} \qquad x_2 = \begin{bmatrix} 1-i \\ 2 \\ -2i \end{bmatrix} \qquad x_3 = \begin{bmatrix} 1+i \\ 2 \\ 2i \end{bmatrix}$$

Since

$$\det(\lambda I - A^T) = \det(\lambda I - A)$$

the eigenvalues of A^T coincide with those of A. The eigenvectors, however, are generally different. Denoting by y_j the eigenvector of A^T corresponding to λ_j, it may be written that

$$A^T y_j = \lambda_j y_j$$

or

$$y_j^T A = \lambda_j y_j^T$$

That is why y_j^T is referred to as the *left eigenvector* of A, corresponding to λ_j. For the matrix from Example 1.21 the left eigenvectors, corresponding to $\lambda_1 = 1$, $\lambda_2 = 2 + 4i$ and $\lambda_3 = 2 - 4i$, are respectively

$$y_1^T = [-2, 1, 1]$$
$$y_2^T = [10, -3 - i, -4 + 2i]$$
$$y_3^T = [10, -3 + i, -4 - 2i]$$

The eigenvectors x_j, where $j = 1, 2, ..., n$, are frequently referred to as *right eigenvectors*.

The eigenvalues of some classes of matrices are determined easily. If, for instance, the matrix is diagonal or triangular, the eigenvalues are its diagonal elements.

Let $J_\lambda^{(k)}$ be a matrix of kth order in the form

$$J_\lambda^{(k)} = \begin{bmatrix} \lambda & 1 & & & \mathbf{0} \\ & \lambda & 1 & & \\ & & \ddots & \ddots & \\ & & & \lambda & 1 \\ \mathbf{0} & & & & \lambda \end{bmatrix} \tag{1.75}$$

Then $J_\lambda^{(k)}$ has λ as its eigenvalue with multiplicity k. The eigenvectors of $J_\lambda^{(k)}$ are multiples of $[1, 0, ..., 0]^T$. A matrix in the form (1.75) is called a *Jordan block*.

A square matrix A is called *defective* if it has an eigenvalue λ of multiplicity k having less than k linearly independent eigenvectors. The Jordan block (1.75) is an example of a defective matrix. It should be noted that there are theoretical and computational difficulties associated with defective matrices, as will be shown later in this section.

Let

$$p(\lambda) = \lambda^n + a_n \lambda^{n-1} + \cdots + a_2 \lambda + a_1$$

be a polynomial with roots $\mu_1, \mu_2, ..., \mu_n$. Then the matrix

$$C_p = \begin{bmatrix} 0 & 1 & 0 & \cdots & 0 \\ 0 & 0 & 1 & \cdots & 0 \\ \vdots & \vdots & \vdots & & \vdots \\ -a_1 & -a_2 & -a_3 & \cdots & -a_n \end{bmatrix}$$

has $p(\lambda)$ as its characteristic polynomial. Therefore the eigenvalues of C_p are $\mu_1, \mu_2, ..., \mu_n$. This matrix is referred to as a *companion matrix* of the polynomial $p(\lambda)$.

If

$$Ax = \lambda x$$

and P is a nonsingular $n \times n$ matrix, then

$$(P^{-1}AP)P^{-1}x = \lambda P^{-1}x \tag{1.76}$$

The matrices A and $P^{-1}AP$ are called *similar*. It follows from (1.76) that the similar matrices have the same eigenvalues. Hence, the eigenvalues of a matrix are invariant to similarity transformations. The eigenvectors, however, are multiplied by P^{-1}.

Several computational methods for finding eigenvalues and eigenvectors consist in the determination of a similarity transformation, which reduces a matrix A of a general form to a matrix $\tilde{A} = P^{-1}AP$ of some special form. This special form is chosen so that the eigenvalue problem may be easily solved. If, for example, the matrix is reduced to triangular form then the diagonal elements of this form are the eigenvalues of the original matrix.

If the eigenvalues $\lambda_1, \lambda_2, ..., \lambda_k$ of a matrix A are pairwise different then its eigenvectors $x_1, x_2, ..., x_k$ are linearly independent, $k \leq n$. This implies that the matrix A may be reduced to diagonal form using similarity transformation. In fact, from

$$Ax_j = \lambda_j x_j \qquad j = 1, 2, ..., n$$

it follows that

$$AX = XD \tag{1.77}$$

where

$$X = [x_1, x_2, ..., x_n] \qquad \text{and} \qquad D = \text{diag}(\lambda_1, \lambda_2, ..., \lambda_n)$$

Since the columns of X are linearly independent, $\text{rank}(X) = n$ and the inverse X^{-1} exists. That is why (1.77) may be represented as

$$X^{-1}AX = D = \text{diag}(\lambda_1, \lambda_2, ..., \lambda_n)$$

Example 1.22

Consider the matrix

$$A = \begin{bmatrix} 33 & 16 & 72 \\ -24 & -10 & -57 \\ -8 & -4 & -17 \end{bmatrix}$$

whose eigenvalues are $\lambda_1 = 1$, $\lambda_2 = 2$ and $\lambda_3 = 3$. The matrix formed by the corresponding eigenvectors is

$$X = \begin{bmatrix} -15 & -16 & -4 \\ 12 & 13 & 3 \\ 4 & 4 & 1 \end{bmatrix}$$

and

$$X^{-1}AX = \text{diag}(1, 2, 3)$$

The more general result is that a given matrix may be reduced to diagonal form if it is not defective, since in this case its eigenvectors are linearly independent. This means that a matrix may have multiple eigenvalues, but nevertheless it may be possible to reduce it to diagonal form. That is why the nondefective matrices are also called *diagonalizable*.

If the matrix is defective, it cannot be reduced to a diagonal form by similarity transformation. In this case there exist unique numbers $\lambda_1, \lambda_2, ..., \lambda_r \in \lambda(A)$ and unique positive numbers $n_1, n_2, ..., n_r$; $n_1 + n_2 + \cdots + n_r = n$, such that A is similar to the matrix

$$\text{diag}(J_{\lambda_1}^{(n_1)}, J_{\lambda_2}^{(n_2)}, ..., J_{\lambda_r}^{(n_r)}) \tag{1.78}$$

The form (1.78) is referred to as the *Jordan canonical form* of A. It is unique to the ordering of the Jordan blocks. A typical Jordan form is as follows

$$\begin{bmatrix} -2 & 1 & & & & & \\ 0 & -2 & & & & \mathbf{0} & \\ & & -2 & 1 & 0 & & \\ & & 0 & -2 & 1 & & \\ & & 0 & 0 & -2 & & \\ & & & & & 4 & 1 \\ & \mathbf{0} & & & & 0 & 4 \end{bmatrix} \tag{1.79}$$

Note that the same eigenvalues may occur in several blocks of the Jordan form.

The whole number of the linearly independent eigenvectors of A is equal to the number of the blocks of its Jordan form.

If X is the matrix which reduces A to its Jordan form (1.78), then $x_1, x_2, \ldots, x_{n_1}$ satisfy

$$Ax_1 = \lambda_1 x_1$$
$$Ax_{j+1} = \lambda_1 x_{j+1} + x_j \qquad j = 1, 2, \ldots, n_1 - 1$$

and similar relations hold for the vectors, corresponding to the other blocks. The vectors x_j, for $j = 2, 3, \ldots, n_1$, are called *principal vectors* of the matrix A.

The polynomials

$$p_i(\lambda) = \det(\lambda I - J_{\lambda_i}^{(n_i)}) = (\lambda - \lambda_i)^{n_i}$$

are referred to as *elementary divisors* of A. They divide the characteristic polynomial of A which may be written as

$$p(\lambda) = p_1(\lambda)p_2(\lambda) \ldots p_r(\lambda)$$

The Jordan form is diagonal only when $n_1 = n_2 = \cdots = n_r = 1$ in which case every divisor $p_i(\lambda)$ is linear. That is why if A is a defective matrix it has nonlinear elementary divisors.

Matrices in the form (1.79), in which some eigenvalues occur in more than one block, are called *derogatory matrices*. In this case several linearly independent eigenvectors correspond to a given λ_i.

In practice, the general similarity transformations have the following serious drawback. Although the matrix P must be nonsingular, it may be ill conditioned with respect to inversion, which makes it difficult to form $P^{-1}AP$. In fact, using the results from Section 1.4, it may be shown that

$$\mathrm{fl}(P^{-1}AP) = P^{-1}AP + E$$

where

$$\| E \|_2 \approx \varepsilon \| P^{-1} \|_2 \| A \|_2 \| P \|_2 = \varepsilon \, \mathrm{cond}_2(P) \| A \|_2$$

Thus large errors may be introduced in the evaluation of $P^{-1}AP$ if P is ill conditioned.

Example 1.23

If

$$A = \begin{bmatrix} 11.3 & -32.0 & 132.4 \\ -9.4 & 27.2 & -111.85 \\ -3.2 & 9.2 & -37.9 \end{bmatrix}$$

$$P = \begin{bmatrix} -6.5 & 8.0 & -54.0 \\ 6.0 & -5.5 & 43.5 \\ 2.0 & -2.0 & 15.0 \end{bmatrix}$$

and we use chopped arithmetic with $\varepsilon = 2^{-15} \approx 3 \times 10^{-5}$, then we obtain (up to three decimal digits)

$$\mathrm{fl}(P^{-1}AP) = \begin{bmatrix} 0.107 & -0.006\,84 & 0.0540 \\ -0.007\,32 & 0.207 & -0.0518 \\ -0.001\,95 & 0.001\,95 & 0.286 \end{bmatrix}$$

The exact answer is

$$P^{-1}AP = \mathrm{diag}(0.1, 0.2, 0.3)$$

so that

$$\| \mathrm{fl}(P^{-1}AP) - P^{-1}AP \|_2 \approx 0.0775$$

This large error is due to the ill conditioning of P as far as

$$\mathrm{cond}_2(P) = 6.63 \times 10^3$$

To avoid the difficulties related to the evaluation of $P^{-1}AP$, it is appropriate to use similarity transformations that are well conditioned, such as the unitary transformations. If the matrix U is unitary, then $U^{-1} = U^H$, $\mathrm{cond}_2(U) = 1$ and

$$\mathrm{fl}(U^H AU) = U^H AU + E$$

where $\| E \|_2 \approx \varepsilon \| A \|_2$, so that the error is as small as possible.

Note, however, that if the similarity transformations are restricted to the class of unitary transformations, then the final form will be more complicated than the diagonal form.

Let x be an eigenvector of A, corresponding to the eigenvalue λ, i.e.

$$Ax = \lambda x$$

Constructing a unitary matrix U such that

$$\tilde{x} = U^H x = [g, \ 0, \ ..., \ 0]^T$$

where $| g | = \| x \|_2 > 0$. Then

$$U^H AU\tilde{x} = \tilde{x}\lambda \tag{1.80}$$

Equation (1.80) implies that the matrix $U^H AU$ has the form

$$U^H AU = \begin{bmatrix} \lambda & a_{12} \\ 0 & \\ \vdots & A_{22} \\ 0 & \end{bmatrix}$$

Thus it is possible to reduce the matrix A into a block-triangular form using unitary transformations.

Example 1.24

The matrix

$$A = \begin{bmatrix} 6.8 & -1.6 & 7.2 & -2.8 \\ -11.4 & 7.8 & -7.6 & 0.4 \\ -19.0 & 8.0 & -19.0 & 6.0 \\ -22.4 & 6.8 & -27.6 & 11.4 \end{bmatrix}$$

has an eigenvalue $\lambda = 5$ with corresponding eigenvector

$$x = [2,\ 1,\ -3,\ -7]^T$$

Using an elementary reflection, chosen to introduce zeros in the last three positions of x, we obtain (up to three significant digits)

$$U = \begin{bmatrix} -0.252 & -0.126 & 0.378 & 0.882 \\ -0.126 & 0.987 & 0.0380 & 0.0887 \\ 0.378 & 0.0380 & 0.886 & -0.266 \\ 0.882 & 0.0887 & -0.266 & 0.379 \end{bmatrix}$$

$$U^T x = [-7.94,\ 0,\ 0,\ 0]^T$$

$$U^T A U = \begin{bmatrix} 5.00 & 11.7 & -42.1 & -9.72 \\ 0 & 10.3 & -16.1 & -8.45 \\ 0 & 5.85 & -9.69 & -4.91 \\ 0 & -0.430 & 0.751 & 1.39 \end{bmatrix}$$

Let the matrix A have eigenvalues $\lambda_1, \lambda_2, ..., \lambda_n$. It may be proved that there exists a unitary matrix U such that $U^H A U$ is an upper triangular matrix with diagonal elements $\lambda_1, \lambda_2, ..., \lambda_n$. This triangular matrix is called the *Schur form* of the matrix A and the columns of U are referred to as *Schur vectors*. If the matrix A is Hermitian, i.e. $A^H = A$, then it may be reduced by unitary similarity transformations to a diagonal form.

Example 1.25

If

$$A = \begin{bmatrix} -6 & 2 & 8 \\ 3 & -5 & -7 \\ -9 & -4 & 1 \end{bmatrix}$$

then (up to three significant digits)

$$U = \begin{bmatrix} 0.250 - 0.574i & -0.285 + 0.415i & -0.0905 - 0.588i \\ -0.285 + 0.467i & 0.0699 - 0.267i & -0.120 \ \ -0.781i \\ 0.522 - 0.187i & 0.781 - 0.247i & -0.0223 - 0.145i \end{bmatrix}$$

is a unitary matrix and

$$U^H A U = \begin{bmatrix} -2.88 + 4.93i & 11.0 + 4.34i & 0.0215 + 3.71i \\ 0 & -2.88 - 4.93i & -1.56 + 6.84i \\ 0 & 0 & -4.24 + 0i \end{bmatrix}$$

If the original matrix is real, then the use of unitary transformations may introduce complex elements in $U^H A U$. That is why, if the computations involve real matrices, it is desirable to use real unitary (i.e. orthogonal) transformations. If A has complex eigenvalues, then it is clear that it cannot be reduced to triangular form using orthogonal transformations. In this case A is reduced to quasitriangular form, in which the 2×2 diagonal blocks have complex conjugate eigenvalues. This quasitriangular form is called the *real Schur form* of the matrix A. If A is symmetric, it may be reduced by orthogonal similarity transformations to a diagonal form.

Example 1.26

If

$$A = \begin{bmatrix} -6 & 2 & 8 \\ 3 & -5 & -7 \\ -9 & -4 & 1 \end{bmatrix}$$

then (up to three decimal digits)

$$U = \begin{bmatrix} 0.161 & 0.787 & 0.595 \\ 0.0612 & -0.610 & 0.790 \\ -0.985 & 0.0911 & 0.147 \end{bmatrix}$$

is orthogonal and

$$U^T A U = \begin{bmatrix} 1.67 & 3.93 & 7.92 \\ -11.4 & -7.43 & 0.539 \\ 0 & 0 & -4.24 \end{bmatrix}$$

The eigenvalues of the 2×2 diagonal block are $-2.88 \pm 4.93i$. (Compare with Example 1.25.)

1.9.2 Sensitivity of Eigenvalues and Eigenvectors

The numerically stable computational methods for solving the eigen-value problem yield a set of approximated eigenvalues of a matrix A, which are the exact eigenvalues of a slightly perturbed matrix $A + E$. In this way the effect of rounding errors on the eigenvalues computation may be expressed by some small change in the given matrix. That is why the accuracy of the computed eigenvalues depends on their sensitivity to perturbations in the matrix elements. If small perturbations in the ele-ments of A lead to large changes in some of its eigenvalues then the cor-responding eigenvalue problem is ill conditioned. If the eigenvalue problem is ill conditioned, it is difficult to obtain exact results for the eigenvalues.

Example 1.27

(Davis and Moler, 1978). The matrix

$$A = \begin{bmatrix} -149 & -50 & -154 \\ 537 & 180 & 546 \\ -27 & -9 & -25 \end{bmatrix}$$

has eigenvalues $\lambda_1 = 1$, $\lambda_2 = 2$ and $\lambda_3 = 3$. If the element a_{22} is changed from 180 to 180.01, then the eigenvalues become (up to eight decimal digits)

$$\lambda_1 = 0.207\,265\,65 \qquad \lambda_2 = 2.300\,834\,90 \qquad \lambda_3 = 3.501\,899\,44$$

Consider the effect on the eigenvalues of perturbations in A in the form $A + \alpha E$, where E is a fixed arbitrary matrix with $\| E \|_2 = 1$ and α is a variable parameter. Assume first that the matrix A is nondefective, i.e. that it has a full system of linearly independent eigenvectors x_j, where $j = 1, 2, ..., n$.

The simplest measure of the sensitivity of the eigenvalues is the condition number of the eigenvector matrix,

$$\text{cond}_2(X) = \| X \|_2 \| X^{-1} \|_2$$

According to the *Bauer–Fike theorem*, changes in the elements of A may lead to $\text{cond}_2(X)$ times larger changes in the eigenvalues of A.

For the matrix A from Example 1.27

$$X = \begin{bmatrix} 0.316\ 23 & 0.404\ 06 & -0.139\ 14 \\ -0.948\ 69 & -0.909\ 14 & 0.973\ 38 \\ 0 & -0.101\ 02 & -0.178\ 89 \end{bmatrix}$$

$$X^{-1} = \begin{bmatrix} 411.10 & 135.98 & 420.58 \\ -267.29 & -89.09 & -277.19 \\ 150.93 & 50.31 & 150.93 \end{bmatrix}$$

and

$$\| X \|_2 = 1.709 \qquad \| X^{-1} \|_2 = 754.1$$

so that

$$\text{cond}_2(X) = 1288.94$$

In this way, the change of a_{22} from 180 to 180.01 could cause a change in the eigenvalues of size 12.89. The actual change of λ_1 is in fact much less than the one predicted by the theorem.

Because of its dependence on the spectral norm of A, $\text{cond}_2(X)$ is referred to as the *spectral condition number*. Since $\text{cond}_2(X)$ characterizes the overall sensitivity of the eigenvalues of A, this number may be considered as a condition number of A with respect to the eigenvalue problem.

If the matrix A is real and symmetric (or complex and Hermitian) then the eigenvector matrix X is orthogonal (respectively unitary) and $\text{cond}_2(X) = 1$. That is why the eigenvalue problem for a symmetric matrix is always well conditioned. In accordance with the *Wielandt–Hoffman theorem*, if A is symmetric, then

$$\sqrt{\sum_{j=1}^{n} (\delta\lambda_j)^2} \leq n\alpha$$

where $\delta\lambda_j$ is the change in λ_j, owing to a perturbation αE in the matrix A. Clearly, $|\delta\lambda_j| \leq n\alpha$, for $j = 1, 2, ..., n$.

The investigation of the sensitivity of the individual eigenvalues involves the consideration of the right and left eigenvectors of A.

Let $\lambda_j(\alpha)$ be an eigenvalue of $A + \alpha E$ which satisfies $\lambda_j(0) = \lambda_j$. Then for small α the following is fulfilled:

$$\lambda_j(\alpha) = \lambda_j + \frac{d\lambda_j}{d\alpha}\alpha + \cdots$$

Using the equation

$$(A + \alpha E)x_j(\alpha) = \lambda_j(\alpha)x_j(\alpha)$$

where $x_j(\alpha)$ is the corresponding eigenvector, it may be shown that the first derivative of λ_j is given by

$$\frac{d\lambda_j}{d\alpha} = \frac{y_j^H E x_j}{y_j^H x_j}$$

If the right and left eigenvectors are normalized so that $\| x_j \|_2 = \| y_j \|_2 = 1$, then

$$\left| \frac{d\lambda_j}{d\alpha} \right| = \frac{| y_j^H E x_j |}{| y_j^H x_j |} \leq \frac{1}{| y_j^H x_j |}$$

In this way, up to first-order terms relative to the small parameter α, the following is fulfilled:

$$| \lambda_j(\alpha) - \lambda_j | \leq \text{cond}(\lambda_j) | \alpha | \tag{1.81}$$

where

$$\text{cond}(\lambda_j) = \frac{1}{| y_j^H x_j |}$$

is called *condition number of the eigenvalue* λ_j.

Equation (1.81) shows that if the perturbations in A are of order α then the eigenvalue λ_j may be perturbed by $| \alpha | \text{cond}(\lambda_j)$. Thus if $\text{cond}(\lambda_j)$ is large, then the eigenvalue problem for λ_j is ill conditioned.

The overall condition number $\text{cond}_2(X)$ and the individual condition numbers $\text{cond}(\lambda_j)$ are related by

$$\text{cond}(\lambda_j) \leq \text{cond}_2(X) \qquad j = 1, 2, \ldots, n$$

For the matrix from Example 1.27 the individual condition numbers are

$$\text{cond}(\lambda_1) = 603.64 \qquad \text{cond}(\lambda_2) = 395.24 \qquad \text{cond}(\lambda_3) = 219.29$$

so that the first eigenvalue is more sensitive than the others.

The condition numbers $\text{cond}_2(X)$ and $\text{cond}(\lambda_j)$, where $j = 1, 2, \ldots, n$, are invariant to unitary similarity transformations of A. The individual condition numbers may be computed without using the eigenvectors as shown in Chan, Feldman and Parlett (1977).

A drawback of the individual condition numbers is that they are not invariant to diagonal similarity transformations.

Example 1.28

(Golub and Wilkinson, 1976). If

$$A = \begin{bmatrix} 2 & 1 \\ 1 & 2 \end{bmatrix}$$

then

$$\lambda_1 = 3 \qquad \lambda_2 = 1$$

$$x_1^H = \frac{1}{\sqrt{2}} [1, 1] \qquad y_1^H = \frac{1}{\sqrt{2}} [1, 1]$$

$$\text{cond}(\lambda_1) = 1$$

so that the eigenvalue λ_1 is well conditioned. However, for the transformed matrix

$$D^{-1}AD = \begin{bmatrix} 2 & d \\ d^{-1} & 2 \end{bmatrix} \qquad \text{where } D = \begin{bmatrix} 1 & 0 \\ 0 & d \end{bmatrix}$$

one has

$$x_1^H = \frac{1}{\sqrt{(1 + d^2)}} [d, 1] \qquad y^H = \frac{1}{\sqrt{(1 + d^2)}} [1, d]$$

$$\text{cond}(\lambda_1) = \frac{1 + d^2}{2d}$$

Hence $\text{cond}(\lambda_1)$ may be made arbitrarily large by taking d sufficiently large or sufficiently small. Clearly, the large value of $\text{cond}(\lambda_1)$ obtained in this way is unrealistic. For this example, when $\text{cond}(\lambda_1)$ is large, then $\| D^{-1}AD \|_2 \gg \| A \|_2$. That is why, in practice, the values of $\text{cond}(\lambda_j)$ of the matrix $D^{-1}AD$ are meaningful when D is chosen so that $\| D^{-1}AD \|_2$ is minimized. The process of searching for a diagonal matrix D, such that $\| D^{-1}AD \|_2$ is small, is called *balancing of a matrix*.

For a symmetric matrix $y_j^H x_j = 1$ and $\text{cond}(\lambda_j) = 1$ and from (1.81) it also follows that the eigenvalue problem is well conditioned. When the matrix is not symmetric the vectors x_j and y_j may be almost orthogonal and the sensitivity of the eigenvalues may be extremely large.

Example 1.29

If

$$A = \begin{bmatrix} a & 1 \\ 0 & a + \alpha \end{bmatrix}$$

then $\lambda_1 = a$ and $\lambda_2 = a + \alpha$ and

$$x_1^H = [1, 0] \qquad y_1^H = \frac{1}{\sqrt{(1 + \alpha^2)}} [-\alpha, 1]$$

$$x_2^H = \frac{1}{\sqrt{(1 + \alpha^2)}} [1, \alpha] \qquad y_2^H = [0, 1]$$

In this way

$$y_1^H x_1 = -\frac{\alpha}{\sqrt{(1+\alpha^2)}} \qquad y_2^H x_2 = \frac{\alpha}{\sqrt{(1+\alpha^2)}}$$

These two quantities tend to zero with $\alpha \to 0$, so that $\text{cond}(\lambda_1)$ and $\text{cond}(\lambda_2)$ may become arbitrarily large. When $\alpha = 0$, the matrix A is defective, i.e. it has a nonlinear elementary divisor, and the condition numbers may be considered as infinite. This shows that if the eigenvalues are sensitive, i.e. if the condition numbers are large, then the matrix should be near to a matrix having multiple eigenvalues (see Wilkinson (1965, ch. 2)).

The eigenvalue problem for multiple eigenvalues, corresponding to a nonlinear elementary divisor, is usually ill conditioned. In general, if λ is a multiple eigenvalue of multiplicity k which corresponds to a nonlinear elementary divisor, then perturbations in A of order α may result in a perturbation in λ of order $\alpha^{1/k}$.

Example 1.30

Let

$$J_{-1}^{(5)} = \begin{bmatrix} -1 & 1 & 0 & 0 & 0 \\ 0 & -1 & 1 & 0 & 0 \\ 0 & 0 & -1 & 1 & 0 \\ 0 & 0 & 0 & -1 & 1 \\ 0 & 0 & 0 & 0 & -1 \end{bmatrix}$$

If the element in the position $(5, 1)$ is replaced by $\alpha = 10^{-5}$ then the characteristic equation becomes

$$(\lambda + 1)^5 = \alpha$$

and the multiple eigenvalue $\lambda = -1$ is replaced by five different eigenvalues, listed below up to eight significant digits

$$\lambda_1 \ = -0.900\,000\,00 + 0.000\,000\,00i$$
$$\lambda_{2,3} = -0.969\,098\,30 \pm 0.095\,105\,652i$$
$$\lambda_{4,5} = -1.080\,901\,7 \pm 0.058\,778\,525i$$

In practice, the multiplicities of the eigenvalues and the degrees of the elementary divisors should be discovered from the computed eigenvalues. That is why the exact determination of the eigenvalue multiplicity and the accurate computation of the Jordan form of a matrix may present a difficult problem.

Sensitivity analysis in the presence of multiple eigenvalues may be done using the *Gershgorin theorem* which is stated as follows.

Theorem 1.1

Let A be a $n \times n$ complex matrix. Then each eigenvalue of A lies in one of the circles (called *Gershgorin circles*) in the complex plane, centred at a_{ii} and of radius

$$\sum_{\substack{j=1 \\ j \neq i}}^{n} |a_{ij}|, \text{ i.e.}$$

$$|\lambda - a_{ii}| \leq \sum_{j \neq i} |a_{ij}| \qquad i = 1, 2, \ldots, n$$

Analogical results may be obtained considering A^{T} instead of A.

Example 1.31

If

$$A = \begin{bmatrix} -1 & 2 \times 10^{-5} & -1 \times 10^{-5} \\ -1 \times 10^{-5} & -1 & -1 \times 10^{-5} \\ 3 \times 10^{-5} & 2 \times 10^{-5} & -2 \end{bmatrix}$$

then the Gershgorin circles C_1, C_2, C_3 of A are centred at -1, -1, -2 and their radii are respectively 3×10^{-5}, 2×10^{-5} and 5×10^{-5}. Thus all eigenvalues of A lie in the union of two circles of radii 3×10^{-5} and 5×10^{-5} centred at -1 and -2.

Note that the theorem does not state in which circles the eigenvalues lie. It is possible that the eigenvalues lie in some of the Gershgorin circles, leaving the others empty, i.e. it is not always true that every circle contains an eigenvalue.

If k Gershgorin circles of the matrix A are disjoint from the other circles, then exactly k eigenvalues of A lie in the union of these k circles. In particular, if any Gershgorin circle is isolated from the other circles, then it contains precisely one eigenvalue of A. Thus, in Example 1.31 the circles C_1 and C_2 are disjoint from C_3. That is why two eigenvalues of A are in the union of C_1 and C_2 and the circle C_3 contains exactly one eigenvalue.

To obtain better approximations to the eigenvalues, the radii of

Gershgorin circles may be decreased using diagonal similarity transformations. For the above example, if $D = \text{diag }(1, 1, 2 \times 10^{-5})$, then $\tilde{A} = DAD^{-1}$ has the form

$$\tilde{A} = \begin{bmatrix} -1 & 2 \times 10^{-5} & -5 \times 10^{-1} \\ -1 \times 10^{-5} & -1 & -5 \times 10^{-1} \\ 6 \times 10^{-10} & 4 \times 10^{-10} & -2 \end{bmatrix}$$

The Gershgorin circles $\tilde{C}_1$, $\tilde{C}_2$, $\tilde{C}_3$ for $\tilde{A}$ are centred again at -1, -1, -2 but their radii are respectively

$$0.500\ 002,\ 0.500\ 001 \text{ and } 10 \times 10^{-10}$$

The circle $\tilde{C}_3$ is disjoint from $\tilde{C}_1$ and $\tilde{C}_2$ and therefore contains exactly one eigenvalue of $\tilde{A}$ and hence of A. The radius of the circle $\tilde{C}_3$ is 10^5 times smaller than that of C_3, which makes it possible to localize more precisely the corresponding eigenvalue. In fact, the eigenvalues of A (up to 15 decimal digits) are

$$\lambda_{1,2} = -1.000\ 000\ 000\ 249\ 998 \pm 0.000\ 014\ 142\ 277\ 046i$$
$$\lambda_3 \ = -1.999\ 999\ 999\ 500\ 004$$

so that

$$|\lambda_3 - a_{33}| = 4.999\ 96 \times 10^{-10} < 10 \times 10^{-10}$$

This technique may be used to reduce the size of any isolated Gershgorin circle. For this purpose, the corresponding element of the diagonal transformation is chosen so that the radius of the circle is as small as possible, at the same time keeping the rest of the circles sufficiently small, so that the circle remains isolated.

The analysis of the eigenvectors' sensitivity shows that the perturbation in an eigenvector x_j depends on the conditioning of the corresponding eigenvalue λ_j and the separation of λ_j from the other eigenvalues. Thus, even if an eigenvalue is well conditioned, the corresponding eigenvector may be very sensitive if this eigenvalue is close to the others.

Example 1.32

If

$$A = \begin{bmatrix} 4.99 & 1.00 & -0.99 \\ -2.99 & 1.00 & 0.99 \\ 2.98 & 1.00 & 1.02 \end{bmatrix}$$

then

$$\lambda_1 = 2 \qquad \lambda_2 = 2.01 \qquad \lambda_3 = 3$$
$$\text{cond}(\lambda_1) \approx 3.464$$

and

$$x_1 = [1, -2, 1]^T$$

If the element a_{11} is changed to 5.00, then (up to five significant digits)

$$x_1 = [0.719\ 40, \ -1.1794, \ 1]^T$$

Although some of the eigenvectors may be sensitive to perturbations in the elements of the matrix, it is possible that the subspace spanned by them is relatively insensitive. In this respect it is important to estimate the sensitivity of a set of Schur vectors, corresponding to a cluster of eigenvalues.

Let

$$U^H A U = \begin{bmatrix} T_{11} & T_{12} \\ 0 & T_{22} \end{bmatrix} \begin{matrix} \}l \\ \}n-l \end{matrix}$$
$$\underbrace{\phantom{T_{11}}}_{l} \quad \underbrace{\phantom{T_{12}}}_{n-l}$$

be the Schur decomposition of a complex $n \times n$ matrix A, and partition the unitary matrix U in the form

$$U = [\underbrace{U_1,}_{l} \ \underbrace{U_2}_{n-l}]$$

The columns of U_1 span an invariant subspace of A.

Suppose now that the matrix A is perturbed, which leads to a new matrix

$$\hat{A} = A + E$$

Let the Schur decomposition of $\hat{A}$ be

$$\hat{U}^H \hat{A} \hat{U} = \begin{bmatrix} \hat{T}_{11} & \hat{T}_{12} \\ 0 & \hat{T}_{22} \end{bmatrix}$$

and let $\hat{U}$ be partitioned as

$$\hat{U} = [\hat{U}_1, \ \hat{U}_2]$$

It is reasonable to expect that if $\| E \|$ is small, then the subspace spanned by $\hat{U}_1$ will lie near to the subspace spanned by U_1. In accordance with the previous considerations, we may presume that the sensitivity of the subspace spanned by U_1 will depend on the distance between $\lambda(T_{11})$ and $\lambda(T_{22})$.

The *separation* between the matrices T_{11} and T_{22} is defined as

$$\text{sep}(T_{11}, T_{22}) = \min_{X \neq 0} \frac{\| T_{11}X - XT_{22} \|_F}{\| X \|_F}$$

and $U^H E U$ is partitioned as

$$U^H E U = \begin{bmatrix} E_{11} & E_{12} \\ E_{21} & E_{22} \end{bmatrix}$$

The following result, due to Stewart (1973a), gives a precise bound on the sensitivity of U_1.

Theorem 1.2

If

$$\delta = \text{sep}(T_{11}, T_{22}) - \| E_{11} \|_2 - \| E_{22} \|_2 > 0$$

and

$$\| E_{21} \|_2 (\| T_{12} \|_2 + \| E_{12} \|_2) \leq \tfrac{1}{4} \delta^2 \tag{1.82}$$

then there exists a complex $(n - l) \times l$ matrix P which satisfies

$$\| P \|_2 \leq \frac{2}{\delta} \| E_{21} \|_2$$

such that the columns of

$$\hat{U}_1 = (U_1 + U_2 P)(I + P^H P)^{-1/2}$$

form an orthogonal basis for a subspace that is invariant for $A + E$.

Example 1.33

Let

$$A = T = \begin{bmatrix} T_{11} & T_{12} \\ 0 & T_{22} \end{bmatrix}$$

$$T_{11} = \begin{bmatrix} 2 & -6 \\ 0 & -3 \end{bmatrix} \quad T_{12} = \begin{bmatrix} 5 & -4 & 8 \\ 7 & 1 & 4 \end{bmatrix} \quad T_{22} = \begin{bmatrix} 5 & 3 & -9 \\ 0 & 1 & -7 \\ 0 & 0 & 2.1 \end{bmatrix}$$

Since the matrix A is already in Schur form, it follows that

$$U_1 = \begin{bmatrix} 1 & 0 \\ 0 & 1 \\ 0 & 0 \\ 0 & 0 \\ 0 & 0 \end{bmatrix}$$

In this case

$$\text{sep}(T_{11}, T_{22}) \approx 5.22 \times 10^{-3}$$

Suppose that the matrix A is perturbed so that

$$E_{21} = 10^{-7} \begin{bmatrix} 1 & 1 \\ 1 & 1 \\ 1 & 1 \end{bmatrix}$$

and $E_{11} = E_{12} = E_{22} = 0$. Hence $\| E_{21} \|_2 \approx 2.45 \times 10^{-7}$ and the condition (1.82) is satisfied.

From the Schur decomposition of

$$A + E = \begin{bmatrix} T_{11} & T_{12} \\ E_{21} & T_{22} \end{bmatrix}$$

we find (up to three decimal digits) that

$$\hat{U}_1 = \begin{bmatrix} 1.00 & 0.136 \times 10^{-4} \\ -0.136 \times 10^{-4} & 1.00 \\ -0.101 \times 10^{-4} & 0.121 \times 10^{-4} \\ 0.710 \times 10^{-5} & -0.865 \times 10^{-5} \\ -1.00 \times 10^{-6} & 0.116 \times 10^{-5} \end{bmatrix}$$

In this way

$$\| \hat{U}_1 - U_1 \|_2 \approx 2.37 \times 10^{-5} \approx \frac{\| E_{21} \|_2}{\text{sep}(T_{11}, T_{22})}$$

Further discussion on the eigenvalue/eigenvector sensitivity and its estimation can be found in Wilkinson (1984), Van Loan (1987) and Bai, Demmel and McKenney (1989).

EXERCISES

1.49 Show that if λ is an eigenvalue of A then for any consistent matrix norm

$$|\lambda| \leq \| A \|$$

1.50 Prove that every square nth order matrix A satisfies its characteristic equation

$$A^n + a_n A^{n-1} + \cdots + a_1 I = 0$$

(see Equation (1.74)). This assertion is the so-called *Caley–Hamilton theorem*.

1.51 Show that the eigenvalues of the block-triangular matrix

$$\begin{bmatrix} T_{11} & T_{12} \\ 0 & T_{22} \end{bmatrix}$$

are the eigenvalues of T_{11} and T_{22}.

1.52 Prove that every real symmetric matrix is orthogonally similar to a diagonal matrix, and that the eigenvalues of a positive definite (non-negative definite) matrix are positive (non-negative).

1.53 Prove that if A is a symmetric matrix with eigenvalues $\lambda_1, \lambda_2, ..., \lambda_n$ then

$$\| A \|_2 = \max\{| \lambda_i | : i = 1, 2, ..., n\}$$

$$\| A \|_F = \sqrt{\sum_{i=1}^{n} \lambda_i^2}$$

1.54 Show that the moduli of the eigenvalues of an orthogonal matrix are equal to 1.

1.55 Show that if λ is an eigenvalue of A then λ^k is an eigenvalue of A^k for $k \geq 0$.

1.56 Show that the eigenvector matrix of

$$\begin{bmatrix} \lambda & 1 \\ 0 & \lambda + \varepsilon \end{bmatrix}$$

has a condition number of order $1/\varepsilon$ for small positive ε.

1.57 Prove that there is only one eigenvector corresponding to a Jordan block.

1.58 Prove that

$$\text{sep}(B, C) + \| E \| + \| F \| \geq \text{sep}(B + E, C + F)$$
$$\geq \text{sep}(B, C) - \| E \| - \| F \|$$

1.59 Prove that

$$\text{sep}(XBX^{-1}, YCY^{-1}) \geq \frac{\text{sep}(B, C)}{\text{cond}(X)\text{cond}(Y)}$$

1.10 COMPUTING THE SCHUR AND JORDAN FORMS

1.10.1 Reduction into Hessenberg Form

The Schur form of a matrix cannot be determined, in general, by a finite sequence of transformations. That is why it is necessary to use *iterative methods* in which the subdiagonal elements of the matrix are made smaller and smaller using an (infinite) sequence of similarity transformations. After a number of such transformations the subdiagonal elements become so small that they may be neglected, and the resulting matrix may be accepted as triangular. Later we describe the so-called *QR algorithm*, which is used for the iterative reduction of a matrix into Schur form by orthogonal similarity transformations. Because of its remarkable numerical properties, the QR algorithm is the heart of many computational methods related to the solution of the eigenvalue problem.

The iterative implementation of similarity transformations on a dense matrix has the drawback that it requires a large number of operations. If, for example, A and U are real $n \times n$ matrices and U is orthogonal, then $2n^3$ flops are required to compute the matrix $A' = U^T A U$. If the iterative procedure requires only one similarity transformation per eigenvalue, then $2n^4$ flops will be necessary to accomplish the full reduction. To reduce the computations, the given matrix is preliminarily transformed to some simplified form which has a large number of zeros. In particular, using orthogonal transformations, a real matrix may be reduced into *Hessenberg form*

$$H = \begin{bmatrix} h_{11} & h_{12} & \cdots & h_{1n} \\ h_{21} & h_{22} & \cdots & h_{2n} \\ 0 & h_{32} & \cdots & h_{3n} \\ \vdots & \vdots & \ddots & \vdots \\ 0 & 0 & \cdots h_{n,n-1} & h_{nn} \end{bmatrix}$$

This reduction may be accomplished by implementing elementary reflections.

The matrix which is to be reduced into Hessenberg form, initially looks as shown below for $n = 5$,

$$A = \begin{bmatrix} x & x & x & x & x \\ x & x & x & x & x \\ \textcircled{x} & x & x & x & x \\ \textcircled{x} & x & x & x & x \\ \textcircled{x} & x & x & x & x \end{bmatrix} \qquad (1.83)$$

Let an elementary reflection

$$U_1 = \begin{bmatrix} 1 & 0 \\ 0 & U_1' \end{bmatrix}$$

be constructed, such that $U_1 A$ has zeros in the positions circled in (1.83).

The right multiplication by the matrix $U_1^T = U_1$ leaves the first column of $U_1 A$ unchanged. Hence $A_2 = U_1 A U_1$ has the form

$$A_2 = \begin{bmatrix} x & x & x & x & x \\ x & x & x & x & x \\ 0 & x & x & x & x \\ 0 & \textcircled{x} & x & x & x \\ 0 & \textcircled{x} & x & x & x \end{bmatrix} \tag{1.84}$$

At the second stage we determine a reflection

$$U_2 = \begin{bmatrix} I_2 & 0 \\ 0 & U_2' \end{bmatrix}$$

such that $U_2 A_2$ acquires zeros in the circled positions in (1.84). The zero elements in the first column of A_2 remain unchanged after the similarity transformation accomplished with the matrix U_2. That is why $A_3 = U_2 A_2 U_2$ has the form

$$A_3 = \begin{bmatrix} x & x & x & x & x \\ x & x & x & x & x \\ 0 & x & x & x & x \\ 0 & 0 & x & x & x \\ 0 & 0 & \textcircled{x} & x & x \end{bmatrix} \tag{1.85}$$

The final transformation with

$$U_3 = \begin{bmatrix} I_3 & 0 \\ 0 & U_3' \end{bmatrix}$$

zeros the single circled element in (1.85), so that the matrix

$$A_4 = U_3 A_3 U_3 = \begin{bmatrix} x & x & x & x & x \\ x & x & x & x & x \\ 0 & x & x & x & x \\ 0 & 0 & x & x & x \\ 0 & 0 & 0 & x & x \end{bmatrix}$$

is in upper Hessenberg form.

The reduction of a real matrix into Hessenberg form requires approximately $\frac{5}{3} n^3 + \frac{1}{2} n^2$ flops. If the elementary reflections are accumulated (this is necessary when finding the eigenvectors), then an additional $\frac{2}{3} n^3$ flops are required.

This algorithm has favourable numerical properties. Using the error analysis for construction and implementation of elementary reflections (see Section 1.7), it may be shown that the computed Hessenberg form $\bar{A}_{n-1}$ satisfies

$$\bar{A}_{n-1} = U^{\mathrm{T}}(A + \Delta A)U \qquad (1.86)$$

where U is orthogonal and

$$\| \Delta A \|_{\mathrm{F}} \leq (6n^2 + 48n - 120)\varepsilon \| A \|_{\mathrm{F}} \qquad (1.87)$$

Equations (1.86) and (1.87) show that the matrix $\bar{A}_{n-1}$ may be considered as the exact result of an orthogonal similarity transformation of the matrix $A + \Delta A$, where $\| \Delta A \|_{\mathrm{F}}$ is small in comparison to $\| A \|_{\mathrm{F}}$. This proves that the algorithm is numerically stable.

Example 1.34

If

$$A = \begin{bmatrix} -3 & 9 & -4 & 1 \\ 5 & 7 & 2 & 8 \\ 4 & -2 & -1 & 6 \\ -9 & 5 & -4 & 3 \end{bmatrix}$$

then (up to three decimal digits) we obtain

$$U_1 = \begin{bmatrix} 1 & 0 & 0 & 0 \\ 0 & -0.453 & -0.362 & 0.815 \\ 0 & -0.362 & 0.910 & 0.203 \\ 0 & 0.815 & 0.203 & 0.543 \end{bmatrix}$$

$$A_2 = \begin{bmatrix} -3.00 & -1.81 & -6.70 & 7.06 \\ -11.00 & -2.09 & -4.73 & -1.26 \\ 0 & 4.98 & -0.372 & -1.50 \\ 0 & 4.27 & -1.68 & 11.50 \end{bmatrix}$$

$$U_2 = \begin{bmatrix} 1 & 0 & 0 & 0 \\ 0 & 1 & 0 & 0 \\ 0 & 0 & -0.759 & -0.651 \\ 0 & 0 & -0.651 & 0.759 \end{bmatrix}$$

$$A_3 = \begin{bmatrix} -3.00 & -1.81 & 0.48 & 9.72 \\ -11.00 & -2.09 & 4.41 & 2.12 \\ 0 & -6.56 & 3.08 & -5.70 \\ 0 & 0 & -5.52 & 8.01 \end{bmatrix}$$

In cases when A is symmetric, the orthogonal transformations keep its symmetry and the above algorithm yields a tridiagonal matrix. This algorithm, however, may be modified to take advantage of the symmetry of A (see for details Stewart (1973b, ch. 7)). As a result the orthogonal reduction of a symmetric matrix into tridiagonal form may be done using only $\frac{2}{3}n^3$ flops.

1.10.2 The QR Algorithm

Let $A_1 = A$ be a complex $n \times n$ matrix. The *unshifted QR algorithm* produces a sequence of similar matrices $A_1, A_2, A_3, \ldots$ in the following way. The matrix A_k, where $k = 1, 2, \ldots$, is decomposed as

$$A_k = Q_k R_k$$

where Q_k is unitary and R_k is upper triangular. Next, a matrix A_{k+1} is computed, such that

$$A_{k+1} = R_k Q_k$$

Since

$$R_k = Q_k^H A_k$$

it follows that the matrix

$$A_{k+1} = Q_k^H A_k Q_k$$

is unitary similar to A_k. Also

$$A_{k+1} = (Q_1 Q_2 \ldots Q_k)^H A_1 Q_1 Q_2 \ldots Q_k$$

It may be shown that under some mild restrictions the sequence $A_1, A_2, A_3, \ldots$, converges to an upper triangular matrix. Thus the QR iteration

For $k = 1, 2, \ldots$
  $A = QR$ (QR decomposition)
  $A \leftarrow RQ$

may be used to find the Schur form of a matrix.

Example 1.35

Let the unshifted QR algorithm be applied to the matrix

$$A = \begin{bmatrix} -9 & -5 & -2 & -1 \\ 44 & 23 & 12 & 8 \\ -26 & -13 & -9 & -5 \\ -38 & -19 & -10 & -10 \end{bmatrix}$$

whose eigenvalues are $-3, -2, -1, 1$.

After five iterations we obtain (up to three significant digits)

$$A_6 = \begin{bmatrix} -0.292 \times 10^1 & 0.411 \times 10^0 & 0.748 \times 10^0 & -0.488 \times 10^2 \\ -0.241 \times 10^{-1} & 0.100 \times 10^1 & -0.218 \times 10^0 & 0.563 \times 10^2 \\ 0.101 \times 10^0 & -0.106 \times 10^{-1} & -0.209 \times 10^1 & -0.131 \times 10^2 \\ 0.335 \times 10^{-3} & -0.350 \times 10^{-4} & 0.213 \times 10^{-3} & -0.985 \times 10^0 \end{bmatrix}$$

The next five iterations yield

$$A_{11} = \begin{bmatrix} -0.299 \times 10^1 & 0.429 \times 10^0 & 0.671 \times 10^0 & 0.471 \times 10^2 \\ 0.151 \times 10^{-2} & 0.100 \times 10^1 & -0.271 \times 10^0 & -0.563 \times 10^2 \\ 0.162 \times 10^{-1} & -0.175 \times 10^{-2} & -0.201 \times 10^1 & -0.183 \times 10^2 \\ -0.146 \times 10^{-5} & 0.157 \times 10^{-6} & -0.132 \times 10^4 & -0.100 \times 10^1 \end{bmatrix}$$

Continuing in this way we obtain after 30 iterations the matrix

$$A_{31} = \begin{bmatrix} -0.300 \times 10^1 & 0.424 \times 10^0 & 0.655 \times 10^0 & 0.469 \times 10^2 \\ 0.461 \times 10^{-6} & 0.100 \times 10^1 & -0.276 \times 10^0 & -0.562 \times 10^2 \\ 0.501 \times 10^{-5} & -0.975 \times 10^{-6} & -0.200 \times 10^1 & 0.192 \times 10^2 \\ 0 & 0 & -0.134 \times 10^{-10} & -0.100 \times 10^1 \end{bmatrix}$$

which may be accepted as upper triangular with the eigenvalues of A on the diagonal.

The drawback of the unshifted QR algorithm is that its convergence is relatively slow. To accelerate the convergence the algorithm is modified as follows. From the elements of the matrix A_k a scalar s_k is determined, which is called an *origin shift* (when the iterative process converges, s_k approaches an eigenvalue of A). After that the new approximation A_{k+1} is obtained using the QR decomposition of $A_k - s_k I$,

$$A_k - s_k I = Q_k R_k$$
$$A_{k+1} = R_k Q_k + s_k I$$

Thus

$$A_{k+1} = Q_k^H (A_k - s_k I) Q_k + s_k I = Q_k^H A_k Q_k$$

It may be proved that under a suitable choice of the shift s_k, it is possible to make the off-diagonal elements in the last row of A_k approach zero very quickly. The rest of the subdiagonal elements may also approach zero, although more slowly. If the shift is close to an eigenvalue of A, then the convergence is very rapid. (The proof of the convergence of the QR algorithm is beyond the scope of this book and may be found, for instance, in Golub and Van Loan, (1983, ch. 7).)

When applied to a full matrix, the QR algorithm has a short-coming in that it requires a large number of computational operations, since each iteration requires approximately n^3 flops. If, however, A_1 is an upper Hessenberg matrix, then so are the next matrices A_k, where $k = 2, 3, ...$, and the computations may be arranged so that each iteration requires a multiple of n^2 flops.

The accelerated convergence of the QR algorithm may be achieved without having to subtract the shift from the diagonal of A_k and then having to restore it explicitly. This leads to the *implicit shift techniques* whose main application is to compute the real Schur form of a matrix with complex eigenvalues.

The implicit shift techniques are based on the properties of the so-called *unreduced Hessenberg matrices* (an upper Hessenberg matrix is unreduced if its subdiagonal elements are nonzero).

Assume that a single step of the QR algorithm with a shift s is applied to an unreduced Hessenberg matrix H. This gives an unreduced upper Hessenberg matrix

$$M = Q^H H Q$$

where Q is the unitary part of the QR decomposition of $H - sI$,

$$H - sI = QR$$

Since R is upper triangular, the partition

$$Q = [q_1, q_2, ..., q_n]$$

yields

$$r_{11} q_1 = QRe_1 = (H - sI)e_1$$

where e_1 is the first column of the unit matrix. Thus the first column of Q is a multiple of the first column of $H - sI$. From the equation

$$QM = HQ$$

and from the fact that M is unreduced, it also follows that

$$q_{j+1} = \frac{1}{m_{j+1,j}} \left(H q_j - \sum_{i=1}^{j} m_{ij} q_i \right) \qquad j = 1, 2, ..., n - 1 \qquad (1.88)$$

Since Q is unitary, $q_j^H q_j = 1$ and $q_j^H q_i = 0$ for $i \neq j$. That is why the multiplication of (1.88) by q_i^H gives

$$m_{ij} = q_i^H H q_j \qquad i = 1, 2, \ldots, j$$

In this way m_{ij} and q_{j+1} are determined up to the signs by the first column of Q. This property of the unreduced Hessenberg matrices makes it possible to avoid the explicit computation of the whole matrix $H - sI$ and the restoration of the shift, since in order to determine the first column of Q, it is necessary to know only the first column of $H - sI$. In practice, the determination of the matrix M may be done in the following way, which is more efficient from the computational point of view:

1. Find a unitary matrix V whose first column coincides with the first column of Q.
2. Reduce $V^H H V$ into upper Hessenberg form $M' = U^H V^H H V U$.

In fact, the reduction of $V^H H V$ into Hessenberg form, using a unitary transformation with a matrix U, is performed in such a way that the first columns of Q and $Q' = VU$ coincide. Then from the properties of the unreduced Hessenberg matrices it follows that $Q = Q'$ and $M = M'$.

The matrix V may be determined as follows. If the first column of $H - sI$ is

$$h_1 = [h_{11} - s, h_{21}, 0, \ldots, 0]^T$$

and V is chosen so that

$$V^H h_1 = \pm \| h_1 \|_2 e_1$$

then it is fulfilled that

$$V e_1 = \pm \frac{1}{\| h_1 \|_2} h_1$$

and the first column of V is the same as that of Q up to the sign.

The implicit shift technique may be used to accomplish two QR steps at once. This makes it possible to avoid complex arithmetic in the case of two complex conjugate shifts.

Let a QR step with a shift s_1 be applied to the matrix H to obtain

$$H_2 = Q_1^H H Q_1$$

and a second QR step with a shift s_2 be applied to H_2 to find

$$H_3 = Q_2^H H_2 Q_2 = Q_2^H Q_1^H H Q_1 Q_2$$

Consider the following algorithm.

1. Determine a unitary matrix U_0 with the same first column as Q_1Q_2.
2. Reduce $U_0^H H U_0$ into an upper Hessenberg matrix M.

As in the case of a single shift, this algorithm reduces H into H_3 without explicit use of s_1 and s_2.

To obtain the matrix U_0, let R_1 and R_2 be the upper triangular matrices in the QR decomposition of $H - s_1 I$ and $H_2 - s_2 I$, respectively. Since $H_2 = Q_1^H H Q_1$ it follows that

$$Q_1 Q_2 R_2 R_1 = (H - s_1 I)(H - s_2 I)$$

Since $R_2 R_1$ is upper triangular, the first column of $Q_1 Q_2$ is a multiple of the first column of $(H - s_1 I)(H - s_2 I)$. That is why U_0 may be chosen as an elementary reflection which reduces the first column of $(H - s_1 I)(H - s_2 I)$ into the vector $[x, 0, ..., 0]^T$.

If H is an upper Hessenberg matrix, then only the first three elements in the first column of $(H - s_1 I)(H - s_2 I)$ are nonzero. These elements are

$$h_{10} = h_{11}^2 + h_{12}h_{21} - (s_1 + s_2)h_{11} + s_1 s_2$$
$$h_{20} = h_{21}[h_{11} + h_{22} - (s_1 + s_2)]$$
$$h_{30} = h_{21}h_{32}$$

The shifts s_1 and s_2 are usually taken as eigenvalues of the trailing 2×2 principal submatrix of H,

$$\begin{bmatrix} h_{n-1,n-1} & h_{n-1,n} \\ h_{n,n-1} & h_{nn} \end{bmatrix} \tag{1.89}$$

Considering the characteristic polynomial of (1.89), it may be shown that when such a choice of shifts is made, the following is fulfilled:

$$s_1 + s_2 = h_{n-1,n-1} + h_{nn}$$
$$s_1 s_2 = h_{nn}h_{n-1,n-1} - h_{n-1,n}h_{n,n-1}$$

That is why if H is real, then the numbers h_{10}, h_{20} and h_{30} are also real, even when the eigenvalues of (1.89) are complex.

Hence U_0 is an orthogonal matrix and the reduction of $U_0^T H U_0$ into Hessenberg form may be done in real arithmetic. In this way the double-shifted implicit technique achieves in real arithmetic the effect of two single QR steps with shifts s and $\bar{s}$, respectively. Since these steps would be accomplished in complex arithmetic, this leads to a significant saving in computations.

The reduction of $U_0^T H U_0$ into Hessenberg form is done as follows. Since only three elements in the first column of $(H - s_1 I)(H - s_2 I)$ are nonzero, U_0 has the form

$$U_0 = \begin{bmatrix} U_0' & 0 \\ 0 & I_{n-3} \end{bmatrix}$$

where U_0' is an elementary reflection of third order. That is why $U_0 A U_0$ looks as shown below for $n = 6$

$$\begin{bmatrix} x & x & x & x & x & x \\ x & x & x & x & x & x \\ \widehat{x} & x & x & x & x & x \\ \widehat{x} & x & x & x & x & x \\ 0 & 0 & 0 & x & x & x \\ 0 & 0 & 0 & 0 & x & x \end{bmatrix} \tag{1.90}$$

To reduce $U_0 A U_0$ into Hessenberg form, an elementary reflection U_1 is chosen in the form

$$U_1 = \begin{bmatrix} 1 & 0 & 0 \\ 0 & U_1' & 0 \\ 0 & 0 & I_{n-4} \end{bmatrix}$$

where U_1' is again an elementary reflection of third order so that the elements of $U_1 U_0 H U_0$, encircled in (1.90), are zero. Then $U_1 U_0 H U_0 U_1$ will look as follows

$$\begin{bmatrix} x & x & x & x & x & x \\ x & x & x & x & x & x \\ 0 & x & x & x & x & x \\ 0 & \widehat{x} & x & x & x & x \\ 0 & \widehat{x} & x & x & x & x \\ 0 & 0 & 0 & 0 & x & x \end{bmatrix} \tag{1.91}$$

This process continues by choosing U_2 so that the circled elements are zeroed in (1.91), and so on. At each stage the pattern of two nonzero elements below the subdiagonal moves one position ahead toward the lower right-hand corner of the array. At the final stage the matrix has the form

$$
\begin{bmatrix}
x & x & x & x & x & x \\
x & x & x & x & x & x \\
0 & x & x & x & x & x \\
0 & 0 & x & x & x & x \\
0 & 0 & 0 & x & x & x \\
0 & 0 & 0 & \textcircled{x} & x & x
\end{bmatrix}
$$

To eliminate the last nonzero element below the subdiagonal, one may use an elementary reflection of second order or a rotation in the $(n-1, n)$-plane.

The double QR step with implicit shifts requires approximately $4n^2$ flops.

This technique may be used iteratively to find the real Schur form of a Hessenberg matrix. In the presence of a real eigenvalue the element $h_{n,n-1}^{(k)}$ will converge quadratically to zero. When this element becomes very small, it may be accepted as zero, which yields the matrix

$$
\begin{bmatrix}
x & x & x & x & x \\
x & x & x & x & x \\
0 & x & x & x & x \\
0 & 0 & x & x & x \\
0 & 0 & 0 & 0 & \lambda_n
\end{bmatrix}
$$

The quantity $\lambda_n = h_{nn}^{(k)}$ is an approximate eigenvalue of H. The rest of the eigenvalues of H are the eigenvalues of the leading principal submatrix of order $n - 1$. Thus the problem is *deflated* and at the next stage the algorithm is applied on a smaller matrix.

In the presence of complex conjugate eigenvalues it is possible to accomplish two implicit complex conjugate shifts. In this case the element $h_{n-1,n-2}^{(k)}$ will converge quadratically to zero. Thus we obtain the matrix

$$
\begin{bmatrix}
x & x & x & x & x \\
x & x & x & x & x \\
0 & x & x & x & x \\
0 & 0 & 0 & x & x \\
0 & 0 & 0 & x & x
\end{bmatrix}
$$

The complex conjugate eigenvalues of H are the eigenvalues of the trailing 2×2 submatrix.

If in the final quasitriangular matrix there occur 2×2 blocks with real eigenvalues, they may be reduced into 2×2 upper triangular blocks implementing plane rotations.

It is possible that, with the exception of $h_{n,n-1}^{(k)}$ or $h_{n-1,n-2}^{(k)}$, some other subdiagonal element of H may also approach zero during the iterative process. In this case H will take the form

$$H = \begin{bmatrix} H_{11} & H_{12} \\ 0 & H_{22} \end{bmatrix}$$

and the problem decouples into two smaller problems, involving the upper Hessenberg matrices H_{11} and H_{22}.

A subdiagonal element of H may be considered as small if

$$|h_{r,r-1}^{(k)}| \leq \varepsilon \|H\|$$

Owing to the use of orthogonal transformations, the acceptance of $h_{r,r-1}^{(k)}$ as zero will correspond to a change in the original matrix of order $\varepsilon \|H\|$. This, however, will lead to large errors if H has some eigenvalues of order $\varepsilon \|H\|$ or less. That is why in practice $|h_{r,r-1}^{(k)}|$ is set to zero if

$$|h_{r,r-1}^{(k)}| \leq \varepsilon(|h_{r-1,r-1}^{(k)}| + |h_{rr}^{(k)}|)$$

If two consecutive subdiagonal elements of the matrix become sufficiently small simultaneously, it is possible to take an additional advantage (for details see Wilkinson (1965, ch. 8)).

The double-shifted QR algorithm described so far is not always convergent. If, for example,

$$H = \begin{bmatrix} 0 & 0 & 1 \\ 1 & 0 & 0 \\ 0 & 1 & 0 \end{bmatrix}$$

then s_1 and s_2 are the eigenvalues of

$$\begin{bmatrix} 0 & 0 \\ 1 & 0 \end{bmatrix}$$

and are equal to zero. Hence $Q_1 = Q_2 = H$, $R_1 = R_2 = I_3$ and $H_2 = H_3 = H$, i.e. the matrix remains unchanged. That is why, if there is no convergence after 10 or 20 iterations, the algorithm is accomplished with a pair of randomly chosen shifts. After that, the process continues in the usual way. For this purpose the shifts are determined by

$$s_1 + s_2 = 1.5s$$
$$s_1 s_2 = s^2$$

where

$$s = |a_{n,n-1}| + |a_{n-1,n-2}|$$

These additional shifts are sufficient to destroy any special relationship among the elements of H, which may prevent the convergence.

Assuming that p steps are necessary to find an eigenvalue, the total number of flops, required by the QR algorithm, is

$$p \sum_{j=1}^{n} 5j^2 \approx \tfrac{5}{3}pn^3$$

Usually $p \approx 2$ so that the determination of the eigenvalues requires approximately $\tfrac{10}{3}n^3$ flops. Taking into account the initial reduction into Hessenberg form, $5n^3$ flops are necessary to find the eigenvalues of a dense matrix. To obtain the Schur form $T = Q^{\mathrm{T}}AQ$, the orthogonal transformations at each step must be applied to the whole matrix. This requires approximately $7n^3$ flops to find T (including the initial reduction to Hessenberg form) and $6n^3$ flops to accumulate Q.

The QR algorithm is numerically stable. It may be shown that the computed real Schur form is always orthogonally similar to a matrix $A + E$, where

$$\|E\|_{\mathrm{F}} \leq k_1 n\varepsilon \|A\|_{\mathrm{F}}$$

and k_1 is a small constant.

Example 1.36

If

$$H = \begin{bmatrix} -7 & 3 & -2 & -1 & 4 \\ 6 & 8 & -4 & 2 & -5 \\ 0 & -3 & 9 & 6 & 1 \\ 0 & 0 & -1 & 2 & 7 \\ 0 & 0 & 0 & 5 & -8 \end{bmatrix}$$

then the first four iterations of the double-shifted QR algorithm yield (up to three significant digits)

$$\begin{bmatrix} 0.124 \times 10^2 & -0.143 \times 10^1 & -0.311 \times 10^1 & 0.280 \times 10^1 & 0.233 \times 10^1 \\ 0.333 \times 10^0 & -0.814 \times 10^1 & -0.554 \times 10^0 & -0.283 \times 10^1 & 0.566 \times 10^1 \\ 0 & 0.294 \times 10^0 & 0.375 \times 10^1 & 0.120 \times 10^1 & -0.321 \times 10^1 \\ 0 & 0 & -0.446 \times 10^1 & 0.665 \times 10^1 & -0.222 \times 10^1 \\ 0 & 0 & 0 & -0.463 \times 10^{-21} & -0.107 \times 10^2 \end{bmatrix}$$

so that the $(5, 5)$ element may be accepted as an eigenvalue of H and the problem is deflated. The next three iterations lead to

$$
\begin{bmatrix}
0.771 \times 10^1 & 0.778 \times 10^1 & 0.235 \times 10^1 & -0.388 \times 10^1 & 0.864 \times 10^0 \\
0.948 \times 10^1 & -0.341 \times 10^1 & -0.211 \times 10^1 & -0.146 \times 10^1 & -0.614 \times 10^1 \\
0 & 0.464 \times 10^{-19} & 0.460 \times 10^1 & -0.492 \times 10^1 & 0.337 \times 10^1 \\
0 & 0 & 0.666 \times 10^0 & 0.576 \times 10^1 & 0.169 \times 10^1 \\
0 & 0 & 0 & 0 & -0.107 \times 10^2
\end{bmatrix}
$$

The 2×2 leading diagonal block has real eigenvalues so it may be reduced to an upper triangular form by using a single plane rotation. The second 2×2 diagonal block has complex eigenvalues $5.18 \pm 1.72i$ and it cannot be further reduced. The final real Schur form is

$$
\begin{bmatrix}
0.124 \times 10^2 & -0.170 \times 10^1 & 0.932 \times 10^0 & -0.408 \times 10^1 & -0.242 \times 10^1 \\
0 & -0.808 \times 10^1 & -0.302 \times 10^1 & 0.749 \times 10^0 & -0.571 \times 10^1 \\
0 & 0 & 0.460 \times 10^1 & -0.492 \times 10^1 & 0.337 \times 10^1 \\
0 & 0 & 0.666 \times 10^0 & 0.576 \times 10^1 & 0.169 \times 10^1 \\
0 & 0 & 0 & 0 & -0.107 \times 10^2
\end{bmatrix}
$$

The accuracy of the computed eigenvalues depends on their sensitivity to perturbations in A. That is why the eigenvalues may be determined with large errors if their condition numbers are large.

In practice, it is appropriate to scale the matrix before reduction into Hessenberg form in order to reduce its norm. This may improve the accuracy of the computed eigenvalues. For this purpose a permutation matrix P and a diagonal matrix D are determined so that the original matrix is reduced to

$$
B = D^{-1}P^{T}APD = \begin{bmatrix}
B_{11} & B_{12} & B_{13} \\
0 & B_{22} & B_{23} \\
0 & 0 & B_{33}
\end{bmatrix}
$$

where B_{11} and B_{33} are upper triangular and the matrix B_{22} is balanced, i.e. its rows and columns have approximately equal norms. In this way the eigenvalues of A are the diagonal elements of the matrices B_{11} and B_{33} and the eigenvalues of the matrix B_{22}.

The elements of D are chosen as exact powers of the machine base to avoid introduction of rounding errors. The scaling is usually done with respect to $\|.\|_1$ to reduce the computations.

There is a class of matrices which does not require balancing. For example, the symmetric matrices are always balanced with respect to $\|.\|_2$.

The implicit shift technique may be also used for computing the eigenvalues of a symmetric tridiagonal matrix. In this case the matrices $A_2, A_3, \ldots$ are also tridiagonal, so that the modification of the QR algorithm requires only cn flops, where c is a small constant. If in the symmetric case the shift s is chosen equal to a_{nn}, then the iterative process converges cubically.

1.10.3 Ordering the Real Schur Form and Block Diagonalization

The eigenvalues of the Schur form, obtained by the double-shifted QR algorithm, are generally not ordered in any special way on the diagonal. In some cases, however, the eigenvalues must appear in a specific order, say, for example if the appearance of nearly equal eigenvalues in clusters is required.

The desired ordering of the real Schur form may be achieved using orthogonal similarity transformations. Assume first that it is necessary to interchange the eigenvalues of the 2×2 block

$$T = \begin{bmatrix} \lambda_1 & t_{12} \\ 0 & \lambda_2 \end{bmatrix}$$

where $\lambda_1 \neq \lambda_2$. In this case it is possible to find an orthogonal matrix Q such that

$$Q^T T Q = \begin{bmatrix} \lambda_2 & x \\ 0 & \lambda_1 \end{bmatrix}$$

For this purpose we may use the eigenvector of T corresponding to λ_2. From the equation

$$Tx = \lambda_2 x \tag{1.92}$$

we obtain

$$x = \begin{bmatrix} t_{12} \\ \lambda_2 - \lambda_1 \end{bmatrix}$$

Now, if a plane rotation R is constructed so that the second component of Rx is zero, then it follows from (1.92) that

$$(RTR^T)(Rx) = \lambda_2(Rx)$$

and the matrix RTR^T will, by necessity, have the form

$$RTR^T = \begin{bmatrix} \lambda_2 & t_{12} \\ 0 & \lambda_1 \end{bmatrix}$$

Hence $Q = R^T$.

The interchanging of adjacent eigenvalues is more complicated when the Schur form has a 2×2 block on the diagonal which corresponds to complex eigenvalues. In such a case it is necessary to have a method for interchanging two consecutive blocks, the first one of order

$l_1 \leq 2$ and the second one of order $l_2 \leq 2$. This may be done by iterative implementation of the implicitly shifted QR step. Specifically, suppose that the first block is used to determine an implicit shift. An arbitrary QR step may be performed on both blocks to eliminate the uncoupling between them. Then a sequence of implicit QR steps, using the previously determined shift, is applied on both blocks. Now a block of size l_1, having the eigenvalues of the first block, will emerge in the lower right-hand-side corner of the array after a specific number of iterations.

    A Fortran program for interchanging consecutive 1×1 and 2×2 blocks of the real Schur form is given in Stewart (1976) (see also Flamm and Walker (1982)).

    An arbitrary ordering of the real Schur form may be achieved by symmetric interchanging of adjacent pairs of eigenvalues.

    If the Schur form of a matrix is appropriately reordered, it may be reduced simply to a block diagonal form implementing non-orthogonal transformations.

    Suppose that the real Schur form of the matrix A is partitioned in the form

$$T = Q^T A Q = \begin{bmatrix} T_{11} & T_{12} & \cdots & T_{1q} \\ & T_{22} & \cdots & T_{2q} \\ & & \ddots & \vdots \\ \mathbf{0} & & & T_{qq} \end{bmatrix}$$

where the blocks T_{ii} have no common eigenvalues. Then it is possible to reduce T into block diagonal form

$$X^{-1} T X = \text{diag}(T_{11}, \ T_{22}, \ \ldots, \ T_{qq})$$

    To derive an algorithm for block diagonalization, consider for simplicity the case when

$$T = \begin{bmatrix} T_{11} & T_{12} \\ 0 & T_{22} \end{bmatrix}$$

We seek the transformation matrix X in the form

$$X = \begin{bmatrix} I & P \\ 0 & I \end{bmatrix}$$

where the identity matrices are of the same order as T_{11} and T_{22} respectively.

    The inverse of X is given by

$$X^{-1} = \begin{bmatrix} I & -P \\ 0 & I \end{bmatrix}$$

Hence

$$X^{-1}TX = \begin{bmatrix} T_{11} & T_{11}P - PT_{22} + T_{12} \\ 0 & T_{22} \end{bmatrix}$$

and the problem for determining X reduces to the solution of the *Sylvester equation*

$$T_{11}P - PT_{22} = -T_{12}$$

This equation has a unique solution for P if and only if the matrices T_{11} and T_{22} have no common eigenvalues. Because T_{11} and T_{22} are upper quasitriangular, the Sylvester equation is easily solved by back substitution (details are given in Section 4.2).

Since

$$\text{cond}_F(X) = \| X \|_F \| X^{-1} \|_F = n + \| P \|_F^2$$

the matrix X will be ill conditioned whenever $\| P \|_F$ is large.

In the general case, the matrix T can be reduced to a block diagonal form by zeroing its superdiagonal blocks in the order $T_{12}, T_{13}, T_{23}, \ldots, T_{1q}, \ldots, T_{q-1,q}$. The following algorithm overwrites T with $X^{-1}TX = \text{diag}(T_{11}, \ldots, T_{qq})$ and Q with QX.

Algorithm 1.21 Block diagonalization of an upper block triangular matrix

For $j = 2, 3, \ldots, q$
   For $i = 1, 2, \ldots, j - 1$
      Solve $T_{ii}P - PT_{jj} = -T_{ij}$
      $T_{ij} \leftarrow 0$
      For $k = j + 1, j + 2, \ldots, q$
         $T_{ik} \leftarrow T_{ik} - PT_{jk}$
      For $k = 1, 2, \ldots, q$
         $Q_{kj} \leftarrow Q_{ki}P + Q_{kj}$

This algorithm requires cn^3 flops, where c is a constant of order unity.

If some of the blocks have close eigenvalues, then the corresponding matrix P will have large elements and the transformation matrix X will be ill conditioned. It is possible in such a case for the block diagonal form to be computed with large errors. That is why the matrix T should be partitioned so that the condition number of X is made as small as possible. An algorithm for such block diagonalization is proposed in Bavely and Stewart (1979).

Example 1.37

If the matrix

$$T = \begin{bmatrix} 7 & -2 & -4 & 3 & 1 \\ & 3 & -8 & 5 & -1 \\ & & 3.01 & 4 & -8 \\ & \mathbf{0} & & -6 & -3 \\ & & & & 2 \end{bmatrix}$$

is partitioned so that

$$T_{11} = \begin{bmatrix} 7 & -2 \\ 0 & 3 \end{bmatrix} \qquad T_{22} = \begin{bmatrix} 3.01 & 4 & -8 \\ & -6 & -3 \\ \mathbf{0} & & 2 \end{bmatrix}$$

then (up to four decimal digits)

$$P = \begin{bmatrix} -400 & -178.1 & 3734 \\ -800 & -356.1 & 7469 \end{bmatrix}$$

and $\text{cond}_F(X) = 7.070 \times 10^7$.

If the partitioning of T is done as

$$T_{11} = \begin{bmatrix} 7 & -2 & -4 \\ & 3 & -8 \\ \mathbf{0} & & 3.01 \end{bmatrix} \qquad T_{22} = \begin{bmatrix} -6 & -3 \\ 0 & 2 \end{bmatrix}$$

so that the eigenvalues 3 and 3.01 appear in the same block, then

$$P = \begin{bmatrix} -0.5136 & 38.61 \\ -0.9502 & 77.77 \\ -0.4440 & 9.239 \end{bmatrix}$$

and $\text{cond}_F(X) = 7.630 \times 10^3$.

1.10.4 Determination of the Jordan Structure of a Matrix

The determination of the eigenvalue multiplicity and the degree of an elementary divisor (i.e. the dimension of the corresponding Jordan block) is a difficult task in the presence of rounding errors. The problem is complicated for derogatory matrices when, for a given eigenvalue more than one eigenvector, i.e. more than one Jordan block, corresponds. We shall describe briefly one of the possible methods for numerical determination of the Jordan form of a matrix proposed by Ruhe (1970) and Kågström and Ruhe (1980).

The reduction of A into Jordan form involves computation of the

Schur decomposition $T = Q^H A Q$, determination of the numerical multiplicities of the eigenvalues, block diagonalization and ascertaining the Jordan structure of each diagonal block.

First the matrix A is transformed into a triangular form using the QR algorithm. If A has one or more multiple eigenvalues, then the resulting triangular matrix will have some diagonal elements close to each other. To obtain these eigenvalues in adjacent positions, they are sorted so that

$$| \lambda_{k-1} - \lambda_k | = \min_{j < k} | \lambda_j - \lambda_k | \qquad k = n, \ldots, 2$$

The reordering of the eigenvalues is done by a sequence of unitary transformations, each of them interchanging two adjacent diagonal elements. The maximal number of swappings is $\frac{1}{2} n(n - 1)$ but in practice this number is rarely larger than n. After the sorting is done, close eigenvalues will possibly appear together. Thus we obtain an upper triangular matrix $T_1 = U^H T U$, where U is the product of implemented unitary transformations.

The next stage of the algorithm is to decide which of the eigenvalue approximations form groups corresponding to multiple eigenvalues. This is done using the Gershgorin circles. For a certain group of eigenvalues it is attempted to isolate the corresponding circles from the remaining circles for all small perturbations of the matrix. When a group is found, whose eigenvalues are isolated from the rest of the eigenvalues but not from each other, the mean of the diagonal elements is taken as a numerically multiple eigenvalue. The mean is chosen for the following reason. If a matrix with multiple eigenvalues is perturbed, then the mean of a corresponding group of eigenvalues is perturbed by the same amount as the matrix itself, while the individual eigenvalues may be perturbed by fractional powers of the perturbation (Wilkinson, 1965, ch. 2). After the grouping we obtain

$$T_1 = \begin{bmatrix} T_{11} & T_{12} & \ldots & T_{1q} \\ & T_{22} & \ldots & T_{2q} \\ & & \ddots & \vdots \\ \text{\Large 0} & & & T_{qq} \end{bmatrix}$$

where the blocks T_{ii}, for $i = 1, \ldots, q$, correspond to different numerically multiple eigenvalues.

The block diagonalization of the matrix T_1 may be accomplished by Algorithm 1.21. As a result we find

$$T_2 = X^{-1} T_1 X = \text{diag}(T_{11}, T_{22}, \ldots, T_{qq})$$

Now each T_{ii} is supposed to have one numerically multiple eigenvalue and we can approximate it by the mean λ_i of the diagonal elements. Thus

$$T_{ii} = \lambda_i I + N_i$$

where N_i is the strictly upper triangular part of T_{ii}.

The Jordan structure of an $n \times n$ matrix C in the form

$$C = \lambda I + N$$

may be determined using the fact that the order of a Jordan block

$$J = \begin{bmatrix} \lambda & 1 & & & & \\ & \lambda & & 1 & & \mathbf{0} \\ & & \ddots & & \ddots & \\ & & & \lambda & & 1 \\ & \mathbf{0} & & & & \lambda \end{bmatrix}$$

is the smallest integer p for which

$$(J - \lambda I)^P = 0$$

This equation shows that the matrix N is nilpotent, i.e. $N^l = 0$, where l is at most equal to n.

If the Jordan form of C has several blocks, then their orders may be recovered from the ranks of the powers of N.

The integers are defined as

$$m_k = n - \operatorname{rank}(N^k) \qquad k = 0, 1, \ldots, n$$

It may be shown that the difference $m_k - m_{k-1}$ is equal to the number of the blocks in the Jordan form of C that have order k or greater. In particular, the quantity $m_1 - m_0 = m_1$ gives the number of the Jordan blocks.

Example 1.38

Let the matrix C have the Jordan form

$$J = \begin{bmatrix} \lambda & 1 & 0 & & & \\ 0 & \lambda & 1 & & \mathbf{0} & \\ 0 & 0 & \lambda & & & \\ & & & \lambda & 1 & \\ & & & 0 & \lambda & \\ & \mathbf{0} & & & & \lambda & 1 \\ & & & & & 0 & \lambda \end{bmatrix}$$

For this matrix

$$m_1 = n - \text{rank}(N^1) = 3$$
$$m_2 = n - \text{rank}(N^2) = 6$$
$$m_3 = n - \text{rank}(N^3) = 7$$

(Since $N^3 = 0$ the next powers of N are also zero matrices and are not considered.) Hence

$$m_1 - m_0 = 3 \qquad m_2 - m_1 = 3 \qquad m_3 - m_2 = 1$$

In fact, J has three blocks whose orders are greater than 1, three blocks whose orders are greater than or equal to 2 and one block whose order is equal to 3.

In this way, if we know the ranks of the matrices N^k, $k = 1, 2, \ldots, n$, it is possible to determine the Jordan structure of C.

The numbers m_k, $k = 1, 2, \ldots$, may be found without computing the powers of the matrix N. For this purpose, it is possible to exploit the QR decomposition with column pivoting of N.

Let N be decomposed as $NP_1 = Q_1 R_1$, where P_1 is a permutation matrix, Q_1 is unitary and R_1 is upper triangular. Since N is of rank $n - m_1$, the last m_1 rows of R_1 are zero. Using the QR decomposition of N we obtain the matrix

$$N^{(1)} = Q_1^H N Q_1 = R_1 P_1^T Q_1 = \begin{bmatrix} N_{11}^{(1)} & N_{12}^{(1)} \\ 0 & 0 \end{bmatrix} \begin{matrix} \}n - m_1 \\ \}m_1 \end{matrix}$$

which retains the zero rows of N.

Next we decompose the upper left submatrix $N_{11}^{(1)}$, which is transformed into

$$Q_2^H N_{11}^{(1)} Q_2 = \begin{bmatrix} N_{11}^{(2)} & N_{12}^{(2)} \\ 0 & 0 \end{bmatrix} \begin{matrix} \}n - m_2 \\ \}m_2 - m_1 \end{matrix}$$

We repeat these operations until we obtain the matrix $M = W^H N W$, where W is unitary and M has the form shown below for $n = 7$, $m_1 = 3$, $m_2 = 6$ and $m_3 = 7$.

Further on the superdiagonal blocks of M may be reduced to upper triangular form, and the final transformation into a Jordan form will require implementation of non-unitary transformations.

In this way, the key moment in finding the Jordan structure of a matrix is the accurate determination of the rank of the matrices $N^{(1)}$, $N_{11}^{(1)}$, $N_{11}^{(2)}$ and so on.

EXERCISES

1.60 Give a detailed description of the algorithm which reduces a matrix into Hessenberg form.

1.61 Perform an error analysis of the reduction into Hessenberg form.

1.62 Write an algorithm for reduction into Hessenberg form by using plane rotations.

1.63 Give a detailed description of an algorithm implementing the double QR step with implicit shifts.

1.64 Develop an algorithm which orders the eigenvalues along the diagonal of the Schur form, so that the eigenvalues with negative real parts appear first.

1.65 Expand the Sylvester equation $AX + XB = C$, where A and B are square matrices, as a linear system of equations for the elements of X.

1.66 Give an efficient version of the QR algorithm for symmetric tridiagonal matrices.

1.11 THE GENERALIZED EIGENVALUE PROBLEM

1.11.1 Generalized Eigenvalues and Eigenvectors

Let A and B be $n \times n$ matrices. The set of all matrices of the form $A - \lambda B$, where λ is an arbitrary complex scalar, is referred to as a *pencil*. The *generalized eigenvalue problem* consists in determining those values of λ for which the equation

$$Ax = \lambda Bx \qquad (1.93)$$

has a nontrivial solution. Such values of λ are called eigenvalues of the pencil $A - \lambda B$ or *generalized eigenvalues*. The solution x of (1.93) is an eigenvector of $A - \lambda B$. The eigenvalues of $A - \lambda B$ are the roots of the characteristic equation

$$\det(A - \lambda B) = 0$$

When $B = I_n$ the generalized eigenvalue problem is reduced to the usual eigenvalue problem for A. If B is nonsingular, then (1.93) is equivalent to

$$B^{-1}Ax = \lambda x$$

so that the eigenvalues of $A - \lambda B$ are eigenvalues of $B^{-1}A$. They are also eigenvalues of $AB^{-1} = B(B^{-1}A)B^{-1}$.

The generalized eigenvalue problem has two unusual features. First, both sides of Equation (1.93) may be equal to zero. This happens when A and B are singular and there is a nonzero vector z, such that $Az = Bz = 0$. In this case every number λ is an eigenvalue. Later we shall assume that the pencil $A - \lambda B$ is *regular*, i.e. $\det(A - \lambda B) \neq 0$. Secondly, if B is singular, then the degree of the characteristic polynomial $\det(A - \lambda B)$ is less than n, i.e. the pencil $A - \lambda B$ has less than n eigenvalues. In such a case, if A is nonsingular, equation (1.93) is equivalent to

$$A^{-1}Bx = \lambda^{-1}x = \mu x$$

and the missing eigenvalues correspond to zero eigenvalues of the reciprocal problem $B - \mu A$. In this way the missing eigenvalues of $A - \lambda B$ may be considered as infinite eigenvalues.

Example 1.39

If

$$A = \begin{bmatrix} 1 & 2 \\ 0 & 0 \end{bmatrix} \quad \text{and} \quad B = \begin{bmatrix} 1 & 1 \\ 0 & 0 \end{bmatrix}$$

then $\det(A - \lambda B) \equiv 0$ and every λ is an eigenvalue of $A - \lambda B$.

If

$$A = \begin{bmatrix} 1 & -1 \\ 0 & 2 \end{bmatrix} \quad B = \begin{bmatrix} 0 & 1 \\ 0 & 0 \end{bmatrix}$$

then $\det(A - \lambda B) = 2$ and the pencil $A - \lambda B$ has two infinite eigenvalues.

If B is well conditioned with respect to inversion, then the solution of the generalized eigenvalue problem may be done using the QR algorithm for the equivalent problem $B^{-1}Ax = \lambda x$. However, if B is ill conditioned, then $B^{-1}A$ cannot be computed accurately. Similarly, if A is ill conditioned, it is not possible to work with the reciprocal problem $A^{-1}B$.

To avoid the explicit forming of $B^{-1}A$ or $A^{-1}B$ in solving the generalized eigenvalue problem, the matrices A and B may be reduced to some simple forms, from which the solution is easily found. It may be proved, that for given $n \times n$ matrices A and B, there exist unitary matrices Q and Z such that QAZ and QBZ are both upper triangular. This reduction is called *generalized Schur decomposition*. Using this decomposition the eigenvalues of $A - \lambda B$ may be found as ratios of the diagonal elements of A and B,

$$\lambda_i = \frac{a_{ii}}{b_{ii}}$$

since the eigenvalues of $A - \lambda B$ and $Q(A - \lambda B)Z$ are the same.

Further on we describe the *QZ algorithm* of Moler and Stewart (1973) for reducing A and B in triangular form. Owing to the use of orthogonal, instead of unitary transformations, this algorithm produces a matrix A which is quasitriangular, i.e. it has 2×2 blocks on the diagonal.

The reduction of A and B into triangular form is done as follows. First, A is reduced to an upper Hessenberg form and B is reduced to an upper triangular form. Then the effect of one double-shifted QR step on AB^{-1} is simulated by equivalent orthogonal transformations on the problem $A - \lambda B$. The iterative implementation of this algorithm reduces A to a quasitriangular form.

1.11.2 Reduction to Hessenberg-triangular Form

The reduction of $A - \lambda B$ into Hessenberg-triangular form is accomplished by elementary orthogonal transformations. First, B is reduced to an upper triangular form and the transformations are applied to A. Then A and B look as shown below for $n = 5$:

$$
\begin{bmatrix}
x & x & x & x & x \\
x & x & x & x & x \\
x & x & x & x & x \\
x & x & x & x & x \\
\langle x \rangle & x & x & x & x
\end{bmatrix}
\qquad
\begin{bmatrix}
x & x & x & x & x \\
0 & x & x & x & x \\
0 & 0 & x & x & x \\
0 & 0 & 0 & x & x \\
0 & 0 & 0 & 0 & x
\end{bmatrix}
$$

Let a plane rotation Q_{45} be constructed in the $(4, 5)$-plane, such that $Q_{45}A$ has a zero in the indicated position. Then $Q_{45}B$ will have the form

$$\begin{bmatrix} x & x & x & x & x \\ 0 & x & x & x & x \\ 0 & 0 & x & x & x \\ 0 & 0 & 0 & x & x \\ 0 & 0 & 0 & (\widehat{x}) & x \end{bmatrix}$$

The nonzero entry arising in position $(5,4)$ may be zeroed by a plane rotation Z_{45} in the $(4,5)$-plane. As a result $Q_{45}BZ_{45}$ is upper triangular. The matrix $Q_{45}AZ_{45}$ will have the form

$$\begin{bmatrix} x & x & x & x & x \\ x & x & x & x & x \\ x & x & x & x & x \\ (\widehat{x}) & x & x & x & x \\ 0 & x & x & x & x \end{bmatrix}$$

After that a rotation Q_{34} in the $(3,4)$-plane is computed, such that $Q_{34}Q_{45}BZ_{45}$ will look as follows

$$\begin{bmatrix} x & x & x & x & x \\ 0 & x & x & x & x \\ 0 & 0 & x & x & x \\ 0 & 0 & (\widehat{x}) & x & x \\ 0 & 0 & 0 & 0 & x \end{bmatrix}$$

Now it is possible to choose a rotation Z_{34} in the $(3,4)$-plane, so that $Q_{34}Q_{45}BZ_{45}Z_{34}$ will again be upper triangular.

In general, the zeros are introduced in A by left multiplication with plane rotations $Q_{i,i+1}$, which introduce nonzero subdiagonal elements in B. These elements are immediately zeroed by right multiplication with appropriate plane rotations $Z_{i,i+1}$.

The reduction of a generalized eigenvalue problem into a Hessenberg-triangular form is numerically stable and requires $\frac{17}{3}n^3$ flops. If the transformations $Z_{i,i+1}$ are accumulated (this is necessary in finding eigenvectors), an additional $\frac{3}{2}n^3$ flops are required. The matrix Q is not required in computing the eigenvectors.

Example 1.40

Let

$$A = \begin{bmatrix} 2 & -8 & 9 & -4 \\ -4 & -6 & 1 & 2 \\ 7 & 3 & -7 & -5 \\ 5 & -1 & -3 & 4 \end{bmatrix} \qquad B = \begin{bmatrix} -5 & 1 & -2 & -4 \\ -8 & 3 & -9 & -7 \\ 7 & -6 & 5 & -3 \\ -2 & 4 & 8 & -1 \end{bmatrix}$$

Using the above algorithm, we obtain (up to three decimal digits)

$$Q = \begin{bmatrix} -0.420 & -0.671 & 0.587 & -0.168 \\ -0.504 & 0.0684 & -0.485 & -0.712 \\ -0.755 & 0.321 & -0.009\ 40 & 0.572 \\ -0.007\ 58 & 0.665 & 0.648 & -0.372 \end{bmatrix}$$

$$Z = \begin{bmatrix} 1 & 0 & 0 & 0 \\ 0 & 0.761 & -0.530 & 0.373 \\ 0 & -0.628 & -0.459 & 0.629 \\ 0 & 0.162 & 0.713 & 0.682 \end{bmatrix}$$

$$QAZ = \begin{bmatrix} 5.12 & 11.6 & -3.58 & -3.82 \\ -8.23 & 1.80 & -0.780 & 2.92 \\ 0 & 8.75 & 6.14 & 0.298 \\ 0 & 0 & -0.252 & -4.67 \end{bmatrix}$$

$$QBZ = \begin{bmatrix} 11.9 & -9.59 & 3.04 & 6.12 \\ 0 & 5.27 & 6.31 & -2.42 \\ 0 & 0 & -2.63 & 3.09 \\ 0 & 0 & 0 & -9.08 \end{bmatrix}$$

1.11.3 The QZ Algorithm

To obtain an analogue of the QR algorithm for the generalized eigenvalue problem, suppose that A and B are reduced as described above and let B be nonsingular. If A is an unreduced upper Hessenberg matrix, then $C = AB^{-1}$ is also an unreduced upper Hessenberg matrix. Assume that a double QR step is applied to C to obtain the upper Hessenberg matrix $C' = QCQ^{\mathrm{T}}$. Instead of forming C and using the QR algorithm, we may accomplish the following procedure. Suppose we may find orthogonal matrices Q' and Z with the following properties:

1. Q' has the some first row as Q.
2. $A' = Q'AZ$ is upper Hessenberg.
3. $B' = Q'BZ$ is upper triangular.

Then

$$A'B'^{-1} = Q'AZZ^TB^{-1}Q'^T = Q'CQ'^T$$

is an upper Hessenberg matrix and according to the properties of un-reduced Hessenberg matrices (see Section 1.10) it follows that $Q' = Q$ and $C' = A'B'^{-1}$.

Assume that the above procedure is applied iteratively to produce the sequences $A_1, A_2, A_3, \ldots$ and $B_1, B_2, B_3, \ldots$. If we set $C_k = A_k B_k^{-1}$ then the sequence $C_1, C_2, C_3, \ldots$ is exactly the sequence which would be obtained by the double-shifted QR algorithm. The shifts may be chosen so that C_k approaches a quasitriangular form and since B_k is upper triangular, it follows that $A_k = C_k B_k$ should also approach a quasi-triangular form. The matrices C_k are not formed so instead of them we work directly with A_k and B_k.

The first row of the matrix Q is the same as the first row of the elementary reflection Q_0 that introduces zeros in positions $(2, 1)$ and $(3, 1)$ of $(C - s_1 I)(C - s_2 I)$, where s_1 and s_2 are the shifts. This reflection depends only on the first two columns of C, which are given by

$$[c_1, c_2] = [a_1, a_2] \begin{bmatrix} b_{11} & b_{12} \\ 0 & b_{22} \end{bmatrix}^{-1}$$

where a_1 and a_2 are the first two columns of A.

As in the implicit QR algorithm, the shifts s_1 and s_2 are taken as the roots of

$$\det(\hat{A} - s\hat{B})$$

where

$$\hat{A} = \begin{bmatrix} a_{n-1,n-1} & a_{n-1,n} \\ a_{n,n-1} & a_{nn} \end{bmatrix} \quad \text{and} \quad \hat{B} = \begin{bmatrix} b_{n-1,n-1} & b_{n-1,n} \\ 0 & b_{nn} \end{bmatrix}$$

The shifts s_1 and s_2 are not computed explicitly, which means complex arithmetic can be avoided.

To find A' and B' we form first the matrices $Q_0 A$ and $Q_0 B$. If we find orthogonal matrices U and V, such that $U Q_0 A V$ is upper Hessenberg, $U Q_0 B V$ is upper triangular and $U Q_0$ has the same first row as Q_0, then $U' = U Q_0$ has the same first row as Q, i.e. $A' = U Q_0 A V$ and $B' = U Q_0 B V$. The matrices U and V may be determined in the following way.

The matrix Q_0 has the form

$$Q_0 = \begin{bmatrix} U_0 & 0 \\ 0 & I_{n-3} \end{bmatrix}$$

where U_0 is an elementary reflection of third order. Hence $Q_0 A$ and $Q_0 B$

have the form shown below for $n = 6$:

$$Q_0 A = \begin{bmatrix} x & x & x & x & x & x \\ x & x & x & x & x & x \\ x & x & x & x & x & x \\ 0 & 0 & x & x & x & x \\ 0 & 0 & 0 & x & x & x \\ 0 & 0 & 0 & 0 & x & x \end{bmatrix} \qquad Q_0 B = \begin{bmatrix} x & x & x & x & x & x \\ x^2 & x & x & x & x & x \\ x^1 & x^1 & x & x & x & x \\ 0 & 0 & 0 & x & x & x \\ 0 & 0 & 0 & 0 & x & x \\ 0 & 0 & 0 & 0 & 0 & x \end{bmatrix}$$

First $Q_0 B$ is reduced to an upper triangular form using an elementary reflection Z_0, which introduces zeros in the positions denoted by superscript 1, and a rotation Z_0' which introduces a zero in the position denoted by superscript 2. Then $Q_0 A Z_0 Z_0'$ and $Q_0 B Z_0 Z_0'$ will have the form

$$\begin{bmatrix} x & x & x & x & x & x \\ x & x & x & x & x & x \\ \tilde{(x)} & x & x & x & x & x \\ \tilde{(x)} & x & x & x & x & x \\ 0 & 0 & 0 & x & x & x \\ 0 & 0 & 0 & 0 & x & x \end{bmatrix} \qquad \begin{bmatrix} x & x & x & x & x & x \\ 0 & x & x & x & x & x \\ 0 & 0 & x & x & x & x \\ 0 & 0 & 0 & x & x & x \\ 0 & 0 & 0 & 0 & x & x \\ 0 & 0 & 0 & 0 & 0 & x \end{bmatrix}$$

Next an elementary reflection Q_1 is constructed to introduce zeros in the positions encircled above. Then $Q_1 Q_0 A Z_0 Z_0'$ and $Q_1 Q_0 B Z_0 Z_0'$ will look as follows:

$$\begin{bmatrix} x & x & x & x & x & x \\ x & x & x & x & x & x \\ 0 & x & x & x & x & x \\ 0 & x & x & x & x & x \\ 0 & 0 & 0 & x & x & x \\ 0 & 0 & 0 & 0 & x & x \end{bmatrix} \qquad \begin{bmatrix} x & x & x & x & x & x \\ 0 & x & x & x & x & x \\ 0 & x & x & x & x & x \\ 0 & x & x & x & x & x \\ 0 & 0 & 0 & 0 & x & x \\ 0 & 0 & 0 & 0 & 0 & x \end{bmatrix}$$

In this way the unwanted nonzero elements in A and B have been moved to the lower, right-hand corners. The reduction continues working with submatrices of the fifth order, which have the same form as the original matrices.

This algorithm requires about $13n^2$ flops. In comparison, the double QR step requires approximately $5n^2$ flops, so that the QZ algorithm, applied to two matrices, requires 2.6 times more computations than the QR algorithm.

Applying a sequence of QZ steps to A and B it is possible to reduce A to a quasitriangular form while maintaining B in triangular form. If the element $a_{n,n-1}^{(k)}$ becomes negligible, then the problem may be deflated. If some other subdiagonal element of A_k, say $a_{l,l-1}^{(k)}$, becomes negligible, then additional computations may be saved by working with the rows from l to n.

The result of the algorithm described is an upper triangular matrix $\tilde{B}$ and an upper quasitriangular matrix $\tilde{A}$, which has no two consecutive nonzero subdiagonal elements. In this way the original problem is decomposed into 1×1 and 2×2 subproblems. The eigenvalues of the 1×1 problems are the ratios of the corresponding diagonal elements of $\tilde{A}$ and $\tilde{B}$. The eigenvalues of the 2×2 problems may be computed as roots of quadratic equations and they may be complex even for real A and B.

The QZ algorithm is numerically stable, the computed generalized Schur form being exact for matrices $A + E$ and $B + F$, where $\|E\| \leq c_1 \varepsilon \|A\|$ and $\|F\| \leq c_2 \varepsilon \|B\|$.

Example 1.41

If the QZ algorithm is applied to the matrices from Example 1.40, one obtains (up to three significant digits)

$$\tilde{A} = \begin{bmatrix} -10.6 & -9.48 & -2.46 & 4.36 \\ 3.93 & -9.50 & -0.224 & -2.15 \\ 0 & 0 & 6.82 & -0.136 \\ 0 & 0 & 0 & -4.61 \end{bmatrix}$$

$$\tilde{B} = \begin{bmatrix} -5.05 & 12.6 & -6.20 & -7.42 \\ 0 & -3.90 & -1.96 & -0.723 \\ 0 & 0 & -8.40 & 1.32 \\ 0 & 0 & 0 & -9.08 \end{bmatrix}$$

so that the eigenvalues of $A - \lambda B$ are

$\lambda_{1,2} = 1.01 \pm 2.45i$
$\lambda_3 = -0.813$
$\lambda_4 = 0.508$

Before applying the QZ algorithm, it is appropriate to balance the matrices A and B. This may improve the accuracy of the computed generalized eigenvalues and eigenvectors. An algorithm for balancing the generalized eigenvalue problem is presented by Ward (1981).

Let

$$QAZ = \begin{bmatrix} A_{11} & A_{12} \\ 0 & A_{22} \end{bmatrix} \begin{matrix} \}l \\ \}n-l \end{matrix} \qquad QBZ = \begin{bmatrix} B_{11} & B_{12} \\ 0 & B_{22} \end{bmatrix} \begin{matrix} \}l \\ \}n-l \end{matrix}$$
$$\underbrace{\phantom{A_{11}}}_{l} \underbrace{\phantom{A_{12}}}_{n-l} \qquad\qquad \underbrace{\phantom{B_{11}}}_{l} \underbrace{\phantom{B_{12}}}_{n-l}$$

be the matrices of the generalized Schur decomposition of the pencil $A - \lambda B$. Let the corresponding partitioning of the unitary matrices Q, Z be

$$Q = [\,\underbrace{Q_1,}_{l} \ \underbrace{Q_2}_{n-l}\,] \qquad Z = [\,\underbrace{Z_1,}_{l} \ \underbrace{Z_2}_{n-l}\,]$$

The columns of the matrices Q_1 and Z_1 span a *deflating pair of subspaces* for the generalized eigenvalue problem $Ax = \lambda Bx$. The deflating subspaces are the generalization of invariant subspaces for the usual eigenvalue problem $Ax = \lambda x$.

In applications of the generalized Schur decomposition it is often important to know the sensitivity of specific deflating subspaces to perturbations in A and B. Suppose that the matrices in the pencil $A - \lambda B$ are perturbed as $\hat{A} = A + E$ and $\hat{B} = B + F$. The upper triangular matrices in the Schur decomposition of $\hat{A} - \lambda \hat{B}$ are represented as

$$\hat{Q}\hat{A}\hat{Z} = \begin{bmatrix} \hat{A}_{11} & \hat{A}_{12} \\ 0 & \hat{A}_{22} \end{bmatrix} \qquad \hat{Q}\hat{B}\hat{Z} = \begin{bmatrix} \hat{B}_{11} & \hat{B}_{12} \\ 0 & \hat{B}_{22} \end{bmatrix}$$

where

$$\hat{Q} = [\hat{Q}_1, \ \hat{Q}_2] \qquad \hat{Z} = [\hat{Z}_1, \ \hat{Z}_2]$$

If $\| E \|$ and $\| F \|$ are small, we may expect that the space spanned by $\hat{Q}_1$ (and $\hat{Z}_1$ respectively) will be close to the space spanned by Q_1 (and Z_1 respectively).

Defining the function dif as

$$\text{dif}(A_{11}, B_{11}; A_{22}, B_{22}) = \min_{[X,Y] \neq 0} \frac{\left\| \begin{bmatrix} A_{11}X - YA_{22} \\ B_{11}X - YB_{22} \end{bmatrix} \right\|_F}{\| [X, Y] \|_F}$$

and with

$$QEZ = \begin{bmatrix} E_{11} & E_{12} \\ E_{21} & E_{22} \end{bmatrix} \qquad QFZ = \begin{bmatrix} F_{11} & F_{12} \\ F_{21} & F_{22} \end{bmatrix}$$

the following theorem, due to Stewart (1973a), is an analogue to Theorem 1.2.

Theorem 1.3

If

$$\delta = \mathrm{dif}(A_{11}, B_{11};\ A_{22}, B_{22}) - \| [E_{11}, F_{11}] \|_F - \| [E_{22}, F_{22}] \|_F > 0$$

and

$$\| [E_{12}, F_{12}] \|_F \| [E_{21}, F_{21}] \|_F < \frac{\delta^2}{4}$$

then there exist complex $(n - l) \times l$ matrices X, Y satisfying

$$\| [X, Y] \|_F \leq \| [E_{21}, F_{21}] \|_F / \delta$$

such that the columns of the matrices

$$\hat{Q}_1 = Q_1 + Q_2 X \qquad \hat{Z}_1 = Z_1 + Z_2 Y$$

span a deflating pair of subspaces for the generalized eigenvalue problem $(A + E)x = \lambda(B + F)x$.

Theorem 1.3 shows that if dif has small value then the corresponding deflating subspaces may be sensitive to perturbations in the pencil. The function dif is thus a generalization of the sep-function for the usual eigenproblem.

The function dif has also the property (Stewart, 1972)

$$\mathrm{dif}(A_{11}, I;\ A_{22}, I) \leq \mathrm{sep}(A_{11}, A_{22})$$

This inequality shows that the sensitivity of the deflating subspaces for the generalized eigenproblem $Ax = \lambda I x$ is higher or equal to the sensitivity of the invariant subspaces for the eigenvalue problem $Ax = \lambda x$.

1.11.4 Ordering the Generalized Schur form

In some applications it is necessary to reorder the triangular forms of the matrices A and B, obtained by the QZ algorithm, so that the eigenvalues of $A - \lambda B$ appear in a specific order. Analogically to the reordering of the Schur form of a matrix, the reordering of the pencil $A - \lambda B$ may be done by successive implementation of orthogonal transformations, which interchange two adjacent generalized eigenvalues.

Suppose first that it is necessary to interchange the eigenvalues of a 2×2 pencil $A - \lambda B$, where

$$A = \begin{bmatrix} a_{11} & a_{12} \\ 0 & a_{22} \end{bmatrix} \qquad B = \begin{bmatrix} b_{11} & b_{12} \\ 0 & b_{22} \end{bmatrix}$$

Using orthogonal matrices Q and Z it is possible to reduce A and B to

$$QAZ = \begin{bmatrix} a_{11}^* & a_{12}^* \\ 0 & a_{22}^* \end{bmatrix} \qquad QBZ = \begin{bmatrix} b_{11}^* & b_{12}^* \\ 0 & b_{22}^* \end{bmatrix} \qquad (1.94)$$

where

$$a_{11}^* b_{22} = a_{22} b_{11}^* \qquad (1.95)$$

i.e. the generalized eigenvalue, determined by (a_{11}^*, b_{11}^*), is the same as that determined by (a_{22}, b_{22}). To find Q and Z we may use the eigenvector of $A - \lambda B$ corresponding to the second eigenvalue. From

$$(b_{22}A - a_{22}B)x = 0$$

we obtain

$$x = \begin{bmatrix} a_{12}b_{22} - a_{22}b_{12} \\ a_{11}b_{22} - a_{22}b_{11} \end{bmatrix}$$

provided that $a_{11}b_{22} \neq a_{22}b_{11}$, i.e. that the eigenvalues of $A - \lambda B$ are different.

Let a plane rotation R_1 be constructed such that the second component of $R_1 x$ is equal to zero. Then it follows from

$$(b_{22}AR_1^T - a_{22}BR_1^T)R_1 x = 0$$

that the matrix $b_{22}AR_1^T - a_{22}BR_1^T$ will have the form

$$\begin{bmatrix} 0 & x \\ 0 & 0 \end{bmatrix}$$

Hence the first columns of AR_1^T and BR_1^T are proportional. That is why if we construct a plane rotation R_2 to zero the $(2, 1)$-element of $R_2 AR_1^T$, then the matrices $R_2 AR_1^T$ and $R_2 BR_1^T$ will have the form shown in (1.94) and the condition (1.95) will be fulfilled. Thus, $Q = R_2$ and $Z = R_1^T$.

The reordering of the generalized Schur form in the presence of 2×2 blocks on the diagonal of A is more complicated. To interchange two adjacent blocks of the pencil $A - \lambda B$, one of them of dimension 2×2, it is possible to use iteratively QZ steps with shifts, determined as eigenvalues of the first block. To remove the uncoupling between the blocks, it is necessary to perform first a QZ step with a randomly chosen shift.

A program for ordering the generalized Schur form of a pencil is presented in Van Dooren (1982) (see also the remarks in Petkov, Christov and Konstantinov (1984b)).

EXERCISES

1.67 Verify that for nonsingular Q and Z the eigenvalues of $A - \lambda B$ and $Q(A - \lambda B)Z$ are the same.

1.68 Give a detailed description of the algorithm for reduction into Hessenberg-triangular form.

1.69 Perform an error analysis of the reduction into Hessenberg-triangular form.

1.70 Give a detailed description of an algorithm implementing the double QZ step with implicit shifts.

1.71 Develop an algorithm for reordering the generalized Schur form in the presence of complex generalized eigenvalues.

1.72 Write an efficient algorithm for solving the generalized eigenvalue problem $Ax = \lambda Bx$ when A is symmetric and B is positive definite.

1.12 THE SINGULAR VALUE DECOMPOSITION

In this section we consider the singular value decomposition of rectangular matrices, which allows us to determine the rank of a matrix and to solve the linear least squares problem in a reliable way.

Let A be a real $n \times m$ matrix. There exist an $n \times n$ orthogonal matrix U and an $m \times m$ orthogonal matrix V, such that $U^T A V$ takes one of the forms

$$U^T A V = \begin{bmatrix} \Sigma \\ 0 \end{bmatrix} \quad \text{if } n \geq m$$

$$U^T A V = [\Sigma, \ 0] \quad \text{if } n \leq m \tag{1.96}$$

Here

$$\Sigma = \text{diag}(\sigma_1, \ \sigma_2, \ ..., \ \sigma_l)$$

where $l = \min\{n, \ m\}$ and

$$\sigma_1 \geq \sigma_2 \geq \cdots \geq \sigma_l \geq 0$$

The decomposition (1.96) is called the *singular value decomposition* (SVD) of A. The scalars $\sigma_1, \sigma_2, ..., \sigma_l$ are the *singular values* of A. The columns of U are the *left singular vectors* of A and the columns of V the *right singular vectors* of A.

We shall assume further that $n \geq m$, the extensions to the case $n < m$ being obvious.

It follows from (1.96) that

$$V^T A^T A V = \Sigma^2$$

Hence the quantities $\sigma_1^2, \sigma_2^2, ..., \sigma_m^2$ are the eigenvalues of A^TA, arranged in descending order. The left singular vectors $u_1, ..., u_n$ are the eigenvectors of AA^T, while the right singular vectors $v_1, v_2, ..., v_m$ are the eigenvectors of A^TA. The singular values of A are the same as those of A^T.

The singular values have much in common with the eigenvalues of a symmetric matrix. In fact, if A is symmetric, then the singular values are precisely the moduli of the eigenvalues of A. For a positive definite matrix the singular values coincide with the eigenvalues.

Example 1.42

If

$$A = \begin{bmatrix} -7 & 11 & -3 \\ 8 & -2 & -5 \\ 4 & -13 & 6 \\ -9 & -16 & -8 \\ -6 & 7 & 1 \end{bmatrix}$$

then (up to three decimal digits)

$$\sigma_1 = 24.6 \qquad \sigma_2 = 16.7 \qquad \sigma_3 = 9.77$$

$$U = \begin{bmatrix} 0.468 & 0.407 & 0.063 & -0.751 & -0.219 \\ -0.127 & -0.281 & 0.801 & -0.288 & 0.425 \\ -0.528 & -0.321 & -0.493 & -0.591 & 0.160 \\ -0.624 & 0.764 & 0.155 & 0.0258 & 0.044 \\ 0.311 & 0.262 & -0.295 & 0.0594 & 0.863 \end{bmatrix}$$

$$V = \begin{bmatrix} -0.108 & -0.888 & 0.448 \\ 0.993 & -0.071 & 0.0982 \\ 0.0555 & -0.455 & -0.889 \end{bmatrix}$$

Let A be an $n \times m$ matrix with singular values $\sigma_1 \geqq \sigma_2 \geqq \cdots \geqq \sigma_m$. Then

$$\| A \|_2 = \sigma_1$$

and

$$\| A \|_F^2 = \sigma_1^2 + \sigma_2^2 + \cdots + \sigma_m^2$$

In this way the 2-norm of a matrix is its largest singular value, i.e. the square root of the largest eigenvalue of A^TA or AA^T. The square of the Frobenius norm is the sum of squares of the singular values.

Let the elements of A be accurate up to quantities of order α. Suppose that, having the computed singular value decomposition of A (1.96), it is found that

$$\sigma_{r+1}^2 + \sigma_{r+2}^2 + \cdots + \sigma_m^2 < \alpha^2$$

Then if we set

$$\Sigma' = \text{diag}(\sigma_1, \sigma_2, \ldots, \sigma_r, 0, \ldots, 0)$$

and

$$A' = U \begin{bmatrix} \Sigma' \\ 0 \end{bmatrix} V^{\mathrm{T}}$$

it is fulfilled that

$$\| A - A' \|_{\mathrm{F}} = \sqrt{(\sigma_{r+1}^2 + \cdots + \sigma_m^2)} < \alpha \tag{1.97}$$

Thus A' is a matrix of rank r lying in α-neighbourhood of A with respect to the Frobenius norm. In such a case it is justified to say that A is near to a matrix of rank r.

This property of the singular value decomposition makes it possible to determine the rank of a matrix in the presence of uncertainty in its elements, that is the *numerical rank*. This may be done by setting to zero all singular values which are less than a tolerance α, reflecting the errors in the elements of A. The rank, obtained in such a way, is the exact rank of a matrix A' which fulfils (1.97).

An important property of singular values is that they are well conditioned with respect to perturbations in the matrix. More precisely, the singular values of $A + E$ differ from the corresponding singular values of A with a quantity which is at most $\| E \|_2$.

Example 1.43

The singular values of

$$A = \begin{bmatrix} 4 & 8 & -3 \\ -7 & 1 & 5 \\ -6 & 3 & -2 \\ 9 & -1 & -8 \end{bmatrix}$$

are (up to eight decimal digits)

$$\sigma_1 = 16.018\ 451$$
$$\sigma_2 = 8.892\ 830\ 3$$
$$\sigma_3 = 4.829\ 781\ 1$$

If A is perturbed by

$$E = 10^{-5} \begin{bmatrix} 1 & 1 & 1 \\ 1 & 1 & 1 \\ 1 & 1 & 1 \\ 1 & 1 & 1 \end{bmatrix}$$

then the singular values become

$$\tilde{\sigma}_1 = 16.018\ 452$$
$$\tilde{\sigma}_2 = 8.892\ 837\ 9$$
$$\tilde{\sigma}_3 = 4.829\ 770\ 2$$

so that

$$|\tilde{\sigma}_i - \sigma_i| < \|E\|_2 = 3.46 \times 10^{-5} \qquad i = 1, 2, 3$$

Singular value decomposition may be used to solve the linear least squares problem $Ax \cong b$.

Let A be an $n \times m$ matrix of rank $r < m$ which is decomposed as

$$A = U \begin{bmatrix} \Sigma & 0 \\ 0 & 0 \end{bmatrix} V^{\mathrm{T}}$$

where

$$\Sigma = \mathrm{diag}(\sigma_1, \sigma_2, \ldots, \sigma_r)$$

and

$$\sigma_1 \geqq \sigma_2 \geqq \cdots \geqq \sigma_r > 0$$

Then the vector x, determined by

$$x = V \begin{bmatrix} \Sigma^{-1} & 0 \\ 0 & 0 \end{bmatrix} U^{\mathrm{T}} b \tag{1.98}$$

minimizes $\|b - Ax\|_2^2$. Moreover, if $\|b - Ax'\|_2^2$ is also minimal and $x' \neq x$, then $\|x\|_2 < \|x'\|_2$. Hence the least squares solution (1.98) is also the minimum norm solution.

As shown above, the singular values of A may be found by computing the eigenvalues of $A^{\mathrm{T}}A$. In this way, the problem for determining the singular values may be reduced to the ordinary eigenvalue problem. This, however, is not to be recommended for the following reasons. If A is rounded to t decimal digits, then the singular value of A of order $10^{-s}\|A\|_2$ will be affected in the worst case in its $(t-s)$th digit. On the other hand, the same singular value corresponds to an eigenvalue of $A^{\mathrm{T}}A$ of order $10^{-2s}\|A\|_2^2$. If $A^{\mathrm{T}}A$ is rounded to t digits, this eigenvalue will be determined with accuracy $t - 2s$ digits. In particular, if $2s > t$, then it is possible that there will be no accurate digit in the computed eigenvalue of $A^{\mathrm{T}}A$, although the corresponding singular value may have several accurate digits.

Example 1.44

The matrix

$$A = \begin{bmatrix} 3.049 & 2.951 \\ 2.951 & 3.049 \end{bmatrix}$$

has singular values

$$\sigma_1 = 6 \qquad \sigma_2 = 0.098$$

On the other hand the matrix $A^{\mathrm{T}}A$, rounded to four decimal digits, is

$$\begin{bmatrix} 18.00 & 18.00 \\ 18.00 & 18.00 \end{bmatrix}$$

and its eigenvalues are 36 and 0. In this way the information about the smallest singular value of A was lost in computing $A^{\mathrm{T}}A$.

To avoid the numerical difficulties related to the use of the matrix $A^{\mathrm{T}}A$, it is possible to implement the following approach:

1. Use orthogonal equivalent transformations to reduce A into bidiagonal form. These transformations leave the singular values of A unchanged.
2. Considering the implicitly shifted QR algorithm for finding the eigenvalues of $A^{\mathrm{T}}A$, derive an iterative procedure to reduce the bidiagonal matrix into diagonal form using orthogonal equivalent transformations.

We refer the reader to Dongarra *et al.* (1979, ch. 11), where the corresponding algorithm is described in detail. This algorithm is numerically stable. If Σ, U and V are the computed quantities, then for $n \geq m$

$$U^{\mathrm{T}}(A + E)V = \begin{bmatrix} \Sigma \\ 0 \end{bmatrix}$$

for some matrix E, satisfying

$$\| E \|_2 \leq f(n, m)\varepsilon \| A \|_2$$

where $f(n, m)$ is a slowly growing function of n and m. Hence the computed singular values differ from the exact ones by no more than

$f(n, m)\varepsilon \| A \|_2$. Since $\| A \|_2 = \sigma_1$, the large singular values will be computed with high relative accuracy. If the matrix has singular values of order $\sigma_1 \varepsilon$, their computed approximations may not be accurate at all.

EXERCISES

1.73 Verify (1.97).

1.74 Prove that

$$\text{cond}_2(A) = \frac{\sigma_{\max}(A)}{\sigma_{\min}(A)}$$

where $\sigma_{\min}(A)$ and $\sigma_{\max}(A)$ are the minimum and maximum singular values of the matrix A.

1.75 Prove that the numerical rank of a matrix A remains unchanged under any perturbation E satisfying $\| E \|_2 < \sigma_{\min}(A)$.

1.76 Show that the pseudo-inverse of

$$A = U \begin{bmatrix} \Sigma \\ 0 \end{bmatrix} V^{\mathsf{T}}$$

is given by

$$A^+ = V[\Sigma^{-1}, \ 0] U^{\mathsf{T}}$$

1.77 Verify that (1.98) is the minimum norm solution of the equation $Ax = b$.

1.78 Give an algorithm which reduces a rectangular matrix into bidiagonal form using elementary reflections.

NOTES AND REFERENCES

A detailed exposition of the material presented in this chapter, may be found in Golub and Van Loan (1983), Stewart (1973b), Wilkinson (1965) and Horn and Johnson (1986).

Error analysis of the arithmetic operations is considered in depth in Wilkinson (1963). A very good introduction to this subject are the corresponding chapters from Forsythe and Moler (1967) and Vandergraft (1978).

A clear presentation of the solution of linear algebraic equations is given by Forsythe and Moler (1967) and Forsythe, Malcolm and Moler (1977). The solution of the linear least squares problem is considered in Lawson and Hanson (1974) and Stewart (1973b). High quality Fortran programs for solving linear equations and least squares problems are contained in the program package LINPACK (Dongarra *et al.*, 1979). This package also contains programs for estimating the condition number of a matrix.

A profound exposition of the eigenvalue problem and its computational solution is given in Wilkinson (1965). Clear descriptions of the QR and QZ algorithms may be found in Golub and Van Loan (1983) and Stewart (1973b). The singular value decomposition is discussed in Lawson and Hanson (1974) and Stewart (1973b). Fortran codes for solving the ordinary and generalized eigenvalue problem for different types of matrices are contained in the package EISPACK (Smith *et al.*, 1976; Garbow *et al.*, 1977). Programs for singular value decomposition may be found in LINPACK and EISPACK.

The programs from LINPACK and EISPACK are implemented in MATLAB† – a powerful interactive system for matrix computations, developed by C.B. Moler (1982). The capabilities of MATLAB range from the solution of linear equations and linear least squares problems, through symmetric and nonsymmetric eigenvalue problems to the singular value decomposition. Because of its ability to perform computations with different precisions, MATLAB provides an opportunity to examine the effect of rounding errors on matrix algorithms. The personal computer version PC-MATLAB† (Moler, Little and Bangert, 1987) of this program contains the so-called Control System Toolbox – a collection of algorithms that implement common control system analysis and design techniques.

The appearance of advanced computer architectures in recent years is stimulating the development of efficient matrix algorithms utilizing some kind of parallelism. We refer the reader to the book of Ortega (1988) for an introduction to this subject, which is an area of intensive research at present.

† MATLAB and PC-MATLAB are trademarks of MathWorks, Inc.

2

Linear Control Systems

In this chapter we review the main topics in the analysis and design of linear control systems described in state–space. Since it is not possible to give a systematic exposition of linear system theory in a single chapter we consider only the basic analysis and design problems and some of the methods for their solution. It should be noted that the methods presented in this chapter are mainly of theoretical interest and the computational solution of the corresponding problems is discussed in the succeeding chapters. Profound consideration of linear system theory and the physical interpretation of its results may be found in the books referenced at the end of the chapter.

2.1 STATE–SPACE DESCRIPTION

A *linear continuous time-invariant control system* is described by the set of differential and algebraic equations written in the vector–matrix form

$$\dot{x}(t) = Ax(t) + Bu(t) \qquad x(t_0) = x_0 \tag{2.1}$$

$$y(t) = Cx(t) \tag{2.2}$$

where $x(t)$ is an n-vector of *states*, $u(t)$ is an m-vector of *inputs (controls)* and $y(t)$ is an r-vector of *outputs*. The dimension n of the state vector $x(t)$ is referred to as the *order* of the system and it is usually fulfilled that $m \leq n$, $r \leq n$. The matrices A, B and C have real constant elements and are of dimension $n \times n$, $n \times m$ and $r \times n$ respectively. The matrix A is referred to as a *state matrix*, B as an *input (control) matrix* and C as an *output matrix*. In present day applications, the order of the system is at most several hundreds so that the matrices A, B and C may be considered as dense.

Equation (2.1) is referred to as the *state equation* and equation (2.2) is referred to as the *output equation*.

The state x_0 of the system at the moment t_0 contains the whole information, which along with the information about the behaviour of the input on the interval (t_0, t), allows us to predict the behaviour of the output at every $t \geq t_0$. The state vector $x(t)$ belongs to an n-dimensional linear vector space which is referred to as *state–space*.

Henceforth we shall assume without loss of generality that $t_0 = 0$.

Equations (2.1) and (2.2) are known as the *state–space description* of a linear continuous-time system or as the *state–space model*. Later we shall identify a linear continuous-time system with Equations (2.1) and (2.2) or with the ordered triple (A, B, C).

If $m = 1$ $(r = 1)$ the system is called *single input* (*single output*). If $m > 1$ $(r > 1)$ the system is called *multi-input* (*multi-output*).

In the case when $u(t) \equiv 0$ the system is said to be *unforced*.

Representations of the type given by (2.1) and (2.2) arise from the descriptions of control systems which are inherently linear, or from the linearization of nonlinear vector equations

$$\dot{x}(t) = f[x(t), u(t)] \tag{2.3}$$

$$y(t) = g[x(t)] \tag{2.4}$$

for small variations of the state, input and output variables.

The description given by (2.1) and (2.2) is not unique. The same system may be described by different state–space models depending on the choice of state, input and output variables. For instance, changing the state variables by the nonsingular transformation

$$x(t) = S\tilde{x}(t)$$

we obtain a new model

$$\dot{\tilde{x}}(t) = \tilde{A}\tilde{x}(t) + \tilde{B}u(t)$$
$$y(t) = \tilde{C}\tilde{x}(t)$$

where

$$\tilde{A} = S^{-1}AS \qquad \tilde{B} = S^{-1}B \qquad \tilde{C} = CS$$

The use of an appropriate model may simplify greatly the solution of the corresponding linear control problem.

In practice, Equations (2.1) and (2.2) are obtained from the mathematical description of the individual system components, or from empirical data for the input–output behaviour of the system. In both cases the system matrices are obtained with some errors, reflecting the tolerances in system components or the errors in the given data. That is why, instead of Equations (2.1) and (2.2), we have in fact the description

$$\dot{x}(t) = (A + \Delta A)x(t) + (B + \Delta B)u(t)$$
$$y(t) = (C + \Delta C)x(t)$$

where $\Delta A, \Delta B$ and ΔC are error matrices. The errors in state–space descriptions can have a significant effect on the solution of linear control problems in the case of high sensitivity to perturbations in the data.

For some linear control systems the state, input and output vectors are defined only at fixed time moments

$$t = k\Delta t$$

where $k = 0, 1, 2, \ldots$, and Δt is a constant time interval. These systems are called *linear discrete-time systems* and the constant Δt is said to be a *sampling interval*. A *linear discrete time-invariant control system* is described by the set of difference and algebraic equations of the form

$$x_{k+1} = Ax_k + Bu_k \tag{2.5}$$

$$y_k = Cx_k \tag{2.6}$$

where x_k, u_k and y_k are the state vector, input (control) vector and output vector, respectively, at the time $t = k\Delta t$ and where A, B and C are real constant matrices.

Equations (2.5) and (2.6) may arise in the description of inherently discrete-time systems or in the discretization of continuous-time models of the type given by (2.1) and (2.2). In the latter case it is often assumed that the input vector $u(t)$ is constant over the interval $(k - 1)\Delta t < t \le k\Delta t$. The discretization of continuous-time models is considered in the next section.

In some cases, the description of a linear control system is obtained in the form

$$E\dot{x}(t) = Ax(t) + Bu(t)$$
$$y(t) = Cx(t)$$

or

$$Ex_{k+1} = Ax_k + Bu_k$$
$$y_k = Cx_k$$

where E is possibly a singular matrix. Systems of this type are called *singular* or *descriptor systems*. Note that when the matrix E is nonsingular its inversion may be undesirable because of ill conditioning.

The problem of control of a linear system is in synthesizing the input in such a way that the output behaves in the desired fashion. To achieve this, the input is produced by a special device called a *controller*. If the input is prescribed as some function of the time, the system is called *open loop*. Since a control system is usually subject to unpredictable disturbances, it is preferable to use information about the actual output when implementing the control. In such a case there is a *feedback* of information from the output to the controller and the system is called *closed loop*.

The *analysis* of a linear control system consists in the study of the properties of state–space models of types given by (2.1) and (2.2) or (2.5) and (2.6). We consider several analysis problems of linear control systems in Sections 2.2–2.5.

The aim of the *design* of a linear control system is to find the appropriate structure and parameters of the controller to ensure the desired behaviour of the output. Some of the design problems of linear systems are discussed in Section 2.6–2.8.

The analysis and design of singular systems require specific methods which are beyond the scope of this book. We refer the reader to the papers of Luenberger (1978), Cobb (1981, 1984), Bender and Laub (1987a,b), Chu (1988), Sincovec *et al.* (1981) and the references therein for detailed consideration of singular systems analysis and design.

EXERCISES

2.1 Show that the differential equation

$$\frac{d^n y(t)}{dt^n} + a_n \frac{d^{n-1} y(t)}{dt^{n-1}} + \cdots + a_1 y(t) = u(t)$$

may be represented in the form of (2.1) and (2.2), where

$$A = \begin{bmatrix} 0 & 1 & 0 & \cdots & 0 \\ 0 & 0 & 1 & \cdots & 0 \\ \vdots & \vdots & \vdots & & \vdots \\ 0 & 0 & 0 & \cdots & 1 \\ -a_1 & -a_2 & -a_3 & \cdots & -a_n \end{bmatrix} \qquad b = \begin{bmatrix} 0 \\ 0 \\ \vdots \\ 0 \\ 1 \end{bmatrix} \qquad c = [1 \ 0 \ \cdots \ 0]$$

Hint: use the *phase variables*

$$x_1(t) = y(t)$$

$$x_2(t) = \frac{dy(t)}{dt}$$

$$\vdots$$

$$x_n(t) = \frac{d^{n-1} y(t)}{dt^{n-1}}$$

2.2 Give an interpretation of the conditions $m > n, r > n$.

2.3 Find a linear approximation of the form of (2.1) and (2.2) to the nonlinear model of (2.3) and (2.4) by linearizing (2.3) and (2.4) about nominal state $x^*(t)$, nominal input $u^*(t)$ and nominal output $y^*(t)$.

2.4 Find a state–space representation for which the matrix A is in diagonal form.

2.2 SOLUTION OF STATE EQUATIONS

Consider an unforced continuous-time linear system described by the homogeneous differential equation

$$\dot{x}(t) = Ax(t) \qquad x(0) = x_0 \tag{2.7}$$

where $x(t)$ is the n-vector of states and A is an $n \times n$ constant matrix. The solution of (2.7) is given by

$$x(t) = e^{At}x_0 \tag{2.8}$$

where e^{At} is the *matrix exponential* defined by the convergent matrix power series

$$e^{At} \equiv \sum_{k=0}^{\infty} (At)^k/k! = I_n + At/1! + (At)^2/2! + \cdots$$

The matrix exponential has the following properties:

1. $e^{A(t+s)} = e^{At}e^{As}$
2. e^{At} is nonsingular for each t
3. $(e^{At})^{-1} = e^{-At}$

4. $\dfrac{d}{dt}(e^{At}) = Ae^{At} = e^{At}A$

The matrix e^{At} is also referred to as the *state transition matrix* in control theory since it relates the state $x(t)$ to the state x_0. If $x(s)$ is the state vector at the moment s then the state vector at every moment t is determined by

$$x(t) = e^{A(t-s)}x(s)$$

Consider next a system described by the nonhomogeneous equation

$$\dot{x}(t) = Ax(t) + Bu(t) \qquad x(0) = x_0 \tag{2.9}$$

where $u(t)$ is the m-vector of inputs and B is an $n \times m$ matrix. The solution of (2.9) is

$$x(t) = e^{At}x_0 + \int_0^t e^{A(t-s)}Bu(s)\, ds$$

Thus, it is apparent that the computation of the matrix exponential e^{At} is a basic problem in the solution of continuous-time state equations.

One of the possible ways to compute e^{At} is to transform the matrix A into some simpler form $\tilde{A} = X^{-1}AX$ (X is nonsingular), such that $e^{\tilde{A}t}$ is easy to obtain explicitly, and then to find the exponential from

$e^{At} = Xe^{\tilde{A}t}X^{-1}$. If, for instance, the matrix A is nondefective (see Section 1.9) it may be decomposed as

$$A = X\tilde{A}X^{-1} \qquad \tilde{A} = \text{diag}(\lambda_1, \lambda_2, ..., \lambda_n)$$

where λ_i, $i = 1, ..., n$, are the eigenvalues of A, and X is the matrix of corresponding eigenvectors x_i, $i = 1, ..., n$, and therefore

$$e^{At} = X \, \text{diag}(e^{\lambda_1 t}, e^{\lambda_2 t}, ..., e^{\lambda_n t})X^{-1}$$

Example 2.1

If

$$A = \begin{bmatrix} -9 & 17 & -17 & 12 \\ -5 & 9 & -11 & 8 \\ -5 & 11 & -13 & 8 \\ -4 & 10 & -10 & 5 \end{bmatrix}$$

then

$$X = \begin{bmatrix} -3 & 1 & -1 & 2 \\ -1 & 1 & -2 & 2 \\ -1 & 2 & -3 & 2 \\ -2 & 2 & -2 & 1 \end{bmatrix} \qquad \tilde{A} = \begin{bmatrix} -1 & & & 0 \\ & -2 & & \\ & & -2 & \\ 0 & & & -3 \end{bmatrix}$$

and

$$e^{At} = \begin{bmatrix} -3 & 1 & -1 & 2 \\ -1 & 1 & -2 & 2 \\ -1 & 2 & -3 & 2 \\ -2 & 2 & -2 & 1 \end{bmatrix} \begin{bmatrix} e^{-t} & & & 0 \\ & e^{-2t} & & \\ & & e^{-2t} & \\ 0 & & & e^{-3t} \end{bmatrix}$$

$$\times \begin{bmatrix} 1 & -3 & 3 & -2 \\ 3 & -8 & 7 & -4 \\ 3 & -7 & 6 & -4 \\ 2 & -4 & 4 & -3 \end{bmatrix}$$

Let the matrix X^{-1} be represented as

$$X^{-1} = \begin{bmatrix} y_1^T \\ y_2^T \\ \vdots \\ y_n^T \end{bmatrix}$$

where y_i, $i = 1, 2, ..., n$, are the left eigenvectors of A. Then the solution of (2.7) may be written as

$$x(t) = \sum_{i=1}^{n} c_i e^{\lambda_i t} x_i \tag{2.10}$$

where $c_i = y_i^T x_0$.

Equation (2.10) shows that the system response may be expressed as a linear combination of elementary motions along the eigenvectors of A in the form $e^{\lambda_i t} x_i$. These motions are called *modes* and the eigenvector matrix X is referred to as a *modal matrix*.

It is clear that the eigenvalues λ_i, $i = 1, 2, ..., n$, of A have an essential role in the system behaviour. These values are also said to be *poles* of the system.

In case of complex conjugate eigenvalues $\sigma \pm i\omega$ of A, instead of the corresponding complex conjugate eigenvectors we may use the real and imaginary parts of the complex eigenvector associated with $\sigma + i\omega$. This leads to the appearance of a 2×2 block

$$\begin{bmatrix} \sigma & \omega \\ -\omega & \sigma \end{bmatrix}$$

in the matrix $\tilde{A}$ so that $e^{\tilde{A}t}$ will have the block

$$e^{\sigma t} \begin{bmatrix} \cos \omega t & \sin \omega t \\ -\sin \omega t & \cos \omega t \end{bmatrix}$$

in the same position. This allows us to avoid complex arithmetic in the computation of e^{At}.

In the case of a defective matrix A, we may use its Jordan form

$$J = S^{-1} A S = \text{diag}(J_1, J_2, ..., J_q)$$

where $J_k = \lambda_k I_{m_k} + N_{m_k}$ is a Jordan block of dimension m_k,

$$N_{m_k} = \begin{bmatrix} 0 & I_{m_k - 1} \\ 0 & 0 \end{bmatrix}$$

In this case

$$e^{At} = S e^{Jt} S^{-1} = S \, \text{diag}(e^{J_1 t}, e^{J_2 t}, ..., e^{J_q t}) S^{-1}$$

where

$$e^{J_k t} = e^{\lambda_k t} \begin{bmatrix} 1 & t/1! & t^2/2! & \cdots & t^p/p! \\ & 1 & t/1! & \cdots & t^{p-1}/(p-1)! \\ & & 1 & \ddots & \vdots \\ & \mathbf{0} & & \ddots & t/1! \\ & & & & 1 \end{bmatrix} \quad p = m_k - 1$$

Thus if the matrix A is defective, the system response will contain terms of the form $e^{\lambda_k t}$, $te^{\lambda_k t}$, $t^2 e^{\lambda_k t}$, ...

Example 2.2

If

$$
A = \begin{bmatrix}
-10 & 9 & 4 & -9 & 6 \\
-3 & 1 & 3 & -4 & 3 \\
5 & -7 & 0 & 3 & -2 \\
5 & -6 & -1 & 3 & -3 \\
-3 & 4 & -1 & -2 & -1
\end{bmatrix}
$$

then

$$
S = \begin{bmatrix}
2 & -2 & -1 & 2 & 1 \\
1 & 0 & -2 & 2 & 1 \\
0 & 2 & -2 & 1 & 1 \\
-1 & 2 & -1 & 0 & 1 \\
0 & -1 & 1 & -1 & 1
\end{bmatrix}
\qquad
J = \left[\begin{array}{ccc|cc}
-1 & 1 & 0 & & \\
0 & -1 & 1 & & \mathbf{0} \\
0 & 0 & -1 & & \\
\hline
 & \mathbf{0} & & -2 & 1 \\
 & & & 0 & -2
\end{array}\right]
$$

and

$$
e^{At} = \begin{bmatrix}
2 & -2 & -1 & 2 & 1 \\
1 & 0 & -2 & 2 & 1 \\
0 & 2 & -2 & 1 & 1 \\
-1 & 2 & -1 & 0 & 1 \\
0 & -1 & 1 & -1 & 1
\end{bmatrix}
\left[\begin{array}{ccc|cc}
e^{-t} & te^{-t} & t^2 e^{-t}/2 & & \\
0 & e^{-t} & te^{-t} & & \mathbf{0} \\
0 & 0 & e^{-t} & & \\
\hline
 & \mathbf{0} & & e^{-2t} & te^{-2t} \\
 & & & 0 & e^{-2t}
\end{array}\right]
$$

$$
\times \begin{bmatrix}
1 & -2 & 2 & -1 & 0 \\
2 & -3 & 1 & 1 & -1 \\
4 & -5 & 0 & 3 & -2 \\
3 & -3 & -1 & 3 & -2 \\
1 & -1 & 0 & 1 & 0
\end{bmatrix}
$$

Let us now consider an unforced discrete-time system described by the homogeneous difference equation

$$ x_{k+1} = Ax_k \qquad k = 0, 1, 2, \ldots $$

The solution of this equation for $k \geq 1$ is

$$ x_k = A^k x_0 $$

Therefore, the transition matrix of a discrete system is the corresponding power of the state matrix.

In the case of a nonhomogeneous difference equation

$$x_{k+1} = Ax_k + Bu_k$$

the solution is given by

$$x_k = A^k x_0 + \sum_{i=0}^{k-1} A^{k-i-1} Bu_i$$

The computation of A^k may be done also by employing the diagonal (or Jordan) form of A.

The solution of difference state equations has much in common with the solution of differential equations. Once e^{Ah} is obtained for some $h = \Delta t$, Equation (2.7) reduces to the difference equation

$$x[(k+1)h] = e^{Ah} x(kh)$$

This yields an alternative way to (2.8) to compute the solution of (2.7).

We turn now to the discretization of a continuous-time system described by the equations

$$\begin{aligned} \dot{x}(t) &= Ax(t) + Bu(t) \\ y(t) &= Cx(t) \end{aligned} \tag{2.11}$$

In the case of a digital control, the input $u(t)$ is usually generated as a piecewise constant function

$$u(t) = u(kh) \qquad kh \le t < (k+1)h \qquad k = 0, 1, 2, \dots$$

where $h = \Delta t$ is the sampling interval.

Since the solution of (2.11) for $t \ge kh$ is

$$x(t) = e^{A(t-kh)} x(kh) + \int_{kh}^{t} e^{A(t-s)} Bu(s) \, ds$$

$$y(t) = Cx(t)$$

the state vector at the moment $t = (k+1)h$ obeys

$$x[(k+1)h] = e^{Ah} x(kh) + \left(\int_{0}^{h} e^{As} \, ds \right) Bu(kh)$$

Setting

$$x_k = x(kh) \qquad u_k = u(kh) \qquad \text{and} \qquad y_k = y(kh)$$

we obtain the discrete-time system

$$\begin{aligned} x_{k+1} &= A_d x_k + B_d u_k \\ y_k &= Cx_k \end{aligned} \tag{2.12}$$

where

$$A_d = e^{Ah} \quad \text{and} \quad B_d = \left(\int_0^h e^{As} \, ds \right) B$$

The state and output vectors of the discrete-time model (2.12) at step k are the same as the state and output vectors of the continuous-time system (2.11) at the moment $t = kh$.

In this way the problem of finding the discrete-time model of a continuous-time system reduces to the computation of the matrix exponential and its integral over the sampling interval.

EXERCISES

2.5 Show that

$$e^{\lambda I + A} = e^{\lambda} e^A$$

2.6 Show that for any nonzero integer q

$$e^A = (e^{A/q})^q$$

2.7 Prove that

$$\int_0^h e^{As} \, ds = Ih + Ah^2/2! + A^2h^3/3! + \cdots$$

2.8 Find e^{At} for

$$A = \begin{bmatrix} A_{11} & A_{12} \\ 0 & A_{22} \end{bmatrix}$$

2.9 Prove that if

$$A = \text{diag}(A_1, A_2, ..., A_p)$$

then

$$e^{At} = \text{diag}(e^{A_1 t}, e^{A_2 t}, ..., e^{A_p t})$$

2.10 Find an analytical expression of e^{At} when A is an upper triangular matrix.

2.11 Prove that

$$\det(e^{At}) = e^{t \, \text{tr}(A)}$$

where $\text{tr}(A)$ is the trace of A.

2.12 Prove that

$$e^{(A+B)t} = e^{At} e^{Bt}$$

if and only if A and B commute (i.e. $AB = BA$).

2.13 Prove that the solution of the matrix differential equation

$$\dot{X}(t) = AX(t) + X(t)B \qquad X(0) = C$$

is

$$X(t) = e^{At}Ce^{Bt}$$

2.14 Prove that

$$e^{(A+E)t} - e^{At} = \int_0^t e^{A(t-s)}Ee^{(A+E)s}\,ds$$

Hint: Use the matrix differential equations

$$\dot{X}(t) = AX(t) \qquad X(0) = I$$
$$\dot{Y}(t) = (A+E)Y(t) \qquad Y(0) = I$$

2.3 STABILITY

A linear time-invariant system described by the homogeneous differential equation

$$\dot{x}(t) = Ax(t) \qquad x(0) = x_0 \tag{2.13}$$

is said to be *asymptotically stable* if $x(t) \to 0$ as $t \to \infty$ for every x_0. If for each x_0 there exists a constant $c < \infty$ such that $\| x(t) \| < c$ as $t \to \infty$ the system is said to be *stable*. If $\| x(t) \| \to \infty$ as $t \to \infty$ for some x_0 the system is said to be *unstable*.

The property of asymptotic stability is fundamental to the correct functioning of a control system of type (2.1), since it guarantees boundedness of the state $x(t)$ provided that the input $u(t)$, $t \geq 0$, is also bounded.

The stability of a given system may be analyzed without solving the state equation, by exploiting the properties of the state matrix. Methods suitable for this purpose are referred to as *stability criteria*. In this section we present several stability criteria for linear time-invariant systems. Their implementation and numerical properties are discussed later in Section 4.1.

The following theorem has a fundamental role in the stability analysis of continuous-time linear systems.

Theorem 2.1

The system (2.13) is asymptotically stable if and only if the eigenvalues λ_k of A have negative real parts:

$$\text{Re}(\lambda_k) < 0 \qquad k = 1, 2, \ldots, n$$

The proof of the theorem is easily done for the case of a diagonalizable matrix A. From the representation (2.10), valid for nondefective matrices, it is seen that $x(t) \to 0$ as $t \to \infty$ for each x_0 if and only if the eigenvalues of A lie in the left-half complex plane. In case of a defective state matrix the proof involves the Jordan form of A.

It may be shown that if all eigenvalues of A have non-positive real parts but $\text{Re}(\lambda_k) = 0$ for some k, then the system's response remains bounded for every x_0 (i.e. the system is stable) if and only if the elementary divisor corresponding to λ_k is linear. Otherwise, the system is stable in this case only if every pure imaginary λ_k (if any) appears in a Jordan block of dimension 1×1. If the elementary divisor corresponding to a multiple pure imaginary λ_k is nonlinear, then the system is unstable. The system is also unstable, if there exists λ_k with $\text{Re}(\lambda_k) > 0$.

If the eigenvalues of a matrix A satisfy the conditions of Theorem 2.1 then A is referred to as a *stability matrix*. Here the distance $\min\{-\text{Re}(\lambda_k): k = 1, 2, ..., n\}$ of the spectrum of A to the imaginary axis is known as the *stability margin* of A.

Example 2.3

If

$$A = \begin{bmatrix} 9 & 10 & 8 & 6 & 4 \\ -21 & -21 & -16 & -12 & -8 \\ 15 & 14 & 11 & 9 & 6 \\ -4 & -4 & -4 & -3 & 0 \\ -2 & -2 & -2 & -3 & -5 \end{bmatrix}$$

then the corresponding system is asymptotically stable since the eigenvalues of A are

$$\lambda_1 = -1 \qquad \lambda_2 = -1 \qquad \lambda_{3,4} = -2 \pm i \qquad \lambda_5 = -3$$

Example 2.4

If

$$A = \begin{bmatrix} -56 & 124 & -92 & 8 & 58 \\ -40 & 90 & -69 & 6 & 46 \\ -26 & 60 & -48 & 4 & 34 \\ -14 & 33 & -27 & 1 & 22 \\ -5 & 12 & -10 & 0 & 9 \end{bmatrix}$$

then the system is unstable, since in the Jordan form of A,

$$J = \begin{bmatrix} -2 & & & & \\ & -1 & & & 0 \\ & & -1 & & \\ 0 & & & 0 & 1 \\ & & & 0 & 0 \end{bmatrix}$$

there is a 2×2 block corresponding to the eigenvalue $\lambda = 0$. If the Jordan form of A were $J = \text{diag}(-2, -1, -1, 0, 0)$ then the system would be stable.

Since the eigenvalues of A are roots of the characteristic equation

$$\det(\lambda I - A) = 0$$

it is possible to study the stability of A without determining its eigenvalues, using the coefficient of the characteristic polynomial $\det(\lambda I - A)$. For this purpose one can apply the classical *stability criteria of Routh–Hurwitz* and *Lienard–Chipart*. The principal disadvantage of this approach is that, in the case of multiple pure imaginary eigenvalues, it is not possible to determine the stability of the system using only the information supplied by the characteristic equation. In fact, the matrices whose Jordan forms are

$$\begin{bmatrix} \lambda & & & \\ & \lambda & & 0 \\ & & \ddots & \\ 0 & & & \lambda \end{bmatrix} \quad \begin{bmatrix} \lambda & 1 & & & \\ & \lambda & 1 & & 0 \\ & & \ddots & \ddots & \\ & & & \ddots & 1 \\ 0 & & & & \lambda \end{bmatrix}$$

have the same characteristic polynomials, but for $\text{Re}(\lambda) = 0$ the first one is a stability matrix while the second one is not.

An investigation of the stability of (2.13) without determining the eigenvalues of A may be done also by the *second method of Lyapunov*. Applying this method to linear time-invariant systems leads to the following theorem.

Theorem 2.2

The system (2.13) is asymptotically stable if and only if for any positive definite matrix Q the solution P of the *Lyapunov matrix equation*

$$A^T P + PA + Q = 0 \tag{2.14}$$

is also positive definite.

The proof of this theorem may be found, for instance, in Lancaster (1969, ch. 8).

The stability analysis by the second method of Lyapunov involves solution of the Lyapunov equation for some positive definite Q (e.g. the identity matrix) and a check of the positive definiteness of P. The latter is determined by the signs of the eigenvalues of P (they must be positive) or by the *Sylvester criterion*: the matrix P is positive definite if and only if

$$p_{11} > 0, \det \begin{bmatrix} p_{11} & p_{12} \\ p_{12} & p_{22} \end{bmatrix} > 0, ..., \det(P) > 0$$

The computational solution of the Lyapunov equation is considered in detail in Section 4.2.

Example 2.5

If

$$A = \begin{bmatrix} 9 & 10 & 8 & 6 & 4 \\ -21 & -21 & -16 & -12 & -8 \\ 15 & 14 & 11 & 9 & 6 \\ -4 & -4 & -4 & -3 & 0 \\ -2 & -2 & -2 & -3 & -5 \end{bmatrix}$$

and $Q = I_5$, the solution of the Lyapunov equation (up to four decimal digits) is

$$P = \begin{bmatrix} 52.75 & 48.61 & 47.59 & 31.47 & 21.20 \\ 48.61 & 45.37 & 44.45 & 29.19 & 19.68 \\ 47.59 & 44.45 & 43.93 & 28.65 & 19.32 \\ 31.47 & 29.19 & 28.65 & 19.43 & 12.86 \\ 21.20 & 19.68 & 19.32 & 12.86 & 8.757 \end{bmatrix}$$

Since the eigenvalues of P are

$\lambda_1 = 168.8$
$\lambda_2 = 0.6068$
$\lambda_3 = 0.5047$
$\lambda_4 = 0.1446$
$\lambda_5 = 0.1342$

it follows that A is a stability matrix (compare with Example 2.3).

The so-called *localization methods* allow us to determine domains in the complex plane in which the eigenvalues of a matrix are situated.

If these domains lie entirely in the left-half complex plane then the corresponding system is asymptotically stable.

Consider as an illustration the implementation of the Gershgorin theorem (see Section 1.9) for stability analysis of linear systems. If the diagonal elements a_{ii} of the matrix A are negative and the radii

$$r_i = \sum_{\substack{j=1 \\ j \neq i}}^{n} |a_{ij}|$$

of the corresponding Gershgorin circles fulfil

$$r_i < |a_{ii}| \qquad i = 1, 2, ..., n$$

then all circles lie in the left-half complex plane. This is sufficient for the system to be asymptotically stable.

Matrices with the property

$$|a_{ii}| > \sum_{\substack{j=1 \\ j \neq i}}^{n} |a_{ij}|$$

are called *diagonally dominant*. In this way if the diagonal elements of a matrix are negative and this matrix is diagonally dominant, then the corresponding system is asymptotically stable.

Example 2.6

If

$$A = \begin{bmatrix} -4.7 & 1.5 & -2.3 & -0.4 \\ 0.9 & -2.9 & 1.6 & 0.1 \\ -0.7 & 3.2 & -5.9 & -0.8 \\ 1.4 & 2.5 & -0.9 & -6.1 \end{bmatrix}$$

then the radii of the Gershgorin circles are

$$r_1 = 4.2 \qquad r_2 = 2.6 \qquad r_3 = 4.7 \qquad r_4 = 4.8$$

and hence A is a stability matrix. Indeed, the eigenvalues of A are (up to four decimal digits)

$$\lambda_1 = -8.261 \qquad \lambda_2 = -5.895 \qquad \lambda_3 = -3.641 \qquad \lambda_4 = -1.803$$

If there is a Gershgorin circle which is isolated from the other circles (i.e. it contains exactly one eigenvalue) and if this circle lies entirely in the right-half complex plane, then the system is unstable. If, however, there is a circle which is crossed by the imaginary axis, then nothing can be said about the stability of the system and further investigation is needed. This approach may be used relative to both A and A^{T} thus enlarging the domain of its applicability.

A more general localization criterion is given by a theorem due to Ostrowski. According to this theorem, each eigenvalue of the matrix A lies in one of the circles

$$| \lambda - a_{ii} | \leq P_i^{\alpha} Q_i^{1-\alpha} \qquad i = 1, 2, \ldots, n$$

where

$$P_i = \sum_{\substack{j=1 \\ j \neq i}}^{n} | a_{ij} | \qquad Q_i = \sum_{\substack{j=1 \\ j \neq i}}^{n} | a_{ji} |$$

and $0 \leq \alpha \leq 1$ (for proof see Marcus and Minc (1964, ch. 3)). Setting $\alpha = 1$ the Ostrowski theorem reduces to the Gershgorin theorem.

Consider now the discrete-time linear system

$$x_{k+1} = A x_k \tag{2.15}$$

This system is said to be asymptotically stable if $x_k \to 0$ as $k \to \infty$ for every x_0. If for each x_0 there exists a constant $c < \infty$ such that $\| x_k \| < c$ as $k \to \infty$, the system is said to be stable. If $\| x_k \| \to \infty$ as $k \to \infty$ for some x_0 the system is unstable.

For discrete systems we have the following result.

Theorem 2.3

The system (2.15) is asymptotically stable if and only if the eigenvalues of A have modulus less than unity.

The proof is done taking into account that the solution of (2.15) has the form

$$x_k = A^k x_0$$

and that $A^k \to 0$ if and only if $| \lambda_i | < 1$, $i = 1, 2, \ldots, n$. If there is some λ_i with $| \lambda_i | > 1$, then the system is unstable. It is also unstable if there is an eigenvalue of unit modulus whose elementary divisor is nonlinear. Finally, if all eigenvalues of A have moduli not greater than 1 but there is λ_k with $| \lambda_k | = 1$, then the system is stable only if the corresponding elementary divisor is linear.

A matrix A with eigenvalues satisfying the condition

$$| \lambda_i | < 1 \qquad i = 1, 2, \ldots, n$$

is called a *convergent matrix*.

Example 2.7

If

$$
A = \begin{bmatrix}
1.0 & 0.9 & 0.8 & 0.6 & 0.4 & 0.2 \\
-1.9 & -1.8 & -1.6 & -1.2 & -0.8 & -0.4 \\
1.7 & 1.7 & 1.5 & 1.2 & 0.8 & 0.4 \\
-1.2 & -1.2 & -1.1 & -1.0 & -0.6 & -0.3 \\
0.4 & 0.4 & 0.4 & 0.4 & 0.1 & 0.1 \\
0.1 & 0.1 & 0.1 & 0.1 & 0.2 & 0.0
\end{bmatrix}
$$

then system (2.15) is asymptotically stable since the eigenvalues of A are

$$\lambda_1 = -0.1 \qquad \lambda_2 = -0.1 \qquad \lambda_3 = -0.1$$
$$\lambda_4 = 0.1 \qquad \lambda_5 = -0.1i \qquad \lambda_6 = 0.1i$$

Hence A is a convergent matrix.

Example 2.8

If

$$
A = \begin{bmatrix}
2.7 & -8.4 & 15.7 & -9.5 & -15.4 & 39.5 \\
1.1 & -2.0 & 1.3 & 5.5 & -9.4 & 2.3 \\
2.2 & -3.6 & -3.0 & 17.0 & -15.2 & -9.8 \\
0.1 & 5.4 & -19.3 & 25.5 & 2.2 & -49.1 \\
-0.2 & 4.8 & -14.8 & 17.0 & 5.0 & -37.4 \\
-0.9 & 4.8 & -9.9 & 7.5 & 7.8 & -24.5
\end{bmatrix}
$$

then the corresponding discrete-time system is stable since in the Jordan form of A,

$$
J = \begin{bmatrix}
1 & & & & & \\
& 1 & & & \mathbf{0} & \\
& & 1 & & & \\
& & & 0.8 & & \\
& \mathbf{0} & & & 0.4 & \\
& & & & & -0.5
\end{bmatrix}
$$

the eigenvalues with unit modulus appear in 1×1 blocks.

The stability of a discrete-time system may be investigated by the methods intended for continuous systems in the following way. The matrix A_c is defined by

$$A_c = (A - I)(A + I)^{-1}$$

provided the matrix $A + I$ is nonsingular (i.e. A has no eigenvalue $\lambda = -1$). Then the eigenvalues μ of A_c satisfy

$$\mu = \frac{\lambda - 1}{\lambda + 1} \tag{2.16}$$

The transformation (2.16) maps the unit circle in the λ-plane into the left half of the μ-plane.

Since

$$\lambda = \frac{1 + \mu}{1 - \mu} \tag{2.17}$$

it follows that $|\lambda_i| < 1$ if and only if $\text{Re}(\mu_i) < 0$, $i = 1, 2, \ldots, n$, i.e. A is a convergent matrix if and only if A_c is a stability matrix.

If we know the characteristic polynomial of A,

$$\det(\lambda I - A) = \lambda^n + a_n \lambda^{n-1} + \cdots + a_1 \tag{2.18}$$

the stability of (2.15) may be analyzed by implementing the Routh–Hurwitz criterion to the polynomial obtained from (2.18) using the substitution (2.17). Another possibility is to apply directly the *criterion of Schur–Cohn* (Jury, 1974, ch. 3) to (2.18). The latter is a discrete analogue to the Routh–Hurwitz criterion.

The implementation of the second method of Lyapunov to the discrete-time system (2.15) leads to the following result.

Theorem 2.4

The system (2.15) is asymptotically stable if and only if for any positive definite matrix Q the solution P of the *discrete Lyapunov matrix equation*

$$A^T P A - P + Q = 0 \tag{2.19}$$

is positive definite.

The proof of the theorem may be found in Barnett and Cameron (1985, ch. 4).

The computational solution of the discrete Lyapunov equation is considered in Section 4.2.

Example 2.9

If

$$A = \begin{bmatrix}
1.0 & 0.9 & 0.8 & 0.6 & 0.4 & 0.2 \\
-1.9 & -1.8 & -1.6 & -1.2 & -0.8 & -0.4 \\
1.7 & 1.7 & 1.5 & 1.2 & 0.8 & 0.4 \\
-1.2 & -1.2 & -1.1 & -1.0 & -0.6 & -0.3 \\
0.4 & 0.4 & 0.4 & 0.4 & 0.1 & 0.1 \\
0.1 & 0.1 & 0.1 & 0.1 & 0.2 & 0.0
\end{bmatrix}$$

and $Q = I_6$, the solution of (2.19) (up to four decimal digits) is

$$P = \begin{bmatrix}
10.17 & 8.882 & 7.937 & 6.340 & 4.089 & 2.055 \\
8.882 & 9.610 & 7.695 & 6.159 & 3.968 & 1.995 \\
7.937 & 7.695 & 7.882 & 5.517 & 3.546 & 1.784 \\
6.340 & 6.159 & 5.517 & 5.452 & 2.844 & 1.433 \\
4.089 & 3.968 & 3.546 & 2.844 & 2.864 & 0.9172 \\
2.055 & 1.995 & 1.784 & 1.433 & 0.9172 & 1.464
\end{bmatrix}$$

The eigenvalues of P are

$$\lambda_1 = 32.36 \qquad \lambda_2 = 1.050 \qquad \lambda_3 = 1.034$$
$$\lambda_4 = 1.003 \qquad \lambda_5 = 1.001 \qquad \lambda_6 = 1.000$$

so that A is a convergent matrix (compare with Example 2.7).

To investigate the stability of (2.15) it is possible also to apply some localization method. The discrete-time system is stable if the domains containing the eigenvalues lie in the unit circle.

The stability of a linear system may be strongly affected by perturbations of the system parameters owing to manufacturing tolerances, influence of external factors, measurement errors and so on. It is possible that small variations in the parameters may cause a stable system to turn into an unstable one and vice versa. Systems which remain stable for relatively large perturbations of their parameters are referred to as *robust* with respect to the stability.

Consider the continuous system (2.13) and suppose that the coefficients of the characteristic polynomial of A,

$$\det(\lambda I - A) = \lambda^n + a_n\lambda^{n-1} + \cdots + a_1$$

vary so that

$$b_i \le a_i \le c_i \qquad i = 1, 2, \ldots, n$$

The following theorem gives necessary and sufficient conditions which allow us to determine whether any polynomial

$$\lambda^n + \tilde{a}_n\lambda^{n-1} + \cdots + \tilde{a}_1 \qquad \tilde{a}_i \in [b_i, c_i] \qquad (2.20)$$

is stable (i.e. its zeros lie in the left-half complex plane).

Theorem 2.5

The polynomials (2.20) are stable if and only if the four polynomials

$$\lambda^n + b_n\lambda^{n-1} + b_{n-1}\lambda^{n-2} + c_{n-2}\lambda^{n-3}$$
$$+ c_{n-3}\lambda^{n-4} + b_{n-4}\lambda^{n-5} + b_{n-5}\lambda^{n-6} + \cdots$$
$$\lambda^n + b_n\lambda^{n-1} + c_{n-1}\lambda^{n-2} + c_{n-2}\lambda^{n-3}$$
$$+ b_{n-3}\lambda^{n-4} + b_{n-4}\lambda^{n-5} + c_{n-5}\lambda^{n-6} + \cdots$$
$$\lambda^n + c_n\lambda^{n-1} + b_{n-1}\lambda^{n-2} + b_{n-2}\lambda^{n-3}$$
$$+ c_{n-3}\lambda^{n-4} + c_{n-4}\lambda^{n-5} + b_{n-5}\lambda^{n-6} + \cdots$$
$$\lambda^n + c_n\lambda^{n-1} + c_{n-1}\lambda^{n-2} + b_{n-2}\lambda^{n-3}$$
$$+ b_{n-3}\lambda^{n-4} + c_{n-4}\lambda^{n-5} + c_{n-5}\lambda^{n-6} + \cdots$$

are stable.

For a proof of this theorem see Kharitonov (1978).

Theorem 2.5 gives a simple way to analyze the robust stability of a linear system provided that the bounds on the coefficients of the characteristic polynomial are known.

Example 2.10

Let

$$\det(\lambda I - A) = \lambda^4 + a_4\lambda^3 + a_3\lambda^2 + a_2\lambda + a_1$$

and

$$a_1 \in [19, 27] \qquad a_2 \in [40, 55]$$
$$a_3 \in [20, 38] \qquad a_4 \in [6, 12] \qquad (2.21)$$

The four polynomials

$$\lambda^4 + 6\lambda^3 + 20\lambda^2 + 55\lambda + 27$$
$$\lambda^4 + 6\lambda^3 + 38\lambda^2 + 55\lambda + 19$$
$$\lambda^4 + 12\lambda^3 + 20\lambda^2 + 40\lambda + 27$$
$$\lambda^4 + 12\lambda^3 + 38\lambda^2 + 40\lambda + 19$$

are stable and hence the characteristic polynomial is stable for every a_1, a_2, a_3 and a_4 which satisfy (2.21).

A more interesting problem is to find conditions on the matrix bounds B, C ($B \le C$) such that any matrix A with $B \le A \le C$ is stable (here $B \le C$ means $b_{ij} \le c_{ij}$ and so on). Unfortunately, necessary and sufficient conditions for solving this problem are not known up to now. Some results based on sufficient conditions may be found in Argoun (1986) and Yedavalli (1986).

EXERCISES

2.15 Prove Theorems 2.1 and 2.3 using the Schur form of A.

2.16 Prove that if the solution of the equation

$$A^{\mathrm{T}}P + PA + 2\sigma P + Q = 0$$

is positive definite for a positive definite Q, then the eigenvalues of A have real parts less than $-\sigma$.

2.17 Prove that the solution of Lyapunov equation (2.14) satisfies

$$\| P \| \ge \frac{\| Q \|}{2 \| A \|}$$

2.18 Prove that if A is stable, Q is positive definite and P is the positive definite solution of Lyapunov equation (2.14), then the stability margin of A is not less than

$$\frac{1}{2 \| P \|_2 \| Q^{-1} \|_2}$$

2.19 Prove that e^A is a convergent matrix if and only if A is a stability matrix.

2.20 Derive discrete versions of the results from Exercises 2.16 and 2.18.

2.21 Prove that if the system (2.15) is asymptotically stable and Q is positive definite, then

$$\sum_{k=0}^{\infty} x_k^{\mathrm{T}} Q x_k = x_0^{\mathrm{T}} P x_0$$

where P is the positive definite solution of the discrete Lyapunov equation (2.19).

2.22 Find the solutions of the Lyapunov equations (2.14) and (2.19), by using the Jordan form of A.

2.23 Derive a discrete analogue of Theorem 2.5 using the transformation (2.16).

2.4 CONTROLLABILITY AND OBSERVABILITY

The linear control system

$$\dot{x}(t) = Ax(t) + Bu(t) \qquad x(0) = x_0 \tag{2.22}$$

is said to be *completely controllable* (or simply *controllable*) if for any initial state x_0 and any prescribed final state x_f there exist a finite time t_f and a control $u(t)$, $0 \le t \le t_f$, such that $x(t_f) = x_f$. Otherwise, the system is uncontrollable, although it may be possible to transfer certain states to any desired final states.

In broad terms, the controllability means that any initial state can be transferred to any final state in a finite time by a suitable choice of control. For a time-invariant system the final state can be taken to be the null vector.

If the system (2.22) is controllable the pair (A, B) is also said to be controllable.

The following theorem, due to Kalman, represents an algebraic criterion for controllability of linear time-invariant systems.

Theorem 2.6

The system (2.22) is completely controllable if and only if the $n \times nm$ *controllability matrix*

$$P = [B, AB, ..., A^{n-1}B]$$

has rank n. If rank $B = l$, this condition reduces to

$$\text{rank}\,[B, AB, ..., A^{n-l}B] = n$$

For proof of this theorem see Kalman, Falb and Arbib (1969, ch. 2).

Example 2.11

If

$$A = \begin{bmatrix} -5 & 4 & 0 & -2 \\ 0 & -5 & 6 & 0 \\ 2 & -6 & 4 & 4 \\ 1 & -2 & 0 & 4 \end{bmatrix} \quad \text{and} \quad B = \begin{bmatrix} -2 & 3 \\ 4 & 2 \\ -1 & -3 \\ 1 & -2 \end{bmatrix}$$

then the system (2.22) is controllable since the controllability matrix

$$P = \begin{bmatrix} -2 & 3 & 24 & -3 & -212 & -79 & 804 & 297 \\ 4 & 2 & -26 & -28 & -38 & -16 & 598 & 212 \\ -1 & -3 & -28 & -26 & 68 & 22 & 284 & 94 \\ 1 & -2 & -6 & -9 & 52 & 17 & 72 & 21 \end{bmatrix}$$

$$\quad\; B \qquad\quad AB \qquad\quad A^2B \qquad\quad A^3B$$

has four linearly independent columns, i.e. its rank is equal to 4. Also, since rank $B = 2$, it is fulfilled that rank $[B, AB, A^2B] = 4$. Moreover, for this system rank $[B, AB] = 4$.

The controllability of a system is not affected by nonsingular transformations of the state and the input. In fact, if we set

$$x(t) = S\tilde{x}(t) \qquad \text{and} \qquad u(t) = T\tilde{u}(t)$$

for some nonsingular S and T, we obtain

$$\dot{\tilde{x}}(t) = \tilde{A}\tilde{x}(t) + \tilde{B}\tilde{u}(t) \qquad\qquad (2.23)$$

where

$$\tilde{A} = S^{-1}AS \qquad \text{and} \qquad \tilde{B} = S^{-1}BT$$

and hence

$$\text{rank}[\tilde{B}, \tilde{A}\tilde{B}, ..., \tilde{A}^{n-1}\tilde{B}] = \text{rank}\, S^{-1}[B, AB, ..., A^{n-1}B]$$

$$\times \begin{bmatrix} T & & 0 \\ & T & \\ & & \ddots \\ 0 & & T \end{bmatrix}$$

$$= \text{rank}[B, AB, ..., A^{n-1}B]$$

If the matrix A has distinct eigenvalues $\lambda_1, \lambda_2, ..., \lambda_n$, then using the nonsingular transformation

$$x(t) = X\tilde{x}(t)$$

where the matrix X consists of the corresponding eigenvectors of A, the system (2.22) is reduced to the form (2.23) with

$$\tilde{A} = X^{-1}AX = \text{diag}(\lambda_1, \lambda_2, ..., \lambda_n)$$
$$\tilde{B} = X^{-1}B$$

Using Theorem 2.6 it may be shown that the pair $(\tilde{A}, \tilde{B})$ is completely controllable if and only if all rows of the matrix $\tilde{B}$ are nonzero.

In cases when A is diagonalizable but has multiple eigenvalues, it is necessary in addition that the rows of $\tilde{B}$, which correspond to equal eigenvalues, be linearly independent. For a single-input system to be controllable, this means that if the state matrix is diagonalizable it must have distinct eigenvalues.

Example 2.12

Let

$$A = \begin{bmatrix} 5 & -16 & 20 & -12 \\ 8 & -21 & 22 & -8 \\ 6 & -14 & 12 & 0 \\ 2 & -4 & 2 & 3 \end{bmatrix} \quad \text{and} \quad B = \begin{bmatrix} 2 & 5 \\ 3 & -1 \\ 1 & 0 \\ -2 & -4 \end{bmatrix}$$

The eigenvalues of A are

$$\lambda_1 = -1 \qquad \lambda_2 = -1 \qquad \lambda_3 = 0 \qquad \lambda_4 = 1$$

and the matrix of corresponding eigenvectors is

$$X = \begin{bmatrix} 10 & 8 & 4 & 1 \\ 8 & 9 & 6 & 2 \\ 4 & 6 & 5 & 2 \\ 1 & 2 & 2 & 1 \end{bmatrix}$$

The matrices of the transformed system are

$$\tilde{A} = X^{-1}AX = \begin{bmatrix} -1 & 0 & 0 & 0 \\ 0 & -1 & 0 & 0 \\ 0 & 0 & 0 & 0 \\ 0 & 0 & 0 & 1 \end{bmatrix} \qquad \tilde{B} = X^{-1}B = \begin{bmatrix} 0 & 11 \\ -3 & -31 \\ 11 & 48 \\ -18 & -49 \end{bmatrix}$$

The system is controllable since the first and second rows of $\tilde{B}$ are linearly independent and the rest are nonzero.

In the general case when A is not diagonalizable (i.e. it is defective), the system matrices are reduced to

$$\tilde{A} = J = S^{-1}AS \qquad \tilde{B} = S^{-1}B$$

where J is the Jordan form of A, and S consists of the eigenvectors and principal eigenvectors of A. In this case, the system is controllable if and only if the rows of $\tilde{B}$, corresponding to the last rows of the Jordan blocks containing the same eigenvalue, are linearly independent. If the matrix A is nonderogatory (i.e. it does not have two or more Jordan blocks with the same eigenvalue), this condition is replaced by the weaker condition that the rows of $\tilde{B}$ corresponding to the last rows of the Jordan blocks be nonzero.

For a single-input system to be controllable, the above conditions mean that the matrix A must be nonderogatory and the components of B corresponding to the last rows of the Jordan blocks must be nonzero.

Example 2.13

Let

$$A = \begin{bmatrix} -1 & -4 & -10 & 16 \\ -3 & -5 & -4 & 10 \\ -2 & -8 & -17 & 28 \\ -2 & -6 & -11 & 19 \end{bmatrix} \quad \text{and} \quad B = \begin{bmatrix} 45 & 23 \\ -38 & 5 \\ 27 & 11 \\ 8 & 10 \end{bmatrix}$$

Using a nonsingular transformation with

$$S = \begin{bmatrix} 2 & -5 & 6 & 10 \\ -1 & 4 & -6 & -6 \\ 2 & -5 & 4 & 5 \\ 1 & -2 & 1 & 2 \end{bmatrix}$$

the state matrix A is reduced to the Jordan form

$$J = S^{-1}AS = \begin{bmatrix} -1 & 1 & 0 & 0 \\ 0 & -1 & 0 & 0 \\ 0 & 0 & -1 & 1 \\ 0 & 0 & 0 & -1 \end{bmatrix}$$

The system is uncontrollable, since the second and fourth rows of

$$\tilde{B} = \begin{bmatrix} -2 & 1 \\ -1 & -3 \\ 4 & -9 \\ 2 & 6 \end{bmatrix}$$

are linearly dependent.

Several other algebraic criteria exist for complete controllability of linear time-invariant systems, some of which are presented below.

Theorem 2.7

The system (2.22) is completely controllable if and only if the *controllability Grammian*

$$W_t = \int_0^t e^{As}BB^T e^{A^T s}\, ds$$

is positive definite for $t > 0$.

For a proof of this theorem see Kalman (1963).

Theorem 2.8

The system (2.22) is completely controllable if and only if the $n^2 \times n(n + m - 1)$ matrix

$$\begin{bmatrix} I_n & & & & & & & B \\ -A & I_n & & & 0 & & B & \\ & -A & \ddots & & & B & & \\ 0 & \ddots & -A & I_n & B & & 0 \\ & & & -A & B & & & \end{bmatrix}$$

has rank n^2.

For a proof of this theorem see Rosenbrock (1970, ch. 2).

Theorem 2.9

The system (2.22) is completely controllable if and only if

$$\text{rank}\,[A - \lambda_i I, \ B] = n \qquad i = 1, 2, \ldots, n$$

where λ_i, $i = 1, 2, \ldots, n$, are the eigenvalues of A.

For a proof of this theorem see Hautus (1969).

If system (2.22) is uncontrollable then, using a nonsingular transformation of the state, it may be reduced to the form

$$\begin{bmatrix} \dot{x}_1(t) \\ \dot{x}_2(t) \end{bmatrix} = \begin{bmatrix} A_{11} & A_{12} \\ 0 & A_{22} \end{bmatrix} \begin{bmatrix} x_1(t) \\ x_2(t) \end{bmatrix} + \begin{bmatrix} B_1 \\ 0 \end{bmatrix} u(t) \qquad (2.24)$$

where the pair (A_{11}, B_1) is controllable and the vector $x_1(t)$ has a dimension

$$d = \text{rank}\,[B, AB, \ldots, A^{n-1}B] < n$$

The vector $x_2(t)$ contains the state components which are completely uncontrollable. The zeros of $\det(\lambda I - A_{22})$ are referred to as *uncontrollable poles* of the system. If A_{22} is a stability matrix, the system is said to be *stabilizable*. Thus, the uncontrollable modes of a stabilizable system are asymptotically stable. If the system is asymptotically stable or completely controllable, it is stabilizable.

Example 2.14

If

$$
A = \begin{bmatrix} 1 & -10 & 44 & -101 & 103 \\ 7 & -10 & 27 & -82 & 90 \\ 3 & 3 & 12 & -65 & 70 \\ 2 & 2 & 8 & -44 & 48 \\ 1 & 1 & 5 & -22 & 21 \end{bmatrix} \quad \text{and} \quad B = \begin{bmatrix} 25 & 21 \\ 18 & 23 \\ 9 & 21 \\ 6 & 14 \\ 3 & 7 \end{bmatrix}
$$

the system (2.22) is uncontrollable since

$$d = \text{rank}\,[B, AB, ..., A^4 B] = 3$$

Using the nonsingular transformation

$$x(t) = S\tilde{x}(t)$$

$$
S = \begin{bmatrix} 5 & 4 & 3 & 2 & 1 \\ 4 & 4 & 3 & 2 & 1 \\ 3 & 3 & 3 & 2 & 1 \\ 2 & 2 & 2 & 2 & 1 \\ 1 & 1 & 1 & 1 & 1 \end{bmatrix}
$$

the system is reduced to

$$\dot{\tilde{x}}(t) = \tilde{A}\tilde{x}(t) + \tilde{B}u(t)$$

where

$$
\tilde{A} = S^{-1}AS = \left[\begin{array}{ccc:cc} -4 & 2 & 8 & -3 & 5 \\ 3 & -7 & -4 & 1 & 4 \\ 2 & 5 & -6 & -6 & -2 \\ \hdashline & & & 1 & 3 \\ & \mathbf{0} & & -2 & -4 \end{array}\right] \quad \text{and}
$$

$$
\tilde{B} = S^{-1}B = \left[\begin{array}{cc} 7 & -2 \\ 2 & 4 \\ -6 & 5 \\ \hdashline & \mathbf{0} \end{array}\right]
$$

The system is stabilizable, since the eigenvalues of the matrix

$$
A_{22} = \begin{bmatrix} 1 & 3 \\ -2 & -4 \end{bmatrix}
$$

are

$$\lambda_1 = -1 \qquad \lambda_2 = -2$$

The linear system

$$\dot{x}(t) = Ax(t) + Bu(t) \qquad x(0) = x_0$$
$$y(t) = Cx(t) \tag{2.25}$$

is said to be *completely observable* (or simply *observable*) if for any initial state x_0 there exists a finite time t_f such that a knowledge of $u(t)$ and $y(t)$ for $0 \leq t \leq t_f$ is sufficient to determine x_0. The control $u(t)$ here may be assumed to be equal to zero on the whole interval $[0, t_f]$ since it does not affect the property of system (2.25) to be observable.

In broad terms, the observability means that it is possible to determine the state of a system by measuring only the output. If the system (2.23) is observable, we will say that the pair (C, A) is also observable.

The following is an analogue of Theorem 2.6.

Theorem 2.10

The system (2.25) is completely observable if and only if the $rn \times n$ *observability matrix*

$$Q = \begin{bmatrix} C \\ CA \\ CA^2 \\ \vdots \\ CA^{n-1} \end{bmatrix}$$

has rank n.

For proof of this theorem see Kalman, Falb and Arbib (1969, ch. 2).

Example 2.15

If

$$A = \begin{bmatrix} -5 & -3 & 4 \\ 2 & -6 & 1 \\ 7 & 3 & -4 \end{bmatrix} \quad \text{and} \quad C = \begin{bmatrix} 2 & 1 & 4 \\ 3 & -5 & -2 \end{bmatrix}$$

then the system (2.25) is observable since the observability matrix

$$
Q = \left[
\begin{array}{ccc}
2 & 1 & 4 \\
3 & -5 & -2 \\
\hdashline
20 & 0 & -7 \\
-39 & 15 & 15 \\
\hdashline
-149 & -81 & 108 \\
330 & 72 & -201
\end{array}
\right]
\begin{array}{l}
C \\ \\
CA \\ \\
CA^2
\end{array}
$$

has three linearly independent rows.

The observability matrix Q is identical to the transpose of the controllability matrix P associated with the pair (A^T, C^T). Conversely, the controllability matrix is identical to the transpose of the observability matrix associated with the pair (B^T, A^T). This duality between controllability and observability allows us to reformulate easily all controllability criteria for the case of observability analysis.

If the system (2.25) is not completely observable, then by using a nonsingular transformation of the state vector, it may be reduced to the form

$$
\begin{bmatrix} \dot{x}_1(t) \\ \dot{x}_2(t) \end{bmatrix} = \begin{bmatrix} A_{11} & 0 \\ A_{21} & A_{22} \end{bmatrix} \begin{bmatrix} x_1(t) \\ x_2(t) \end{bmatrix} + \begin{bmatrix} B_1 \\ B_2 \end{bmatrix} u(t)
$$

$$
y(t) = \begin{bmatrix} C_1 & 0 \end{bmatrix} \begin{bmatrix} x_1(t) \\ x_2(t) \end{bmatrix}
$$

where the pair (C_1, A_{11}) is completely observable and $x_1(t)$ has dimension equal to rank Q. If A_{22} is a stability matrix the system is said to be *detectable*. Thus the unobservable modes of a detectable system are asymptotically stable.

Considering the general case of an uncontrollable and unobservable system, we have the following result.

Theorem 2.11 (Kalman decomposition)

The system (2.25) can be reduced by using a nonsingular transformation of the state to the form

$$
\begin{bmatrix} \dot{x}_1(t) \\ \dot{x}_2(t) \\ \dot{x}_3(t) \\ \dot{x}_4(t) \end{bmatrix} = \begin{bmatrix} A_{11} & A_{12} & A_{13} & A_{14} \\ 0 & A_{22} & 0 & A_{24} \\ 0 & 0 & A_{33} & A_{34} \\ 0 & 0 & 0 & A_{44} \end{bmatrix} \begin{bmatrix} x_1(t) \\ x_2(t) \\ x_3(t) \\ x_4(t) \end{bmatrix} + \begin{bmatrix} B_1 \\ B_2 \\ 0 \\ 0 \end{bmatrix} u(t)
$$

$$y(t) = [0 \quad C_2 \quad 0 \quad C_4] \begin{bmatrix} x_1(t) \\ x_2(t) \\ x_3(t) \\ x_4(t) \end{bmatrix}$$

where the four subsystems are respectively as follows:

1. Completely controllable but unobservable.
2. Completely controllable and completely observable.
3. Uncontrollable and unobservable.
4. Completely observable but uncontrollable.

For proof of this theorem see Kalman (1963).

The definitions and criteria for controllability and observability of continuous-time systems carry over the discrete-time systems

$$\begin{aligned} x_{k+1} &= Ax_k + Bu_k \\ y_k &= Cx_k \end{aligned} \tag{2.26}$$

with minor modifications, except for the following. If the matrix A is singular, then the possibility of reaching the state $x_f = 0$ from any initial state x_0 (controllability to the origin) is not the same as reaching any state $x_f \neq 0$ from $x_0 = 0$ (controllability from the origin). If, for instance, the matrix A is nilpotent, i.e. $A^q = 0$ for some q, then any x_0 can be driven to $x_q = 0$ with the input $u_k = 0$, $k = 0, 1, ..., q - 1$. It may be shown that the requirement for full rank of the matrix

$$P = [B, AB, ..., A^{n-1}B]$$

is equivalent to the controllability from the origin. A discrete-time system which is controllable from the origin is said to be *completely reachable*. The matrix P in the case of a discrete-time system is often called a *reachability matrix*. Analogically, instead of observability of a discrete-time system we speak about *reconstructability*, both terms being equivalent, if A is nonsingular. The system (2.26) is completely reconstructable if and only if the *reconstructability matrix*

$$Q = \begin{bmatrix} C \\ CA \\ \vdots \\ CA^{n-1} \end{bmatrix}$$

has rank n.

For unity with the continuous-time case, later we shall identify controllability with reachability and observability with reconstructability.

EXERCISES

2.24 Prove that the pair

$$A = \begin{bmatrix} 0 & 1 & 0 & \dots & 0 \\ 0 & 0 & 1 & \dots & 0 \\ \vdots & \vdots & \vdots & & \vdots \\ 0 & 0 & 0 & \dots & 1 \\ -a_1 & -a_2 & -a_3 & \dots & -a_n \end{bmatrix} \qquad b = \begin{bmatrix} 0 \\ 0 \\ \vdots \\ 0 \\ 1 \end{bmatrix}$$

is controllable.

2.25 Find conditions for complete controllability of the pair

$$A = \begin{bmatrix} a_{11} & a_{12} & \dots & a_{1n} \\ a_{21} & a_{22} & \dots & a_{2n} \\ & a_{32} & \dots & a_{3n} \\ & \mathbf{0} & \ddots & \vdots \\ & & a_{n,n-1} & a_{nn} \end{bmatrix} \qquad b = \begin{bmatrix} b_1 \\ 0 \\ 0 \\ \vdots \\ 0 \end{bmatrix}$$

2.26 Prove that a single-input nth order system with real matrices A and b is completely controllable for each nonzero b if and only if one of the following two conditions hold: (1) $n = 1$, or (2) $n = 2$ and A has no real eigenvalues.

2.27 Prove that the pair (A, B) is uncontrollable if and only if there exists a row vector $w \neq 0$ such that

$$wA = \lambda w \qquad wB = [0 \quad 0 \quad \dots \quad 0]$$

2.28 Prove that the minimum number of inputs for complete controllability is equal to the largest number of Jordan blocks containing the same eigenvalue.

2.29 Prove that if the pair (A, B) is controllable and

$$S^{-1}AS = \begin{bmatrix} A_{11} & A_{12} \\ A_{21} & A_{22} \end{bmatrix} \qquad S^{-1}B = \begin{bmatrix} B_1 \\ 0 \end{bmatrix}$$

then the pair (A_{22}, A_{21}) is also controllable.

2.30 Prove that if the pair (A, B) is controllable, then so is the pair $(A - BK, B)$. *Hint*: use Theorem 2.9.

2.31 Show by an example that if the pair (A, B) is controllable and the pair (C, A) is observable then the pair $(C, A - BK)$ may be unobservable.

2.32 Prove that if the pair (C, A) is detectable and the Lyapunov equation $A^{\mathrm{T}}P + PA + C^{\mathrm{T}}C = 0$ has a symmetric positive semidefinite solution P then A is a stability matrix.

2.5 CANONICAL FORMS

A completely controllable and observable linear system can be reduced via nonsingular transformations of the state, input and output into a certain *equivalent system* having special structure of the matrices A, B and C. This system, characterized by a large number of specified entries (e.g. zeros or units) in given positions, is said to be a *canonical form* of the original system. A given system may be reduced into different canonical forms which are characterized by different dispositions of the specified elements.

The reduction into canonical form allows us to reveal the structural properties of the system and to simplify the solution of design problems. A rigorous definition of the notion of a canonical form may be found in Kalman, Falb and Arbib (1969, ch. 2).

Consider first the case of a single-input, single-output system

$$\dot{x}(t) = Ax(t) + bu(t)$$
$$y(t) = cx(t)$$
(2.27)

where A is an $n \times n$ matrix, b is a column n-vector and c is a row n-vector.

It is already known (see Exercise 2.1) that a system described by the differential equation

$$\frac{\mathrm{d}^n y(t)}{\mathrm{d}t^n} + a_n \frac{\mathrm{d}^{n-1} y(t)}{\mathrm{d}t^{n-1}} + \cdots + a_1 y(t) = u(t)$$

can be represented in the form

$$\dot{\tilde{x}}(t) = \tilde{A}\tilde{x}(t) + \tilde{b}u(t)$$
$$y(t) = \tilde{c}\tilde{x}(t)$$
(2.28)

where

$$\tilde{A} = \begin{bmatrix} 0 & 1 & 0 & \cdots & 0 \\ 0 & 0 & 1 & \cdots & 0 \\ \vdots & \vdots & \vdots & & \vdots \\ -a_1 & -a_2 & -a_3 & \cdots & -a_n \end{bmatrix} \qquad \tilde{b} = \begin{bmatrix} 0 \\ 0 \\ \vdots \\ 1 \end{bmatrix} \qquad \text{and}$$

$$\tilde{c} = [1 \quad 0 \quad \cdots \quad 0]$$

The matrix $\tilde{A}$ is in companion form.

Now it may be shown that the system (2.27) can be reduced into the form (2.28) using nonsingular transformation $x(t) = S\tilde{x}(t)$, except that the matrix $\tilde{c}$ will not have a special structure, and provided that the controllability matrix

$$P = [b, \; Ab, \; \ldots, \; A^{n-1}b]$$

has rank n. In fact, since controllability is not altered by similarity transformations, it follows that the matrix

$$\tilde{P} = [\tilde{b}, \; \tilde{A}\tilde{b}, \; \ldots, \; \tilde{A}^{n-1}\tilde{b}] = S^{-1}P$$

must have rank n so that the matrix

$$S = P\tilde{P}^{-1}$$

is nonsingular.

The representation (2.28), in which the matrix $\tilde{c} = cS$ has no special form, is called a *controllable canonical form* of the system (2.27). The controllable canonical form is useful for several reasons. In particular, the characteristic polynomial of A is determined directly from

$$\det(\lambda I - A) = \det(\lambda I - \tilde{A}) = \lambda^n + a_n\lambda^{n-1} + \cdots + a_1.$$

As was pointed out in Section 2.4, a necessary condition for complete controllability of the system (2.27) is that the matrix A be cyclic (nonderogatory).

If the coefficients of the characteristic polynomial of A are known, the matrix $\tilde{P}^{-1}$ is easily found from

$$\tilde{P}^{-1} = \begin{bmatrix} a_2 & a_3 & \cdots & a_n & 1 \\ a_3 & a_4 & \cdots & 1 & \\ \vdots & \vdots & \ddots & & \\ a_n & 1 & & \mathbf{0} & \\ 1 & & & & \end{bmatrix}$$

In this case the matrix $\tilde{A}$ may be written directly, without computing $S^{-1}AS$.

The transformation into controllable canonical form may also be found from

$$S^{-1} = \begin{bmatrix} s_1 \\ s_1 A \\ s_1 A^2 \\ \vdots \\ s_1 A^{n-1} \end{bmatrix}$$

where s_1 is the nth row of P^{-1}.

Example 2.16

If

$$A = \begin{bmatrix} -8 & 4 & -4 & -13 \\ 7 & 2 & -4 & 18 \\ 8 & -1 & 0 & 17 \\ 3 & 1 & -2 & 7 \end{bmatrix} \quad \text{and} \quad b = \begin{bmatrix} -2 \\ 2 \\ 2 \\ 1 \end{bmatrix}$$

then

$$\det(\lambda I - A) = \lambda^4 - \lambda^3 - 3\lambda^2 + \lambda + 2$$

$$P = \begin{bmatrix} -2 & 3 & -7 & 4 \\ 2 & 0 & 7 & 9 \\ 2 & -1 & 7 & 5 \\ 1 & -1 & 4 & 0 \end{bmatrix} \qquad \tilde{P}^{-1} = \begin{bmatrix} 1 & -3 & -1 & 1 \\ -3 & -1 & 1 & 0 \\ -1 & 1 & 0 & 0 \\ 1 & 0 & 0 & 0 \end{bmatrix}$$

$$S = \begin{bmatrix} 0 & -4 & 5 & -2 \\ 4 & 1 & -2 & 2 \\ 3 & 2 & -3 & 2 \\ 0 & 2 & -2 & 1 \end{bmatrix}$$

and

$$\tilde{A} = S^{-1}AS = \begin{bmatrix} 0 & 1 & 0 & 0 \\ 0 & 0 & 1 & 0 \\ 0 & 0 & 0 & 1 \\ -2 & -1 & 3 & 1 \end{bmatrix} \quad \text{and} \quad \tilde{b} = S^{-1}b = \begin{bmatrix} 0 \\ 0 \\ 0 \\ 1 \end{bmatrix}$$

Exploiting the duality between controllability and observability it is possible to show that the system (2.27) can be reduced into *observable canonical form*

$$\dot{x}(t) = \tilde{A}\tilde{x}(t) + \tilde{b}u(t)$$
$$y(t) = \tilde{c}\tilde{x}(t)$$

$$\tilde{A} = \begin{bmatrix} 0 & 0 & \cdots & 0 & -a_1 \\ 1 & 0 & \cdots & 0 & -a_2 \\ 0 & 1 & \cdots & 0 & -a_3 \\ \vdots & \vdots & & \vdots & \vdots \\ 0 & 0 & \cdots & 1 & -a_n \end{bmatrix} \qquad \tilde{c} = \begin{bmatrix} 0 & 0 & \cdots & 0 & 1 \end{bmatrix}$$

provided that

$$\text{rank} \begin{bmatrix} c \\ cA \\ \vdots \\ cA^{n-1} \end{bmatrix} = n$$

In this case the matrix $\tilde{b}$ is not in a special form.

Consider now the completely controllable multi-input multi-output system

$$\dot{x}(t) = Ax(t) + Bu(t)$$
$$y(t) = Cx(t) \tag{2.29}$$

and assume without loss of generality that the matrix $B = [b_1, \ldots, b_m]$ is of full rank, i.e. $\text{rank}(B) = m$. Since the controllability matrix

$$P = [b_1, \ldots, b_m, Ab_1, \ldots, Ab_m, \ldots, A^{n-1}b_1, \ldots, A^{n-1}b_m] \tag{2.30}$$

has rank n, it is possible to select n linearly independent columns in (2.30). If these columns are chosen in the order in which they occur from left to right (this includes by assumption all columns of B), we obtain, after reordering, the matrix

$$S = [b_1, \ldots, A^{p_1-1}b_1, \ldots, b_m, \ldots, A^{p_m-1}b_m] \qquad p_1 + p_2 + \cdots + p_m = n$$

The integers $p_1, p_2, \ldots, p_m$ are referred to as *Kronecker invariants* of the pair (A, B) or *controllability indices* if $p_1 \geq p_2 \geq \cdots \geq p_m$. The integer $p = \max\{p_i: i = 1, 2, \ldots, m\}$ is often called the *controllability index of the system*.

A transformation matrix T which reduces the system (2.29) into canonical form is obtained in the following way. Let f_i, $i = 1, 2, \ldots, m$ be the σ_ith row of S^{-1}, where

$$\sigma_i = \sum_{j=1}^{i} p_j$$

Then

$$T = \begin{bmatrix} f_1 \\ \vdots \\ f_1 A^{p_1-1} \\ \vdots \\ f_m \\ \vdots \\ f_m A^{p_m-1} \end{bmatrix}$$

Using the transformation $\tilde{x}(t) = Tx(t)$, the system is reduced into *Luenberger canonical form*

$$\dot{\tilde{x}}(t) = \tilde{A}\tilde{x}(t) + \tilde{B}u(t)$$
$$y(t) = \tilde{C}\tilde{x}(t)$$

where $\tilde{A} = TAT^{-1}$ and $\tilde{B} = TB$ have the following structure respectively

$$\tilde{A} = \left[\begin{array}{ccccc|ccccc|ccccc}
0 & 1 & 0 & \ldots & 0 & 0 & 0 & \ldots & 0 & & 0 & 0 & \ldots & 0 \\
0 & 0 & 1 & \ldots & 0 & 0 & 0 & \ldots & 0 & & 0 & 0 & \ldots & 0 \\
\vdots & \vdots & \vdots & & \vdots & \vdots & \vdots & & \vdots & \cdots & \vdots & \vdots & & \vdots \\
0 & 0 & 0 & \ldots & 1 & 0 & 0 & \ldots & 0 & & 0 & 0 & \ldots & 0 \\
x & x & x & \ldots & x & x & x & \ldots & x & & x & x & \ldots & x \\\hline
0 & 0 & & \ldots & 0 & 0 & 1 & 0 & \ldots & 0 & 0 & 0 & \ldots & 0 \\
0 & 0 & & \ldots & 0 & 0 & 0 & 1 & \ldots & 0 & 0 & 0 & \ldots & 0 \\
\vdots & \vdots & & & \vdots & \vdots & \vdots & \vdots & & \vdots & \vdots & \vdots & & \vdots \\
0 & 0 & & \ldots & 0 & 0 & 0 & 0 & \ldots & 1 & 0 & 0 & \ldots & 0 \\
x & x & & \ldots & x & x & x & x & \ldots & x & x & x & \ldots & x \\\hline
 & & \vdots & & & & & \vdots & & & & & \vdots & \\\hline
0 & 0 & & \ldots & 0 & 0 & 0 & & \ldots & 0 & 0 & 1 & 0 & \ldots & 0 \\
0 & 0 & & \ldots & 0 & 0 & 0 & & \ldots & 0 & 0 & 0 & 1 & \ldots & 0 \\
\vdots & \vdots & & & \vdots & \vdots & \vdots & & & \vdots & \vdots & \vdots & \vdots & & \vdots \\
0 & 0 & & \ldots & 0 & 0 & 0 & & \ldots & 0 & 0 & 0 & 0 & \ldots & 1 \\
x & x & & \ldots & x & x & x & & \ldots & x & x & x & x & \ldots & x
\end{array}\right]
\begin{array}{l} \\ \\ \\ \\ \leftarrow \sigma_1 \\ \\ \\ \\ \\ \leftarrow \sigma_2 \\ \\ \\ \\ \\ \\ \leftarrow \sigma_m \end{array}$$

$$\underleftrightarrow{\quad p_1 \quad} \underleftrightarrow{\quad p_2 \quad} \underleftrightarrow{\quad p_m \quad}$$

$$(2.31)$$

$$\tilde{B} = \left[\begin{array}{cccc}
0 & 0 & \ldots & 0 \\
0 & 0 & \ldots & 0 \\
\vdots & \vdots & & \vdots \\
0 & 0 & \ldots & 0 \\
1 & x & \ldots & x \\\hline
0 & 0 & \ldots & 0 \\
0 & 0 & \ldots & 0 \\
\vdots & \vdots & & \vdots \\
0 & 0 & \ldots & 0 \\
0 & 1 & \ldots & x \\\hline
 & \vdots & & \\\hline
0 & 0 & \ldots & 0 \\
0 & 0 & \ldots & 0 \\
\vdots & \vdots & & \vdots \\
0 & 0 & \ldots & 0 \\
0 & 0 & \ldots & 1
\end{array}\right]
\begin{array}{l} \\ \\ \\ \\ \leftarrow \sigma_1 \\ \\ \\ \\ \\ \leftarrow \sigma_2 \\ \\ \\ \\ \\ \\ \leftarrow \sigma_m \end{array}$$

Note that the matrix $\tilde{A}$ contains m companion matrices, each of dimension p_i, along the leading diagonal. The largest companion form is of dimension p – the controllability index of the system. It may be shown that some of the nonspecified (x) entries in $\tilde{A}$ and $\tilde{B}$ must also be zero, but this is not important for most applications. The matrix $\tilde{C} = CT^{-1}$ has no special form in this case.

Example 2.17

If

$$A = \begin{bmatrix} 0 & -6 & 2 & -3 & 10 \\ 1 & -17 & 7 & -10 & 27 \\ -4 & -11 & 4 & -7 & 21 \\ -3 & -7 & 2 & -4 & 14 \\ 1 & -10 & 4 & -6 & 16 \end{bmatrix} \quad \text{and} \quad B = \begin{bmatrix} 1 & 0 \\ 2 & 0 \\ 0 & 3 \\ 0 & 2 \\ 1 & 0 \end{bmatrix}$$

then the linearly independent columns of the controllability matrix from left to right are

$$[b_1, \ b_2, \ Ab_1, \ Ab_2, \ A^2b_1]$$

so that

$$\begin{aligned} S &= [b_1, \ Ab_1, \ A^2b_1, \ b_2, \ Ab_2] \\[6pt] &= \begin{bmatrix} 1 & -2 & 5 & 0 & 0 \\ 2 & -6 & 14 & 0 & 1 \\ 0 & -5 & 12 & 3 & -2 \\ 0 & -3 & 8 & 2 & -2 \\ 1 & -3 & 8 & 0 & 0 \end{bmatrix} \end{aligned}$$

and $p_1 = 3$, $p_2 = 2$.

The transformation matrix T is obtained using the third and fifth rows of S^{-1}. In the given case

$$\begin{aligned} f_1 &= [1 \quad -2 \quad 2 \quad -3 \quad 3] \\ f_2 &= [2 \quad -3 \quad 4 \quad -6 \quad 4] \end{aligned}$$

and

$$T = \begin{bmatrix} 1 & -2 & 2 & -3 & 3 \\ 2 & -3 & 2 & -3 & 4 \\ 2 & -2 & 1 & -2 & 3 \\ 2 & -3 & 4 & -6 & 4 \\ 3 & -3 & 3 & -4 & 3 \end{bmatrix}$$

Hence

$$\tilde{A} = TAT^{-1} = \left[\begin{array}{ccc:cc} 0 & 1 & 0 & 0 & 0 \\ 0 & 0 & 1 & 0 & 0 \\ 0 & 2 & 0 & -1 & 0 \\ \hdashline 0 & 0 & 0 & 0 & 1 \\ 1 & 0 & -1 & 0 & -1 \end{array}\right]$$

$$\tilde{B} = TB = \left[\begin{array}{cc} 0 & 0 \\ 0 & 0 \\ 1 & -1 \\ \hdashline 0 & 0 \\ 0 & 1 \end{array}\right]$$

Using a dual procedure, the system (2.29) can be reduced into multivariable observable canonical form. In this case the corresponding integers $q_1, q_2, ..., q_r$ are referred to as *observability indices* and if the system is completely observable then

$$\sum_{i=1}^{r} q_i = n$$

If a given system is not completely controllable it can be reduced to the form (2.24) where the pair (A_{11}, B_1) is in canonical form. Since in the given case

$$\text{rank}[B, \ AB, ..., A^{n-1}B] = d < n$$

the transformation matrix may be obtained by using $n - d$ additional column vectors which are linearly independent of the columns of the controllability matrix.

The canonical forms of linear discrete-time systems are the same as for continuous-time systems. Other canonical forms of linear systems are described in Maroulas and Barnett (1979). Canonical forms relative to orthogonal state transformations are considered later in Section 4.4.

EXERCISES

2.33 Show that the columns of the transformation matrix S, which reduces a single-input system into controllable canonical form, may be computed recursively by

$$s_n = b \qquad s_j = a_{j+1}b + As_{j+1} \qquad j = n - 1, ..., 1$$

Hint: use the identity $AS \equiv S\tilde{A}$.

2.34 Show that in the single-input case

$$P^{-1}AP = \begin{bmatrix} 0 & 0 & \dots & 0 & -a_1 \\ 1 & 0 & \dots & 0 & -a_2 \\ 0 & 1 & \dots & 0 & -a_3 \\ \vdots & \vdots & & \vdots & \vdots \\ 0 & 0 & \dots & 1 & -a_n \end{bmatrix} \qquad P^{-1}b = \begin{bmatrix} 1 \\ 0 \\ 0 \\ \vdots \\ 0 \end{bmatrix}$$

provided $\text{rank}(P) = n$. Try to develop similar canonical form in the multi-input case and compare it with the form (2.31).

2.35 Prove that if for a completely controllable single-input system

$$S = [A^{n-1}b, \ A^{n-2}b, \ \dots, \ Ab, \ b]$$

then

$$S^{-1}AS = \begin{bmatrix} -a_n & 1 & 0 & \dots & 0 \\ -a_{n-1} & 0 & 1 & \dots & 0 \\ \vdots & \vdots & \vdots & & \vdots \\ -a_2 & 0 & 0 & \dots & 1 \\ -a_1 & 0 & 0 & \dots & 0 \end{bmatrix} \qquad S^{-1}b = \begin{bmatrix} 0 \\ 0 \\ \vdots \\ 0 \\ 1 \end{bmatrix}$$

2.36 Find the transformation matrix which reduces the companion form into diagonal form in the case of distinct eigenvalues.

2.37 Derive a procedure for partial reduction of a single-input system, which is not completely controllable, into controllable canonical form.

2.38 Prove that the controllability indices are invariant under nonsingular transformations of the states and inputs.

2.39 Prove that the controllability index of a controllable system is the minimal integer p such that

$$\text{rank}\,[B, \ AB, \ \dots, \ A^{p-1}B] = n$$

2.40 Find matrices K, T and V such that the pair

$$[T(A + BK)T^{-1}, \ TBV]$$

is in *Brunovsky canonical form*

$$\tilde{A} = \begin{bmatrix} A_1 & & & \\ & A_2 & & \mathbf{0} \\ & & \ddots & \\ \mathbf{0} & & & A_m \end{bmatrix} \qquad \tilde{B} = \begin{bmatrix} B_1 \\ B_2 \\ \vdots \\ B_m \end{bmatrix}$$

where the blocks

$$A_i = \begin{bmatrix} 0 & 1 & 0 & \dots & 0 \\ 0 & 0 & 1 & \dots & 0 \\ \vdots & \vdots & \vdots & & \vdots \\ 0 & 0 & 0 & \dots & 1 \\ 0 & 0 & 0 & \dots & 0 \end{bmatrix} \qquad B_i = \begin{bmatrix} & & & 0 & & \\ & & & 0 & & \\ & & & \vdots & & \\ & & & 0 & & \\ 0 & \dots & & 010 & \dots & 0 \\ & & & \uparrow & & \\ & & & i & & \end{bmatrix}$$

have dimensions $p_i \times p_i$ and $p_i \times m$, respectively. *Hint*: use the Luenberger canonical form.

2.6 POLE ASSIGNMENT

Consider the linear control system

$$\dot{x}(t) = Ax(t) + Bu(t)$$
$$y(t) = Cx(t) \tag{2.32}$$

In cases when the state vector $x(t)$ is available, the input vector $u(t)$ may be determined by the *linear control law* (or *linear state feedback*)

$$u(t) = -Kx(t)$$

where the $m \times n$ matrix K is said to be a *feedback matrix* (or *gain matrix*).

The corresponding closed-loop system is described by the equations

$$\dot{x}(t) = (A - BK)x(t)$$
$$y(t) = Cx(t) \tag{2.33}$$

and hence the behaviour of the closed-loop system is determined by the properties of the matrix $A - BK$. For asymptotic stability of the system it is necessary to choose the matrix K so that the closed-loop poles, i.e. the eigenvalues of $A - BK$, lie in the left-half complex plane.

The problem of finding an appropriate feedback matrix K is referred to as the *problem of synthesis of a state regulator*. A particular case of this problem is the *pole assignment problem* which consists in the determination of a feedback matrix K, such that the poles of the closed-loop system are allocated to desired places in the complex plane. The possibility of assigning the closed-loop poles is based on the following fundamental theorem.

Theorem 2.12

Let $\Lambda = \{\lambda_1, \lambda_2, ..., \lambda_n\}$ be an arbitrary set of n complex numbers λ_i, in which every λ_i with $\text{Im}(\lambda_i) \neq 0$ appears in a conjugate pair. Then the matrix pair (A, B) is completely controllable if and only if for every choice of Λ there exists a real matrix K such that $A - BK$ has Λ as its set of eigenvalues.

For proof of this theorem see Wonham (1986, ch. 2).

This result shows that the poles of the closed-loop system (2.33) may be prescribed by a suitable choice of the matrix K at arbitrary places (under the restriction that the complex poles must appear in conjugate pairs) provided that the system (2.32) is completely controllable. In particular, every controllable system can be stabilized by linear state feedback.

Since the poles of the closed-loop system determine the character of the system modes, the pole assignment is also referred to as *modal control*.

In this section we consider some of the available methods for pole assignment. Consider first the single-input system

$$\dot{x}(t) = Ax(t) + bu(t) \tag{2.34}$$

where b is a column n-vector and $u(t)$ is scalar valued. Assume that the system is controllable, i.e. the matrix

$$P = [b, \ Ab, \ ..., \ A^{n-1}b]$$

is of full rank. The linear state feedback is of the form

$$u(t) = -kx(t) \tag{2.35}$$

where k is a row n-vector. The desired position of closed-loop poles may be specified by the coefficients of the closed-loop characteristic polynomial

$$d(\lambda) = \lambda^n + d_n\lambda^{n-1} + \cdots + d_1 \tag{2.36}$$

Notice that under the above mentioned restrictions on the poles, these coefficients are real numbers.

Using a nonsingular transformation

$$x(t) = S\tilde{x}(t)$$

the system (2.34) is reduced to the controllable canonical form

$$\dot{\tilde{x}}(t) = \tilde{A}\tilde{x}(t) + \tilde{b}u(t) \tag{2.37}$$

$$\tilde{A} = S^{-1}AS = \begin{bmatrix} 0 & 1 & 0 & \cdots & 0 \\ 0 & 0 & 1 & \cdots & 0 \\ \vdots & \vdots & \vdots & & \vdots \\ 0 & 0 & 0 & \cdots & 1 \\ -a_1 & -a_2 & -a_3 & \cdots & -a_n \end{bmatrix} \qquad \tilde{b} = S^{-1}b = \begin{bmatrix} 0 \\ 0 \\ \vdots \\ 0 \\ 1 \end{bmatrix}$$

The numbers $a_1, a_2, ..., a_n$ are the coefficients of the open-loop characteristic polynomial

$$\det(\lambda I - A) = \lambda^n + a_n\lambda^{n-1} + \cdots + a_1$$

The control law (2.35) takes the form

$$u(t) = -\tilde{k}\tilde{x}(t) \tag{2.38}$$

where

$$\tilde{k} = kS = [k_1, k_2, ..., k_n]$$

The system (2.37) together with the control law (2.38) is described by

$$\dot{\tilde{x}}(t) = (\tilde{A} - \tilde{b}\tilde{k})\tilde{x}(t)$$

where

$$\tilde{A} - \tilde{b}\tilde{k} = \begin{bmatrix} 0 & 1 & \cdots & 0 \\ 0 & 0 & & 0 \\ \vdots & \vdots & & \vdots \\ 0 & 0 & \cdots & 1 \\ -a_1 - k_1 & -a_2 - k_2 & \cdots & -a_n - k_n \end{bmatrix}$$

Since the matrix $\tilde{A} - \tilde{b}\tilde{k}$ is in companion form its characteristic polynomial is

$$\det(\lambda I - \tilde{A} + \tilde{b}\tilde{k}) = \lambda^n + (a_n + k_n)\lambda^{n-1} + \cdots + a_1 + k_1 \qquad (2.39)$$

By making equal the expressions (2.39) and (2.36) we obtain

$$\tilde{k} = d - a$$
$$d = [d_1, d_2, ..., d_n]$$
$$a = [a_1, a_2, ..., a_n]$$

In this way the feedback matrix which assigns the poles to the desired places is given by

$$k = (d - a)S^{-1}$$

Exploiting the results from Section 2.5 we get

$$k = (d - a)\tilde{P}P^{-1} \qquad \tilde{P} = [\tilde{b}, \tilde{A}\tilde{b}, ..., \tilde{A}^{n-1}\tilde{b}]$$

This result shows that for a single-input system the solution for the feedback matrix is unique.

Notice that since the pair $(A - bk, b)$ is controllable it follows that the matrix $A - bk$ is *cyclic* (nonderogatory), i.e. if there are multiple desired eigenvalues they appear in a single Jordan block in the Jordan form of $A - bk$.

Example 2.18

Let

$$A = \begin{bmatrix} -3 & 7 & -3 & -2 \\ -6 & 12 & -5 & -4 \\ -5 & 10 & -4 & -4 \\ -3 & 6 & -3 & -2 \end{bmatrix} \quad \text{and} \quad b = \begin{bmatrix} 1 \\ 2 \\ 2 \\ 1 \end{bmatrix}$$

and the desired closed-loop poles be

$$-1+i, \quad -1-i, \quad -2, \quad -3$$

Applying a nonsingular transformation with a matrix

$$S = \begin{bmatrix} 0 & 0 & 0 & 1 \\ 0 & 1 & -2 & 2 \\ 0 & 2 & -3 & 2 \\ -1 & 2 & -2 & 1 \end{bmatrix}$$

we obtain

$$\tilde{A} = S^{-1}AS = \begin{bmatrix} 0 & 1 & 0 & 0 \\ 0 & 0 & 1 & 0 \\ 0 & 0 & 0 & 1 \\ 2 & -3 & -1 & 3 \end{bmatrix} \quad \text{and}$$

$$\tilde{b} = S^{-1}b = \begin{bmatrix} 0 \\ 0 \\ 0 \\ 1 \end{bmatrix}$$

Since the desired closed-loop characteristic polynomial is

$$\lambda^4 + 7\lambda^3 + 18\lambda^2 + 22\lambda + 12$$

it is easy to find that

$$\tilde{k} = [2, \ -3, \ -1, \ 3] - [-12, \ -22, \ -18, \ -7]$$
$$= [14, \ 19, \ 17, \ 10]$$

and

$$k = \tilde{k}S^{-1} = [96, \ -119, \ 83, \ -14]$$

In the multi-input case

$$\dot{x}(t) = Ax(t) + Bu(t) \tag{2.40}$$

where B is an $n \times m$ matrix $(1 < m < n)$ the control law

$$u(t) = -Kx(t)$$

which assigns the closed-loop poles at prescribed places, is not unique. This is because the feedback matrix contains nm parameters which satisfy n conditions, corresponding to the n coefficients of the desired characteristic polynomial. This freedom may be exploited to satisfy additional requirements, i.e. to specify the eigenstructure of the matrix $A - BK$.

The simplest way to find the gain matrix in the multi-input case is to reduce the system (2.40) to equivalent single-input system using the *dyadic feedback matrix*

$$K = qk \qquad (2.41)$$

where q is a column m-vector and k is a row n-vector. Since in this case the rank of the matrix K is equal to 1, the corresponding feedback is known also as *unity-rank feedback*. In this way, the pole assignment of the multi-input system reduces to pole assignment of the equivalent single-input system

$$\dot{x}(t) = Ax(t) + bw(t)$$
$$b = Bq \qquad u(t) = qw(t) \qquad (2.42)$$

For a given q, the matrix k may be determined for each set of desired poles if the system (2.42) is controllable. This requires that the matrix A be cyclic, since only in this case does there exist a vector b such that the system (2.42) is controllable.

The use of dyadic feedback leads to loss of freedom in the synthesis, since the abilities of the multi-input system are artificially reduced to those of a single-input system.

A full-rank feedback matrix which yields the desired poles of a multi-input system may be found using the Luenberger canonical form. Implementing the nonsingular transformation

$$\tilde{x}(t) = Tx(t)$$

where the matrix T is chosen as described in Section 2.5, the system (2.40) is reduced to the form

$$\dot{\tilde{x}}(t) = \tilde{A}\tilde{x}(t) + \tilde{B}u(t) \qquad (2.43)$$

where the matrices $\tilde{A}$ and $\tilde{B}$ have the form (2.31).

The control law of system (2.43) is taken as

$$u(t) = -\tilde{K}\tilde{x}(t)$$

where $\tilde{K}$ is an $m \times n$ matrix.

Since the closed-loop state matrix $\tilde{F} = \tilde{A} - \tilde{B}\tilde{K}$ of the transformed system has nontrivial rows only in the positions $\sigma_1, \sigma_2, \ldots, \sigma_m$, $\sigma_i = \sum_{j=1}^{i} p_j$, the desired set of eigenvalues of $\tilde{F}$ may be achieved if the rows of $\tilde{K}$ are chosen so that the under- or over-diagonal blocks of $\tilde{F}$ are zeroed and the last rows of the diagonal blocks are modified in an appropriate way. Then the matrix $\tilde{F}$ takes the form (if the under-diagonal blocks are zeroed)

$$\tilde{F} = \begin{bmatrix} F_{11} & F_{12} & \cdots & F_{1m} \\ & F_{22} & \cdots & F_{2m} \\ & & \ddots & \vdots \\ \text{\Large 0} & & & F_{mm} \end{bmatrix}$$

where each $p_i \times p_i$ matrix F_{ii}, $i = 1, 2, \ldots, m$, is in companion form. The characteristic polynomial of $\tilde{F}$

$$\det(\lambda I - \tilde{F}) = \prod_{i=1}^{m} \det(\lambda I - F_{ii})$$

is a product of the characteristic polynomials of the diagonal blocks. The coefficients of these polynomials are the elements in the last rows of the diagonal blocks taken with opposite signs. These elements are obtained from the desired eigenvalues of the diagonal blocks.

From the rows of the matrices $\tilde{A}$, $\tilde{F}$, $\tilde{B}$ in positions $\sigma_1, \sigma_2, \ldots, \sigma_m$ let the $m \times n$ matrix $\hat{A}$, the $m \times n$ matrix $\hat{F}$ and the $m \times m$ matrix $\hat{B}$, respectively, be constructed. Notice that the matrix $\hat{B}$ is in upper triangular form with units on the diagonal

$$\hat{B} = \begin{bmatrix} 1 & x & x & \cdots & x \\ & 1 & x & \cdots & x \\ & & \ddots & & \vdots \\ & & & 1 & x \\ \text{\Large 0} & & & & 1 \end{bmatrix}$$

and hence is a nonsingular matrix. Then the equation

$$\tilde{B}\tilde{K} = \tilde{A} - \tilde{F}$$

is reduced into

$$\hat{B}\tilde{K} = \hat{A} - \hat{F}$$

so that

$$\tilde{K} = \hat{B}^{-1}(\hat{A} - \hat{F})$$

The matrix $\tilde{K}$ obtained in this way is non-unique since the desired eigenvalues of the closed-loop state matrix may be distributed in a different way between the blocks of $\tilde{F}$.

The feedback matrix of the original system is determined from

$$K = \tilde{K}T$$

Generally, the matrix K is of full rank.

Example 2.19

Let

$$A = \begin{bmatrix} -3 & 11 & -5 & 4 & -26 \\ -5 & 14 & 2 & -4 & -33 \\ 4 & -7 & 6 & -1 & 5 \\ 3 & -7 & 9 & -6 & 9 \\ -3 & 7 & 1 & -3 & -14 \end{bmatrix} \quad \text{and} \quad B = \begin{bmatrix} -3 & 7 \\ -8 & 19 \\ -4 & 10 \\ -4 & 10 \\ -3 & 7 \end{bmatrix}$$

and the desired closed-loop system poles be

$$-1 + i, \quad -1 - i, \quad -2 + i, \quad -2 - i, \quad -3$$

Using a nonsingular transformation with matrix

$$T = \begin{bmatrix} 77 & -102 & 1 & 50 & 127 \\ 52 & -49 & -6 & 34 & 41 \\ 3 & -4 & 0 & 2 & 5 \\ 2 & -2 & 0 & 1 & 2 \\ 1 & 1 & -3 & 4 & -5 \end{bmatrix}$$

we get

$$\tilde{A} = TAT^{-1} = \begin{bmatrix} 0 & 1 & 0 & 0 & 0 \\ -5 & 7 & 125 & -155 & 0 \\ 0 & 0 & 0 & 1 & 0 \\ 0 & 0 & 0 & 0 & 1 \\ -2 & 3 & 49 & -66 & -10 \end{bmatrix}$$

$$\tilde{B} = TB = \begin{bmatrix} 0 & 1 \\ 1 & 0 \\ 0 & 0 \\ 0 & 0 \\ 0 & 1 \end{bmatrix}$$

so that $p_1 = 2$ and $p_2 = 3$.

Representing the closed-loop characteristic polynomial in conformity with the values of Kronecker indices

$$(\lambda^2 + 2\lambda + 2)(\lambda^3 + 7\lambda^2 + 17\lambda + 15)$$

we obtain

$$\tilde{F} = \begin{bmatrix} 0 & 1 & 0 & 0 & 0 \\ -2 & -2 & 125 & -155 & 0 \\ 0 & 0 & 0 & 1 & 0 \\ 0 & 0 & 0 & 0 & 1 \\ 0 & 0 & -15 & -17 & -7 \end{bmatrix}$$

Taking the second and fifth rows of $\tilde{A}$, $\tilde{F}$ and $\tilde{B}$, we get

$$\hat{A} = \begin{bmatrix} -5 & 7 & 125 & -155 & 0 \\ -2 & 3 & 49 & -66 & -10 \end{bmatrix}$$

$$\hat{F} = \begin{bmatrix} -2 & -2 & 125 & -155 & 0 \\ 0 & 0 & -15 & -17 & -7 \end{bmatrix}$$

$$\hat{B} = \begin{bmatrix} 1 & 0 \\ 0 & 1 \end{bmatrix}$$

Hence

$$\tilde{K} = \hat{A} - \hat{F} = \begin{bmatrix} -3 & 9 & 0 & 0 & 0 \\ -2 & 3 & 64 & -49 & -3 \end{bmatrix}$$

and

$$K = \tilde{K}T = \begin{bmatrix} 237 & -135 & -57 & 156 & -12 \\ 93 & -104 & -11 & 69 & 106 \end{bmatrix}$$

Another solution is obtained if we take

$$\tilde{F} = \begin{bmatrix} 0 & 1 & 0 & 0 & 0 \\ -2 & -2 & 0 & 0 & 0 \\ 0 & 0 & 0 & 1 & 0 \\ 0 & 0 & 0 & 0 & 1 \\ -2 & 3 & -15 & -17 & -7 \end{bmatrix}$$

so that

$$\hat{F} = \begin{bmatrix} -2 & -2 & 0 & 0 & 0 \\ -2 & 3 & -15 & -17 & -7 \end{bmatrix}$$

$$\tilde{K} = \begin{bmatrix} -3 & 9 & 125 & -155 & 0 \\ 0 & 0 & 64 & -49 & -3 \end{bmatrix}$$

and

$$K = \begin{bmatrix} 302 & -325 & -57 & 251 & 303 \\ 91 & -161 & 9 & 67 & 237 \end{bmatrix}$$

Consider now the case when the system

$$\dot{x}(t) = Ax(t) + Bu(t)$$

is not completely controllable. Implementing a nonsingular transformation $x(t) = S\tilde{x}(t)$ we obtain

$$\dot{\tilde{x}}(t) = \tilde{A}\tilde{x}(t) + \tilde{B}u(t)$$

where

$$\tilde{A} = \begin{bmatrix} A_{11} & A_{22} \\ 0 & A_{22} \end{bmatrix} \qquad \tilde{B} = \begin{bmatrix} B_1 \\ 0 \end{bmatrix}$$

and the pair (A_{11}, B_1) is controllable. Let the control law be taken as

$$u(t) = -\tilde{K}\tilde{x}(t) = -[K_1, K_2]x(t)$$

where the partition of $\tilde{K}$ conforms to that of $\tilde{A}$. The transformed closed-loop system is described by

$$\dot{\tilde{x}}(t) = \tilde{F}\tilde{x}(t)$$

where

$$\tilde{F} = \begin{bmatrix} A_{11} - B_1 K_1 & A_{12} - B_1 K_2 \\ \hline 0 & A_{22} \end{bmatrix}$$

is an upper block-triangular matrix.

Since

$$\det(\lambda I - \tilde{F}) = \det(\lambda I - A_{11} + B_1 K_1)\det(\lambda I - A_{22})$$

the eigenvalues of the matrix $\tilde{F}$ are those of $A_{11} - B_1 K_1$ and A_{22}. Therefore, the state feedback affects only the controllable part of the system. The controllable poles may be assigned at the desired places using the above considered methods, while the uncontrollable poles are not altered by the state feedback. If the system is stabilizable, A_{22} is a stability matrix and hence it is possible to find a state feedback for which the closed-loop system is asymptotically stable. The matrix K_2 does not affect the closed-loop poles and may be arbitrarily chosen.

Example 2.20

If

$$A \begin{bmatrix} -7 & 9 & -14 & 11 & 3 \\ -4 & -4 & 15 & -11 & 11 \\ 14 & -18 & 29 & -23 & -3 \\ 20 & -20 & 27 & -22 & -10 \\ 6 & -9 & 15 & -12 & 1 \end{bmatrix} \qquad \text{and} \qquad b = \begin{bmatrix} 3 \\ 1 \\ 1 \\ 2 \\ 2 \end{bmatrix}$$

then implementing a transformation with matrix

$$S = \begin{bmatrix} 4 & -4 & 3 & -2 & 2 \\ 4 & -3 & 1 & -1 & 2 \\ 0 & -1 & 1 & -2 & 2 \\ -1 & -1 & 2 & -3 & 2 \\ 2 & -2 & 2 & -2 & 1 \end{bmatrix}$$

we obtain

$$\tilde{A} = S^{-1}AS = \left[\begin{array}{ccc|cc} 0 & 1 & 0 & -1 & 0 \\ 0 & 0 & 1 & 3 & -2 \\ 1 & -2 & 2 & 0 & -3 \\ \hline 0 & 0 & 0 & -4 & -2 \\ 0 & 0 & 0 & 1 & -1 \end{array}\right] \qquad \tilde{b} = S^{-1}b = \left[\begin{array}{c} 0 \\ 0 \\ 1 \\ \hline 0 \\ 0 \end{array}\right]$$

The pair (A, b) is not controllable, but it is stabilizable, since the eigenvalues of the block A_{22} are -2 and -3.

Taking the poles of the controllable part as

$-1 \pm i, \quad -1$

which corresponds to desired characteristic polynomial

$\lambda^3 + 3\lambda^2 + 4\lambda + 2$

we obtain

$k_1 = [3, \ 2, \ 5]$

If the matrix k_2 is chosen as zero matrix we get

$\tilde{k} = [3, \ 2, \ 5, \ 0, \ 0]$

and

$k = \tilde{k}S^{-1} = [24, \ -55, \ 98, \ -79, \ 24]$

The closed-loop poles are $-1 \pm i, \ -1, \ -2, \ -3$.

The problem for pole assignment of a linear discrete-time system

$$x_{k+1} = Ax_k + Bu_k \tag{2.44}$$

is virtually the same as for continuous-time systems. Theorem 2.12 holds also in the discrete-time case. The only difference is that the desired poles must lie in the unit circle which ensures asymptotic stability of the closed-loop system.

An interesting feature of the discrete-time case is that the time response of the closed-loop system may be made to 'die' for n samples, i.e. it is possible to find a state feedback such that $x_n = 0$ for all x_0. Discrete-time systems with this property are called *deadbeat*. The closed-loop system is deadbeat if and only if all poles are equal to zero. According to the Caley–Hamilton theorem, in this case

$$(A - BK)^n = 0$$

i.e. $A - BK$ is a nilpotent matrix and

$$x_n = (A - BK)^n x_0 = 0$$

The following theorem shows that the state of a multi-input discrete-time system can be driven to zero for less than n samples.

Theorem 2.13

If the discrete-time system (2.44) is completely controllable, then there exists a feedback matrix K such that $(A - BK)^p = 0$, where $p = \max\{p_i : i = 1, 2, ..., m\}$ is the controllability index of the system.

Indeed, using the Luenberger canonical form of (2.44) we can find a feedback matrix K such that the transformed closed-loop state matrix is in the Jordan form

$$\tilde{F} = \text{diag}(N_1, N_2, ..., N_m)$$

where the diagonal blocks

$$N_i = \begin{bmatrix} 0 & 1 & 0 & \cdots & 0 \\ 0 & 0 & 1 & \cdots & 0 \\ \vdots & \vdots & \vdots & & \vdots \\ 0 & 0 & 0 & \cdots & 1 \\ 0 & 0 & 0 & \cdots & 0 \end{bmatrix}$$

are of dimension $p_i \times p_i$, $i = 1, 2, ..., m$. Now the proof of the theorem easily follows, taking into account that $\tilde{F}^p = 0$.

EXERCISES

2.41 Prove that in the single-input case the feedback matrix may be determined by *Ackermann's formula*

$$k = [0, ..., 0, 1] [b, Ab, ..., A^{n-1}b]^{-1} P(A)$$

where

$$P(A) = A^n + d_n A^{n-1} + \cdots + d_1 I$$

and $d_1, d_2, \ldots, d_n$ are the coefficients of the desired closed-loop characteristic polynomial. *Hint*: use the Caley–Hamilton theorem for $A - bk$.

2.42 Let the state matrix A of a single-input system have distinct eigenvalues $\lambda_1, \lambda_2, \ldots, \lambda_n$ and denote by X the matrix of the corresponding eigenvectors of A. Prove that the gain matrix k which yields the desired eigenvalues $\mu_1, \mu_2, \ldots, \mu_n$ of $A - bk$ may be found by the *Mayne–Murdoch formula*

$$k = \alpha X^{-1}$$

where

$$\alpha = [\alpha_1, \ \alpha_2, \ \ldots, \ \alpha_n]$$

$$\alpha_j = \frac{\displaystyle\prod_{i=1}^{n} (\mu_i - \lambda_j)}{\beta_j \displaystyle\prod_{\substack{i=1 \\ i \neq j}}^{n} (\lambda_i - \lambda_j)} \qquad j = 1, 2, \ldots, n$$

$$\beta \equiv [\beta_1, \beta_2, \ldots, \beta_n] = X^{-1} b$$

2.43 Derive explicit expressions for the gain matrix elements of a second-order, completely controllable, single-input system using different pole assignment methods.

2.44 Derive a procedure for pole assignment of multi-input systems using the Brunovsky canonical form (see Exercise 2.40).

2.45 If the pair (A, B) is in Luenberger canonical form, find a matrix K such that $A - BK$ has a desired Jordan form J. *Hint*: use the equation $(A - BK)S = SJ$, where S is a nonsingular matrix.

2.46 (*Fundamental theorem of linear state feedback*) Let $f_1, f_2, \ldots, f_q$ be the degrees of the invariant polynomials of the matrix F. Prove that there exists a matrix K such that $A - BK$ is similar to F if and only if

$$\sum_{i=1}^{s} (f_i - p_i) \geqq 0 \qquad \text{for all} \qquad s = 1, 2, \ldots, \min\{q, m\}$$

where $p_1, p_2, \ldots, p_m$ are the controllability indices of the pair (A, B).

2.47 Show that the feedback matrix of a single-input deadbeat discrete-time system may be found as the last row of the matrix

$$[A^{-n}b, \ A^{-n+1}b, \ \ldots, \ A^{-1}b]^{-1}$$

provided A is nonsingular. *Hint*: use Ackermann's formula.

2.7 QUADRATIC OPTIMIZATION

The requirements for the dynamic behaviour of a control system may be formulated as a problem for minimization of a quadratic cost function of states and inputs. The use of such a function for linear systems allows us to obtain an optimal control which can be realized by linear state feedback.

The *linear quadratic optimization problem* for the system

$$\dot{x}(t) = Ax(t) + Bu(t) \qquad x(0) = x_0 \tag{2.45}$$

is to find a control $u^0(t)$ which minimizes the *quadratic performance cost*

$$J = \int_0^\infty [x^T(t)Qx(t) + u^T(t)Ru(t)] \, dt \tag{2.46}$$

where $Q = M^T M$ is a positive semidefinite matrix and R is a positive definite matrix. The matrices Q and R are said to be the *state* and *control-weighting matrices*, respectively.

The solution of the quadratic optimization problem is given by the following theorem.

Theorem 2.14

If the pair (A, B) is stabilizable and the pair (M, A) is detectable, then there exists a unique optimal control $u^0(t)$ which minimizes (2.46) and may be expressed as a linear state feedback

$$u^0(t) = -K^0 x(t) \qquad K^0 = R^{-1} B^T P^0$$

where P^0 is the unique positive semidefinite solution of the *matrix algebraic Riccati equation*

$$A^T P + PA + Q - PBR^{-1}B^T P = 0 \tag{2.47}$$

Furthermore, the closed-loop state matrix $A - BK^0$ is a stability matrix and the minimum value of the performance index J is equal to $x_0^T P^0 x_0$.

For proof of this theorem see Kwakernaak and Sivan (1972, ch. 3).

The matrix Riccati equation (2.47) may have several solutions for P but under the conditions of Theorem 2.14 only one of the solutions is positive semidefinite. If the pair (M, A) is observable, then P^0 is a positive definite matrix.

The choice of the positive (semi)definite solution of (2.47) is determined by the requirement for asymptotic stability of the closed-loop system. In fact, the Riccati equation may be written as a Lyapunov equation with respect to the closed-loop system,

$$(A - BK^0)^T P + P(A - BK^0) + Q + K^{0T} RK^0 = 0$$

where $Q + K^{0T} RK^0$ is at least a positive semidefinite matrix. In accordance with the second method of Lyapunov, if the matrix P^0 is positive semidefinite and the pair $(M, A - BK^0)$ is detectable, where $M^T M = Q + K^{0T} R K^0$, then $A - BK^0$ is a stability matrix (see Exercise 2.32).

Owing to its symmetry, the matrix Riccati equation can be expressed as a system of $n(n + 1)/2$ quadratic algebraic equations for the elements of P.

Example 2.21

Consider a completely controllable system with matrices

$$A = \begin{bmatrix} 0 & 1 \\ 0 & 0 \end{bmatrix} \quad \text{and} \quad b = \begin{bmatrix} 0 \\ 2 \end{bmatrix}$$

The weighting matrices are chosen as

$$Q = \begin{bmatrix} 2 & 0 \\ 0 & 0 \end{bmatrix} = \begin{bmatrix} \sqrt{2} \\ 0 \end{bmatrix} [\sqrt{2},\ 0] \quad R = 2$$

so that $M = [\sqrt{2},\ 0]$ and the pair (M, A) is observable.

The solution of the Riccati equation is in the form

$$P = \begin{bmatrix} p_1 & p_2 \\ p_2 & p_3 \end{bmatrix}$$

Since the pair (A, B) is controllable and the pair (M, A) is observable, the Riccati equation must have a unique positive definite solution. Therefore, according to the Sylvester criterion, the following must be satisfied

$$p_1 > 0 \qquad p_1 p_3 > p_2^2$$

The substitution of A, b, Q, R and P in (2.47) yields

$$\begin{bmatrix} 0 & 0 \\ 1 & 0 \end{bmatrix}\begin{bmatrix} p_1 & p_2 \\ p_2 & p_3 \end{bmatrix} + \begin{bmatrix} p_1 & p_2 \\ p_2 & p_3 \end{bmatrix}\begin{bmatrix} 0 & 1 \\ 0 & 0 \end{bmatrix} + \begin{bmatrix} 2 & 0 \\ 0 & 0 \end{bmatrix}$$

$$- \begin{bmatrix} p_1 & p_2 \\ p_2 & p_3 \end{bmatrix}\begin{bmatrix} 0 \\ 2 \end{bmatrix}\frac{1}{2}[0\ \ 2]\begin{bmatrix} p_1 & p_2 \\ p_2 & p_3 \end{bmatrix} = \begin{bmatrix} 0 & 0 \\ 0 & 0 \end{bmatrix}$$

This matrix equation is equivalent to the following system of nonlinear equations

$$\begin{align} -2p_2^2 + 2 &= 0 \\ p_1 - 2p_2 p_3 &= 0 \\ 2p_2 - 2p_3^2 &= 0 \end{align} \tag{2.48}$$

From the first and third equation in (2.48) we get

$$p_2 = \pm 1 \qquad p_3 = \pm \sqrt{p_2}$$

Since the elements of P must be real, it follows that $p_2 = 1$ and $p_3 = \pm 1$, while the requirement $P > 0$ yields $p_3 = 1$.

From the second equation in (2.48) we obtain $p_1 = 2$ so that

$$P^0 = \begin{bmatrix} 2 & 1 \\ 1 & 1 \end{bmatrix}$$

is positive definite.

The optimal feedback matrix is

$$k^0 = R^{-1} B^{\mathrm{T}} P^0 = [1, 1]$$

and the poles of the optimal closed-loop system (i.e. the eigenvalues of $A - bk^0$) are

$$\lambda_{1,2} = -1 \pm i$$

There exist several versions of the linear quadratic optimization problem which are sometimes used instead of the original formulation. We shall consider the following two versions.

1. *Quadratic performance index with cross-weighting terms*
   Let

$$J = \int_0^\infty [x^{\mathrm{T}}(t)Qx(t) + 2x^{\mathrm{T}}(t)Su(t) + u^{\mathrm{T}}(t)Ru(t)] \, dt$$

under the condition that the matrix $Q - SR^{-1}S^{\mathrm{T}}$ must be positive semidefinite.

This quadratic function may be represented as

$$J = \int_0^\infty \begin{bmatrix} x(t) \\ u(t) \end{bmatrix}^{\mathrm{T}} \begin{bmatrix} Q & S \\ S^{\mathrm{T}} & R \end{bmatrix} \begin{bmatrix} x(t) \\ u(t) \end{bmatrix} \, dt$$

If the pair $(M, A - BR^{-1}S^{\mathrm{T}})$ is detectable, where $M^{\mathrm{T}}M \equiv Q - SR^{-1}S^{\mathrm{T}}$, the optimal control is given by

$$u^0(t) = -K^0 x(t) \qquad K^0 = R^{-1}(B^{\mathrm{T}} P^0 + S^{\mathrm{T}})$$

where P^0 is the unique positive semidefinite solution of the Riccati equation

$$(A - BR^{-1}S^{\mathrm{T}})^{\mathrm{T}} P + P(A - BR^{-1}S^{\mathrm{T}}) \\ + Q - SR^{-1}S^{\mathrm{T}} - PBR^{-1}B^{\mathrm{T}}P = 0$$

2. *Quadratic optimization with a prescribed degree of stability*
In this version the performance index is taken in the form

$$J = \int_0^\infty e^{2\alpha t}[x^T(t)Qx(t) + u^T(t)Ru(t)] \ dt$$

where $\alpha \geq 0$.

The optimal control is obtained as

$$u^0(t) = -K^0 x(t) \qquad K^0 = R^{-1}B^T P^0 \tag{2.49}$$

where P^0 is the positive semidefinite solution of

$$(A + \alpha I)^T P + P(A + \alpha I) + Q - PBR^{-1}B^T P = 0 \tag{2.50}$$

Equation (2.50) prompts that (2.49) is also the optimal control of the system

$$\dot{x}(t) = (A + \alpha I)x(t) + Bu(t) \qquad x(0) = x_0 \tag{2.51}$$

subject to the performance index

$$J = \int_0^\infty [x^T(t)Qx(t) + u^T(t)Ru(t)] \ dt$$

The optimal closed-loop system for (2.51) is

$$\dot{x}(t) = (A + \alpha I - BK^0)x(t)$$

Denoting by $\lambda_i(A)$ the eigenvalues of A, we get

$$\lambda_i(A + \alpha I - BK^0) = \lambda_i(A - BK^0) + \alpha$$

Since $\mathrm{Re}[\lambda_i(A + \alpha I - BK^0)] < 0$ it follows that

$$\mathrm{Re}[\lambda_i(A - BK^0)] + \alpha < 0$$

i.e.

$$\mathrm{Re}[\lambda_i(A - BK^0)] < -\alpha$$

In this way the poles of the optimal closed-loop system

$$\dot{x}(t) = (A - BK^0)x(t)$$

have real parts less than $-\alpha$, i.e. the system has degree of stability at least α.

The Riccati equation (2.47) is closely related to the $2n \times 2n$ *Hamiltonian matrix*

$$H = \begin{bmatrix} A & -BR^{-1}B^T \\ -Q & -A^T \end{bmatrix}$$

which has the interesting property that

$$W^{-1}HW = -H^T \tag{2.52}$$

where

$$W = \begin{bmatrix} 0 & I \\ -I & 0 \end{bmatrix} \qquad W^{-1} = W^T = -W$$

Equation (2.52) shows that the matrices H and $-H^T$ are similar, i.e. they have the same set of eigenvalues. Since on the other hand the eigenvalues of H and H^T are the same, it follows that if λ is an eigenvalue of H, then $-\lambda$ is also an eigenvalue of H with the same multiplicity. Moreover, it is possible to show that n eigenvalues of H coincide with the poles of the optimal closed-loop system. To prove this we construct the matrix

$$T = \begin{bmatrix} I_n & 0 \\ P^0 & I_n \end{bmatrix}$$

and perform the nonsingular transformation

$$T^{-1}HT$$

$$= \begin{bmatrix} I_n & 0 \\ -P^0 & I_n \end{bmatrix} \begin{bmatrix} A & -BR^{-1}B^T \\ -Q & -A^T \end{bmatrix} \begin{bmatrix} I_n & 0 \\ P^0 & I_n \end{bmatrix}$$

$$= \begin{bmatrix} A - BR^{-1}B^T P^0 & -BR^{-1}B^T \\ -(A^T P^0 + P^0 A + Q - P^0 BR^{-1}B^T P^0) & -(A - BR^{-1}B^T P^0)^T \end{bmatrix}$$

$$= \begin{bmatrix} A - BK^0 & -BR^{-1}B^T \\ 0 & -(A - BK^0)^T \end{bmatrix} \tag{2.53}$$

Since the matrix in (2.53) is in a block-triangular form, it follows that the eigenvalues of H are the eigenvalues of $A - BK^0$ together with the eigenvalues of $-(A - BK^0)$.

Consider next the discrete-time system

$$x_{k+1} = Ax_k + Bu_k \tag{2.54}$$

with the quadratic performance index

$$J = \sum_{k=0}^{\infty} (x_k^T Q x_k + u_k^T R u_k) \tag{2.55}$$

where $Q = M^T M$ is positive semidefinite and R is a positive definite matrix.

The following theorem gives the solution of the discrete quadratic optimization problem.

Theorem 2.15

If the pair (A, B) is stabilizable and the pair (M, A) is detectable then the optimal control u_k, $k = 0, 1, 2, \ldots$, which minimizes (2.55) is determined by

$$u_k = K^0 x_k \qquad K^0 = (R + B^T P^0 B)^{-1} B^T P^0 A \qquad (2.56)$$

where P^0 is the unique positive semidefinite solution of the *discrete matrix algebraic Riccati equation*

$$A^T P A - P + Q - A^T P B (R + B^T P B)^{-1} B^T P A = 0 \qquad (2.57)$$

Under the control law (2.56) the optimal discrete-time closed-loop system

$$x_{k+1} = (A - BK^0) x_k$$

is asymptotically stable and the quadratic cost (2.55) has the minimum value

$$J^0 = x_0^T P^0 x_0$$

For a proof of this theorem see Sage and White (1977, ch. 6).

It may be shown that if the pair (M, A) is observable, then P^0 is a positive definite matrix.

Example 2.22

Let

$$A = \begin{bmatrix} 1 & 0 \\ 0 & 0.5 \end{bmatrix} \qquad b = \begin{bmatrix} 1 \\ 0 \end{bmatrix} \qquad Q = \begin{bmatrix} 1 & 0 \\ 0 & 3 \end{bmatrix} \qquad R = 2$$

The pair (A, b) is stabilizable and the pair (M, A), $M = \text{diag}(1, \sqrt{3})$, $M^T M = Q$, is observable, so that the discrete Riccati equation must have a unique positive definite solution.

Substituting

$$P = \begin{bmatrix} p_1 & p_2 \\ p_2 & p_3 \end{bmatrix}$$

into (2.57) we get the following system of nonlinear algebraic equations

$$1 - \frac{p_1^2}{2 + p_1} = 0$$

$$-\frac{p_2}{2} - \frac{p_1 p_2}{2(2 + p_1)} = 0$$

$$-\frac{3}{4} p_3 + 3 - \frac{p_2^2}{4(2 + p_1)} = 0$$

As a solution of the first equation we take the positive root $p_1 = 2$ which yields

$$p_2 = 0 \qquad p_3 = 4$$

The matrix

$$P^0 = \begin{bmatrix} 2 & 0 \\ 0 & 4 \end{bmatrix}$$

is obviously positive definite.

The optimal feedback matrix is $k^0 = [0.5, 0]$ and the poles of the closed-loop system are $\lambda_{1,2} = 0.5$.

The discrete Riccati equation (2.57) may be written in the equivalent form

$$A^T PA - P + Q - A^T PBR^{-1}B^T A^{-T}(P - Q) = 0 \tag{2.58}$$

provided the matrix A is nonsingular. The optimal feedback matrix is then found from

$$K^0 = R^{-1}B^T A^{-T}(P^0 - Q)$$

Equation (2.58) is related to the $2n \times 2n$ *symplectic matrix*

$$S = \begin{bmatrix} A + BR^{-1}B^T A^{-T}Q & -BR^{-1}B^T A^{-T} \\ -A^{-T}Q & A^{-T} \end{bmatrix}$$

This matrix has the property

$$W^{-1}SW = S^{-T} \tag{2.59}$$

where

$$W = \begin{bmatrix} 0 & I_n \\ -I_n & 0 \end{bmatrix}$$

and

$$S^{-1} = \begin{bmatrix} A^{-1} & A^{-1}BR^{-1}B^{\mathrm{T}} \\ QA^{-1} & A^{\mathrm{T}} + QA^{-1}BR^{-1}B^{\mathrm{T}} \end{bmatrix}$$

According to (2.59) the matrices S and $S^{-\mathrm{T}}$ are similar, i.e. if λ is an eigenvalue of S then $1/\lambda$ is also an eigenvalue of S with the same multiplicity (since S is nonsingular then all λ are nonzero).

Let the matrix T be determined as

$$T = \begin{bmatrix} I_n & 0 \\ P^0 & I_n \end{bmatrix}$$

Carrying the nonsingular transformation $T^{-1}ST$ we obtain

$$T^{-1}ST = \begin{bmatrix} A - BR^{-1}B^{\mathrm{T}}A^{-\mathrm{T}}(P^0 - Q) & -BR^{-1}B^{\mathrm{T}}A^{-\mathrm{T}} \\ 0 & A^{-\mathrm{T}} + P^0 BR^{-1}B^{\mathrm{T}}A^{-\mathrm{T}} \end{bmatrix}$$

and using the *matrix inversion lemma*

$$(A + BCD)^{-1} = A^{-1} - A^{-1}B(DA^{-1}B + C^{-1})^{-1}DA^{-1}$$

which holds for A and C being nonsingular, it may be shown that

$$(A^{-1} + A^{-1}BR^{-1}B^{\mathrm{T}}P^0)^{-1} = A - B(R + B^{\mathrm{T}}P^0 B)^{-1}B^{\mathrm{T}}P^0 A = A - BK^0$$

Hence

$$T^{-1}ST = \begin{bmatrix} A - BK^0 & -BR^{-1}B^{\mathrm{T}}A^{-\mathrm{T}} \\ 0 & (A - BK^0)^{-\mathrm{T}} \end{bmatrix} \tag{2.60}$$

Equation (2.60) shows that the eigenvalues of the symplectic matrix S are the eigenvalues of $A - BK^0$ (the poles of the optimal closed-loop discrete system) together with the eigenvalues of $(A - BK^0)^{-1}$.

EXERCISES

2.48 Show that the continuous Riccati equation may be represented as

$$[I_n, \ P]\, Y \begin{bmatrix} I_n \\ P \end{bmatrix} = 0$$

where

$$Y = Y^{\mathrm{T}} = \begin{bmatrix} Q & A^{\mathrm{T}} \\ A & -BR^{-1}B^{\mathrm{T}} \end{bmatrix}$$

2.49 Show that, using a nonsingular transformation $x(t) = T\tilde{x}(t)$, the linear quadratic optimization problem reduces to a problem with matrices

$$\tilde{A} = T^{-1}AT \qquad \tilde{B} = T^{-1}B$$
$$\tilde{Q} = T^T Q T \qquad \tilde{R} = R$$

2.50 Let P be the positive definite solution of the Riccati equation

$$A^T P + PA + C^T C - PBB^T P = 0$$

Prove that if $A = A^T$ and $B = C^T$ then P^{-1} is also a solution of this equation.

2.51 Prove that if the continuous Riccati equation has a positive definite solution and $B^T B = I$ then

$$P = A^T BRB^T + T_1 T_2$$

where

$$T_1 T_1^T = A^T BRB^T A + Q$$
$$T_2 = R_1 B^{-1} \qquad R_1 R_1^T = R$$

2.52 Prove that the solution of the continuous Riccati equation satisfies

$$\| P \| \geq \frac{\| Q \|}{\| A \| + (\| A \|^2 + \| BR^{-1}B^T \| \| Q \|)^{1/2}}$$

2.53 Prove that if

$$A = \begin{bmatrix} A_{11} & A_{12} \\ A_{22} & A_{22} \end{bmatrix}$$

and

$$T = \begin{bmatrix} I & 0 \\ X & I \end{bmatrix}$$

where X is a solution of $A_{22}X - XA_{11} + A_{21} - XA_{12}X = 0$, then

$$T^{-1}AT = \begin{bmatrix} A_{11} + A_{12}X & A_{12} \\ 0 & A_{22} - XA_{12} \end{bmatrix}$$

2.54 Prove that the discrete Riccati equation may be written in the equivalent form

$$Q - P + A^T P(I_n + BR^{-1}B^T P)^{-1}A = 0$$

Hint: use the matrix inversion lemma.

2.8 STATE OBSERVERS

The linear state feedback often cannot be realized directly since it requires that all elements of the state vector be available for measurement. Usually only the output $y(t)$ is known. It is important, therefore, to find an approximation of the state vector, $\hat{x}(t) \cong x(t)$, knowing only the output $y(t)$ and the input $u(t)$ of the system. This approximation may then be used in the state feedback instead of the state vector itself.

The lth order system

$$\dot{z}(t) = Fz(t) + Gy(t) + Hu(t) \qquad z(0) = z_0 \tag{2.61}$$

is said to be the *state observer* (or *Luenberger observer*) of the system

$$\begin{aligned}\dot{x}(t) &= Ax(t) + Bu(t) \qquad x(0) = x_0 \\ y(t) &= Cx(t)\end{aligned} \tag{2.62}$$

if for every initial state x_0 of (2.62) there exists an initial state z_0 of (2.61) such that

$$z(t) = Tx(t) \qquad t \geq 0 \tag{2.63}$$

for every $u(t)$, $t \geq 0$, and some constant $l \times n$ matrix T. The relation (2.63) allows us, by measuring $z(t)$, to obtain an approximation to the state vector $x(t)$.

The following result has a basic role in the design of state observers.

Theorem 2.16

The system (2.61) is a state observer of the system (2.62) if there exist matrices G and T which satisfy the linear matrix equation

$$TA - FT = GC \tag{2.64}$$

and the matrix H is determined from

$$H = TB \tag{2.65}$$

To fulfil the condition (2.63) it is necessary that the initial state of the observer be chosen as

$$z_0 = Tx_0 \tag{2.66}$$

The proof of the theorem follows easily by substitution of (2.63) into (2.61) bearing in mind (2.62).

Equation (2.64) is a particular case of the Sylvester equation (see Section 1.10). It may be shown that for a given matrix G, this equation has a unique solution for T if and only if the matrices A and F have no common eigenvalues.

If the condition (2.66) is not satisfied, then instead of (2.63) there exists the relation

$$z(t) = Tx(t) + e(t)$$

where the error vector $e(t)$ is a solution of the equation

$$\dot{e}(t) = Fe(t) \qquad e(0) = z_0 - Tx_0 \tag{2.67}$$

Consider first the so-called *full-order observer* (or *identical observer*). The system of nth order

$$\dot{\hat{x}}(t) = F\hat{x}(t) + Gy(t) + Hu(t) \qquad \hat{x}(0) = \hat{x}_0 \tag{2.68}$$

is an identical state observer of (2.62) if the condition $\hat{x}(0) = x_0$ leads to $\hat{x}(t) = x(t)$, $t \geqq 0$.

In accordance with (2.64), (2.65) and (2.66) it follows that (2.68) is an identical observer of (2.62) if and only if

$$F = A - KC \qquad G = K \qquad H = B$$

where K is a matrix which will be determined later.

The identical observer is represented in the form

$$\dot{\hat{x}}(t) = A\hat{x}(t) + Bu(t) + K[y(t) - C\hat{x}(t)]$$

This equation shows that the full-order observer is a model of the system (2.62) with an additional input proportional to the error $y(t) - \hat{y}(t)$, where $\hat{y}(t) = C\hat{x}(t)$ is the reconstructed output.

The full-order observer may also be represented in the form

$$\dot{\hat{x}}(t) = (A - KC)\hat{x}(t) + Bu(t) + Ky(t) \tag{2.69}$$

which shows that the stability of the observer is determined by the eigenvalues of $A - KC$.

The reconstruction error $e(t) = x(t) - \hat{x}(t)$ in the case of an identical observer satisfies the equation

$$\dot{e}(t) = (A - KC)e(t) \qquad e(0) = x_0 - \hat{x}_0 \tag{2.70}$$

The error will converge asymptotically to zero, i.e. $e(t) \to 0$ as $t \to \infty$ if and only if the observer is asymptotically stable. That is why the problem of the design of the full-order observer consists in the determination of a matrix K which yields the desired dynamical behaviour of (2.70).

The conditions for existence of this matrix are given by the following theorem.

Theorem 2.17

The eigenvalues of the matrix $A - KC$ may be allocated at arbitrary symmetric places in the complex plane by suitable choice of the matrix K if and only if the pair (C, A) is completely observable.

The proof of this theorem is easily done considering the pair (A^T, C^T) which is controllable provided the pair (C, A) is observable. According to Theorem 2.12 it is always possible to find a matrix K^T which assigns the eigenvalues of $A^T - C^T K^T$ at desired places in the complex plane.

The poles of the full-order observer (the eigenvalues of $A - KC$) have to be situated on the left of the poles of the observed system which will guarantee fast convergence of the error $e(t)$.

Consider the particular case of a single-input, single-output system

$$\dot{x}(t) = Ax(t) + bu(t)$$
$$y(t) = cx(t)$$

where the pair (c, A) is completely observable. Using a nonsingular transformation $x(t) = S\tilde{x}(t)$ this system may be reduced to observable canonical form

$$\dot{\tilde{x}}(t) = \tilde{A}\tilde{x}(t) + \tilde{b}u(t)$$
$$y(t) = \tilde{c}\tilde{x}(t)$$

where

$$\tilde{A} = S^{-1}AS = \begin{bmatrix} 0 & 0 & \cdots & 0 & -a_1 \\ 1 & 0 & \cdots & 0 & -a_2 \\ 0 & 1 & \cdots & 0 & -a_3 \\ \vdots & \vdots & & \vdots & \vdots \\ 0 & 0 & \cdots & 1 & -a_n \end{bmatrix} \quad \begin{array}{l} \tilde{c} = cS = [0, \; 0, \; \cdots, \; 0, \; 1] \\ \tilde{b} = S^{-1}b \end{array}$$

All (n) eigenvalues of $\tilde{A} - \tilde{k}\tilde{c}$, $\tilde{k} = kS$, may be located at arbitrary places by an appropriate choice of $\tilde{k} = [k_1, k_2, \ldots, k_n]$ since

$$\tilde{A} - \tilde{k}\tilde{c} = \begin{bmatrix} 0 & 0 & \cdots & 0 & -a_1 - k_1 \\ 1 & 0 & \cdots & 0 & -a_2 - k_2 \\ 0 & 1 & \cdots & 0 & -a_3 - k_3 \\ \vdots & \vdots & & \vdots & \vdots \\ 0 & 0 & \cdots & 1 & -a_n - k_n \end{bmatrix}$$

and

$$\det(\lambda I - \tilde{A} + \tilde{k}\tilde{c}) = \lambda^n + (a_n + k_n)\lambda^{n-1} + \cdots + (a_2 + k_2)\lambda + a_1 + k_1$$

The synthesis of a full-order observer for a multi-output system can be done in a similar way by reducing the pair (A^T, C^T) into Luenberger canonical form (see Section 2.5) and finding matrix K^T by pole assignment methods.

The implementation of a full-order observer involves construction of an additional nth order linear system whose state components approximate the dynamic behaviour of the corresponding components of $x(t)$. It should be taken into account, however, that r output variables $y(t)$ are available for measurement and for rank $C = r$ each component $y_i(t)$ represents a known linearly independent combination of the state variables. That is why it is possible to use an observer with reduced dimension $n - r$ in order to reconstruct only those linear combinations of the states which are not present in $Cx(t) = y(t)$.

The *reduced-order observer* has the form of (2.61) with $l = n - r$. If conditions (2.64) and (2.65) are fulfilled, it is possible to estimate the state vector by using the vector

$$\hat{x}(t) = \begin{bmatrix} T \\ C \end{bmatrix}^{-1} \begin{bmatrix} z(t) \\ y(t) \end{bmatrix}$$

provided the matrix $\begin{bmatrix} T \\ C \end{bmatrix}$ is nonsingular. It may be shown that this condition can always be satisfied by suitable choice of the matrices F and G if the system (2.62) is completely observable.

If (2.66) is fulfilled, then $\hat{x}(t) = x(t)$ and the state is reconstructed exactly. In the general case $(z_0 \neq Tx_0)$, we have

$$\hat{x}(t) = x(t) - \begin{bmatrix} T \\ C \end{bmatrix}^{-1} \begin{bmatrix} e(t) \\ 0 \end{bmatrix} \tag{2.71}$$

where the $(n - r)$-vector $e(t)$ is a solution of (2.67). The desired asymptotic behaviour of $e(t)$ can be achieved by an appropriate choice of eigenvalues of F (the observer poles).

We will sketch two of the available methods for synthesis of reduced-order observers.

In the first method, the matrices F and G are set and for asymptotic stability it is required that the eigenvalues of F have negative real parts. For given F and G the Sylvester equation (2.64) is solved for T using some of the methods considered later in Section 4.2. To ensure a unique solution of (2.64) it is necessary to choose F so that its eigenvalues are different from those of A. The matrix T must satisfy the condition

$$\text{rank} \begin{bmatrix} T \\ C \end{bmatrix} = n$$

If this condition is not fulfilled it is necessary to set other F and G and the procedure is repeated again.

Example 2.23

Consider a system of the form (2.62) with matrices

$$A = \begin{bmatrix} 1 & 0 & -2 \\ 0 & 2 & 0 \\ -3 & 1 & 0 \end{bmatrix} \quad B = \begin{bmatrix} 1 & 0 \\ 0 & -1 \\ 0 & 3 \end{bmatrix} \quad C = \begin{bmatrix} 0 & 1 & 2 \\ 0 & 0 & 1 \end{bmatrix}$$

The eigenvalues of A are

$$\lambda_1 = 2 \qquad \lambda_2 = 3 \qquad \lambda_3 = -2$$

Since rank $C = 2$, the reduced-order observer is of the first order. With given matrices

$$F = -5 \qquad \text{and} \qquad G = [9, -10]$$

we obtain

$$T = [1, 1, 2]$$

so that

$$H = [1, \; 5] \qquad \begin{bmatrix} T \\ C \end{bmatrix}^{-1} = \begin{bmatrix} 1 & -1 & 0 \\ 0 & 1 & -2 \\ 0 & 0 & 1 \end{bmatrix}$$

It follows from (2.71) that

$$x_1(t) = \hat{x}_1(t) + e(t)$$
$$x_2(t) = \hat{x}_2(t)$$
$$x_3(t) = \hat{x}_3(t)$$

where the error $e(t) = e^{-5t}e_0$ is a solution of

$$e(t) = -5e(t) \qquad e_0 = z_0 - x_1(0) - x_2(0) - 2x_3(0)$$

In the second method for synthesis of reduced-order observers we shall use the nonsingular transformation

$$\tilde{x}(t) = Sx(t) \qquad S = \begin{bmatrix} C \\ D \end{bmatrix}$$

where the matrix D is chosen from the condition

$$\text{rank} \begin{bmatrix} C \\ D \end{bmatrix} = n$$

Under this transformation the system (2.62) is reduced to

$$\dot{x}_1(t) = A_{11}x_1(t) + A_{12}x_2(t) + B_1u(t)$$
$$\dot{x}_2(t) = A_{21}x_1(t) + A_{22}x_2(t) + B_2u(t)$$
$$y(t) = x_1(t)$$

where

$$SAS^{-1} \equiv \begin{bmatrix} A_{11} & A_{12} \\ A_{21} & A_{22} \end{bmatrix} \qquad SB \equiv \begin{bmatrix} B_1 \\ B_2 \end{bmatrix}$$

$$CS^{-1} = [I_r, \; 0]$$

and x_1 is an r-vector, and x_2 is an $(n-r)$-vector.

In the given case it is necessary to obtain an estimation of the unmeasurable vector $x_2(t)$. The state observer is taken in the form (2.61) with

$$F = A_{22} - T_1 A_{12}$$
$$G = A_{22}T_1 - T_1 A_{11} + A_{21} - T_1 A_{12} T_1$$
$$H = B_2 - T_1 B_1$$

and the matrix T_1 is determined later.

If, for estimation of $x_2(t)$, we use

$$\hat{x}_2 x_2(t) = T_1 y(t) + z(t)$$

then the estimation error

$$e(t) = x_2(t) - \hat{x}_2(t)$$

will satisfy the equation

$$\dot{e}(t) = (A_{22} - T_1 A_{12}) e(t)$$

Since the pair (C, A) is observable the same is valid for the pair (A_{12}, A_{22}) (cf. Exercise 2.29). Hence it is always possible to find a matrix T_1 such that the observer poles (the eigenvalues of $A_{22} - T_1 A_{12}$) are located at desired places. Since

$$T = [-T_1, I_{n-r}]$$

the matrix $\begin{bmatrix} T \\ C \end{bmatrix}$ is always nonsingular and

$$\begin{bmatrix} \hat{x}_1(t) \\ \hat{x}_2(t) \end{bmatrix} = \begin{bmatrix} 0 & I_r \\ I_{n-r} & T_1 \end{bmatrix} \begin{bmatrix} z(t) \\ y(t) \end{bmatrix}$$

This method does not require the eigenvalues of A and F to be distinct.

Example 2.24

For the system from Example 2.23 let the matrix D be chosen as

$$D = [1 \quad 0 \quad 0]$$

Then

$$S = \begin{bmatrix} 0 & 1 & 2 \\ 0 & 0 & 1 \\ 1 & 0 & 0 \end{bmatrix} \qquad S^{-1} = \begin{bmatrix} 0 & 0 & 1 \\ 1 & -2 & 0 \\ 0 & 1 & 0 \end{bmatrix}$$

$$A_{11} = \begin{bmatrix} 4 & -8 \\ 1 & -2 \end{bmatrix} \qquad A_{12} = \begin{bmatrix} -6 \\ -3 \end{bmatrix}$$

$A_{21} = [0, \ -2] \qquad A_{22} = 1$

$B_1 = \begin{bmatrix} 0 & 5 \\ 0 & 3 \end{bmatrix} \qquad B_2 = [1 \quad 0]$

For $F = -5$ the elements of the matrix $T = [t_1, \quad t_2]$ must satisfy

$A_{22} - T_1 A_{12} = 1 + 6t_1 + 3t_2 = -5$

Choosing arbitrary $t_1 = 0$ we obtain $t_2 = -2$. Hence

$G = [2, \ 4] \qquad H = [1, \ 6] \qquad T = [0, \ 2, \ 1]$
$\hat{x}_2(t) = z(t) - 2y_2(t)$

The basic implementation of the Luenberger observer is for the approximation of linear state feedback using the estimation $\hat{x}(t)$ instead of the unavailable state vector $x(t)$.

Consider the linear state feedback

$$u(t) = -Kx(t) \tag{2.72}$$

where the matrix K is determined so as to ensure the desired dynamics of the closed-loop system

$\dot{x}(t) = (A - BK)x(t)$

Using the estimation $\hat{x}(t)$, obtained by a reduced-order observer in the form (2.61), the control law (2.72) is realized as

$$u(t) = -K\hat{x}(t) = -Bz(t) - Dy(t) \tag{2.73}$$

where

$$[E, \quad D] = \begin{bmatrix} T \\ C \end{bmatrix}^{-1}$$

The extended $(2n - r)$th order system which incorporates the observer is described by the equation

$$\begin{bmatrix} \dot{x}(t) \\ \dot{z}(t) \end{bmatrix} = \begin{bmatrix} A - BDC & -BE \\ GC - HDC & F - HE \end{bmatrix} \begin{bmatrix} x(t) \\ z(t) \end{bmatrix} \tag{2.74}$$

If the conditions (2.64) and (2.65) are satisfied then using the non-singular transformation

$$\begin{bmatrix} x(t) \\ e(t) \end{bmatrix} = \begin{bmatrix} I_n & 0 \\ -T & I_{n-r} \end{bmatrix} \begin{bmatrix} x(t) \\ z(t) \end{bmatrix}$$

the system (2.74) is reduced to

$$\begin{bmatrix} \dot{x}(t) \\ \dot{e}(t) \end{bmatrix} = \begin{bmatrix} A - BK & -BE \\ 0 & F \end{bmatrix} \begin{bmatrix} x(t) \\ e(t) \end{bmatrix} \tag{2.75}$$

Equation (2.75) shows that the poles of the extended system are the eigenvalues of $A - BK$ together with the eigenvalues of F. That is why the state observer does not affect the poles of the closed-loop system with state feedback, but only adjoins its poles.

In this way, for a completely controllable and completely observable system with r linearly independent outputs, it is possible to construct an additional $(n - r)$th order system such that all $(2n - r)$ poles of the extended system are located at the desired places.

In the case under consideration, the state of the system is estimated in order to realize a linear function Kx of the state, whose dimension m is usually less than n. It might be expected that there is an observer of dimension less than $n - r$ which produces an appropriate estimation of the desired function. In particular, it may be shown that a scalar linear function of the state may always be estimated by an observer of order $q - 1$, where q is the observability index of the system.

Equation (2.73) shows that when approximating linear state feedback by using an observer, the input vector $u(t)$ becomes a function of the output vector $y(t)$ and the observer state vector $z(t)$. That is why the additional dynamic system can be taken directly in the form of a *dynamic compensator*

$$\dot{z}(t) = \hat{F}z(t) + \hat{G}y(t)$$
$$u(t) = -Ez(t) - Dy(t)$$

where the matrices $\hat{F}$, $\hat{G}$, D and E are chosen so as to ensure the desired dynamics of the extended system. It may be shown that, in particular, the prescribed position of the system poles can be achieved by a compensator of order less than $n - r$. This, however, may be done at the price of losing the ability to obtain an estimation of the state vector.

The state observer of a discrete-time system

$$x_{k+1} = Ax_k + Bu_k$$
$$y_k = Cx_k$$

is taken in the form

$$z_{k+1} = Fz_k + Gy_k + Hu_k$$

where the matrices F, G and H again satisfy the equations (2.64) and (2.65). The eigenvalues of F must lie in the unit circle which guarantees convergence to zero of the estimation error. In contrast to the continuous-time case, the estimation error may be forced to zero in a finite number of steps choosing the eigenvalues of F as zeros. Observers with such properties are called *deadbeat observers*. Using the observable canonical form it may be shown that the full-order deadbeat observer of a single-output system will produce an exact estimate of the state vector in n steps. In the multi-output case it is possible to obtain an exact estimate in q steps, where q is the observability index of the system.

In some cases, it is desirable to ensure the prescribed dynamics of a closed-loop linear system without the implementation of an additional dynamic system. This is done by using an *output feedback*

$$u(t) = K_0 y(t)$$

where K_0 is an $m \times r$ matrix.

It should be stressed that the desired location of all closed-loop poles can be achieved by output feedback only for special matrices B and C. Specifically, it may be shown that for a completely controllable and completely observable linear system with rank$(B) = m$ and rank$(C) = r$, 'almost every' desired spectrum of $A + BK_0C$ is achievable provided $m + r \geq n + 1$. For some systems, however, it is not possible to allocate more than $\max\{m, r\}$ poles arbitrarily close to $\max\{m, r\}$ prescribed numbers, using only output feedback. That is why in certain cases it is impossible even to stabilize the closed-loop system by output feedback.

EXERCISES

2.55 Prove Theorem 2.16.

2.56 Is it possible to construct a stable observer if the system is not completely observable? Give an example.

2.57 Derive a procedure for the synthesis of a full-order observer using the Luenberger canonical form of the pair (A^T, C^T).

2.58 Find an analytical solution of Equation (2.64) for the case $r = 1$, reducing the pair (C, A) to observable canonical form.

2.59 Derive a procedure for the synthesis of reduced-order observers for case of rank$(C) < r$.

2.60 Find the value of the quadratic performance index

$$J = \int_0^\infty [x^T(t)Qx(t) + u^T(t)Ru(t)] \, dt$$

for the case when the optimal control is realized using a state observer.

2.61 Show that the time response of a discrete-time system incorporating a full-order observer can be driven to zero in $p + q$ steps, where p is the controllability index and q is the observability index of the system.

NOTES AND REFERENCES

A vast amount of literature exists on the analysis and design of linear control systems described in the state space. A good introduction to this subject may be found in Barnett and Cameron (1985), Chen (1984), Kailath (1980), Layton (1976), Ogata (1987), Owens (1981), Sinha (1984), Wolovich (1974) and Zadeh and Desoer (1963). An advanced treatment of linear system theory is presented in Brockett (1970), Desoer (1970), Kalman, Falb and Arbib (1969), Rosenbrock (1970) and Wonham (1986).

The theory and application of discrete-time systems is discussed in Åström and Wittenmark (1986), Cadzow (1973), Franklin and Powell (1980), Kuo (1980) and Leigh (1985).

A detailed exposition of stability theory and the second method of Lyapunov is given in Barnett and Storey (1970), Lehnigk (1966) and Willems (1970).

The controllability and observability properties and the canonical representations of linear systems are studied in Kalman, Falb and Arbib (1969), Luenberger (1967), Brunovsky (1970) and Popov (1972).

The pole assignment problem of linear systems is considered in detail in Wonham (1967), Fallside (1977), Patel and Munro (1982) and Porter and Crossley (1972).

The theory of linear optimal control is presented in depth in Anderson and Moore (1971) and Kwakernaak and Sivan (1972).

The design of state observers is considered in Luenberger (1966, 1971) and O'Reilly (1983). If the system is subject to noises whose statistical characteristics are known, it is possible to obtain optimal estimation of the state vector. This is done by using a special observer, which is called *Kalman filter* (see Anderson and Moore, 1979).

3

Solution of State Equations

This chapter is devoted to the computational solution of the state equations of linear systems. Emphasis is put on the computation of the matrix exponential in the continuous-time case. Since the elements of the state matrix are not known exactly, it is necessary to study first the sensitivity of the exponential to perturbations in the data. Then we consider several ways of finding the matrix exponential including series methods, ordinary differential equation methods and matrix decomposition methods. The error analysis for these methods makes it possible to estimate the error in the solution of state equations. A closely related problem to the computation of the matrix exponential is the determination of a discrete-time model for a continuous-time system.

3.1 SENSITIVITY OF THE MATRIX EXPONENTIAL

In practice, the elements of the state matrix A are known with some error which leads to an error in the computed exponential e^{At}. The methods attempting to compute e^{At} also give an approximation to the exact matrix exponential due to rounding errors. Hence from the backward error analysis point of view, instead of e^{At} we obtain the exact exponential of a perturbed matrix $A + E$. The size of the elements of E depends on the error in A and on the numerical properties of the method used in the computation of e^{At}. In this way there arises the problem of estimating the perturbation $H(t) = e^{(A+E)t} - e^{At}$ of e^{At} where only a bound $\|E\| \leq \gamma$ for the norm of the perturbation E is usually known. As a measure of $H(t)$ we take the norm

$$h(t) = \| H(t) \|_2 = \| e^{(A+E)t} - e^{At} \|_2$$

Sometimes the exponential problem is ill conditioned, i.e. a small perturbation in A may lead to a large change in $H(t)$. For such a problem it is not possible to obtain an exact answer even if a numerically stable method is implemented for computation of e^{At}.

Example 3.1

The exact matrix exponential of

$$A = \begin{bmatrix} 48 & -49 & 50 & 49 \\ 0 & -2 & 100 & 0 \\ 0 & -1 & -2 & 1 \\ -50 & -50 & 50 & -52 \end{bmatrix}$$

for $t = 1$ is (up to five decimal digits)

$$e^A = \begin{bmatrix} 10.285 & -10.015 & 10.150 & 10.015 \\ -112.78 & 106.15 & -99.246 & -106.01 \\ -3.3834 & 3.2480 & -3.2480 & -3.2480 \\ -119.55 & 112.78 & -106.01 & -112.64 \end{bmatrix}$$

If the $(1, 2)$-element of A is changed to -48.996 we obtain

$$e^{A+E} = \begin{bmatrix} 8.8665 & -8.7208 & 9.3086 & 8.7385 \\ -110.10 & 103.73 & -98.808 & -103.71 \\ -3.1962 & 3.0788 & -3.1723 & -3.0832 \\ -115.75 & 109.34 & -104.89 & -109.33 \end{bmatrix}$$

so that a relative perturbation

$$\| E \|_2 / \| A \|_2 = 3.263 \times 10^{-5}$$

leads to a relative change in the result equal to

$$\| e^{A+E} - e^A \|_2 / \| e^A \|_2 = 2.55 \times 10^{-2}$$

Thus the exponential problem for the given A is ill conditioned.

When the matrices A and E commute, i.e. $AE = EA$, it is easy to obtain a bound on the relative perturbation of the exponential since

$$e^{(A+E)t} - e^{At} = e^{At}(e^{Et} - I_n)$$

and thus

$$\| H(t) \|_2 / \| e^{At} \|_2 \leq e^{\gamma t} - 1 \leq \gamma t e^{\gamma t} \qquad \gamma = \| E \|_2$$

In the general case $AE \neq EA$ the sensitivity analysis of e^{At} becomes considerably more difficult. To obtain an expression for $H(t)$ we can exploit the property that the matrix exponential is a solution of the homogeneous matrix differential equation

$$\dot{X}(t) = AX(t) \qquad X(0) = I_n \tag{3.1}$$

The perturbed problem (3.1) is presented in the form

$$\dot{Y}(t) = (A + E)Y(t) \qquad Y(0) = I_n \tag{3.2}$$

Subtracting (3.1) from (3.2) we find that $H(t) = Y(t) - X(t)$ is the solution of the nonhomogeneous matrix differential equation

$$\dot{H}(t) = AH(t) + EY(t) \qquad H(0) = 0 \tag{3.3}$$

The solution of (3.3) is expressed explicitly as

$$H(t) = \int_0^t e^{A(t-s)} E e^{(A+E)s}\, ds$$

In this way we have

$$h(t) \leqq \gamma \int_0^t \| e^{A(t-s)} \|_2 \, \| e^{(A+E)s} \|_2 \, ds$$

$$\leqq \gamma \int_0^t f(t-s)[h(s) + f(s)]\, ds \tag{3.4}$$

where $f(t)$ is an upper bound for $\| e^{At} \|_2$, i.e.

$$\| e^{At} \|_2 \leqq f(t) \tag{3.5}$$

Hence two basic problems arise in obtaining the estimate (3.4):

1. Find a bound (3.5) for the norm of the matrix exponential.
2. Find the maximal solution of integral inequality (3.4).

3.1.1 Bounds for the Matrix Exponential

An easy way to find a bound on $\| e^{At} \|$ is to use the power series for e^{At}. Thus

$$\| e^{At} \|_2 \leqq f_1(t) = \sum_{k=0}^{\infty} \| At \|_2^k / k! = e^{\| A \|_2 t} \tag{3.6}$$

Another way to bound $\| e^{At} \|_2$ is based on the concept of a *logarithmic norm* $\mu(A)$ which is defined as the maximum eigenvalue of the matrix $(A + A^H)/2$ for a complex matrix A (in the real case $A^H = A^T$). Some of the properties of the logarithmic norm are

$$|\mu(A)| \leqq \| A \|_2$$
$$\mu(A + B) \leqq \mu(A) + \mu(B) \tag{3.7}$$
$$\alpha(A) \leqq \mu(A) \qquad \text{with } \alpha(A) = \mu(A) \text{ if and only if } A^H A = AA^H$$

where $\alpha(A)$ is the *spectral abscissa* of A, i.e. the maximum real part of the eigenvalues of A. It may be proved that

$$\| e^{At} \|_2 \leqq f_2(t) = e^{\mu(A)t} \tag{3.8}$$

where the bound is attained if and only if A is *normal*, i.e. $A^H A = AA^H$.

A bound on $\| e^{At} \|$ may be obtained also by using the Jordan form of A. Let $J = S^{-1}AS$ be the Jordan canonical form of A and $m \geq 1$ be the dimension of the maximum block in J. We suppose that the matrix S is chosen so that the condition number

$$\text{cond}(S) = \| S \|_2 \| S^{-1} \|_2$$

is minimized.

Since $e^{At} = Se^{Jt}S^{-1}$ we obtain

$$\| e^{At} \|_2 \leq \| S \|_2 \| S^{-1} \|_2 \| e^{Jt} \|_2 = \text{cond}(S)\| e^{Jt} \|_2 \tag{3.9}$$

Now it is possible to prove that

$$\| e^{Jt} \|_2 \leq \| e^{J_m(\alpha)t} \|_2 \tag{3.10}$$

where $J_m(\alpha) = \alpha(A)I_m + N_m$

$$N_m = \begin{bmatrix} 0 & 1 & 0 & \dots & 0 \\ 0 & 0 & 1 & \dots & 0 \\ . & . & . & \dots & . \\ . & . & . & \dots & 1 \\ 0 & 0 & 0 & \dots & 0 \end{bmatrix}$$

Hence

$$\| e^{Jt} \|_2 \leq e^{\alpha(A)t} \| e^{N_m t} \|_2 \tag{3.11}$$

It is shown in Kågström (1977) that

$$\| e^{N_m t} \|_2 \leq e^{d_m t} \qquad d_m = \cos \frac{\pi}{m+1} \tag{3.12}$$

Another estimate is obtained from

$$\| e^{N_m t} \|_2 = \left\| \sum_{k=0}^{m-1} (N_m t)^k / k! \right\|_2$$

$$\leq \sum_{k=0}^{m-1} \| N_m \|_2^k t^k / k! = \sum_{k=0}^{m-1} t^k / k! \tag{3.13}$$

Combining (3.9)–(3.13) we get

$$\| e^{At} \|_2 \leq f_3(t) = \text{cond}(S)\, e^{[\alpha(A) + d_m]t} \tag{3.14}$$

and

$$\| e^{At} \|_2 \leq f_4(t) = \text{cond}(S)\, e^{\alpha(A)t} \sum_{k=0}^{m-1} t^k / k! \tag{3.15}$$

Sometimes it is convenient to use a bound on $\|e^{At}\|$ based on the Schur decomposition of A

$$T = U^H A U = D + N$$

where U is unitary, D is diagonal and N is a strictly upper triangular matrix. It is assumed that the matrix U is chosen so that the norm of the matrix N, $\nu = \|N\|_2$, is minimized.

Denoting by $l \geq 1$ the *index of nilpotency* of N

$$l = \min\{s: N^s = 0\}$$

the following estimate for $\|e^{At}\|$ is obtained

$$\|e^{At}\|_2 = \|U e^{(D+N)t} U^H\|_2 = \|e^{(D+N)t}\|_2 \leq e^{\alpha(A)t} \|e^{Nt}\|_2$$

Since

$$\|e^{Nt}\|_2 \leq \sum_{k=0}^{l-1} \|N\|_2^k t^k / k! = \sum_{k=0}^{l-1} (\nu t)^k / k!$$

we have

$$\|e^{At}\|_2 \leq f_5(t) = e^{\alpha(A)t} \sum_{k=0}^{l-1} (\nu t)^k / k! \tag{3.16}$$

The estimates (3.6), (3.8), (3.14), (3.15) and (3.16) may be written in the form

$$\|e^{At}\|_2 \leq f(t)$$

where

$$f(t) = c_0 e^{\beta t} \sum_{k=0}^{p-1} (\nu_0 t)^k / k! = c_0 e^{\beta t} f_0(t) \tag{3.17}$$

Here $f(t)$ is one of $f_1(t), \ldots, f_5(t)$ and the values of the constants c_0, β, ν_0 and p are given in Table 3.1.

Table 3.1

	Power series (3.6)	Log norm (3.8)	Jordan (3.14)	Jordan (3.15)	Schur (3.16)
c_0	1	1	cond(S)	cond(S)	1
β	$\|A\|_2$	$\mu(A)$	$\alpha(A) + d_m$	$\alpha(A)$	$\alpha(A)$
ν_0	0	0	0	1	ν
p	—	—	—	m	l

Example 3.2

(Van Loan, 1977). If

$$A = \begin{bmatrix} -1 + \delta & 4 \\ 0 & -1 - \delta \end{bmatrix} \qquad \delta = 10^{-6}$$

then (3.6), (3.8), (3.14), (3.15) and (3.16) yield respectively

$$\| e^{At} \|_2 \leq f_1(t) = e^{4.236t}$$
$$\| e^{At} \|_2 \leq f_2(t) = e^{(-1 + \sqrt{(4 + \delta^2)})t}$$
$$\| e^{At} \|_2 \leq f_3(t) = f_4(t) = 4 \times 10^6 \times e^{(-1 + \delta)t}$$
$$\| e^{At} \|_2 \leq f_5(t) = (1 + 4t)e^{(-1 + \delta)t}$$

The comparison of these bounds for different values of t is shown in Table 3.2.

It is evident that the bound based on the power series is not relevant to this case since it increases with t while the actual norm decreases. Since the behaviour of e^{At} for $t \to \infty$ is determined by the sign of $\alpha(A)$, the log norm bound may or may not be informative depending on the sign of $\mu(A)$. It is seen from the table that the bounds based on Jordan and Schur forms decay precisely when e^{At} decays. The use of the Schur form gives better results when the eigensystem of A is ill conditioned, i.e. when cond(S) is large. It is possible, however, that in some cases the bounds based on Jordan form are better than those based on Schur form. In general, the implementation of a corresponding bound depends on the particular matrix.

A disadvantage of the estimates exploiting Jordan and Schur forms is that they require knowledge of the transformations which minimize cond(S), and $\| N \|_2$ respectively. In practice, to obtain Jordan and Schur forms, we use methods which do not necessarily produce

Table 3.2

t	$\| e^{At} \|_2$	Power series (3.6)	Log norm (3.8)	Jordan (3.14), (3.15)	Schur (3.16)
0	1.0×10^0	1.0×10^0	1.0×10^0	4.0×10^6	1.0×10^0
5	1.4×10^{-1}	1.6×10^9	1.5×10^2	2.7×10^4	1.4×10^{-1}
10	1.8×10^{-3}	2.5×10^{18}	2.2×10^4	1.8×10^2	1.9×10^{-3}
15	1.8×10^{-5}	3.9×10^{27}	3.3×10^6	1.2×10^0	1.9×10^{-5}
20	1.6×10^{-7}	6.2×10^{36}	4.9×10^8	8.2×10^{-3}	1.7×10^{-7}
25	1.4×10^{-9}	9.8×10^{45}	7.2×10^{10}	5.6×10^{-5}	1.4×10^{-9}
30	1.1×10^{-11}	1.6×10^{55}	1.1×10^{13}	3.7×10^{-7}	1.1×10^{-11}

transformations of the desired property. That is why the estimates obtained are usually pessimistic.

3.1.2 Perturbation Bounds for the Matrix Exponential

Substituting $f(t)$ from (3.6) into (3.4) for the case of the power series bound we get

$$h(t) \leq \gamma \int_0^t e^{\|A\|_2(t-s)} e^{\|A+E\|_2 s} \, ds \tag{3.18}$$

The integration of the right-hand side of (3.18) gives

$$h(t) \leq \gamma (e^{\|A+E\|_2 t} - e^{\|A\|_2 t}) / (\|A+E\|_2 - \|A\|_2)$$

Using the inequality $\|A+E\|_2 \leq \|A\|_2 + \|E\|_2$ we find

$$h(t) \leq e^{\|A\|_2 t} (e^{\gamma t} - 1) = h_1(t) \tag{3.19}$$

Exploiting the properties of the log norm it may be shown in the same way that

$$h(t) \leq e^{\mu(A)t} (e^{\gamma t} - 1) = h_2(t) \tag{3.20}$$

Similarly, the implementation of the estimate (3.14) based on the Jordan form yields

$$h(t) \leq c_0 e^{[\alpha(A) + d_m]t} (e^{\gamma c_0 t} - 1) = h_3(t) \tag{3.21}$$

The estimates $f_4(t)$ and $f_5(t)$, based on Jordan and Schur forms respectively, may be employed in a unified manner producing the following result. Let $\rho_p(\omega)$, $\omega > 0$, be the unique positive root of the equation

$$\rho^p = \omega \sum_{k=0}^{p-1} \rho^k$$

Then

$$h(t) \leq c_0 e^{\alpha(A)t} \left[b_m(\gamma c_0) e^{\rho_m(\gamma c_0)t} - \sum_{k=0}^{m-1} t^k / k! \right] = h_4(t) \tag{3.22}$$

in the case of (3.15) and

$$h(t) \leq e^{\alpha(A)t} \left[b_l(\gamma/\nu) e^{\nu \rho_l(\gamma/\nu)t} - \sum_{k=0}^{l-1} (\nu t)^k / k! \right] = h_s(t) \tag{3.23}$$

in the case of (3.16), where

$$b_p(\omega) = \left[\frac{1+\omega}{\rho_p(\omega)} \right]^{p-1}$$

In (3.22) and (3.23) $\rho_p(\omega)$ may be replaced by $\bar{\rho}_p(\omega)$

$$\rho_p(\omega) < \bar{\rho}_p(\omega) = 1 + \omega - \omega(1+\omega)^{-p}$$

Example 3.3

Consider the change in e^{At}

$$A = \begin{bmatrix} -3.5 & 0.5 & 0.5 & -0.5 \\ 0 & -4.0 & 1.0 & 0 \\ -1.0 & 0 & -5.0 & 0 \\ 0.5 & -0.5 & 1.5 & -3.5 \end{bmatrix}$$

due to a perturbation

$$E = \begin{bmatrix} -2.5 \times 10^{-4} & 5.0 \times 10^{-5} & -3.5 \times 10^{-4} & -2.5 \times 10^{-4} \\ 4.0 \times 10^{-4} & 0 & 2.0 \times 10^{-4} & 2.0 \times 10^{-4} \\ 1.5 \times 10^{-4} & -5.0 \times 10^{-5} & 1.5 \times 10^{-4} & 1.5 \times 10^{-4} \\ 1.0 \times 10^{-4} & 0 & 2.0 \times 10^{-4} & 1.0 \times 10^{-4} \end{bmatrix}$$

In the given case $\|A\|_2 = 5.71$, $\mu(A) = -3.15$ and $\gamma = \|E\|_2 = 7.65 \times 10^{-4}$. The matrix A is similar to the Jordan block

$$J = \begin{bmatrix} -4 & 1 & 0 & 0 \\ 0 & -4 & 1 & 0 \\ 0 & 0 & -4 & 1 \\ 0 & 0 & 0 & -4 \end{bmatrix}$$

so that

$$c_0 = 5.55 \qquad \alpha(A) = -4 \qquad m = 4 \qquad d_m = 0.81$$

From the Schur form of A we obtain

$$l = 4 \qquad v = \|N\|_2 = 2.36$$

In this way the estimates (3.19)–(3.23) become

$$h_1(t) = e^{5.71t}(e^{0.000\,765\,t} - 1)$$
$$h_2(t) = e^{-3.15t}(e^{0.000\,765\,t} - 1)$$
$$h_3(t) = 5.55e^{-3.19t}(e^{0.004\,25\,t} - 1)$$
$$h_4(t) = 5.55e^{-4t}[48.0e^{0.276t} - (1 + t + t^2/2 + t^3/6)]$$
$$h_5(t) = e^{-4t}\{371.0e^{0.329t} - [1 + 2.36t + (2.36t)^2/2 + (2.36t)^3/6]\}$$

These bounds are compared for different values of t in Table 3.3.

The comments made after Example 3.2 hold for this example also. The estimate based on power series bound is misleading since $e^{\|A\|_2 t}$ increases with t while $\|e^{At}\|_2 \to 0$ as $t \to \infty$. The estimate implementing log norm yields the best result for the above values of t because of the negative sign of $\mu(A)$. For this example, the estimates utilizing the Jordan form give better results than the estimates based on

Table 3.3

t	$\|e^{(A+E)t} - e^{At}\|_2$	Power series (3.19)	Log norm (3.20)	Jordan (3.21)	Jordan (3.22)	Schur (3.23)
0	0	0	0	0	2.6×10^2	3.7×10^2
1	1.6×10^{-5}	0.2×10^0	3.3×10^{-5}	9.7×10^{-4}	6.2×10^0	9.3×10^0
2	6.7×10^{-7}	1.4×10^2	2.8×10^{-6}	8.0×10^{-5}	1.4×10^{-1}	2.3×10^{-1}
3	2.7×10^{-8}	6.4×10^4	1.8×10^{-7}	5.0×10^{-6}	3.3×10^{-3}	5.5×10^{-3}
4	1.2×10^{-9}	2.6×10^7	1.0×10^{-8}	2.7×10^{-7}	7.6×10^{-5}	1.3×10^{-4}
5	4.8×10^{-11}	9.7×10^9	5.6×10^{-10}	1.4×10^{-8}	1.7×10^{-6}	3.2×10^{-6}
6	1.8×10^{-12}	3.5×10^{12}	2.9×10^{-11}	6.9×10^{-10}	4.0×10^{-8}	7.9×10^{-8}
7	6.0×10^{-14}	1.2×10^{15}	1.5×10^{-12}	3.3×10^{-11}	9.3×10^{-10}	1.9×10^{-9}
8	1.9×10^{-15}	4.3×10^{17}	7.2×10^{-14}	1.6×10^{-12}	2.2×10^{-11}	4.9×10^{-11}
9	5.8×10^{-17}	1.5×10^{20}	3.5×10^{-15}	7.3×10^{-14}	5.2×10^{-13}	1.2×10^{-12}
10	1.7×10^{-18}	5.0×10^{22}	1.7×10^{-16}	3.3×10^{-15}	1.3×10^{-14}	3.2×10^{-14}

Schur form as a result of the low condition number of the transformation into Jordan form. This comparison of log norm, Jordan and Schur estimates is, however, valid for moderate values of t.

For t asymptotically large the Jordan estimate (3.22) is best with behaviour of order $e^{-3.724t}$. The Schur estimate (3.23) is closely behind ($\approx e^{-3.671t}$), while the Jordan estimate (3.21) ($\approx e^{-3.186t}$) is slightly better than the log norm estimate (3.20) ($\approx e^{-3.149t}$).

EXERCISES

3.1 Verify (3.7).
3.2 Verify (3.8).
3.3 Verify (3.12).
3.4 Compare the bounds for the matrix exponential of a normal matrix.
3.5 Find the perturbation bounds for the matrix exponential of a diagonalizable matrix.
3.6 Verify (3.22) and (3.23).
3.7 Compute the perturbation bounds for e^A, where A is the matrix from Example 3.1. Comment on the results.

3.2 SERIES METHODS

The series methods for computing the matrix exponential e^A are derived by the direct application to matrices of standard approximation techniques for the scalar function e^a. The commonly used series methods involve truncated Taylor series or Padé approximations.

3.2.1 Taylor Series

By definition, the matrix exponential of A is determined as

$$e^A = I_n + A/1! + A^2/2! + \ldots = \sum_{k=0}^{\infty} A^k/k! \qquad (3.24)$$

The terms of the series (3.24) may be computed recursively, the kth term being A/k times the $(k-1)$th term. The summation of the series in floating-point arithmetic is done up to the moment when the addition of a new term does not alter the sum. That is, if

$$S_l = \sum_{k=0}^{l} A^k/k!$$

is the matrix computed in floating-point arithmetic, then there exists a number L such that $\mathrm{fl}(S_L) = \mathrm{fl}(S_{L+1})$. Hence $\mathrm{fl}(S_L)$ is the approximation of e^A we search for. The following algorithm computes $\mathrm{fl}(S_L)$ and stores it in the array S.

Set $S = I_n$, $R = I_n$
For $k = 1, 2, \ldots$
   $R \leftarrow AR/k$
   If $\mathrm{fl}(S + R) = S$, quit
   $S \leftarrow S + R$

The algorithm requires $(L + 1)n^3$ flops, the necessary array storage being proportional to $3n^2$.

This algorithm may produce unsatisfactory results even in the scalar case because of rounding errors.

Example 3.4

Let the exponential of

$$A = \begin{bmatrix} -43 & 42 \\ -28 & 27 \end{bmatrix}$$

be computed in floating-point arithmetic with $\varepsilon = 2^{-15} \approx 3 \times 10^{-5}$. After $L = 44$ terms, the algorithm yields (up to six decimal digits)

$$e^A \approx \begin{bmatrix} 16.1782 & -53.3721 \\ -20.2168 & -79.2988 \end{bmatrix}$$

The correct exponential is

$$e^A = \begin{bmatrix} -0.735\ 758 & 1.103\ 64 \\ -0.735\ 758 & 1.103\ 64 \end{bmatrix}$$

so that the computed approximation has wrong signs in three components, i.e. the relative error in the result is at least of order 1.

Considering the intermediate results, it may be observed that the two matrices

$$A^{15}/15! = \begin{bmatrix} -0.100\ 355 \times 10^7 & 0.100\ 416 \times 10^7 \\ -0.669\ 008 \times 10^6 & 0.669\ 440 \times 10^6 \end{bmatrix}$$

$$A^{16}/16! = \begin{bmatrix} 0.940\ 864 \times 10^6 & -0.941\ 344 \times 10^6 \\ 0.627\ 232 \times 10^6 & -0.627\ 600 \times 10^6 \end{bmatrix}$$

have elements of nearly equal magnitudes but of opposite signs. Since we use arithmetic with relative precision of only 10^{-5}, the elements of these intermediate results have absolute errors of order 10, i.e. larger than the final result. Thus the large error in the final result is due to the catastrophic cancellation in floating-point arithmetic (see Section 1.4). Clearly the truncation of the series is not important in this case.

If we use arithmetic with $\varepsilon = 1.4 \times 10^{-17}$, after $L = 44$ terms we obtain

$$e^A \approx \begin{bmatrix} -0.734\ 172 & 1.102\ 05 \\ -0.734\ 701 & 1.102\ 58 \end{bmatrix}$$

so that there are at least two accurate digits in the result.

The truncation and rounding errors associated with the Taylor series approach increase with the norm of the matrix A. That is why it is desirable to decrease the matrix norm to some prescribed level. For this purpose we may use the following techniques (Ward, 1977).

The first technique attempts to minimize the matrix norm over all possible translations, i.e. we search for a scalar s such that $\|A - sI_n\|$ is minimized. Since it is difficult to compute the translation which produces the exact minimum, we can use an approximation to this translation determined by computing the mean of the eigenvalues of A. The sum of the eigenvalues is equal to the trace of A,

$$\text{tr}(A) = \sum_{i=1}^{n} a_{ii}$$

and consequently the mean is given by $\text{tr}(A)/n$. In this way, we obtain

$$A' = A - (\text{tr}(A)/n)I_n$$

and

$$e^A = e^{\text{tr}(A)/n} e^{A'}$$

The second technique attempts to minimize the 1-norm of A' over all possible diagonal similarity transformations. This may be done by balancing the matrix A' as described in Section 1.10, finding a permutation matrix P and a diagonal matrix D_0 so that the 1-norm of

$$A'' = D_0^{-1} P^T A' P D_0$$

is minimal. To avoid introducing rounding errors, the elements of D_0 are chosen as integer powers of the machine arithmetic base.

The third technique assures that the matrix norm is bounded by unity. It employs the fact that if A is divided by a nonzero scalar q then $e^{A/q}$ should be raised to the qth power to obtain e^A, i.e.

$$e^A = (e^{A/q})^q$$

For this purpose we determine an integer $m > 0$ such that

$$2^{m-1} \leq \| A'' \|_1 \leq 2^m$$

The matrix A'' is then scaled by the factor 2^{-m} resulting in a matrix with the desired property. Then we have

$$A''' = 2^{-m} A''$$

and

$$e^A = e^{\operatorname{tr}(A)/n} P D_0 (e^{A'''})^{2^m} D_0^{-1} P^{\mathrm{T}}$$

The computation of the matrix $(e^{A'''})^{2^m}$ can now be carried out by squaring $e^{A'''}$ m times. This technique is referred to as *scaling and squaring*.

In this way we obtain the following algorithm, which for a given $n \times n$ matrix A computes the exponential e^A and stores it in the array S.

Algorithm 3.1 Computation of the matrix exponential by Taylor series expansion using scaling and squaring

Set $s = \operatorname{tr}(A)/n$
$A \leftarrow A - sI_n$
Determine a diagonal matrix D_0 and a permutation matrix P such that
    $D_0^{-1} P^{\mathrm{T}} A P D_0$ is balanced.
Determine $m \geq 0$ such that $2^{-m} \| -A \|_1 \leq 1$
Set $S = I_n$, $R = I_n$
$A \leftarrow A/2^m$
For $k = 1, 2, \ldots$
    $R \leftarrow AR/k$
    If $\mathrm{fl}(S + R) = S$, terminate the loop
    $S \leftarrow S + R$
If $m > 0$
    For $k = 1, 2, \ldots, m$
    $S \leftarrow S^2$
$S \leftarrow e^s P D_0 S D_0^{-1} P^{\mathrm{T}}$

This algorithm requires approximately $(L + m + 1)n^3$ flops. The necessary array storage is proportional to $3n^2$.

Consider the numerical properties of the algorithm. We shall analyze first the errors arising in the computation of e^B

$$B = 2^{-m} D_0^{-1} P^{\mathrm{T}} (A - sI_n) P D_0$$

The Taylor series expansion of e^B is represented as

$$e^B = S_L + T_{L+1}$$

where

$$S_L = \sum_{k=0}^{L} B^k/k!$$

and

$$T_{L+1} = \sum_{k=L+1}^{\infty} B^k/k!$$

is the truncation error matrix.

Using the error analysis of the basic matrix operations from Section 1.4 and utilizing the inequality $\|B\|_1 \leq 1$ it may be shown that the computed finite series $\overline{S_L}$ satisfies

$$\overline{S_L} = S_L + F_L$$

where

$$\|F_L\|_1 < e(L + n + 1)\varepsilon \approx 2.72(L + n + 1)\varepsilon \tag{3.25}$$

In this way we have that

$$\overline{e^B} = \overline{S_L} = e^B + F_L - T_{L+1}$$

The analysis of the truncation error yields

$$\|T_{L+1}\|_1 < e(e - 1)\varepsilon \cong 4.68\varepsilon \tag{3.26}$$

Thus

$$\overline{e^B} = e^B + F$$

where

$$\|F\|_1 < 2.72(L + n + 2.72)\varepsilon$$

Since

$$\|e^B\|_1 \geq e^{-\|B\|_1} \geq \sum_{k=0}^{\infty} (-1)^k/k! = 1/e > 0.36$$

it follows that the relative error in $\overline{e^B}$ is bounded as

$$\|\overline{e^B} - e^B\|_1/\|e^B\|_1 \leq 7.56(L + n + 2.72)\varepsilon \tag{3.27}$$

Now the problem is to compute the exponential $(\overline{e^B})^{2^m}$ by repeated squaring. Unfortunately, the rounding errors in $(\overline{e^B})^{2^m}$ are usually small relative to $\|\overline{e^B}\|_1^{2^m}$ rather than to $\|(\overline{e^B})^{2^m}\|_1$. Consequently, it is possible that in some cases the relative error in the final result will be large. That is why the best we can do is to monitor the error during the squaring process. This is done as follows.

For $1 \leq k \leq m$ we obtain

$$\mathrm{fl}\,[(\overline{\mathrm{e}^B})^{2^k}] = \{\mathrm{fl}\,[(\overline{\mathrm{e}^B})^{2^{k-1}}]\}^2 + H_k$$

where

$$\| H_k \|_1 \leq n\varepsilon \,\| \,\mathrm{fl}\,[(\overline{\mathrm{e}^B})^{2^{k-1}}]\,\|_1^2$$

If G_k is defined by

$$G_k = \mathrm{fl}\,[(\overline{\mathrm{e}^B})^{2^k}] - (\mathrm{e}^B)^{2^k}$$

the above expression yields the recurrent relation

$$\| G_k \|_1 \leq 2 \,\| (\mathrm{e}^B)^{2^{k-1}} \|_1 \| G_{k-1} \|_1 + \| G_{k-1} \|_1^2 + n\varepsilon \,\| \,\mathrm{fl}\,[(\overline{\mathrm{e}^B})^{2^{k-1}}]\,\|_1^2$$

$$(3.28)$$

where

$$\| G_0 \|_1 = \| \,\overline{\mathrm{e}^B} - \mathrm{e}^B \,\|_1$$

The value of $\| (\mathrm{e}^B)^{2^k} \|_1$, $0 \leq k < m-1$, can be approximated by the value of $\| \,\mathrm{fl}\,[(\overline{\mathrm{e}^B})^{2^k}]\,\|_1$ which it is possible to compute during the squaring. Thus a bound on G_m, the absolute error in computing $(\mathrm{e}^B)^{2^m}$, is approximated by the recurrence relation (3.28) with $\| (\mathrm{e}^B)^{2^{k-1}} \|_1$ replaced by $\| \,\mathrm{fl}\,[(\overline{\mathrm{e}^B})^{2^{k-1}}]\,\|_1$ for $1 \leq k \leq m$ and

$$\| G_0 \|_1 < 7.56(L + n + 2.72)\varepsilon$$

The final stage of the algorithm is to transform backwards the matrix $\mathrm{fl}\,[(\overline{\mathrm{e}^B})^{2^m}]$. The only rounding errors at this stage arise in the multiplication of the matrix

$$PD_0\mathrm{fl}\,[(\overline{\mathrm{e}^B})^{2^m}]D_0^{-1}P^{\mathrm{T}}$$

by the scalar e^s. These errors are negligible, but the effect of this stage on the absolute error G_m must be taken into account. It follows from the definition of G_m that the computed exponential fulfils

$$\overline{\mathrm{e}^A} = \mathrm{e}^s PD_0\,[(\mathrm{e}^B)^{2^m} + G_m]D_0^{-1}P^{\mathrm{T}}$$

Thus the bound on the relative error

$$\| \,\overline{\mathrm{e}^A} - \mathrm{e}^A \,\|/\| \,\mathrm{e}^A \,\|$$

in the final result may be approximated as

$$\mathrm{e}^s \| D_0 \|_1 \| G_m \|_1 \| D_0^{-1} \|_1 / \| \,\overline{\mathrm{e}^A} \,\|_1$$

Example 3.5

Let

$$A = \begin{bmatrix} 4 & 2 & 0 \\ 1 & 4 & 1 \\ 1 & 1 & 4 \end{bmatrix}$$

The mean of the eigenvalues is

$$s = \text{tr}(A)/n = 4$$

and

$$A' = A - sI_3 = \begin{bmatrix} 0 & 2 & 0 \\ 1 & 0 & 1 \\ 1 & 1 & 0 \end{bmatrix}$$

Since the matrix A' is balanced, $P = D_0 = I_3$ and $A'' = A'$. From

$$\| 2^{-m} A'' \|_1 \leq 1$$

we find that $m = 2$ and

$$B = 2^{-m} A'' = \begin{bmatrix} 0 & 0.5 & 0 \\ 0.25 & 0 & 0.25 \\ 0.25 & 0.25 & 0 \end{bmatrix}$$

Using arithmetic with $\varepsilon = 2^{-56} \approx 1.39 \times 10^{-17}$ we obtain $L = 15$ and (up to 17 decimal digits)

$$\overline{e^B} = \begin{bmatrix} 0.106\,877\,427\,894\,764\,60 \times 10^1 & 0.516\,431\,458\,346\,888\,96 \times 10^0 \\ 0.289\,973\,495\,876\,241\,09 \times 10^0 & 0.110\,053\,204\,565\,044\,26 \times 10^1 \\ 0.289\,973\,495\,876\,241\,09 \times 10^0 & 0.321\,731\,262\,579\,037\,71 \times 10^0 \end{bmatrix}$$

$$\begin{bmatrix} 0.635\,155\,334\,055\,932\,51 \times 10^{-1} \\ 0.258\,215\,729\,173\,444\,48 \times 10^0 \\ 0.103\,701\,651\,224\,484\,93 \times 10^1 \end{bmatrix}$$

Utilizing the bound (3.27) we get

$$\| \overline{e^B} - e^B \|_1 / \| e^B \|_1 < 2.17 \times 10^{-15}$$

the actual relative error being of order ε.

The double squaring of $\overline{e^B}$ and the subsequent multiplication by

e^4 yield

$$\overline{e^A} = \begin{bmatrix} 147.866\ 622\ 446\ 370\ 15 & 183.765\ 138\ 646\ 368\ 43 \\ 127.781\ 085\ 523\ 182\ 48 & 183.765\ 138\ 646\ 368\ 42 \\ 127.781\ 085\ 523\ 182\ 48 & 163.679\ 601\ 723\ 180\ 76 \end{bmatrix}$$

$$\begin{bmatrix} 71.797\ 032\ 399\ 996\ 549 \\ 91.882\ 569\ 323\ 184\ 216 \\ 111.968\ 106\ 246\ 371\ 88 \end{bmatrix}$$

Since $\| G_0 \| = 0.78 \times 10^{-15}$ the monitoring of the error gives

$$\| G_1 \|_1 \leq 2 \| \overline{e^B} \|_1 \| G_0 \|_1 + \| G_0 \|_1^2 + n\varepsilon \| \overline{e^B} \|_1^2 = 0.32 \times 10^{-14}$$

$$\| G_2 \|_1 \leq 2 \| (\overline{e^B})^2 \|_1 \| G_1 \|_1 + \| G_1 \|_1^2 + n\varepsilon \| (\overline{e^B})^2 \|_1^2 = 0.22 \times 10^{-13}$$

Hence

$$e^s \| G_2 \|_1 / \| \overline{e^A} \|_1 = 0.23 \times 10^{-14}$$

which implies that at least 15 decimal digits are accurate in e^A. The actual number of the accurate digits is approximately 16.

It is worthwhile making some comments on the scaling and squaring approach. Difficulties may arise during the squaring process if A is a stability matrix such that the norm of its exponential grows before it decays, as shown in Figure 3.1, and 2^{-m} is under the 'hump', but 1 is beyond it (see Exercise 3.10). For such matrices one has

$$\| e^A \|_1 \ll \| e^{A/2^m} \|_1^{2^m} \tag{3.29}$$

On the other hand, the error analysis of the squaring process reveals that because of cancellation, rounding errors of order $\varepsilon \| e^B \|_1^{2^m}$,

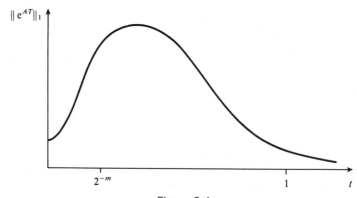

Figure 3.1

$B = A/2^m$ can be expected in the computed exponential of A. It follows from (3.29) that in this case

$$\| e^A \|_1 = \| (e^B)^{2^m} \|_1 \ll \| e^B \|_1^{2^m}$$

and hence the error may be large with respect to $\| e^A \|_1$. Consequently any algorithm which tries to pass the 'hump' by repeated squaring may lead to a large relative error in the computed exponential.

In this way a large error estimate, obtained by the recurrence relation (3.28), could be the result of any of the following three causes (Moler and Van Loan, 1978):

1. The error estimate is an overestimate of the true error, which is actually small. The computed exponential has low relative error, but the estimate is very pessimistic.
2. The true error is large due to cancellation in going over the 'hump', but the exponent is not very sensitive. In this case it is necessary to use another algorithm, which may produce a more accurate result.
3. The underlying problem is inherently sensitive and no algorithm can be expected to produce a more accurate result.

The right distinction among these three situations is a difficult problem which requires analysis of the sensitivity of the exponential and investigations of the 'hump'.

3.2.2 Padé Approximations

Let

$$f(x) = s_0 + s_1 x + s_2 x^2 + \cdots$$

be a power series with $s_0 \neq 0$ and let $n(x)$ and $d(x)$ be polynomials of degree p, q respectively,

$$n(x) = \sum_{k=0}^{p} a_k x^k \qquad d(x) = \sum_{k=0}^{q} b_k x^k$$

The ratio $n(x)/d(x)$ is referred to as (p, q) *Padé approximation* to $f(x)$ if the terms $x^0, x^1, ..., x^{p+q}$ are cancelled out in $f(x)d(x) - n(x)$. The sum $p + q$ is the *order* of the Padé approximation.

It may be proved that the coefficients $a_0, a_1, ..., a_p; b_0, b_1, ..., b_q$ always exist and the ratio $n(x)/d(x)$ is unique.

The (p, q) Padé approximation to the matrix exponential e^A is defined by

$$R_{pq}(A) = [D_{pq}(A)]^{-1} N_{pq}(A)$$

where

$$N_{pq}(A) = \sum_{k=0}^{p} a_k A^k \qquad a_k = (p+q-k)! \, p! / [(p+q)! \, k! \, (p-k)!]$$

$$D_{pq}(A) = \sum_{k=0}^{p} b_k (-A)^k \qquad b_k = (p+q-k)! \, q! / [(p+q)! \, k! \, (q-k)!]$$

The nonsingularity of the matrix $D_{pq}(A)$ is assured if the parameters p and q are sufficiently large or if the eigenvalues of A have negative real parts (i.e. A is a stability matrix). The choice of p and q is done to achieve the prescribed accuracy of the approximation.

For a given accuracy it is preferable to use *diagonal Padé approximations* ($p = q$). In fact, suppose that $p > q$. The evaluation of the approximation $R_{pq}(A)$ which is of order $p + q$ requires approximately pn^3 flops. However, the same work is necessary to compute the approximation $R_{pp}(A)$ which is of order $2p > p + q$. Similar arguments hold for the case $p < q$.

The Padé approximations can be used when the norm of A is not very large. Similarly to the use of Taylor series, rounding errors may affect these approximations very much. When implementing diagonal approximations for large p, the matrix $D_{pp}(A)$ approaches the series for $e^{-A/2}$ while $N_{pp}(A)$ tends to the series for $e^{A/2}$. That is why these matrices may be computed with large errors because of cancellation. Also, the denominator matrix $D_{pq}(A)$ may be ill conditioned with respect to inversion. If, for example, the eigenvalues $\lambda_1, \lambda_2, ..., \lambda_n$ of the matrix A are widely spread, i.e. $|\lambda_1| \gg |\lambda_n|$, then

$$\text{cond}_2 [D_{qq}(A)] \approx \text{cond}_2(e^{-A/2}) \geqq e^{(|\lambda_1| - |\lambda_n|)/2}$$

is very large.

In the case of nondiagonal approximation, if A is a stability matrix, then the computed approximations with $p > q$ tend to have larger rounding errors due to cancellation, while the computed approximations with $p < q$ tend to have larger rounding errors due to the ill conditioning of the matrix $D_{pq}(A)$. In addition, the diagonal Padé approximations to e^{At} have the desirable property of being bounded for $t \to \infty$ which is not valid for $p \neq q$.

Example 3.6

If the diagonal Padé approximations are computed for

$$A = \begin{bmatrix} -43 & 42 \\ -28 & 27 \end{bmatrix}$$

using floating-point arithmetic with $\varepsilon = 2^{-15} \approx 3 \times 10^{-5}$ then the most accurate results are obtained for $p = 9$. In this case we have (to six significant digits)

$$R_{99}(A) = \begin{bmatrix} -0.747\,070 & 1.090\,52 \\ -0.747\,147 & 1.090\,64 \end{bmatrix}$$

so that only two digits are accurate in the final result (compare with Example 3.4). The condition number of the denominator matrix is

$$\text{cond}_2\,[D_{99}(A)] = 6634$$

If the computations are performed with precision $\varepsilon = 2^{-56} \approx 1.4 \times 10^{-17}$ then

$$R_{99}(A) = \begin{bmatrix} -0.735\,814 & 1.103\,69 \\ -0.735\,795 & 1.103\,67 \end{bmatrix}$$

so that at least four digits are accurate in the computed approximation.

The number of the necessary diagonal Padé approximations is determined by the prescribed accuracy. It is shown by Moler and Van Loan (1978) that if

$$\| A \|_1 \leq \tfrac{1}{2}$$

then in the absence of rounding errors there exists a matrix E such that

$$R_{pp}(A) = e^{A+E}$$
$$EA = AE$$
$$\| E \|_1 \leq \delta \| A \|_1$$
$$\delta = 2^{3-2p}(p!)^2 / [\,(2p)!(2p+1)!\,]$$

The relative error in the computed approximation is then determined by

$$\| R_{pp}(A) - e^A \|_1 / \| e^A \|_1 = \| e^A (e^E - I_n) \|_1 / \| e^A \|_1$$
$$\leq \| E \|_1 e^{\| E \|_1} \leq \delta \| A \|_1 e^{\delta \| A \|_1}$$

Now we may choose p so that the relative error is less than some tolerance. Since A has errors of order $\varepsilon \| A \|_1$ it is useless to set $\delta < \varepsilon$.

If the norm of A is large, we may use the techniques implemented in Algorithm 3.1. Specifically, we may determine a scalar s, a diagonal matrix D_0, a permutation matrix P and an integer m, such that

$$\| B \|_1 \leqq 1 \qquad B = 2^{-m} D_0^{-1} P^{\mathrm{T}} (A - sI_n) PD_0$$

The exponential of B is then computed by diagonal Padé approximations. The coefficients in the powers of B are computed recursively by

$$c_k = c_{k-1} \frac{p - k + 1}{k(2p - k + 1)} \qquad k = 1, 2, ..., p \qquad c_0 = 1$$

This leads to the following algorithm, which for a given $n \times n$ matrix A and a positive integer p computes the Padé approximation to e^A and stores it in the array R.

Algorithm 3.2 Computation of the matrix exponential by Padé approximation using scaling and squaring

Set $s = \operatorname{tr}(A)/n$

$\quad A \leftarrow A - sI_n$

Determine a diagonal matrix D_0 and a permutation matrix P such that $D_0^{-1} P^{\mathrm{T}} APD_0$ is balanced

Determine $m \geqq 0$ such that $2^{-m} \| A \|_1 \leqq 1$

$\quad A \leftarrow A/2^m$

Set $\;c = 1$
Set $\;D = I_n, N = I_n, Q = I_n$
For $\;k = 1, 2, ..., q$

$$c \leftarrow c \, \frac{p - k + 1}{k(2p - k + 1)}$$

$\quad Q \leftarrow AQ$
$\quad N \leftarrow N + cQ$
$\quad D \leftarrow D + (-1)^k cQ$

Solve $DR = N$ for R using Gaussian elimination with partial pivoting

$\quad$ If $m > 0$
$\quad$ For $k = 1, 2, ..., m$
$\quad\quad R \leftarrow R^2$
$R \leftarrow e^s PD_0 RD_0^{-1} P^{\mathrm{T}}$

This algorithm requires approximately $(p + m + \frac{1}{3})n^3$ flops. The necessary array storage is proportional to $4n^2$.

Algorithm 3.2 is generally more efficient than Algorithm 3.1 which uses Taylor approximations. In the absence of rounding errors, when $\|A\|$ is small, Padé approximations require about one half of the work for the same accuracy. As $\|A\|$ grows this advantage decreases because of the larger amount of scaling needed which is equal for both methods.

The rounding error properties of Algorithm 3.2 are similar to those of Algorithm 3.1.

Consider first the effect of rounding errors on the computation of Padé approximations to e^B,

$$B = 2^{-m} D_0^{-1} P^T (A - sI_n) P D_0$$

The pth diagonal Padé approximation fulfils

$$e^B = D_p^{-1} N_p + E_T(p) \tag{3.30}$$

where

$$N_p = \sum_{k=0}^{p} c_k B^k \quad , \quad D_p = \sum_{k=0}^{p} c_k(-B)^k$$

and $E_T(p)$ is the truncation error matrix which is a function of p.

An analysis of the errors in computing e^B yields

$$\|\overline{e^B} - e^B\|_1 / \|e^B\|_1 < (38.1n + 207.8)\varepsilon$$
$$+ [1 + (31.9n + 55)\varepsilon] \|E_T(p)\|_1 / \|e^B\|_1$$

This estimate of the relative error shows how to select the appropriate diagonal Padé approximation for a particular computer. Clearly, we should determine p so that $\|E_T(p)\|_1 / \|e^B\|_1$ is of order ε.

The error analysis for the squaring of the matrix e^B is the same as in the use of Taylor series. The initial error, however, in the given case is bounded by $(38.1n + 208.8)\varepsilon$.

Example 3.7

Let

$$A = \begin{bmatrix} 4 & 2 & 0 \\ 1 & 4 & 1 \\ 1 & 1 & 4 \end{bmatrix}$$

and the computations be performed with relative precision $\varepsilon = 2^{-56} \approx 1.4 \times 10^{-17}$. Since $\|E_T(p)\|_1 / \|e^B\|_1 \approx \varepsilon$ for $p = 9$, the ninth-order diagonal Padé approximation is implemented. Using Algorithm

3.2 we obtain (up to 17 decimal digits)

$$\overline{e^A} = \begin{bmatrix} 147.866\ 622\ 446\ 370\ 15 & 183.765\ 138\ 646\ 368\ 43 \\ 127.781\ 085\ 523\ 182\ 49 & 183.765\ 138\ 646\ 368\ 43 \\ 127.781\ 085\ 523\ 182\ 49 & 163.679\ 601\ 723\ 180\ 77 \end{bmatrix}$$

$$\begin{bmatrix} 71.797\ 032\ 399\ 996\ 548 \\ 91.882\ 569\ 323\ 184\ 216 \\ 111.968\ 106\ 246\ 371\ 88 \end{bmatrix}$$

The accuracy estimate indicates a minimum of 14 digits accurate in the 1-norm of e^A. The actual number of accurate digits is approximately 16 (compare with Example 3.5).

EXERCISES

3.8 Verify (3.25) and (3.26).

3.9 Prove that the condition number of the linear system $DR = N$ from Algorithm 3.2 does not exceed 5.06.

3.10 Investigate the 'hump' of $\|e^{At}\|_1$ for

$$A = \begin{bmatrix} -5 & 0 & -3 & 5 \\ 3 & -2 & 3 & -5 \\ 3 & 0 & -7 & 1 \\ -5 & 0 & 1 & 1 \end{bmatrix}$$

Compare $\|e^A\|_1$ and $\|e^{A/2^m}\|_1^{2^m}$ for $m = 4$.

3.11 Analyze the relative truncation error of the Padé approximation of e^A as a function of the order p for the matrix A from Example 3.7.

3.12 Write computer programs implementing Algorithms 3.1 and 3.2. Incorporate monitoring of the error during the squaring.

3.13 Compare the accuracy of Algorithms 3.1 and 3.2 ($p = 9$) in computing the exponential of the matrix A from Example 3.1.

3.3 ORDINARY DIFFERENTIAL EQUATION METHODS

Since the state equation of a continuous system is a particular case of the general nonlinear ordinary differential equation (ODE)

$$\dot{x}(t) = f(x(t), u(t), t) \qquad x(0) = x_0$$

its solution may be found by some of the methods for numerical integration of an ODE.

A disadvantage of this approach is the large volume of computational operations required, since it does not make use of the special nature of the problem. The implementation of an ODE solver, however, is indispensible in case of a time-dependent state matrix or in the presence of a forcing term in the state equation.

The numerical methods for solving an ODE

$$\dot{x}(t) = f(x(t), t) \qquad x(0) = x_0 \tag{3.31}$$

are discrete methods, i.e. they compute a sequence of approximations $x_k \approx x(t_k)$ for a set of points

$$t_{k+1} = t_k + h_k \qquad k = 0, 1, ..., N - 1$$

The quantity h_k is called *step size* and it may vary from step to step. In general, the approximation x_{k+1} is computed using l previous values $x_k, x_{k-1}, ..., x_{k-l+1}$. If $l = 1$ the corresponding method is called *single step* and if $l > 1$ this is a *multistep method*. Examples of single step ODE methods are the Euler and classical Runge–Kutta methods.

The errors in the numerical solution of an ODE are a result of discretization (this error is also called truncation error) and rounding errors. The *local discretization error* is the error made at a given step provided that the previous values are exact and there are no rounding errors in the computations. This error is determined in the following way. Let the function $y_k(t)$ satisfy

$$\dot{y}_k(t) = f(y_k(t), t) \qquad y_k(t_k) = x_k$$

Then the local error d_k is given by

$$d_k = x_{k+1} - y_k(t_{k+1})$$

In this way the local error is the difference between the computed approximation and the theoretical solution defined by the same initial condition at the point t_k. Note, however, that $y_k(t_{k+1})$ itself may be far from the exact solution at the point t_{k+1}.

The *global discretization error* e_k at step k is the difference between the computed approximation x_k (neglecting the rounding errors) and the exact solution $x(t_k)$ determined by the initial condition x_0, i.e.

$$e_k = x_k - x(t_k)$$

Clearly, the global error is a function of all local errors made at the previous steps.

Practically all present-day codes for solving ODEs control the local error at each step and do not even attempt to control the global error directly.

The rounding-error properties of the existing ODE methods are not studied in depth, partly because the accuracy requested from ODE solvers is usually less than the full machine precision.

The *order* of a numerical method for solving ODEs is defined in terms of the local discretization error in the following way. We say that the method is of an order p if there exists a number c so that

$$\| d_k \| \leq c h_k^{p+1}$$

The number c may depend on the function f but it must not depend on k or on the step size h_k. The Euler method, for example, is of order $p = 1$. The higher the order of the method, the smaller is the global discretization error for a given h_k. The decreasing of the global error by decreasing step size may be done until the moment when the effect of rounding errors becomes prevalent. Hence there is an optimum step size depending on the method, nature of the problem and machine precision used.

The costs of solving an ODE may be split into two parts – the cost of repeated function evaluation, and the remaining part called *overhead*.

The solution of an ODE poses some difficulties in case of stiff equations. A stable differential equation of type (3.31) is said to be *stiff* if the eigenvalues λ_i, $i = 1, 2, ..., n$, of the Jacobi matrix $\partial f / \partial x$ satisfy

$$\max_i \ \mathrm{Re}(-\lambda_i) / \min_i \ \mathrm{Re}(-\lambda_i) \gg 1$$

The stiff equation has a particular solution in the form of a decaying exponential whose time constant is much smaller than the interval on which the solution is considered. Most of the standard ODE methods are not well suited for stiff equations.

Example 3.8

(Forsythe, Malcolm and Moler, 1977). Consider the equation

$$\dot{x}(t) = Ax(t)$$

where

$$A = \begin{bmatrix} 998 & 1998 \\ -999 & -1999 \end{bmatrix}$$

The eigenvalues of A are -1 and -1000 so that the equation is stiff. If

$$x_1(0) = x_2(0) = 1$$

then the solution is

$$x_1(t) = 4e^{-t} - 3e^{-1000t}$$
$$x_2(t) = -2e^{-t} + 3e^{-1000t}$$

The component e^{-1000t} is decaying very quickly therefore the step size chosen is very small. On the other hand after a very small time interval the solution is close to

$$x_1(t) = 4e^{-t}$$
$$x_2(t) = -2e^{-t}$$

where e^{-t} is decaying slowly.

Let the equation be solved by the Euler method. In this case we obtain

$$x_{k+1} = x_k + hAx_k$$

The values of the first components x_{1k} and $x_1(t_k)$ of x_k and $x(t_k)$ for $h = 10^{-2}$ and $k = 1, 2, ..., 5$ are shown in Table 3.4.

Owing to the discretization and the rounding errors the computed solution gradually diverges from the exact one.

Consider in brief some of the numerical methods for solving an ODE. The classical *Runge–Kutta* method of fourth order is a single-step ODE method expressed by

$$x_{k+1} = x_k + \tfrac{1}{6}(c_1 + 2c_2 + 2c_3 + c_4)$$

where

$$c_1 = hf(x_k, t_k)$$
$$c_2 = hf(x_k + \tfrac{1}{2}c_1, t_k + \tfrac{1}{2}h)$$
$$c_3 = hf(x_k + \tfrac{1}{2}c_2, t_k + \tfrac{1}{2}h)$$
$$c_4 = hf(x_k + c_3, t_k + h)$$

This method requires four evaluations of the function $f(x, t)$ per step. The method is simple to use since it is self-starting, the step size can be changed easily and the storage requirements are minimal. With this scheme, however, it is not possible to achieve automatic step size control.

Table 3.4

t_k	x_{1k}	$x_1(t_k)$
0.01	3.096×10^1	3.960
0.02	-2.391×10^2	3.921
0.03	2.191×10^3	3.882
0.04	-1.968×10^4	3.843
0.05	1.772×10^5	3.805

More efficient is the Runge–Kutta–Fehlberg method which is of fifth order and requires six function evaluations per step. Four of these evaluations are implemented in the fourth-order scheme and all six evaluations are used in a fifth-order scheme. Comparison of the results is employed to obtain an estimate of the error, which is used in the step size control. The writing of a good Runge–Kutta code demands attention to many details (Shampine and Watts, 1977).

A disadvantage of the Runge–Kutta method is that if the function is complicated they require a considerable amount of computation. These methods are not suitable for stiff equations.

The multistep methods give better accuracy than single-step methods since they use information for the previously computed values $x_{k-1}, x_{k-2}, \ldots$ and derivatives $f_{k-1}, f_{k-2}, \ldots$.

The *linear multistep methods* can generally be represented by an equation of the form

$$\sum_{i=0}^{l} \alpha_i x_{k-i+1} = h \sum_{i=0}^{l} \beta_i f_{k-i+1} \qquad f_j = f(x_j, t_j) \tag{3.32}$$

where l is the order of the method and α_i and β_i are suitably chosen constants. If $\beta_0 = 0$ the right-hand side of (3.32) does not involve f_{k+1} and the equation can be solved explicitly for x_{k+1}: such methods are called *explicit*. If $\beta_0 \neq 0$ then x_{k+1} is present in the right-hand side of (3.32) through the function $f_{k+1} = f(x_{k+1}, t_{k+1})$. In this case the solution requires the use of an iterative procedure and the method is called *implicit*.

Implementation of the multistep methods involves the use of two multistep schemes at each step. First, an approximation of x_{k+1} is obtained using an explicit method which is referred to as a *predictor*. This approximation is employed to determine the value of f_{k+1}. This value is then implemented in an implicit method which is called a *corrector*, to obtain a more accurate approximation of x_{k+1}. Finally, this corrected value may be substituted in the differential equation to obtain a more accurate value of f_{k+1} and the integration can then be advanced one step further.

An example of the predictor–corrector multistep method is the *Adams–Bashforth* method of fourth-order given by

predictor: $x_{k+1} = x_k + h(55f_k - 59f_{k-1} + 37f_{k-2} - 9f_{k-3})/24$

corrector: $x_{k+1} = x_k + h(9f_{k+1} + 19f_k - 5f_{k-1} + f_{k-2})/24$

Multistep schemes of high order lead to efficient methods for solving ODEs which may be of high accuracy. The main disadvantage of these methods is that special methods are needed to start the integration. For this purpose it is possible to use, for example, Runge–Kutta methods. The necessity of using information about the previous values leads to the requirement for more storage in comparison to the single-step methods.

The most widely used class of linear multistep methods for stiff problems are the so-called *backward differentiation formulas*. The derivative f_{k+1} at the $(k + 1)$th step is approximated by $l + 1$ values x_{k-i+1}, $i = 0, 1, ..., l$, which leads to a linear multistep method of the form

$$\sum_{i=0}^{l} \alpha_i x_{k-i+1} = h\beta_0 f_{k+1} \qquad \alpha_0 = 1$$

For $l = 1$ we obtain the *backward Euler method*, expressed by

$$x_{k+1} = x_k + hf(x_{k+1}, t_{k+1})$$

There is a large variety of computer programs intended to solve nonlinear ODEs. Good programs implementing the methods mentioned in this section may be found in many software libraries such as IMSL, NAG, and so on. Among the good available programs are the subroutines DERKF, DEABM and DEBDF from the DEPAC package (Shampine and Watts, 1979), which are briefly described below.

DERKF is a fifth-order, variable-step Runge–Kutta code. It is the simplest of the three choices, both algorithmically and in usage. DERKF is primarily designed to solve nonstiff and mildly stiff ODEs when derivative evaluations are not expensive. Because DERKF has a very low overhead cost, it will usually result in the least expensive integration when solving problems requiring a modest amount of accuracy and having equations that are not costly to evaluate. The code gives an indication when the problem is stiff.

The subroutine DEABM uses the Adams–Moulton predictor–corrector formulas of variable order (one to twelve) and variable step. It is used when derivative evaluations are expensive, high accuracy is needed or answers at many specific points are required. It also gives a warning in case of stiff equations.

DEBDF is a variable order (one to five), variable-step backward differentiation formula code which implements Gear stiff method (Gear, 1971). It is the most complicated of the three choices. DEBDF is primarily designed to solve stiff ODEs of crude to moderate tolerances. DEBDF will be inefficient compared to DERKF and DEABM on nonstiff problems because it uses much more storage, has a much larger overhead and the low-order formulas will not give high accuracies efficiently.

All the codes in DEPAC use a mixed relative–absolute error

criterion. At each step the error is to be controlled so that

| estimate local error | $\leq$ relative error | solution | + absolute error

for each vector component and the user has control over the type of error test to be performed.

General-purpose ODE solvers may be used to compute the matrix exponential setting $f(x, t) = Ax$. The jth column of e^{At} can be obtained choosing x_0 equal to the jth column of the unit matrix and integrating from 0 to t. This is obviously inefficient since the ODE routine does not take into account the linear, constant coefficient nature of the problem.

Example 3.9

Consider the computation of e^A for

$$A = \begin{bmatrix} 4 & 2 & 0 \\ 1 & 4 & 1 \\ 1 & 1 & 4 \end{bmatrix}$$

The eigenvalues of A are 3, 3 and 6 so that the equation is not stiff.

Using the subroutines DERKF, DEABM and DEBDF with relative and absolute local error tolerance 10^{-6} and precision $\varepsilon = 2^{-24} \approx 5.96 \times 10^{-8}$, we obtain (up to eight decimal places):

DERKF

$$\overline{e^A} = \begin{bmatrix} 147.866\,26 & 183.764\,63 & 71.796\,745 \\ 127.780\,72 & 183.764\,65 & 91.882\,278 \\ 127.780\,72 & 163.679\,12 & 111.967\,81 \end{bmatrix}$$

$$\| \overline{e^A} - e^A \|_1 / \| e^A \|_1 = 2.78 \times 10^{-6}$$

DEABM

$$\overline{e^A} = \begin{bmatrix} 147.866\,17 & 183.764\,65 & 71.796\,730 \\ 127.780\,63 & 183.764\,65 & 91.882\,248 \\ 127.780\,63 & 163.679\,12 & 111.967\,78 \end{bmatrix}$$

$$\| \overline{e^A} - e^A \|_1 / \| e^A \|_1 = 2.75 \times 10^{-6}$$

DEBDF

$$\overline{e^A} = \begin{bmatrix} 147.869\,71 & 183.769\,36 & 71.799\,835 \\ 127.784\,18 & 183.769\,33 & 91.885\,422 \\ 127.784\,19 & 163.683\,81 & 111.970\,99 \end{bmatrix}$$

$$\| \overline{e^A} - e^A \|_1 / \| e^A \|_1 = 2.38 \times 10^{-5}$$

Table 3.5

	10^{-4}	10^{-5}	10^{-6}
DERKF	69	107	169
DEABM	49	60	78
DEBDF	48	65	82

The average number of function evaluations for the three starting vectors and for three different local error tolerances is given in Table 3.5. This data, although a result of a single experiment, shows that the implementation of general-purpose ODE solvers requires approximately $100n^2$ flops for solving a single differential equation $\dot{x}(t) = Ax(t)$ (one function evaluation in this case requires n^2 flops) or $100n^3$ flops for computing e^{At}. For comparison, the Padé approximation method requires $10n^3$ flops to find e^{At} with higher accuracy (see Example 3.7).

Consider a specialization of the general ODE methods for solving the linear differential equation (Moler and Van Loan, 1978)

$$\dot{x}(t) = Ax(t) \qquad x(0) = x_0 \tag{3.33}$$

The fourth-order Runge–Kutta method with a fixed step size in this case produces

$$x_{k+1} = x_k + \tfrac{1}{6}c_1 + \tfrac{1}{3}c_2 + \tfrac{1}{3}c_3 + \tfrac{1}{6}c_4 \tag{3.34}$$

where

$$c_1 = hAx_k \qquad c_2 = hA(x_k + \tfrac{1}{2}c_1) \qquad c_3 = hA(x_k + \tfrac{1}{2}c_2)$$
$$c_4 = hA(x_k + c_3)$$

Thus equation (3.34) reduces to

$$x_{k+1} = (I_n + Ah + \cdots + A^4 h^4/4!)x_k = S_4(Ah)x_k \tag{3.35}$$

so that in the absence of rounding errors this method is equivalent to the Taylor series method of order four. Since the step size is fixed, the representation (3.35) requires the matrix $S_4(Ah)$ to be computed only once. The vector x_{k+1} is then obtained from x_k with one matrix–vector multiplication. The Runge–Kutta scheme (3.34) would require four such multiplications per step.

There is a close connection between single-step ODE methods with fixed step size and the technique of scaling and squaring. In fact, if $h = 1/m$ then

$$x(1) = x(mh) \approx x_m = [S_4(Ah)]^m x_0.$$

In this case Ah is the scaled matrix and its exponential is approximated by $S_4(Ah)$. In the numerical solution of an ODE, however, the matrix $[S_4(Ah)]^m$ is not obtained by repeated squaring even if m is a power of 2.

Variable-step ODE methods for solving (3.33) may be constructed in the following way. To control the step size it is necessary to compute two approximations to x_{k+1}, say $y_{k+1} = S_4(Ah_k)x_k$ and $x_{k+1} = S_5(Ah_k)x_k$. Their difference is used to choose h_k from the condition

$$\| x_{k+1} - y_{k+1} \| \approx \delta \| x_k \|$$

where δ is some prescribed relative local error tolerance. Since

$$x_{k+1} - y_{k+1} = \frac{A^5 h_k^5}{5!} x_k$$

the required step size is given approximately by

$$h_k = \left[\frac{5! \, \delta}{\| A^5 \|} \right]^{1/5}$$

This method is not very efficient if the step size does not change very much, since the matrices $S_4(Ah_k)$ and $S_5(Ah_k)$ must be precomputed at each step. If, however, the problem involves a large 'hump', the necessary step size beyond the 'hump' can be large and the variable-step method will require less work.

The specialization of multistep methods, such as the Adams–Bashforth formulas, for solving equation (3.33) has not found practical application although it might be efficient and may have good numerical properties.

EXERCISES

3.14 Apply Algorithms 3.1 and 3.2 to the solution of the stiff equation from Example 3.8. Compare with the exact solution.

3.15 Apply the standard subroutines for solving ODEs from your local program library to compute the exponential of the matrix A from Example 3.9. Compare with the result obtained in Example 3.7. Analyze the dependence of the global discretization error on the local error tolerance.

3.16 Compare the rounding error properties of the fourth-order Runge–Kutta method with a fixed step size (expression (3.34)) and the Taylor series method of order four (expression (3.35)).

3.17 Develop an algorithm and a computer program for solving the linear differential equation $\dot{x}(t) = Ax(t)$, $x(0) = x_0$, implementing the Adams–Bashforth formulas.

3.18 Write a computer program, based on a Runge–Kutta code, for solving the nonhomogeneous state equation $\dot{x}(t) = Ax(t) + Bu(t)$, $x(0) \equiv x_0$, where $u(t)$ is given in tabular form for $t = kh$, $k = 1, 2, \ldots$. Use some interpolation method to get the values of $u(t)$ for $kh < t < (k+1)h$.

3.4 MATRIX DECOMPOSITION METHODS

If the matrix A is decomposed as

$$A = SBS^{-1} \tag{3.36}$$

then the matrix exponential e^{At} satisfies

$$e^{At} = Se^{Bt}S^{-1}$$

As is pointed out by Moler and Van Loan (1978), all methods for computation of e^{At} which employ a decomposition of the form (3.36) involve two conflicting objectives:

1. Make the matrix B close to a diagonal matrix so that e^{Bt} is easy to compute.
2. Make the matrix S well conditioned so that the errors in computing $Se^{Bt}S^{-1}$ are small.

The use of Jordan canonical form of A puts emphasis on the first objective, while the implementation of Schur form puts emphasis on the second one.

Let us consider the numerical properties of methods for computation of e^{At} employing the basic matrix decompositions.

3.4.1 Diagonalization

As noted in Section 2.2, if the matrix A is nondefective, the exponential e^{At} may be found from

$$e^{At} = X \operatorname{diag}(e^{\lambda_1 t}, e^{\lambda_2 t}, \ldots, e^{\lambda_n t}) X^{-1}$$

where $\lambda_1, \lambda_2, \ldots, \lambda_n$ are the eigenvalues of A, and X is the matrix of the corresponding eigenvectors. This yields the following algorithm for computation of e^{At}.

Determine the eigenvalues λ_i and eigenvectors x_i $(i = 1, 2, ..., n)$ of A

Set $F = X$ diag $(e^{\lambda_1 t}, e^{\lambda_2 t}, ..., e^{\lambda_n t})$

$\qquad = [x_1 e^{\lambda_1 t}, x_2 e^{\lambda_2 t}, ..., x_n e^{\lambda_n t}]$

Solve $e^{At} X = F$ for e^{At}

Since approximately $15n^3$ flops are necessary to determine the eigenvalues and eigenvectors of A, this algorithm requires about $16\frac{1}{3}n^3$ flops. It is easily modified to handle complex conjugate eigenvalues using only real arithmetic.

Unfortunately, the algorithm produces large errors when A is a nearly defective matrix, i.e. when the eigenvectors of A are almost linearly dependent. This situation is indicated by the large value of cond(X).

Example 3.10

(Moler and Van Loan, 1978). If

$$A = \begin{bmatrix} 1 + \delta & 1 \\ 0 & 1 - \delta \end{bmatrix}$$

then $\lambda_1 = 1 + \delta$ and $\lambda_2 = 1 - \delta$

$$X = \begin{bmatrix} 1 & -1 \\ 0 & 2\delta \end{bmatrix}$$

and cond(X) is of order $1/\delta$ $(\delta \neq 0)$. If $\delta = 10^{-5}$ and arithmetic with $\varepsilon = 2^{-19} \approx 1.9 \times 10^{-6}$ is used to compute e^A, we obtain

$$\overline{e^A} = \begin{bmatrix} 2.718\ 307 & 2.861\ 023 \\ 0 & 2.718\ 250 \end{bmatrix}$$

The exact exponential (up to seven decimal digits) is

$$e^A = \begin{bmatrix} 2.718\ 309 & 2.718\ 282 \\ 0 & 2.718\ 255 \end{bmatrix}$$

so that the computed exponential has errors of order 10^5 times the machine precision (as may be expected). It may be shown that in the given case e^A is not very sensitive to changes in A and therefore the above algorithm must be regarded as numerically unstable.

More reliable results in the computation of e^{At} by diagonalization may be obtained taking into account the way of determining the eigenvectors by the QR algorithm. Using the double-shifted QR algorithm (see Section 1.10) we obtain an orthogonal matrix Q and a quasitriangular matrix T such that

$$Q^T A Q = T$$

Next the eigenvectors of A are obtained from $X = QR$, where R is an upper quasitriangular matrix, a solution of

$$TR = RD$$

and D is the diagonal part of T.

In this way the inverse of the eigenvector matrix may be computed from

$$X^{-1} = R^{-1} Q^T \qquad (3.37)$$

The representation (3.37) has two advantages. First, the inversion of X is reduced to the inversion of a triangular matrix, which is associated with smaller errors. Secondly, $\text{cond}(X) = \text{cond}(R)$ so that the estimation of $\text{cond}(X)$ can be done efficiently. If, however, A is defective or nearly defective, the use of (3.37) also leads to large errors.

3.4.2 Jordan Decomposition

In case of a defective matrix A we can rely on the Jordan form of A. Now there are powerful methods for determining the Jordan structure of a general matrix. Such a method is implemented in the algorithm JNF of Kågström and Ruhe (see Section 1.10), which produces the exact Jordan decomposition of $A^* = A + E$

$$J = S^{-1} A^* S$$

where $\| E \|_2 \leq \gamma \| A \|_2$ and γ is small. If the Jordan form of A is well conditioned, then the size of γ is of order ε and $\text{cond}(S)$ is relatively low. The algorithm obtains the unnormalized Jordan form in which the blocks

$$J_k = \begin{bmatrix} \lambda_k & c_1 & & & \text{\Large 0} \\ & \lambda_k & c_2 & & \\ & & \ddots & \ddots & \\ \text{\Large 0} & & & \ddots & c_p \\ & & & & \lambda_k \end{bmatrix} \qquad p = m_k - 1$$

of order m_k have superdiagonal elements different from 1. The exponential of J_k is given by

$$
e^{J_k t} = e^{\lambda_k t}
\begin{bmatrix}
1 & c_1 t & c_1 c_2 t^2/2 & \cdots & c_1 \ldots c_p t^p/p! \\
  & 1 & c_2 t & \cdots & c_2 \ldots c_p t^{p-1}/(p-1)! \\
  & & \textbf{0} & \ddots & \vdots \\
  & & & \ddots & c_p t \\
  & & & & 1
\end{bmatrix}.
\tag{3.38}
$$

Thus we have the following algorithm.

Algorithm 3.3 Computation of e^{At} by Jordan decomposition of A

Compute the Jordan decomposition of A, $A = SJS^{-1}$, $J = \text{diag}\,(J_1, J_2, \ldots, J_r)$ by the algorithm JNF of Kågström and Ruhe
For $k = 1, 2, \ldots, r$

  Compute $e^{J_k t}$ from (3.38)

Compute $e^{At} = S\,\text{diag}(e^{J_1 t}, e^{J_2 t}, \ldots, e^{J_r t})S^{-1}$

Since the determination of Jordan form as described in Section 1.10 requires at most $\frac{7}{12} n^4 + 9n^3$ flops, the full implementation of Algorithm 3.3 will require $\frac{7}{12} n^4 + 11\frac{1}{3} n^3$ flops. If it is necessary to compute only $x(t) = e^{At} x_0$ for various t, the equation $Sv = x_0$ should be solved once (at the cost of $\frac{1}{3} n^3$ flops) and then each $x(t) = Se^{Jt}v$ can be obtained with n^2 flops.

Algorithm 3.3 has good numerical properties for a wide class of problems and may be recommended as one of the best ways for finding the matrix exponential. It may give poor results only when the matrix A has ill-conditioned Jordan form. Fortunately, this situation arises only rarely and can be easily recognized by the large value of cond(S). An important feature of the algorithm of Kågström and Ruhe is that it yields reliable estimates of cond(S) and $\|E\|$. Thus it is possible to compute an a priori estimate of the error in the exponential using the bounds (3.21) or (3.22). A disadvantage of Algorithm 3.3 is that it is relatively expensive.

Example 3.11

For

$$
A = \begin{bmatrix}
4 & 2 & 0 \\
1 & 4 & 1 \\
1 & 1 & 4
\end{bmatrix}
$$

and $\varepsilon = 2^{-56} \approx 1.4 \times 10^{-17}$, the algorithm of Kågström and Ruhe yields

$$J = \begin{bmatrix} 6. & 0 & 0 \\ 0 & 3. & c_1 \\ 0 & 0 & 3. \end{bmatrix} \qquad c_1 = 1.114\ 172\ 029\ 062\ 311\ 1$$

with 17 true decimal digits.

The condition number of the transformation matrix S is about 1.31 so that the errors in computing $Se^{Jt}S^{-1}$ are small. As a final result we get

$$\overline{e^A} = \mathrm{fl}(Se^J S^{-1}) = \begin{bmatrix} 147.866\ 622\ 446\ 370\ 18 & 183.765\ 138\ 646\ 368\ 49 \\ 127.781\ 085\ 523\ 182\ 54 & 183.765\ 138\ 646\ 368\ 53 \\ 127.781\ 085\ 523\ 182\ 53 & 163.679\ 601\ 723\ 180\ 85 \end{bmatrix}$$

$$\begin{bmatrix} 71.797\ 032\ 399\ 996\ 580 \\ 91.882\ 569\ 323\ 184\ 266 \\ 111.968\ 106\ 246\ 371\ 93 \end{bmatrix}$$

and

$$\| \overline{e^A} - e^A \|_1 / \| e^A \|_1 = 4.82 \times 10^{-16}$$

Hence approximately 17 digits are exact in $\overline{e^A}$ (compare with Examples 3.5 and 3.7).

3.4.3 Schur Decomposition

Let the Schur decomposition of A

$$A = QTQ^H$$

where Q is unitary and T is upper triangular, be known. Then

$$e^A = Qe^TQ^H$$

and the problem for determining e^A is reduced to the computation of the exponential of a triangular matrix. For this purpose it is possible to use the recursive algorithm from Parlett (1976), which produces the upper triangular matrix $F = f(T)$ for a given analytic function f. This algorithm is based on the property

$$FT = TF$$

which implies

$$\sum_{k=1}^{j} f_{ik}t_{kj} = \sum_{k=i}^{j} t_{ik}f_{kj} \qquad j > i$$

Provided that $t_{ii} \neq t_{jj}$ it follows from the above identities that

$$f_{ij} = \left[t_{ij}(f_{ii} - f_{jj}) + \sum_{k=i+1}^{j-1} (f_{ik}t_{kj} - t_{ik}f_{kj}) \right] \Big/ (t_{ii} - t_{jj}) \qquad (3.39)$$

In this way it is possible to compute one superdiagonal of F at a time beginning with the diagonal $f(t_{11}), ..., f(t_{nn})$. In applying this to the determination of e^A we come to the following algorithm

Compute the Schur decomposition $A = QTQ^H$ of A

For $i = 1, 2, ..., n$

  Set $f_{ii} = e^{t_{ii}}$

For $l = 1, 2, ..., n - 1$
  For $i = 1, 2, ..., n - l$
    Set $j = i + l$
    Compute f_{ij} by (3.39)
Compute $e^A = QFQ^H$

In this algorithm the computation of $F = e^T$ requires only $n^3/3$ flops. Unfortunately, difficulties may arise when A has confluent or almost confluent eigenvalues.

Example 3.12

If

$$A = \begin{bmatrix} 1.000\,01 & 1 \\ 0 & 0.999\,99 \end{bmatrix}$$

and $\varepsilon = 2^{-19} \approx 1.9 \times 10^{-6}$ then we obtain (up to seven decimal digits)

$$\bar{f}_{11} = 2.718\,307 \qquad \bar{f}_{22} = 2.718\,250$$

and

$$\bar{f}_{12} = \mathrm{fl}\,[(\bar{f}_{11} - \bar{f}_{22})/(t_{11} - t_{22})] = 2.857\,140$$

while the exact values are

$$f_{11} = 2.718\,309 \qquad f_{22} = 2.718\,255 \qquad f_{12} = 2.718\,282$$

The large error in $\bar{f}_{12}$ is due to the cancellation which takes place in computing $\bar{f}_{11} - \bar{f}_{22}$ and the subsequent division by 2×10^{-5}.

To handle the case of multiple or close eigenvalues of A it is possible to use a block version of the above algorithm, see Parlett (1976). This approach, however, does not have any advantage over the use of Jordan form of A.

3.4.4 Block Diagonalization

A compromise between the Jordan and Schur decompositions is the block diagonal decomposition (see Section 1.10) in which the matrix A is reduced to

$$A = X \, \mathrm{diag}(T_1, T_2, ..., T_q) X^{-1}$$

using a well-conditioned transformation matrix X. Each block T_k involves a cluster of nearly confluent eigenvalues. In the algorithm of Bavely and Stewart (1979) the size of each block is made as small as possible while keeping the condition number of X less than some prescribed bound 10^l. This implies that at most roughly l digits will be lost, because of rounding errors, in the computation of e^{At} by

$$e^{At} = X \, \mathrm{diag}(e^{T_1 t}, e^{T_2 t}, ..., e^{T_q t}) X^{-1}$$

A larger l will lead to the loss of more digits, while a smaller l will require more computations to evaluate the exponentials of diagonal blocks.

The computation of $e^{T_k t}$ may be done by taking into account that the matrix T_k is represented in the form

$$T_k = \lambda_k I_{n_k} + N_k$$

where λ_k is the average value of diagonal elements of T_k, and N_k is an upper triangular matrix. If the grouping of eigenvalues in T_k is properly done then N_k should be a nearly nilpotent matrix, i.e.

$$N_k^{n_k} \approx 0$$

where n_k is the order of T_k. Thus $e^{T_k t}$ can be found efficiently from

$$e^{T_k t} = e^{\lambda_k t} e^{N_k t}$$

where $e^{N_k t}$ is computed by a few terms of the Taylor series or by a low-order Padé approximation.

If the matrix A has multiple eigenvalues, it may be expected that the diagonal elements of T_k will be found with errors of order ε^{1/n_k} (see Section 1.9). In such a case the matrix N_k will not be nearly nilpotent. However, the diagonal elements of $N_k^{n_k}$ will be of order ε and may be neglected.

Overall we have the following algorithm.

Algorithm 3.4 Computation of e^{At} by block diagonalization

Compute the block diagonal form $X^{-1}AX = \text{diag}(T_1, T_2, ..., T_q)$ of A
  by Algorithm 1.21

For $k = 1, 2, ..., q$

    Compute $e^{T_k t}$ using Algorithm 3.1 or 3.2

Compute $e^{At} = X \, \text{diag}(e^{T_1 t}, e^{T_2 t}, ..., e^{T_q t}) X^{-1}$

Since at most $\frac{1}{12} n^4 + 15n^3$ flops are necessary to find the block diagonal
form of A, the full implementation of Algorithm 3.4 will require about
$\frac{1}{12} n^4 + 17\frac{1}{3} n^3$ flops. The block diagonalization does not depend on t so
that the calculation of e^{At} for each new t will require at most $2n^3$ flops.

    In this way Algorithm 3.4 gives one of the best approaches to computing the matrix exponential. If, however, the matrix A has large
Jordan blocks then the use of Jordan decomposition might produce
more accurate results. Also, the implementation of block diagonal form
makes it more difficult to estimate the error in the computed
exponential.

Example 3.13

For
$$A = \begin{bmatrix} 4 & 2 & 0 \\ 1 & 4 & 1 \\ 1 & 1 & 4 \end{bmatrix}$$
and $\varepsilon = 2^{-56} \approx 1.4 \times 10^{-17}$, the block diagonal form of A obtained by
Algorithm 1.21 is (up to 17 decimal digits)

$$\overline{X^{-1}AX} = \begin{bmatrix} 6.0 & 0 & 0 \\ 0 & 3.0 & c \\ 0 & 0 & 3.0 \end{bmatrix} \qquad c_1 = 1.154\,700\,538\,379\,251\,5$$

and $\text{cond}_2(X) \approx 1.31$.

    The computed exponential is

$$\overline{e^A} = \begin{bmatrix} 147.866\,622\,446\,370\,11 & 183.765\,138\,646\,368\,42 \\ 127.781\,055\,523\,182\,47 & 183.765\,138\,646\,368\,46 \\ 127.781\,085\,523\,182\,47 & 163.679\,601\,723\,180\,78 \end{bmatrix}$$

$$\begin{matrix} 71.797\,032\,399\,996\,534 \\ 91.882\,569\,323\,184\,222 \\ 111.968\,106\,246\,371\,88 \end{matrix}$$

so that

$$\| \overline{e^A} - e^A \|_1 / \| e^A \|_1 = 1.47 \times 10^{-16}$$

(compare with Example 3.11).

EXERCISES

3.19 Show that if the eigenvalues and eigenvectors of a nondefective $n \times n$ matrix A are known exactly, then the error in evaluating e^{At} by diagonalization satisfies

$$\| \overline{e^{At}} - e^{At} \|_2 \leqq n \, e^{\lambda_{max}(A)t} \, \text{cond}(X)\varepsilon \qquad \lambda_{max}(A) = \max_i | \lambda_i |$$

3.20 Write a computer program implementing Algorithm 3.3. Incorporate an estimation of the error in the computed exponential based on the bounds (3.21) or (3.22). Analyze the performance of the program on matrices with known exponential.

3.21 Develop an algorithm for finding e^{At} exploiting the Schur form of A, the latter being ordered so that the clusters of nearly confluent eigenvalues appear in blocks on the diagonal.

3.22 Write a computer program implementing Algorithm 3.4. Compare the performance of this program and the program utilizing the Jordan decomposition on test examples of high order.

3.5 COMPARISON OF METHODS FOR COMPUTING THE MATRIX EXPONENTIAL

Following from the results presented in Sections 3.2, 3.3 and 3.4, currently there are at least four methods which may be used for reliable computation of e^{At}. Two of them are series methods which employ Taylor series and Padé approximations. Because of the implementation of scaling and squaring, both methods have acceptable numerical properties. They are not numerically stable but they produce a warning when excessive errors are introduced in the solution. Series methods are suitable for matrices with a low norm since in this case the number of squarings is small and the effect of rounding errors is not significant. These methods compute e^A for a single matrix A, the number of necessary operations being from $10n^3$ to $20n^3$. To find e^{At} for k values of t, k times more work is required. Since the method employing Padé approximations is more efficient than that using Taylor series, it should be preferred.

The matrix decomposition methods for computing e^{At} are preferable in cases when the exponential must be calculated for several values of t and/or $\|A\|$ is large. They require more computational operations than the series methods (the number of flops may be of order n^4) but generally they produce more accurate results. The most reliable algorithms employ Jordan decomposition or block diagonalization. The Jordan decomposition method is more expensive than the block diagonalization method, but in some cases it may produce better results. Also, the Jordan decomposition method makes it possible to obtain easily an estimate of the error in the computed exponential.

The choice of method for finding e^{At} should be problem oriented and depends on the implementation details.

3.6 ERRORS IN THE SOLUTION OF STATE EQUATIONS

As was pointed out in Section 2.2, there are two alternative ways to compute the solution of a continuous-time state equation

$$\dot{x}(t) = Ax(t) \qquad x(0) = x_0 \tag{3.40}$$

The first way is for every t to find the exponential e^{At} and then to compute

$$x(t) = e^{At}x_0$$

This approach is efficient when e^{At} is determined by some of the matrix decomposition methods, since the computation of $x(t)$ for each new t does not require much work. When series methods are implemented, e^{At} must be computed for every t and this requires a large number of operations. With t increasing, the norm of At also increases, which leads to increasing the number of scalings and squarings. This in turn causes enlargement of the volume of both work and errors.

The solution of (3.40) is done efficiently by using the difference equation

$$x[(k+1)h] = Fx(kh)$$

where $F = e^{Ah}$ and $h = \Delta t$ is an appropriate time step. In this case the computation of the matrix exponential is done only once, and the state vector is determined by repeated multiplication which costs only n^2 flops at each step. Hence, if e^{Ah} is computed, the solution of (3.40) reduces to the solution of the difference equation

$$x_{k+1} = Fx_k \tag{3.41}$$

The same equation with $F = A$ is solved in the discrete-time case. Thus, the error analysis of the solution of (3.41), presented in this section, holds both for the continuous-time and discrete-time cases.

In the presence of rounding errors, instead of (3.41) we solve the equation

$$\bar{x}_{k+1} = \mathrm{fl}(\bar{F}\bar{x}_k) \qquad \bar{x}_0 = x_0 \tag{3.42}$$

where $\bar{F} = F + H$. For a continuous-time system H is the error in the computed matrix exponential, while for a discrete-time system H is the error in the state matrix.

Equation (3.42) may be represented as

$$\bar{x}_{k+1} = (\bar{F} + G)\bar{x}_k \qquad \bar{x}_0 = x_0 \tag{3.43}$$

where, employing the error bound for matrix multiplication (see Section 1.4), it may be shown that

$$\| G \|_F \le n\varepsilon \|\bar{F}\|_F$$

We are interested in obtaining a bound on the error in the state vector

$$\Delta_k = \bar{x}_k - x_k$$

Subtracting (3.41) from (3.43) we find that Δ_k is the solution of the nonhomogeneous difference equation

$$\Delta_{k+1} = F\Delta_k + (G + H)\bar{x}_k \qquad \Delta_0 = 0 \tag{3.44}$$

The solution of (3.44) is

$$\Delta_k = \sum_{i=1}^{k} F^{k-i}(G + H)\bar{x}_{i-1}$$

Since

$$\bar{x}_k = (\bar{F} + G)^k x_0$$

it follows that

$$\Delta_k = \sum_{i=1}^{k} F^{k-i}(G + H)(\bar{F} + G)^{i-1} x_0 \tag{3.45}$$

Taking the norm of both sides of (3.45) we get

$$\| \Delta_k \|_2 \le (\| G \|_F + \| H \|_F) \sum_{i=1}^{k} \| F^{k-i} \|_F \| (\bar{F} + G)^{i-1} \|_F \| x_0 \|_2$$

If $\| H \|_F \ll \| F \|_F$ we may use the approximate estimate

$$\| \Delta_k \|_2 \le (\| G \|_F + \| H \|_F) \sum_{i=1}^{k} \| \bar{F}^{k-i} \|_F \| \bar{F}^{i-1} \|_F \| x_0 \|_2 \tag{3.46}$$

which involves only known quantities.

The inequality (3.46) makes it possible to estimate, for a given k, the error in the solution owing to the error matrix H and the rounding errors.

The bound (3.46) may be strengthened to

$$\| \Delta_k \|_2 \leq kn(\| G \|_F + \| H \|_F) \| \bar{F} \|_F^{k-1} \| x_0 \|_2 \tag{3.47}$$

Unfortunately, since $\| \bar{F}^i \|_F$ may be very small in comparison to $\| \bar{F} \|_F^i$, the bound (3.47) is sometimes very pessimistic. Also, if $\| \bar{F} \|_F > 1$, the right-hand side of (3.47) will always increase with k, although $\| \Delta_k \|_2$ may decrease after some k.

Example 3.14

Let

$$F = \begin{bmatrix} -1.366 & 0.1349 & 0.2681 & -0.7030 \\ -0.4168 & 1.683 & 0.8159 & 1.602 \\ -0.2825 & -3.478 & 1.188 & 1.289 \\ 0.6504 & 0.7523 & -0.6478 & -0.5045 \end{bmatrix}$$

and

$$H = 10^{-4} \begin{bmatrix} -5 & -8 & 3 & -9 \\ 4 & 2 & -3 & -7 \\ 5 & -8 & -3 & 1 \\ 9 & 2 & 7 & -6 \end{bmatrix}$$

The results of solving $x_{k+1} = Fx_k$ and $\bar{x}_{k+1} = \bar{F}\bar{x}_k$, $\bar{F} = F + H$ for $x_0 = [0.1, 0.3, -0.1, -0.2]^T$, by using arithmetic with $\varepsilon = 2^{-56}$

Table 3.6

k	$\| x_k \|_2$	$\| \Delta_k \|_2$ $= \| \bar{x}_k - x_k \|_2$	$\| \Delta_k \|_2$ (3.46)	$\| \Delta_k \|_2$ (3.47)
1	0.12×10^1	0.15×10^{-3}	0.36×10^{-2}	0.36×10^{-2}
2	0.34×10^1	0.10×10^{-2}	0.18×10^{-1}	0.36×10^{-1}
3	0.65×10^1	0.26×10^{-2}	0.66×10^{-1}	0.27×10^0
4	0.54×10^1	0.93×10^{-2}	0.19×10^0	0.18×10^1
5	0.32×10^1	0.27×10^{-1}	0.41×10^0	0.12×10^2
6	0.16×10^1	0.83×10^{-1}	0.71×10^0	0.71×10^2
7	0.70×10^0	0.19×10^0	0.10×10^1	0.42×10^3
8	0.30×10^0	0.27×10^0	0.11×10^1	0.24×10^4
9	0.13×10^0	0.26×10^0	0.99×10^0	0.14×10^5
10	0.55×10^{-1}	0.20×10^0	0.72×10^0	0.77×10^5

$\approx 1.4 \times 10^{-17}$, along with the estimates (3.46) and (3.47), are shown in Table 3.6.

It is seen that for $k > 8$ the value of $\| \Delta_k \|_2$ becomes greater than $\| x_k \|_2$, i.e. there are no true digits in the elements of the computed solution $\bar{x}_k$.

For this example the estimate (3.46) reflects properly the behaviour of the error, in contrast to (3.47). The estimate (3.46), however, is more expensive to compute than (3.47).

EXERCISES

3.23 Write a computer program for solving the continuous-time state equation $\dot{x}(t) = Ax(t)$, $x(0) = x_0$, utilizing the Jordan or block-diagonal form of A. Avoid for efficiency the explicit computation of e^{At}. Incorporate an error estimation based on the bound (3.46).

3.24 Develop an algorithm for solving the nonhomogeneous state equation $\dot{x}(t) = Ax(t) + Bu(t)$, $x(0) = x_0$, where $u(t) = \text{const}$.

3.25 Develop algorithms for solving the discrete-time state equation $x_{k+1} = Fx_k$ based on the Schur, Jordan and block-diagonal forms of F.

3.26 Develop an efficient algorithm and a computer program for solving the nonhomogeneous difference equation $x_{k+1} = Ax_k + Bu_k$.

3.7 DISCRETIZATION OF CONTINUOUS-TIME SYSTEMS

As was shown in Section 2.2, the determination of the discrete-time model of a continuous-time system requires computation of the matrix exponential and its integral over the sampling interval.

Consider the methods for computing the integral of the matrix exponential. From the definition of e^{At}

$$e^{At} = I_n + At/1! + A^2t^2/2! + \cdots$$

it follows that

$$\int_0^h e^{As} \, ds = I_n h + Ah^2/2! + A^2h^3/3! + \cdots \tag{3.48}$$

The implementation of the series representation (3.48) is associated with the same difficulties as the computation of e^{At} by series methods (see Section 3.2). If the norm of A is large, large errors may be introduced in the computed result. To reduce $\| A \|$ it is possible to scale the elements of A by a scalar q. Then we get (Healey, 1973)

$$\int_0^h e^{As} \, ds = [I_n + e^{A^*h} + \cdots + (e^{A^*h})^{q-1}] \int_0^h e^{A^*s} \, ds \qquad A^* = A/q$$

and hence it is not possible to use repeated squaring for $q = 2^m$ as in the computation of the exponential.

Since

$$A \int_0^h e^{As} \, ds = e^{Ah} - I_n$$

we obtain that

$$\int_0^h e^{As} \, ds = A^{-1}(e^{Ah} - I_n) = (e^{Ah} - I_n)A^{-1} \tag{3.49}$$

whenever the matrix A is nonsingular. The use of (3.49) in the presence of rounding errors is restricted to the cases when A is well conditioned.

If the matrix A is nondefective we may use the decomposition

$$A = X \, \mathrm{diag}(\lambda_1, \lambda_2, \ldots, \lambda_n)X^{-1}$$

where $\lambda_1, \lambda_2, \ldots, \lambda_n$ are the eigenvalues of A, and X is the matrix of corresponding eigenvectors. Hence

$$\int_0^h e^{As} \, ds = X \, \mathrm{diag}(d_1, d_2, \ldots, d_n)X^{-1}$$

where

$$d_k = \begin{cases} (e^{\lambda_k h} - 1)/\lambda_k & \lambda_k \neq 0 \\ h & \lambda_k = 0 \end{cases}$$

In the presence of complex conjugate eigenvalues $\sigma \pm i\omega$ we may use the real 2×2 block

$$\Lambda = \begin{bmatrix} \sigma & \omega \\ -\omega & \sigma \end{bmatrix}$$

in the decomposition of A. Now the matrix

$$D = \int_0^h e^{\Lambda s} \, ds$$

is computed by means of real arithmetic only:

$$D = (\sigma^2 + \omega^2)^{-1} \begin{bmatrix} p & q \\ -q & p \end{bmatrix}$$

where

$$p = \sigma(e^{\sigma h} \cos \omega h - 1) + \omega e^{\sigma h} \sin \omega h$$
$$q = \sigma e^{\sigma h} \sin \omega h - \omega(e^{\sigma h} \cos \omega h - 1)$$

This method may give poor results even when the eigenvector matrix X is well conditioned. Indeed, when $\lambda_k h$ is small, the error in the divided difference $(e^{\lambda_k h} - 1)/\lambda_k$ may become very large as a result of cancellation.

When the matrix A is defective or nearly defective, the integral of e^{As} may be computed exploiting the Jordan decomposition

$$A = S \, \mathrm{diag}(J_1, J_2, ..., J_r)S^{-1}$$

of A, where $J_k = \lambda_k I_{m_k} + N_k$ is an $m_k \times m_k$ Jordan block. It is easy to show that

$$\int_0^h e^{As} \, ds = S \, \mathrm{diag}(M_1, M_2, ..., M_r)S^{-1} \qquad M_k = \int_0^h e^{J_k s} \, ds$$

where

$$M_k = \begin{bmatrix} \mu_0 & \mu_1 & \mu_2 & \cdots & \mu_p \\ & \mu_0 & \mu_1 & \cdots & \mu_{p-1} \\ & & \ddots & & \vdots \\ \mathbf{0} & & & \ddots & \mu_1 \\ & & & & \mu_0 \end{bmatrix} \qquad p = m_k - 1 \qquad (3.50)$$

and

$$\mu_i = \left(\int_0^h s^i e^{\lambda_k s} \, ds \right) \Big/ i! =$$

$$\begin{cases} \dfrac{h^i e^{\lambda_k h}}{\lambda_k} \sum_{l=0}^{i} 1/[(-\lambda_k h)^l (i-l)!] + (-\lambda_k)^{-n-1} & \lambda_k \neq 0 \\ h^{i+1}/(i+1)! & \lambda_k = 0 \end{cases}$$

The implementation of (3.50) again requires the computation of divided differences which is associated with large errors in the case of small λ_k and h because of cancellation.

The matrices of the discrete model can be reliably determined exploiting the following result (Van Loan, 1978). If

$$Q = \begin{bmatrix} A & B \\ 0_{m \times n} & 0_{m \times m} \end{bmatrix}$$

then

$$e^{Qh} = \begin{bmatrix} e^{Ah} & \left(\int_0^h e^{As} \, ds \right) B \\ 0_{m \times n} & I_m \end{bmatrix} \qquad (3.51)$$

Hence the matrices A_d and B_d may be obtained simultaneously using some of the methods for computing e^{Qh}. The only disadvantage of this approach is that it is not efficient for large m.

Example 3.15

Let

$$A = \begin{bmatrix} -1 & -6 & 13 \\ 4 & 10 & -8 \\ 1 & 2 & 3 \end{bmatrix} \qquad B = \begin{bmatrix} -7 & 2 \\ 3 & 11 \\ -9 & -4 \end{bmatrix}$$

and $h = 0.1$. Using arithmetic with $\varepsilon = 2^{-56} \approx 1.4 \times 10^{-17}$ and computing e^{Qh} by Algorithm 3.4, we obtain (up to 15 decimal digits)

$$\bar{A}_d = \begin{bmatrix} 0.845\ 705\ 930\ 572\ 588 & -0.751\ 393\ 655\ 175\ 163 \\ 0.570\ 779\ 960\ 596\ 270 & 2.362\ 962\ 679\ 352\ 71 \\ 0.165\ 147\ 051\ 374\ 620 & 0.330\ 294\ 102\ 749\ 239 \end{bmatrix}$$

$$\begin{bmatrix} 1.727\ 806\ 524\ 993\ 08 \\ -1.111\ 623\ 839\ 558\ 47 \\ 1.326\ 677\ 646\ 266\ 65 \end{bmatrix}$$

$$\bar{B}_d = \begin{bmatrix} -1.466\ 197\ 666\ 085\ 72 & -0.513\ 070\ 927\ 770\ 771 \\ 0.756\ 002\ 499\ 766\ 302 & 2.027\ 345\ 695\ 559\ 31 \\ -1.050\ 351\ 067\ 073\ 79 & -0.294\ 933\ 938\ 474\ 527 \end{bmatrix}$$

The comparison with the exact matrices A_d and B_d shows that

$$\| \bar{A}_d - A_d \|_1 / \| A_d \|_1 = 2.3 \times 10^{-16}$$
$$\| \bar{B}_d - B_d \|_1 / \| B_d \|_1 = 1.2 \times 10^{-15}$$

EXERCISES

3.27 Develop an algorithm for discretization of a continuous-time system employing the block-diagonal form of the state matrix.

3.28 Write a computer program for discretization of a continuous-time system based on the expression (3.51) and utilizing the Jordan decomposition in computing the matrix exponential.

3.29 Perform an analysis of the errors in the computed matrices A_d and B_d of the discrete-time model, obtained by using (3.51).

3.30 Develop a method for finding the matrices of a continuous-time model which is equivalent to a given discrete-time state–space description at the sampling points. Is this model unique?

NOTES AND REFERENCES

The sensitivity of the matrix exponential to perturbations in the data has been investigated in the papers of Kågström (1977) and Van Loan (1977). In Section 3.1 we present improved sensitivity estimates based on the Jordan and Schur forms.

All known methods for the computation of the matrix exponential are discussed in the excellent survey paper of Moler and Van Loan (1978). It is shown in this paper that most methods, for instance the methods exploiting the characteristic polynomial, are numerically unstable.

The implementation of the Taylor series in conjunction with scaling and squaring for determining the matrix exponential is considered in Saha and Boriotti (1981). The error analysis of the Padé approximation method is done by Ward (1977). He proved the important result that the condition number of the denominator matrix of the Padé approximation for a scaled matrix has a very low condition number. Ward's program PADE8 is among the best available programs for computation of a matrix exponential.

A very good introduction to the numerical solution of ordinary differential equations may be found in Forsythe, Malcolm and Moler (1977), Ortega and Poole (1981). The multistep methods for solving an ODE are considered in detail in the books of Gear (1971) and Shampine and Gordon (1975). Useful information about the practical solution of nonstiff and stiff ODEs is contained in Shampine, Watts and Davenport (1976) and Shampine and Gear (1979). The specialization of multistep methods for solving large linear systems of ODEs is considered in Enright (1979).

The matrix decomposition methods for finding the matrix exponential are presented in Moler and Van Loan (1978) and in Golub and Van Loan (1983).

Methods for the computation of the integral of a matrix exponential are considered in Johnson and Philips (1971), Healey (1973) and Van Loan (1978).

4

Stability, Controllability and Observability Computations

In this chapter we consider the computational solution of problems related to stability and controllability/observability analysis of linear control systems. An important question is the numerical behaviour of the corresponding methods presented in Chapter 2. It is shown that few of these methods may actually be used on a computer because of the effect of rounding errors. That is why we emphasize numerically stable methods based on orthogonal transformations. It is demonstrated, however, that even numerically stable methods may produce meaningless results either for ill-conditioned problems or when implemented inappropriately. For this reason we also study the sensitivity of the corresponding problem to perturbations in system matrices.

4.1 STABILITY ANALYSIS

We saw in Section 2.3 that the stability of a linear continuous-time system

$$\dot{x}(t) = Ax(t)$$

or a discrete-time system

$$x_{k+1} = Ax_k$$

where A is an $n \times n$ matrix, can be investigated in several ways. This includes determination of the eigenvalues of A, evaluation of its characteristic polynomial using the second method of Lyapunov, or implementation of some localization method. Historically, the methods based on the characteristic polynomial of the state matrix have been first in favour. This was because of the lack of efficient methods for finding the eigenvalues at that time and to the presence of well-developed techniques for root location analysis of polynomials, such as the Routh–Hurwitz and Schur–Cohn criteria. These techniques have the advantage that they allow easy checking of the stability of a system without determining the roots of the characteristic equation. Unfortunately, the methods for finding the characteristic polynomial of a matrix, which do not require computation of the eigenvalues, are numerically unstable. We shall demonstrate this by the Leverrier–Faddeev method which is still widely used because of its simplicity and ease of programming.

For an $n \times n$ matrix A, the Leverrier–Faddeev method finds the coefficients of the characteristic polynomial of A,

$$\det(\lambda I - A) = \lambda^n + a_n \lambda^{n-1} + \cdots + a_1$$

as follows. Let $R_n = I_n$. Then

$$a_{n-k+1} = -\frac{1}{k} \operatorname{tr}(A R_{n-k+1})$$

$$R_{n-k} = a_{n-k+1} I_n + A R_{n-k+1}$$

for $k = 1, 2, \ldots, n$. This leads to the following simple algorithm.

Algorithm of Leverrier–Faddeev

Set $R = I_n$
For $k = 1, 2, \ldots, n$
   Set $P = AR$

$$a_{n-k+1} = -\frac{1}{k} \operatorname{tr}(P)$$

$$R \leftarrow a_{n-k+1} I_n + P$$

Since $a_n = -\operatorname{tr}(A)$, $a_{n-1} = -\frac{1}{2} \operatorname{tr}(a_n A + A^2)$ and so on, the Leverrier–Faddeev algorithm involves computation of the subsequent powers of the matrix A. We have noted in Section 1.4 that this may cause large errors in the final result because of the severe subtractive cancellations which may occur (see Example 1.5). This danger is serious, especially for matrices of large norm or high order. For this reason the coefficients $a_{n-1}, a_{n-2}, \ldots, a_1$ of the characteristic polynomial are computed by the Leverrier–Faddeev algorithm usually with decreasing accuracy.

Example 4.1

Let

$$A = \begin{bmatrix} -12.9 & 24.6 & -25.2 & 38.7 \\ -8.4 & 15.8 & -16.2 & 25.2 \\ -8.0 & 15.6 & -16.3 & 24.6 \\ -4.0 & 8.0 & -8.4 & 12.4 \end{bmatrix}$$

and assume the above algorithm is used in floating-point arithmetic with accuracy $\varepsilon = 2^{-15} \approx 3.05 \times 10^{-5}$. As a result we obtain (up to five decimal digits)

$\bar{a}_4 = 0.100\ 00 \times 10^1$
$\bar{a}_3 = 0.346\ 80 \times 10^0$
$\bar{a}_2 = 0.101\ 84 \times 10^0$
$\bar{a}_1 = -0.167\ 47 \times 10^{-2}$

while the exact answer is

$a_4 = 0.100\ 00 \times 10^1$
$a_3 = 0.350\ 00 \times 10^0$
$a_2 = 0.500\ 00 \times 10^{-1}$
$a_1 = 0.240\ 00 \times 10^{-2}$

We see that even the sign of the coefficient a_1 is not computed correctly. The polynomial

$$\lambda^4 + \bar{a}_4\lambda^3 + \bar{a}_3\lambda^2 + \bar{a}_2\lambda + \bar{a}_1$$

has roots (up to three decimal digits)

$\bar{\lambda}_1 = -0.719$
$\bar{\lambda}_{2,3} = -0.148 \pm 0.357i$
$\bar{\lambda}_4 = 0.0156$

The exact eigenvalues of A are

$\lambda_1 = -0.4$
$\lambda_2 = -0.3$
$\lambda_3 = -0.2$
$\lambda_4 = -0.1$

In this way, the conclusion that the corresponding continuous-time system is unstable, based on the computed coefficients of the characteristic polynomial, is wrong.

Another candidate as a method for stability analysis is the second method of Lyapunov. The implementation of this method to linear systems requires the solution of the continuous (2.14) or discrete (2.19)

Lyapunov equation. We shall see in the next section that the best available methods for solving the Lyapunov equation involve, as an intermediate step, reduction of the matrix A into Schur form. Hence the eigenvalues of A are already available during the solution of the equation. In addition, when using the second method of Lyapunov, the solution must be checked for positive definiteness, which in turn requires determination of its eigenvalues. That is why use of the Lyapunov method for stability analysis of linear time-invariant systems is inefficient. It should be stressed, however, that the Lyapunov equation has a number of other applications in the analysis and design of control systems, which makes its computational solution very important.

The implementation of localization methods for stability analysis also cannot be generally recommended. These methods give conditions which are only sufficient, i.e. if these conditions are not fulfilled it cannot be said whether the system is stable or not.

In this way, the only reliable technique for stability analysis of a linear system is to find the eigenvalues of the state matrix by some numerically stable method, and then to check their disposition about the stability region. For this purpose, it is appropriate to use the double-shifted QR algorithm (see Section 1.10), which is generally recognized as the best method for computing the eigenvalues of a dense matrix. To reduce the computations, the matrix is preliminarily reduced to Hessenberg form by using orthogonal similarity transformations. The full path for finding the eigenvalues of a general nonsymmetric matrix by the QR algorithm requires approximately $5n^3$ flops and may be realized, for instance, by the EISPACK routines ORTHES and HQR. Using the QR algorithm for the stability analysis of linear systems has several implications which will be discussed in detail.

As pointed out in Section 1.10, the QR algorithm finds the exact eigenvalues of a matrix $A + E$, where

$$\| E \|_F \leq k_1 n \varepsilon \| A \|_F$$

and k_1 is a constant of order unity. On the other hand, the perturbation in a simple eigenvalue λ_i of A, owing to the perturbation E, may be $\text{cond}(\lambda_i)$ times $\| E \|_F$ where $\text{cond}(\lambda_i)$ is the individual condition number of λ_i. Hence we can expect that the eigenvalues of A are determined with absolute errors, which satisfy

$$| \Delta\lambda_i | \leq k_1 n \varepsilon \ \text{cond}(\lambda_i) \| A \|_F \qquad i = 1, 2, ..., n \qquad (4.1)$$

By the way, this bound explains why the smaller eigenvalues are usually computed with higher relative errors. It is more important, however, that if there is an eigenvalue on the imaginary axis (or on the unit circle in the discrete-time case) the corresponding system should be considered as unstable, since insignificant perturbations in A may lead to unstable poles. That is why the notion of stability (in contrast to asymptotic stability) is only of theoretical interest.

In this way, the reliable determination of system stability in the case of simple poles involves the determination of the condition numbers $\text{cond}(\lambda_i)$, $i = 1, 2, ..., n$. As mentioned in Section 1.9, these numbers can be computed efficiently by the algorithm of Chan, Feldman and Parlett (1977). This algorithm involves as a preliminary step the reduction of the matrix A into Schur form by the QR method.

Example 4.2

Let

$$A = \begin{bmatrix} -23.21 & 44.62 & -43.62 & 63.83 & -42.62 \\ 0.36 & -1.62 & 1.12 & -1.48 & 1.12 \\ -8.64 & 18.08 & -19.58 & 27.22 & -18.58 \\ -22.48 & 43.36 & -42.36 & 61.84 & -41.36 \\ -13.12 & 23.04 & -21.04 & 32.16 & -21.04 \end{bmatrix}$$

Using the subroutine ORTHES from EISPACK, which reduces A to upper Hessenberg form and the subroutines QR2NOZ and CONDIT, which realize the algorithm of Chan, Feldman and Parlett, in floating-point arithmetic with $\varepsilon = 2^{-56} \approx 1.39 \times 10^{-17}$, we obtain the eigenvalues (up to 17 decimal digits)

$$\bar{\lambda}_1 = -0.199\,999\,999\,999\,998\,88 \times 10^1$$
$$\bar{\lambda}_2 = -0.100\,000\,000\,000\,000\,84 \times 10^1$$
$$\bar{\lambda}_3 = -0.500\,000\,000\,000\,000\,000 \times 10^0$$
$$\bar{\lambda}_4 = -0.999\,999\,999\,999\,941\,80 \times 10^{-1}$$
$$\bar{\lambda}_5 = -0.100\,000\,000\,000\,067\,64 \times 10^{-1}$$

with corresponding condition numbers (up to four decimal digits)

$$\text{cond}(\lambda_1) = 0.1271 \times 10^3$$
$$\text{cond}(\lambda_2) = 0.1044 \times 10^3$$
$$\text{cond}(\lambda_3) = 0.1349 \times 10^2$$
$$\text{cond}(\lambda_4) = 0.1590 \times 10^3$$
$$\text{cond}(\lambda_5) = 0.1730 \times 10^3$$

Hence, the eigenvalues are computed with absolute errors:

$$|\Delta\lambda_1| \leq 1.38 \times 10^{-12}$$
$$|\Delta\lambda_2| \leq 1.14 \times 10^{-12}$$
$$|\Delta\lambda_3| \leq 1.47 \times 10^{-13}$$
$$|\Delta\lambda_4| \leq 1.73 \times 10^{-12}$$
$$|\Delta\lambda_5| \leq 1.88 \times 10^{-12}$$

where we take in (4.1) for definiteness $k_1 = 1$ and make use of $\| A \|_F = 156.65$. This shows that the corresponding continuous-time system is asymptotically stable since for each i

$$\lambda_i = \bar{\lambda}_i - \Delta\lambda_i$$

has a negative real part. In fact, the exact eigenvalues of A are

$$\lambda_1 = -2$$
$$\lambda_2 = -1$$
$$\lambda_3 = -0.5$$
$$\lambda_4 = -0.1$$
$$\lambda_5 = -0.01$$

so that the actual errors in the eigenvalues are

$$\Delta\lambda_1 = 0.1120 \times 10^{-13}$$
$$\Delta\lambda_2 = -0.8400 \times 10^{-14}$$
$$\Delta\lambda_3 = 0$$
$$\Delta\lambda_4 = 0.5820 \times 10^{-14}$$
$$\Delta\lambda_5 = -0.6764 \times 10^{-14}$$

Since the eigenvalue condition numbers are not invariant to diagonal similarity transformations (see Example 1.28), significant improvement of eigenvalue accuracy may be achieved if the matrix A is preliminarily balanced. Theoretically, the matrix A is balanced if a diagonal similarity transformation D is found such that $\| D^{-1}AD \|_2$ is minimized. This transformation, however, is difficult to find and in practice the balancing is done as described in Section 1.10. Since balancing is not time-consuming (the number of necessary computational operations is proportional to n^2) its use is generally recommended. This can be done, in particular, by the subroutine BALANC from EISPACK.

When A is a defective matrix, the QR algorithm may produce eigenvalues with large errors owing to their high sensitivity (the individual condition numbers of eigenvalues corresponding to nonlinear elementary divisor are theoretically infinite, in the sense that the linear estimation (4.1) is not valid). That is why, in addition to the QR method, in such a case it is appropriate to use some grouping strategy, i.e. a technique which recognizes the multiple eigenvalues. For instance, in the algorithm of Kågström and Ruhe (see Section 1.10), the grouping is done by using the Gershgorin theorem which allows us to find the Jordan structure of a matrix reliably.

Example 4.3

Let the eigenvalues of

$$A = \begin{bmatrix} -35.1 & 27.5 & 8.1 & -37.0 & 63.0 \\ 12.0 & -12.1 & 0.5 & 12.0 & -24.0 \\ -20.0 & 12.0 & 7.9 & -21.5 & 33.0 \\ -16.0 & 8.0 & 8.0 & -18.1 & 25.5 \\ -32.0 & 24.0 & 8.0 & -34.0 & 56.9 \end{bmatrix}$$

be determined by the QR algorithm as realized by the EISPACK subroutines CORTH and COMQR2 in floating-point arithmetic with relative precision $\varepsilon = 2^{-23} \approx 1.19 \times 10^{-7}$ (this is approximately the single precision of most computers). The computed eigenvalues are (to three digits):

$\bar{\lambda}_1 = -0.274$
$\bar{\lambda}_2 = -0.148 + 0.162i$
$\bar{\lambda}_3 = -0.148 - 0.162i$
$\bar{\lambda}_4 = 0.0349 + 0.0967i$
$\bar{\lambda}_5 = 0.0349 - 0.0966i$

so that the corresponding continuous-time system appears to be unstable. On the other hand, using the algorithm of Kågström and Ruhe realized by the subroutine JNF in the same precision, we obtain

$\bar{\lambda}_1 = -0.137$
$\bar{\lambda}_2 = -0.100$
$\bar{\lambda}_3 = -0.100$
$\bar{\lambda}_4 = -0.100$
$\bar{\lambda}_5 = -0.0626$

The actual eigenvalues of A are all equal to -0.1 appearing in a 5×5 Jordan block, so that the system is asymptotically stable. The large errors are a result of the fact that the eigenvalues of a Jordan block are very sensitive to perturbations (cf. Example 1.27). Since the errors in the eigenvalues in the given case are of order $\sqrt[5]{\varepsilon} \approx 0.0412$, we see that the Kågström and Ruhe algorithm produces results which are as accurate as possible, while the QR algorithm itself yields a wrong answer. Hence even a numerically stable method for eigenvalues, such as the QR algorithm, cannot yield reliable results when the eigenvalue problem is ill conditioned. In such a case, the equivalent small perturbation E, which reflects the effect of rounding errors in computing the eigenvalues of A, may transform a stable system with state matrix A to an unstable system with matrix $A + E$. In this way, the stability analysis of a linear system is closely related to the determination of the *distance to an unstable system*, i.e. the size of minimum norm perturbation which causes a stability matrix to become a matrix which has at least one eigenvalue with zero real part. The stability analysis by a numerically stable method will be meaningful only if the distance to an unstable system is sufficiently larger than the norm of the equivalent perturbation introduced by the method itself.

For some systems the distance to an unstable system may be very small as shown by the following example.

Example 4.4

Consider a stable continuous-time system with state matrix

$$A = \begin{bmatrix} -0.5 & 1 & 1 & 1 & 1 & 1 \\ & -0.5 & 1 & 1 & 1 & 1 \\ & & -0.5 & 1 & 1 & 1 \\ & & & -0.5 & 1 & 1 \\ & 0 & & & -0.5 & 1 \\ & & & & & -0.5 \end{bmatrix} \qquad (4.2)$$

If the matrix A is perturbed so that its element in position $(6, 1)$ becomes equal to $\frac{1}{324}$, then the eigenvalues of the resulting matrix become (up to four decimal digits)

$$\lambda_1 = -0.8006$$
$$\lambda_{2,3} = -0.7222 \pm 0.2485i$$
$$\lambda_{4,5} = -0.3775 \pm 0.4120i$$
$$\lambda_6 = 0.000$$

Thus a small perturbation may cause a stable system to become unstable. Notice that in the given case the stability margin

$$\alpha(A) = \min_i \{|\operatorname{Re}(\lambda_i)|\} = 0.5$$

is very large in comparison to the size of the smallest destabilizing perturbation for A. Hence the distance to an unstable system is much more reliable as a measure of stability than the stability margin.

If the distance to an unstable system is relatively large, we shall say that the system is *robust* with respect to the stability in the sense of the definition given in Section 2.3.

The problem of finding the distance to an unstable system is complicated. At present, no explicit expression is known for the lowest bound on the perturbation which makes a stable system become unstable. We shall consider below two methods for estimating this bound, based on the second method of Lyapunov and the Gershgorin theorem respectively.

Consider the linear system

$$\dot{x}(t) = (A + E)x(t)$$

where A is an $n \times n$ stability matrix and E is an $n \times n$ error matrix. In practice, the elements of E are not known exactly and one may only have knowledge of the maximum deviation that can be expected in the entries of A. In this case the entries of E fulfil

$$|e_{ij}| \leqq \delta$$

where δ is the magnitude of the maximum deviation. Perturbations of this type are referred to as *structured perturbations* since the perturbation model structure is known and the bounds on the individual elements of the perturbation matrix are also known. If only the norm bound for the error matrix is known, we shall say that the perturbation is *unstructured*.

For the case of structured perturbations we have the following result.

Theorem 4.1

The matrix $A + E$ is a stability matrix if A is a stability matrix and

$$|e_{ij}|_{max} = \delta < \frac{1}{\sigma_{max}(W)}$$
$$W = (Y + Y^{\mathrm{T}})/2$$
$$Y = |P|V$$

where V is an $n \times n$ matrix whose entries are such that $v_{ij} = 0$ if the perturbation in a_{ij} is known to be zero (i.e. $e_{ij} = 0$), and $v_{ij} = 1$ if the perturbation in a_{ij} is known to be nonzero (i.e. $e_{ij} \neq 0$), and P is an $n \times n$ matrix which satisfies the Lyapunov equation

$$A^{\mathrm{T}}P + PA + 2I_n = 0 \tag{4.3}$$

The proof of this theorem can be found in Yedavalli (1985).

Theorem 4.1 allows us to obtain an upper bound on the entries of perturbations which maintain the stability of a system provided the structure of perturbation is known. If this structure is not known, the matrix V should be taken with all entries equal to unity.

Theorem 4.1 can be implemented by the following algorithm.

Algorithm 4.1 Robust stability analysis by the second method of Lyapunov

Solve $A^T P + PA + 2I_n = 0$ by some of the methods considered in Section 4.2.

For a given V compute

$$Y = |P| V \qquad W = (Y + Y^T)/2$$

Compute the singular values

$\sigma_1, \sigma_2, ..., \sigma_n$ of W

Take δ as $\dfrac{1}{\sigma_{max}(W)}$

Example 4.5

Let

$$A = \begin{bmatrix} 9 & 10 & 8 & 6 & 4 \\ -21 & -21 & -16 & -12 & -8 \\ 15 & 14 & 11 & 9 & 6 \\ -4 & -4 & -4 & -3 & 0 \\ -2 & -2 & -2 & -3 & -5 \end{bmatrix} \tag{4.4}$$

The corresponding continuous-time system is asymptotically stable since the eigenvalues of A are

$$\lambda_1 = -3 \qquad \lambda_{2,3} = -2 \pm i \qquad \lambda_{4,5} = -1$$

Suppose first that the perturbation of A is such that for each $i, j = 1, 2, ..., 5$, it is fulfilled that $|e_{ij}| \leq \delta$. In this case, the matrix V has all its entries equal to 1, the solution of the Lyapunov equation (4.3) is (up to three decimal digits)

$$P = \begin{bmatrix} 105.5 & 97.2 & 95.2 & 62.9 & 42.4 \\ 97.2 & 90.7 & 88.9 & 58.4 & 39.4 \\ 95.2 & 88.9 & 87.9 & 57.3 & 38.6 \\ 62.9 & 58.4 & 57.3 & 38.9 & 25.7 \\ 42.4 & 39.4 & 38.6 & 25.7 & 17.5 \end{bmatrix}$$

and

$$\sigma_{max}(W) = 1585.7$$

Hence the system will maintain its stability if the entries of the perturbation matrix fulfil

$$|e_{ij}|_{max} < 6.3064 \times 10^{-4}$$

Suppose now that only the elements of A in positions $(1, 2)$, $(3, 5)$ and $(5, 3)$ are subject to perturbations. In this case

$$V = \begin{bmatrix} 0 & 1 & 0 & 0 & 0 \\ 0 & 0 & 0 & 0 & 0 \\ 0 & 0 & 0 & 0 & 1 \\ 0 & 0 & 0 & 0 & 0 \\ 0 & 0 & 1 & 0 & 0 \end{bmatrix}$$

and using Algorithm 4.1 we obtain that $A + E$ remains a stability matrix if

$$|e_{12}|, \; |e_{35}|, \; |e_{53}| < 4.539 \times 10^{-3}$$

Another estimate of the magnitude of allowable perturbations can be obtained using the Gershgorin theorem. For this purpose it is appropriate to reduce first the matrix A into Schur form

$$T = U^{H}AU = \begin{bmatrix} t_{11} & t_{12} & \cdots & t_{1n} \\ & t_{22} & \cdots & t_{2n} \\ & \mathbf{0} & \ddots & \vdots \\ & & & t_{nn} \end{bmatrix}$$

implementing a similarity transformation on A with a unitary matrix U, and then considering the Gershgorin circles for the matrix $A + E$. To reduce the radii of Gershgorin circles as much as possible, one may use a diagonal similarity transformation on $U^{H}(A + E)U$. In this way the following result is established:

Theorem 4.2

(Petkov, Christov and Konstantinov (1988b)). The system remains stable if the perturbation satisfies

$$\| E_1 \|_\infty \leqq \min_{k} \{s_k\} \qquad E_1 = U^{H}EU$$

where

$$s_n = | \mathrm{Re}(t_{nn}) |$$

$$s_{n-1} = \frac{| \mathrm{Re}(t_{n-1,n-1}) |}{1 + | t_{n-1,n} |/s_n}$$

$$\vdots$$

$$s_1 = \frac{| \mathrm{Re}(t_{11}) |}{1 + \sum_{j=2}^{n} | t_{1j} |/s_j}$$

With respect to the original perturbation we have

$$\| E \|_F = \| E_1 \|_F \le \delta_1 = \frac{1}{\sqrt{n}} \min_{k} \{s_k\}$$

This result is summarized in Algorithm 4.2.

Algorithm 4.2 Robust stability analysis by the Gershgorin theorem

Reduce A to Schur form $T = U^H A U$ by the QR algorithm

For $k = n, n-1, \ldots, 1$

Compute $s_k = \dfrac{|\operatorname{Re}(t_{kk})|}{1 + \displaystyle\sum_{j=k+1}^{n} |t_{kj}|/s_j}$

Find $\delta_1 = \dfrac{1}{\sqrt{n}} \min_{k} \{s_k\}$

Algorithm 4.2 is suitable for the case of unstructured perturbations. Better results can sometimes be obtained by taking A^T instead of A. It is also possible to use different Schur forms of A and then to take the maximum estimate of the norm of the perturbation E.

Example 4.6

For the matrix A from Example 4.5 we have (up to three decimal digits)

$$T = \begin{bmatrix} -2+i & 0 & -0.114+0.325i & -1.92-0.202i & -44.3+9.16i \\ & -3 & 0.0953-2.67i & -1.62-0.210i & -0.167+0.253i \\ & & -1 & 0.112+1.65i & 8.43+12.7i \\ & & \mathbf{0} & -1 & 9.59-0.413i \\ & & & & -2-i \end{bmatrix}$$

so that

$s_1 = 0.0487$
$s_2 = 0.0506$
$s_3 = 0.0549$
$s_4 = 0.172$
$s_5 = 2.00$

and

$$\delta_1 = \frac{s_1}{\sqrt{5}} = 0.0218$$

Comparing the result with that from Example 4.5, we see that for the case of unstructured perturbations better estimation is obtained by using the Gershgorin theorem. Algorithm 4.1, however, may produce better results in the case of structured perturbations. Both methods give estimations of the distance to an unstable system which are usually pessimistic.

Algorithm 4.2 can obviously be modified for robust stability analysis of discrete-time systems. In this case, one must take $1 - |t_{kk}|$ instead of $|\text{Re}(t_{kk})|$ in Algorithm 4.2.

Example 4.7

Consider a discrete-time system with state matrix

$$A = \begin{bmatrix} 1.0 & 0.9 & 0.8 & 0.6 & 0.4 & 0.2 \\ -1.9 & -1.8 & -1.6 & -1.2 & -0.8 & -0.4 \\ 1.7 & 1.7 & 1.5 & 1.2 & 0.8 & 0.4 \\ -1.2 & -1.2 & -1.1 & -1.0 & -0.6 & -0.3 \\ 0.4 & 0.4 & 0.4 & 0.4 & 0.1 & 0.1 \\ 0.1 & 0.1 & 0.1 & 0.1 & 0.2 & 0.0 \end{bmatrix}$$

whose eigenvalues are

$$\lambda_1 = -0.1 \qquad \lambda_2 = -0.1 \qquad \lambda_3 = -0.1$$
$$\lambda_4 = 0.1 \qquad \lambda_5 = 0.1i \qquad \lambda_6 = -0.1i$$

Using Algorithm 4.2 we obtain

$$s_1 = 0.109$$
$$s_2 = 0.124$$
$$s_3 = 0.353$$
$$s_4 = 0.601$$
$$s_5 = 0.446$$
$$s_6 = 0.900$$

so that

$$\delta_1 < \frac{s_1}{\sqrt{6}} = 0.0445$$

EXERCISES

4.1 Check the stability of the continuous-time system with a state matrix

$$A = \begin{bmatrix} 173.9 & -312.5 & 288.0 & -468.0 \\ 108.0 & -192.1 & 175.5 & -288.0 \\ 116.0 & -208.0 & 191.9 & -312.5 \\ 64.0 & -116.0 & 108.0 & -174.1 \end{bmatrix}$$

by determining the eigenvalues of A using the subroutines ORTHES and HQR from EISPACK. Observe the effect of preliminary balancing of A by the subroutine BALANC. Compare the results with those obtained by the subroutine JNF of Kågström and Ruhe.

4.2 Implement the algorithm of Chan, Feldman and Parlett on matrix (4.4). Comment on the result.

4.3 Show that if the matrix A is diagonal, then according to Theorem 4.1 the distance to an unstable system is estimated as

$$\delta \le \frac{1}{\sigma_{max}(W)}$$
$$W = (Y + Y^T)/2$$
$$Y = |A^{-1}| V$$

4.4 Show that if the matrix A is a Jordan block of dimension n with an eigenvalue λ, then the distance to an unstable system, estimated by Algorithm 4.2, satisfies

$$\|E\|_F \le \frac{1}{\sqrt{n}} \frac{|\lambda|^n}{1 + |\lambda| + |\lambda|^2 + \cdots + |\lambda|^{n-1}}$$

4.5 Write computer programs realizing Algorithms 4.1 and 4.2.

4.6 Implement Algorithms 4.1 and 4.2 on matrix (4.2) and compare the results.

4.2 SOLUTION OF LYAPUNOV AND SYLVESTER EQUATIONS

In this section we consider the computational solution of the Lyapunov matrix equation

$$A^T X + XA = C \tag{4.5}$$

and its discrete counterpart

$$A^T XA - X = C \tag{4.6}$$

with respect to the symmetric matrix X, where A is an $n \times n$ matrix and C is an $n \times n$ symmetric matrix.

Equation (4.5) is a particular case of the Sylvester equation

$$AX + XB = C \tag{4.7}$$

where A is an $n \times n$, B is an $m \times m$, C is an $n \times m$ and X is an $n \times m$ matrix.

It may be shown that the Sylvester equation has a unique solution for X if and only if the matrices A and $-B$ have no eigenvalues in common, i.e.

$$\lambda_i + \mu_j \neq 0 \tag{4.8}$$

where λ_i, $i = 1, 2, ..., n$, are the eigenvalues of A and μ_j, $j = 1, 2, ..., m$, are the eigenvalues of B. With respect to the Lyapunov equation (4.5), condition (4.8) reduces to

$$\lambda_i + \lambda_j \neq 0$$

Thus if A is a stability matrix then (4.5) has a unique solution for every C.

The discrete Lyapunov equation (4.6) has a unique solution for X if and only if the matrix A has not reciprocal eigenvalues, i.e.

$$\lambda_i \lambda_j \neq 1$$

Since C and X are symmetric matrices, Equations (4.5) and (4.6) can be represented as systems of $\frac{1}{2}n(n+1)$ linear equations for the $\frac{1}{2}n(n+1)$ elements of the upper triangular part of X. If, for instance $n = 2$, the Lyapunov equation (4.5) is written as

$$Tx = c \tag{4.9}$$

where

$$T = \begin{bmatrix} 2a_{11} & 2a_{21} & 0 \\ a_{12} & a_{11} + a_{22} & a_{21} \\ 0 & 2a_{12} & 2a_{22} \end{bmatrix}$$

$$x = [x_{11}, x_{12}, x_{22}]^T \qquad c = [c_{11}, c_{12}, c_{22}]^T$$

Hence the Lyapunov equation may be solved directly by a linear system of equations of type (4.9). However, since T is an $\frac{1}{2}n(n+1) \times \frac{1}{2}n(n+1)$ matrix, this approach requires about $n^6/24$ computational operations and storage, proportional to $n^4/16$, which makes it unpractical.

The Sylvester equation (4.7) can also be represented as a system of linear equations. Let the nm-vectors

$$c = \text{vec}(C) = [c_{11}, c_{21}, ..., c_{n1}, c_{12}, c_{22}, ..., c_{nm}]^{\text{T}}$$
$$x = \text{vec}(X) = [x_{11}, x_{21}, ..., x_{n1}, x_{12}, x_{22}, ..., x_{nm}]^{\text{T}}$$

be constructed from the elements of the matrices C and X. Then Equation (4.7) is written as a system of nm equations for the nm unknown elements of X,

$$Tx = c \tag{4.10}$$

where $T = I_m \otimes A + B^{\text{T}} \otimes I_n$ and the $nm \times nm$ matrices

$$I_m \otimes A = \begin{bmatrix} A & & & \\ & A & & \mathbf{0} \\ & & \ddots & \\ \mathbf{0} & & & A \end{bmatrix}$$

$$B^{\text{T}} \otimes I_n = \begin{bmatrix} b_{11}I_n & b_{21}I_n & \cdots & b_{m1}I_n \\ b_{12}I_n & b_{22}I_n & \cdots & b_{m2}I_n \\ \vdots & \vdots & & \vdots \\ b_{1m}I_n & b_{2m}I_n & \cdots & b_{mm}I_n \end{bmatrix}$$

are the Kronecker products of A and B^{T}, respectively, with the unit matrix.

The matrix T is nonsingular if and only if A and $-B$ have no eigenvalues in common since its eigenvalues are $\lambda_i + \mu_j$.

The solution of (4.7) through (4.10) requires approximately $n^3 m^3$ computational operations and storage proportional to $n^2 m^2$ which is not acceptable for large n and m. If, for instance, $n = 25$ and $m = 20$, the matrix T has dimension 500×500.

Later we shall present efficient methods for solving (4.5)–(4.7) which avoid formation of large matrices.

4.2.1 A Perturbation Analysis

When the equation $AX + XB = C$ is solved on a computer having machine precision ε, rounding errors of order $\varepsilon \| A \|_{\text{F}}$, $\varepsilon \| B \|_{\text{F}}$ and $\varepsilon \| C \|_{\text{F}}$ will usually be present in A, B and C. It is important to note that these errors occur before any algorithm is implemented to solve the equation. Hence, in the best case, we can expect that the computed solution will satisfy

$$(A + E)\bar{X} + \bar{X}(B + F) = (C + G) \tag{4.11}$$

where

$$\|E\|_F \leq \varepsilon \|A\|_F$$
$$\|F\|_F \leq \varepsilon \|B\|_F$$
$$\|G\|_F \leq \varepsilon \|C\|_F$$

The accuracy of the solution depends on the sensitivity of X to perturbations in A, B and C. In some cases, small changes in these matrices may cause a large change in the solution and this is independent of both the method used and the properties of the machine arithmetic.

Example 4.8

If

$$A = \begin{bmatrix} 21.6 & -23.46 & 66.3 & -42.46 \\ 8.4 & -7.84 & 25.2 & -17.94 \\ 3.4 & -2.34 & 9.7 & -7.94 \\ 12.2 & -11.82 & 36.6 & -25.17 \end{bmatrix}$$

$$B = \begin{bmatrix} 15.2 & 30.42 \\ -7.8 & -15.61 \end{bmatrix}$$

$$C = \begin{bmatrix} 29.38 & 36.79 \\ 15.22 & 22.63 \\ 10.22 & 17.63 \\ 19.21 & 26.62 \end{bmatrix}$$

then the solution of the Sylvester equation (4.7) is

$$X = \begin{bmatrix} 1 & 1 \\ 1 & 1 \\ 1 & 1 \\ 1 & 1 \end{bmatrix}$$

If the matrix A is perturbed by

$$E = 10^{-7} \begin{bmatrix} 1 & 1 & 1 & 1 \\ 1 & 1 & 1 & 1 \\ 1 & 1 & 1 & 1 \\ 1 & 1 & 1 & 1 \end{bmatrix}$$

then the solution becomes (up to five decimal digits)

$$\bar{X} = \begin{bmatrix} 1.015\,1 & 1.029\,4 \\ 0.985\,53 & 0.971\,80 \\ 0.985\,49 & 0.971\,71 \\ 0.993\,01 & 0.986\,37 \end{bmatrix}$$

Thus a relative change in the norm of the matrix A,

$$\frac{\|E\|_F}{\|A\|_F} = 3.86 \times 10^{-9}$$

leads to a relative change in the solution

$$\frac{\|\bar{X} - X\|_F}{\|X\|_F} = 0.0206$$

i.e. the relative perturbations in the data are magnified about 10^7 times in the solution.

The influence of the perturbations E, F and G on the solution X can be analyzed in the following way. The perturbed Sylvester equation (4.11) is written as

$$(T + Y)\bar{x} = c + g \tag{4.12}$$

where

$$T = I_m \otimes A + B^T \otimes I_n$$
$$Y = I_m \otimes E + F^T \otimes I_n$$
$$\bar{x} = \text{vec}(\bar{X})$$
$$g = \text{vec}(G)$$

We have that

$$\|T\|_2 \leq \|A\|_F + \|B\|_F$$
$$\|Y\|_2 \leq \|E\|_F + \|F\|_F$$
$$\|c\|_2 = \|C\|_F \leq (\|A\|_F + \|B\|_F) \|X\|_F$$
$$\|g\|_2 = \|G\|_F$$

Here we used the easily established inequality

$$\|P \otimes Q\|_2 \leq \|P\|_2 \|Q\|_2$$

Proceeding as in the analysis of conditioning of a system of linear equations (Section 1.6), we obtain that if T is nonsingular, c is nonzero and

$$\|T^{-1}\|_2 \|Y\|_2 < 1$$

then

$$\frac{\|\bar{x} - x\|_2}{\|x\|} \leq \frac{\|T^{-1}\|_2}{1 - \|T^{-1}\|_2 \|Y\|_2} \left(\|Y\|_2 + \frac{\|g\|_2}{\|x\|_2} \right)$$

Since

$$\sigma_{\min}(T) = \min_{x \neq 0} \frac{\|Tx\|_2}{\|x\|_2} = \min_{X \neq 0} \frac{\|AX + XB\|_F}{\|X\|_F} = \text{sep}(A, -B)$$

where $\mathrm{sep}(A, -B)$ is the separation between the matrices A and $-B$ (see Section 1.9), we have that

$$\| T^{-1} \|_2 = \frac{1}{\sigma_{\min}(T)} = \frac{1}{\mathrm{sep}(A, -B)}$$

Note that $\mathrm{sep}(A, -B) > 0$ if and only if A and $-B$ have no eigenvalues in common, i.e. $\lambda_i + \mu_j \neq 0$.

Hence we come to the following statement.

Theorem 4.3

Let the matrices A and $-B$ have no eigenvalues in common, $C \neq 0$ and

$$\frac{\| E \|_F + \| F \|_F}{\mathrm{sep}(A, -B)} \equiv \delta < 1$$

Then the relative perturbation in the solution of the Sylvester equation (4.7), due to perturbations E, F and G in the matrices A, B and C, satisfies

$$\frac{\| \bar{X} - X \|_F}{\| X \|_F} \leq \frac{1}{1 - \delta} \frac{\| E \|_F + \| F \|_F + \dfrac{\| G \|_F}{\| X \|_F}}{\mathrm{sep}(A, -B)} \tag{4.13}$$

In case of ε – relative perturbations in A, B and C we obtain that if

$$\delta = \varepsilon \frac{\| A \|_F + \| B \|_F}{\mathrm{sep}(A, -B)} < 1$$

then

$$\frac{\| \bar{X} - X \|_F}{\| X \|_F} \leq \frac{2\varepsilon}{1 - \delta} \frac{\| A \|_F + \| B \|_F}{\mathrm{sep}(A, -B)} \tag{4.14}$$

The quantity

$$\frac{\| A \|_F + \| B \|_F}{\mathrm{sep}(A, -B)}$$

is considered as a *condition number of the Sylvester equation*.

The bounds (4.13) and (4.14) show that if the separation between A and $-B$ is very small, then the change in $\bar{X}$ can be very large. For the equation from Example 4.8

$$\mathrm{sep}(A, -B) = 8.49 \times 10^{-7}$$

so that the perturbation in X may be about 10^7 times the perturbation in A as shown above. Note that $\mathrm{sep}(A, -B)$ may be very small, although the distance between the eigenvalues of A and B is not. In the above example, the eigenvalues of A are -1, -0.5, -0.2 and -0.01 and those of $-B$ are 0.4 and 0.01 so that the distance between the eigenvalues of A and $-B$ is 0.02, which is appreciably larger than $\text{sep}(A, -B)$.

An algorithm for estimation of $\text{sep}(A, -B)$ without computing the matrix $I_m \otimes A + B^T \otimes I_n$, is proposed in Byers (1983).

For the Lyapunov equation (4.5) we get as a corollary that if A and $-A$ have no eigenvalues in common and

$$\frac{2\varepsilon \| A \|_F}{\text{sep}(A^T, -A)} \equiv \delta < 1$$

then

$$\frac{\| \bar{X} - X \|_F}{\| X \|_F} \leq \frac{4\varepsilon}{1 - \delta} \frac{\| A \|_F}{\text{sep}(A^T, -A)}$$

Consider finally the discrete Lyapunov equation (4.6). This equation is represented as

$$Tx = c$$

where

$$c = \text{vec}(C) \quad \text{and} \quad x = \text{vec}(X)$$

and the matrix

$$T = A^T \otimes A^T - I_{n^2}$$

is nonsingular if and only if the eigenvalues of A obey

$$\lambda_i \lambda_j \neq 1$$

The sensitivity of the discrete Lyapunov equation depends on the quantity

$$\text{sep}_d(A^T, A) \equiv \min_{x \neq 0} \frac{\| Tx \|_2}{\| x \|_2} = \min_{X \neq 0} \frac{\| A^T X A - X \|_F}{\| X \|_F}$$

$$= \sigma_{\min}(A^T \otimes A^T - I_{n^2})$$

Note that $\text{sep}_d(A^T, A) > 0$ provided T is nonsingular. For A being a convergent matrix, one has

$$0 < \text{sep}_d(A^T, A) \leq 1$$

Following similar arguments as in the proof of Theorem 4.3 we obtain that if the matrix A has no reciprocal eigenvalues, $c \neq 0$ and

$$\frac{(2\varepsilon + \varepsilon^2)\| A \|_F^2}{\text{sep}_d(A^T, A)} \equiv \delta < 1$$

then

$$\frac{\|\bar{X} - X\|_{\mathrm{F}}}{\|X\|_{\mathrm{F}}} \leq \frac{\varepsilon}{1 - \delta} \frac{(3 + \varepsilon)\|A\|_{\mathrm{F}}^2 + 1}{\mathrm{sep}_d(A^{\mathrm{T}}, A)}$$

4.2.2 The Bartels–Stewart Algorithm

The Sylvester equation

$$AX + XB = C$$

can be solved efficiently if it is transformed into the equivalent equation

$$\tilde{A}\tilde{X} + \tilde{X}\tilde{B} = \tilde{C} \qquad\qquad (4.15)$$

where

$$\tilde{A} = U^{-1}AU \qquad \tilde{B} = V^{-1}BV$$
$$\tilde{C} = U^{-1}CV \qquad X = U^{-1}XV$$

U and V are certain nonsingular matrices and the matrices $\tilde{A}$ and $\tilde{B}$ have some 'simple' form. This is done by the so-called transformation methods which involve the following four steps:

1. Transform A and B into special form via the similarity transformations $\tilde{A} = U^{-1}AU$ and $\tilde{B} = V^{-1}BV$.
2. Solve $U\tilde{C} = CV$ for $\tilde{C}$.
3. Solve the transformed equation $\tilde{A}\tilde{X} + \tilde{X}\tilde{B} = \tilde{C}$ for $\tilde{X}$.
4. Solve $XV = U\tilde{X}$ for X.

In steps 2 and 4, it is necessary to solve linear systems of equations involving the transformation matrices U and V. If for this purpose we use Gaussian elimination with partial pivoting, then we can expect that the computed solution $\bar{X}$ will be contaminated by relative errors of order

$$\varepsilon[\mathrm{cond}(U) + \mathrm{cond}(V)]$$

Thus in step 1 we are limited to using only similarity transformations which are well conditioned. If, for instance, the matrices A and B are reduced to diagonal forms, which allows Equation (4.15) to be solved easily, then the eigenvector matrices U and V may be ill conditioned which will introduce large errors in the solution.

Example 4.9

Let

$$A = \begin{bmatrix} -2.60 & 12.45 \\ -0.50 & 2.39 \end{bmatrix} \qquad B = \begin{bmatrix} 4.60 & 21.54 \\ -0.60 & -2.59 \end{bmatrix}$$

$$C = \begin{bmatrix} 13.85 & 28.80 \\ 5.89 & 20.84 \end{bmatrix}$$

If the matrices A and B are reduced to diagonal forms $\tilde{A} = U^{-1}AU$ and $\tilde{B} = V^{-1}BV$, respectively, then the matrix $\tilde{X}$ in step 3 is found simply from

$$\tilde{x}_{ij} = \frac{\tilde{c}_{ij}}{\tilde{a}_{ii} + \tilde{b}_{jj}} \qquad i = 1, 2 \qquad j = 1, 2$$

Using floating-point arithmetic with $\varepsilon = 2^{-31} \approx 4.66 \times 10^{-10}$ we obtain (up to 10 decimal digits)

$$\bar{X} = \begin{bmatrix} 0.999\ 592\ 606\ 5 & 0.997\ 550\ 507\ 1 \\ 0.999\ 931\ 941\ 9 & 0.999\ 590\ 881\ 9 \end{bmatrix}$$

while the exact answer is

$$X = \begin{bmatrix} 1.000\ 000\ 000 & 1.000\ 000\ 000 \\ 1.000\ 000\ 000 & 1.000\ 000\ 000 \end{bmatrix}$$

For this example

$$\mathrm{cond}_2(U) = 1.6773 \times 10^6 \qquad \text{and} \qquad \mathrm{cond}_2(V) = 4.9025 \times 10^6$$

so that errors of order

$$\varepsilon[\mathrm{cond}_2(U) + \mathrm{cond}_2(V)] = 0.003$$

appear in the solution.

On the other hand

$$\frac{1}{\mathrm{sep}(A, -B)} = 787.97$$

so that the Sylvester equation is relatively well conditioned. Thus the method used is numerically unstable.

In this way it is preferable to use methods which are based on orthogonal transformation of A and B. This allows us to solve simply and accurately the linear equations in steps 2 and 4. Also, if U and V are orthogonal, then

$$\mathrm{sep}(U^T AU, -V^T BV) = \mathrm{sep}(A, -B)$$

so that the sensitivity of the problem is not increased when we pass on to equation (4.15). This is important for the numerical stability of the corresponding method.

The first method for solving the Lyapunov and Sylvester equations, which rely on implementation of orthogonal transformations, is the algorithm of Bartels and Stewart (1972). It involves orthogonal reduction of A and B into Schur form using the QR method. Equation (4.7) is solved by the Bartels–Stewart algorithm as follows. The matrix A is reduced to lower real Schur form $\tilde{A}$ by an orthogonal transformation with a matrix U

$$\tilde{A} = U^T A U = \begin{bmatrix} \tilde{A}_{11} & & & 0 \\ \tilde{A}_{21} & \tilde{A}_{22} & & \\ \vdots & \vdots & \ddots & \\ \tilde{A}_{p1} & \tilde{A}_{p2} & \cdots & \tilde{A}_{pp} \end{bmatrix}$$

where each matrix $\tilde{A}_{ii}$ is of order one or two. Similarly, B is reduced to upper real Schur form $\tilde{B}$ by using the orthogonal matrix V

$$\tilde{B} = V^T B V = \begin{bmatrix} \tilde{B}_{11} & \tilde{B}_{12} & \cdots & \tilde{B}_{1q} \\ & \tilde{B}_{22} & \cdots & \tilde{B}_{2q} \\ & & \ddots & \vdots \\ 0 & & & \tilde{B}_{qq} \end{bmatrix}$$

where again each $\tilde{B}_{ii}$ is of order at most two. If

$$\tilde{C} = U^T C V = \begin{bmatrix} \tilde{C}_{11} & \cdots & \tilde{C}_{1q} \\ \vdots & & \vdots \\ \tilde{C}_{p1} & \cdots & \tilde{C}_{pq} \end{bmatrix}$$

and

$$\tilde{X} = U^T X V = \begin{bmatrix} \tilde{X}_{11} & \cdots & \tilde{X}_{1q} \\ \vdots & & \vdots \\ \tilde{X}_{p1} & \cdots & \tilde{X}_{pq} \end{bmatrix}$$

then Equation (4.7) is equivalent to

$$\tilde{A}\tilde{X} + \tilde{X}\tilde{B} = \tilde{C}$$

If the partitions of $\tilde{A}, \tilde{B}, \tilde{C}$ and $\tilde{X}$ are conformal, then

$$\tilde{A}_{kk}\tilde{X}_{kl} + \tilde{X}_{kl}\tilde{B}_{ll} = \tilde{C}_{kl} - \sum_{j=1}^{k-1} \tilde{A}_{kj}\tilde{X}_{jl} - \sum_{i=1}^{l-1} \tilde{X}_{ki}\tilde{B}_{il}$$

$$\tag{4.16}$$

$$\begin{cases} k = 1, 2, \ldots, p \\ l = 1, 2, \ldots, q \end{cases}$$

These equations may be solved successively for $\tilde{X}_{11}, \tilde{X}_{21}, ..., \tilde{X}_{p1}, \tilde{X}_{12}, \tilde{X}_{22}, ..., \tilde{X}_{pq}$. The solution of (4.7) is then given by $X = U\tilde{X}V^T$.

The transformation of the matrices A and B into real Schur form can be accomplished by the QR algorithm. For this purpose it is possible to use the subroutines ORTHES, ORTRAN and a shortened version of the subroutine HQR2 from EISPACK. The reduction of A into lower real Schur form is done by reducing the transpose of A into upper Schur form and then transposing the result.

The solution for $\tilde{X}_{kl}$ in (4.16) requires the solution of a matrix equation of the form (4.7), where the matrices $\tilde{A}_{kk}$ and $\tilde{A}_{ll}$ are of order at most two. Hence, the solution of (4.16) can be obtained by solving a linear system of order at most four. For instance, if $\tilde{A}_{kk}$ and $\tilde{B}_{ll}$ and both of order two, then

$$\begin{bmatrix} \tilde{a}_{11} + \tilde{b}_{11} & \tilde{a}_{12} & \tilde{b}_{21} & 0 \\ \tilde{a}_{21} & \tilde{a}_{22} + \tilde{a}_{11} & 0 & \tilde{b}_{21} \\ \tilde{b}_{12} & 0 & \tilde{a}_{11} + \tilde{b}_{22} & \tilde{a}_{12} \\ 0 & \tilde{b}_{12} & \tilde{a}_{21} & \tilde{a}_{22} + \tilde{b}_{22} \end{bmatrix} \begin{bmatrix} \tilde{x}_{11} \\ \tilde{x}_{21} \\ \tilde{x}_{12} \\ \tilde{x}_{22} \end{bmatrix} = \begin{bmatrix} \tilde{d}_{11} \\ \tilde{d}_{21} \\ \tilde{d}_{12} \\ \tilde{d}_{22} \end{bmatrix}$$

where $\tilde{a}_{ij}$, $\tilde{b}_{ij}$ and $\tilde{x}_{ij}$ denote the elements of $\tilde{A}_{kk}$, B_{ll} and $\tilde{X}_{kl}$, respectively, and $\tilde{d}_{ij}$ denotes the elements of the right-hand side of (4.16). The system arising from (4.16) can be solved by Gaussian elimination with partial pivoting by the subroutines SGECO, SGEFA and SGESL (or their double precision counterparts) from LINPACK. Once calculated, the solution $\tilde{X}_{kl}$ may be stored in the locations occupied by $\tilde{C}_{kl}$ which is no longer needed. Thus, we have the following algorithm.

Algorithm 4.3 Bartels–Stewart algorithm for solving the Sylvester equation $AX + XB = C$

Reduce A to lower real Schur form

$$\tilde{A} = U^T A U$$

Reduce B to upper real Schur form

$$\tilde{B} = V^T B V$$

Compute $\tilde{C} = U^T C V$

For $l = 1, 2, ..., q$

$$\tilde{C}_{kl} \leftarrow \tilde{C}_{kl} - \sum_{i=1}^{l-1} \tilde{X}_{ki} \tilde{B}_{il} \qquad k = 1, 2, ..., p$$

For $k = 1, 2, \ldots, p$

$$\tilde{C}_{kl} \leftarrow \tilde{C}_{kl} - \sum_{j=1}^{k-1} \tilde{A}_{kj} \tilde{X}_{jl}$$

Solve $\tilde{A}_{kk} \tilde{X}_{kl} + \tilde{X}_{kl} \tilde{B}_{ll} = \tilde{C}_{kl}$ for $\tilde{X}_{kl}$

Compute $X = U \tilde{C} V^T$

Assuming that the Schur decomposition of A and B requires $13n^3$ and $13m^3$ flops respectively, then the overall solution of (4.7) by Algorithm 4.3 requires

$$13(n^3 + m^3) + \tfrac{5}{2}(nm^2 + mn^2)$$

flops. The necessary storage is proportional to $2n^2 + 2m^2 + nm$ assuming the data is overwritten.

Consider the numerical properties of Algorithm 4.3. In simplified notation the computations are carried out as follows. Having matrices $R = U^T A U$ and $S = V^T B V$, the matrix Y in

$$RY + YS = D \qquad D = U^T C V$$

is essentially obtained by using Gaussian elimination and forward substitution to solve the system

$$Ty = d$$

where $T = I_m \otimes R + S^T \otimes I_n$ is a lower quasitriangular matrix and

$$y = \text{vec}(Y) \qquad d = \text{vec}(D)$$

Using the results from Sections 1.6, 1.7 and 1.10 we have that the computed quantities satisfy

$$
\begin{array}{lll}
\bar{U} = U_1 + E_1 & U_1^T U_1 = I & \| E_1 \|_F \le \mu \\
\bar{V} = V_1 + F_1 & V_1^T V_1 = I & \| F_1 \|_F \le \mu \\
\bar{R} = U_1^T (A + E_2) U_1 & & \| E_2 \|_F \le \mu \| A \|_F \\
\bar{S} = V_1^T (B + F_2) V_1 & & \| F_2 \|_F \le \mu \| B \|_F \\
\bar{D} = U_1^T (C + G_1) V_1 & & \| G_1 \|_F \le \mu \| C \|_F \\
(\bar{T} + G_2) \bar{y} = \bar{d} & & \| G_2 \|_2 \le \mu \| \bar{T} \|_2 \\
\bar{X} = U_1 (\bar{Y} + G_3) V_1^T & & \| G_3 \|_F \le \mu \| \bar{Y} \|_F
\end{array}
$$

where

$$
\begin{aligned}
\bar{T} &= I_m \otimes \bar{R} + \bar{S}^T \otimes I_n \\
\bar{y} &= \text{vec}(\bar{Y}) \\
\bar{d} &= \text{vec}(\bar{D})
\end{aligned}
$$

and μ is a small multiple of the relative machine precision ε. We have assumed in the above relations that the solution of the linear system $\bar{T}y = \bar{d}$ is equivalent to introducing a small perturbation G_2 in the matrix $\bar{T}$.

From this we obtain that

$$(T_1 + \Delta T)\bar{y} = d_1 + \Delta d$$

where

$$
\begin{aligned}
&d_1 = \text{vec}(D_1), \\
&D_1 = U_1^T C V_1 \\
&T_1 = I_m \otimes R_1 + S_1^T \otimes I_n \\
&R_1 = U_1^T A U_1 \\
&S_1 = V_1^T B V_1 \\
&\| \Delta T \|_2 \le (2\mu + \mu^2)(\| A \|_F + \| B \|_F) \\
&\| \Delta d \|_2 \le \mu(\| A \|_F + \| B \|_F)\| X \|_F
\end{aligned}
$$

Hence if

$$\mu(2 + \mu) \frac{\| A \|_F + \| B \|_F}{\text{sep}(A, -B)} \le 1/2$$

then

$$\frac{\| \bar{X} - X \|_F}{\| X \|_F} \le (8\mu + 4\mu^2) \frac{\| A \|_F + \| B \|_F}{\text{sep}(A, -B)} \tag{4.17}$$

The inequality (4.17) shows that the computed solution $\bar{X}$ will be contaminated by errors of order

$$\varepsilon \frac{\| A \|_F + \| B \|_F}{\text{sep}(A, -B)}$$

Comparison of (4.17) and (4.14) reveals that the solution of (4.7) by the Bartels–Stewart algorithm is equivalent to introducing small perturbations in the matrices A, B and C. Hence the algorithm of Bartels and Stewart is numerically stable from a backward error analysis point of view.

A similar analysis shows that the norm of the residual

$$A\bar{X} + \bar{X}B - C$$

satisfies

$$\| A\bar{X} + \bar{X}B - C \|_F \le (8\mu + 4\mu^2)(\| A \|_F + \| B \|_F)\| X \|_F$$

Notice that this bound does not depend on $\text{sep}(A, -B)$ so that the residual is always small relative to $\| X \|_F$.

Example 4.10

Consider the solution of the Sylvester equation $AX + XB = C$ where

$$A = \begin{bmatrix} -5 & 2 & 7 & 1 & -8 \\ 3 & -6 & 2 & -2 & -4 \\ -1 & 4 & -2 & 6 & 3 \\ -4 & -7 & -1 & 5 & -2 \\ -2 & 3 & 1 & 0 & 9 \end{bmatrix}$$

$$B = \begin{bmatrix} -13 & -14 \\ 4 & 5 \end{bmatrix}$$

$$C = \begin{bmatrix} 4 & 7 \\ -52 & -27 \\ -2 & -9 \\ 15 & 36 \\ -3 & -46 \end{bmatrix}$$

The exact solution of this equation is

$$X = \begin{bmatrix} -1 & 2 \\ 3 & 1 \\ -2 & -3 \\ -6 & -4 \\ 1 & -2 \end{bmatrix}$$

The solution obtained by Algorithm 4.3 in floating-point arithmetic with $\varepsilon = 2^{-56} \approx 1.39 \times 10^{-17}$ is accurate to thirteen digits, the relative error being

$$\frac{\| \bar{X} - X \|_F}{\| X \|_F} = 0.661 \times 10^{-14}$$

In the given case

$$\text{sep}(A, -B) = 0.0568$$

so that the error is of order

$$\varepsilon \frac{\| A \|_F + \| B \|_F}{\text{sep}(A, -B)} = 1.02 \times 10^{-14}$$

The relative residual is

$$\frac{\| A\bar{X} + \bar{X}B - C \|_F}{\| X \|_F} = 2.28 \times 10^{-15}$$

which is of order

$$\varepsilon(\| A \|_F + \| B \|_F) = 5.80 \times 10^{-16}$$

Algorithm 4.3 can also be used to solve the Lyapunov equation $A^T X + XA = C$. However, it is possible to take advantage of the symmetry of C and X. Let U be orthogonal and $\tilde{A} = U^T A U$ be in real Schur form. Partition $\tilde{A}, \tilde{C} = U^T C U$ and $\tilde{X} = U^T X U$ in the form

$$\tilde{A} = \begin{bmatrix} \tilde{A}_{11} & \tilde{A}_{12} \\ 0 & \tilde{A}_{22} \end{bmatrix}$$

$$\tilde{C} = \begin{bmatrix} \tilde{C}_{11} & \tilde{C}_{12} \\ \tilde{C}_{21} & \tilde{C}_{22} \end{bmatrix}$$

$$\tilde{X} = \begin{bmatrix} \tilde{X}_{11} & \tilde{X}_{12} \\ \tilde{X}_{21} & \tilde{X}_{22} \end{bmatrix}$$

where $\tilde{A}_{11}$, $\tilde{C}_{11}$ and $\tilde{X}_{11}$ are at most of order two. Then from the equation $\tilde{A}^T \tilde{X} + \tilde{X} \tilde{A} = \tilde{C}$ it follows that

$$\tilde{A}_{22}^T \tilde{X}_{22} + \tilde{X}_{22} \tilde{A}_{22} = \tilde{C}_{22} - \tilde{A}_{12}^T \tilde{X}_{21} - \tilde{X}_{21} \tilde{A}_{12}$$

Hence, once $\tilde{X}_{11}$ and $\tilde{X}_{21}$ have been calculated, the size of the problem can be reduced.

The matrix $\tilde{X}_{21}$ is computed as described above in the case of the Sylvester equation. The matrix $\tilde{X}_{11}$ satisfies the Lyapunov equation

$$\tilde{A}_{11}^T \tilde{X}_{11} + \tilde{X}_{11} \tilde{A}_{11} = \tilde{C}_{11}$$

When $\tilde{A}_{11}$ is of order two, this equation gives a new linear system of order three for the three distinct elements of $\tilde{X}_{11}$.

Saving some operations in solving the Lyapunov equation can be realized in the computation of $\tilde{C} = U^T C U$ and $X = U \tilde{X} U^T$. Let $C = Z + Z^T$ where Z is upper triangular. Then

$$\tilde{C} = U^T C U = U^T Z U + (U^T Z U)^T$$

Thus it is necessary to compute only $U^T Z U$ and since Z is upper triangular, the product ZU can be computed with about half the operations required for the computation of CU.

In this way, we obtain Algorithm 4.4.

Algorithm 4.4 Bartels–Stewart algorithm for solving the Lyapunov equation $A^T X + XA = C$

Reduce A to upper real Schur form

$$\tilde{A} = U^T A U$$

Compute $W = U^T Z U$, where Z is the upper triangular part of C with $z_{ii} = c_{ii}/2$, $i = 1, 2, ..., n$

Compute $\tilde{C} = W + W^T$

For $l = 1, 2, ..., p$

　For $k = l, l + 1, ..., p$

$$\tilde{C}_{kl} \leftarrow \tilde{C}_{kl} - \sum_{i=l}^{k-1} \tilde{A}_{ik}^T \tilde{X}_{il}$$

　　Solve $\tilde{A}_{kk}^T \tilde{X}_{kl} + \tilde{X}_{kl} \tilde{A}_{ll} = \tilde{C}_{kl}$ for $\tilde{X}_{kl}$

　For $j = l + 1, l + 2, ..., p$

$$\tilde{X}_{lj} = \tilde{X}_{jl}^T$$

　　For $i = j, j + 1, ..., p$

$$\tilde{C}_{ij} \leftarrow \tilde{C}_{ij} - \tilde{X}_{il} \tilde{A}_{lj} - \tilde{A}_{li}^T \tilde{X}_{lj}$$

$$\tilde{C}_{ji} = \tilde{C}_{ij}^T$$

Compute $W = U Z U^T$, where Z is the upper triangular part of $\tilde{X}$ with $z_{ii} = \tilde{x}_{ii}/2$, $i = 1, 2, ..., n$

Compute $X = W + W^T$

This algorithm requires approximately $13n^3 + \frac{7}{2}n^3$ flops, where the first term is due to the reduction of A to a real Schur form. The necessary storage is proportional to $3n^2$. The solution error obeys (4.17) for $\|B\|_F = \|A\|_F$ and $\text{sep}(A, -B) = \text{sep}(A^T, -A)$.

Example 4.11

The exact solution of the Lyapunov equation $A^T X + XA = C$, where

$$A = \begin{bmatrix} 5.3 & 12.6 & 14.4 & -19.8 \\ -3.6 & -8.2 & -9.0 & 12.6 \\ 3.2 & 7.6 & 8.7 & -12.0 \\ 1.6 & 4.0 & 4.8 & -6.5 \end{bmatrix}$$

$$C = \begin{bmatrix} 6.6 & 4.9 & 20.2 & -8.8 \\ 4.9 & -14.8 & 15.9 & 23.3 \\ 20.2 & 15.9 & 52.8 & -21.6 \\ -8.8 & 23.3 & -21.6 & -40.4 \end{bmatrix}$$

is

$$X = \begin{bmatrix} 1.0 & 1.0 & 0.0 & 1.0 \\ 1.0 & 2.0 & -1.0 & 1.0 \\ 0.0 & -1.0 & 2.0 & 0.0 \\ 1.0 & 1.0 & 0.0 & 2.0 \end{bmatrix}$$

Implementing Algorithm 4.4 in floating-point arithmetic with $\varepsilon = 2^{-56} \approx 1.39 \times 10^{-17}$, the solution is computed accurately to eleven digits with a relative error

$$\frac{\|\bar{X} - X\|_F}{\|X\|_F} = 0.596 \times 10^{-12}$$

Since

$$\text{sep}(A^T, -A) = 3.63 \times 10^{-5}$$

the error is of order

$$\varepsilon \frac{\|A\|_F}{\text{sep}(A^T, -A)} = 1.47 \times 10^{-11}$$

as predicted by the numerical analysis.

The discrete Lyapunov equation

$$A^T X A - X = C$$

can be solved in a similar way. If the matrices $\tilde{C} = U^T C U$ and $\tilde{X} = U^T X U$ are partitioned according to $\tilde{A} = U^T A U$, where $\tilde{A}$ is in real upper Schur form, then the equation $\tilde{A}^T \tilde{X} \tilde{A} - \tilde{X} = \tilde{C}$ is written as

$$\sum_{i=1}^{k} \tilde{A}_{ik}^T \left(\sum_{j=1}^{l} \tilde{X}_{ij} \tilde{A}_{jl} \right) - \tilde{X}_{kl} = \tilde{C}_{kl} \qquad l = 1, 2, \ldots, p \qquad k = l, l+1, \ldots, p$$

These equations are solved successively for $\tilde{X}_{11}, \tilde{X}_{21}, \ldots, \tilde{X}_{p1}$, $\tilde{X}_{22}, \ldots, \tilde{X}_{p2}, \ldots, \tilde{X}_{pp}$. Each block $\tilde{X}_{kl}$ is obtained by solving a symmetric or unsymmetric discrete Lyapunov equation of order one or two.

Hence we have Algorithm 4.5.

Algorithm 4.5 Solution of the discrete Lyapunov equation $A^T X A - X = C$

Reduce A to upper real Schur form

$$\tilde{A} = U^T A U$$

Compute $\tilde{C} = U^{\mathrm{T}} C U$

For $l = 1, 2, ..., p$
  For $k = l, l + 1, ..., p$

$$\tilde{C}_{kl} \leftarrow \tilde{C}_{kl} - \tilde{A}_{kk}^{\mathrm{T}} \sum_{j=1}^{l-1} \tilde{X}_{kj} \tilde{A}_{jl} - \sum_{i=1}^{k-1} \tilde{A}_{ik}^{\mathrm{T}} \sum_{j=1}^{l} \tilde{X}_{ij} \tilde{A}_{jl}$$

  Solve $\tilde{A}_{kk}^{\mathrm{T}} \tilde{X}_{kl} \tilde{A}_{ll} - \tilde{X}_{kl} = \tilde{C}_{kl}$ for $\tilde{X}_{kl}$

$$\tilde{X}_{lk} = \tilde{X}_{kl}^{\mathrm{T}}$$

Compute $X = U \tilde{X} U^{\mathrm{T}}$

Algorithm 4.5 requires approximately $13n^3 + 4n^3$ flops, where the first term is again due to the reduction to Schur form. The necessary storage is proportional to $3n^2$, provided the data is overwritten. The algorithm is numerically stable, the computed solution satisfying

$$\frac{\| \bar{X} - X \|_{\mathrm{F}}}{\| X \|_{\mathrm{F}}} \leq \frac{k a \varepsilon}{\mathrm{sep}_d(A^{\mathrm{T}}, A)}$$

where $a = \max\{1, \| A \|_{\mathrm{F}}^2\}$ and k is a constant of order 10. The corresponding residual obeys

$$\| A^{\mathrm{T}} \bar{X} A - \bar{X} - C \|_{\mathrm{F}} \leq k a \varepsilon \| X \|_{\mathrm{F}}$$

Example 4.12

If

$$A = \begin{bmatrix} -2.8500 & 6.3000 & -5.7000 & -7.5000 \\ 3.9000 & -7.7000 & 8.0000 & 7.8000 \\ 4.1000 & -8.4000 & 8.6000 & 8.4000 \\ 1.7000 & -3.4000 & 3.2000 & 4.1000 \end{bmatrix}$$

$$C = \begin{bmatrix} 6.5410 & -14.1520 & 13.8230 & 15.6685 \\ -14.1520 & 29.5300 & -28.8820 & -32.7590 \\ 13.8230 & -28.8820 & 28.1400 & 31.7050 \\ 15.6685 & -32.7590 & 31.7050 & 36.1950 \end{bmatrix}$$

then the exact solution of the discrete Lyapunov equation is

$$X = \begin{bmatrix} 0.4000 & -0.2000 & 0.2000 & 0.1000 \\ -0.2000 & 0.2000 & -0.1000 & 0.0000 \\ 0.2000 & -0.1000 & 0.2000 & 0.1000 \\ 0.1000 & 0.0000 & 0.1000 & 0.1000 \end{bmatrix}$$

Using Algorithm 4.5 in floating-point arithmetic with $\varepsilon = 2^{-56} \approx 1.39 \times 10^{-17}$ the solution is obtained accurately to eleven digits. The solution error

$$\frac{\| \bar{X} - X \|_F}{\| X \|_F} = 1.03 \times 10^{-11}$$

is of order

$$\varepsilon \, \frac{\| A \|_F^2}{sep_d(A^T, A)} = 8.42 \times 10^{-11}$$

4.2.3 The Hessenberg–Schur Algorithm

The Bartels–Stewart algorithm for the Sylvester equation $AX + XB = C$ may be streamlined by only reducing the matrix A to Hessenberg form. This allows a substantial decrease in the number of required computational operations whenever $n \gg m$.

Let the matrix A be reduced by orthogonal similarity transformations to upper Hessenberg form

$$\tilde{A} = U^T A U \qquad U^T U = I$$

and the matrix B^T be reduced to real upper Schur form

$$\tilde{B} = V^T B V \qquad V^T V = I$$

The orthogonal reduction of A to Hessenberg form can be accomplished by using elementary reflections in $\frac{5}{3} n^3$ flops, provided the matrix U is not accumulated (see Section 1.10).

In this way we obtain the transformed equation

$$\tilde{A}\tilde{X} + \tilde{X}\tilde{B}^T = \tilde{C}$$

where $\tilde{X} = U^T X V$ and $\tilde{C} = U^T C V$, which may be solved in a manner similar to that in the Bartels–Stewart algorithm.

Assuming $\tilde{b}_{k,k-1}$ is zero it follows that

$$(\tilde{A} + \tilde{b}_{kk} I_n) \tilde{x}_k = \tilde{c}_k - \sum_{j=k+1}^{m} \tilde{b}_{kj} \tilde{x}_j \equiv \tilde{d} \qquad (4.18)$$

where

$$\tilde{X} = [\tilde{x}_1, \tilde{x}_2, ..., \tilde{x}_m]$$
$$\tilde{C} = [\tilde{c}_1, \tilde{c}_2, ..., \tilde{c}_m]$$

Thus $\tilde{x}_k$ can be found from the previously computed $\tilde{x}_{k+1}, ..., \tilde{x}_m$ by solving an $n \times n$ system of linear equations with a Hessenberg coefficient matrix. This system is easily solved by Gaussian elimination with partial pivoting in n^2 flops once the right-hand side is known.

If $\tilde{b}_{k,k-1}$ is nonzero then we have

$$\tilde{A}[\tilde{x}_{k-1}, \tilde{x}_k] + [\tilde{x}_{k-1}, \tilde{x}_k] \begin{bmatrix} \tilde{b}_{k-1,k-1} & \tilde{b}_{k,k-1} \\ \tilde{b}_{k-1,k} & \tilde{b}_{kk} \end{bmatrix}$$

$$= [\tilde{c}_{k-1}, \tilde{c}_k] - \sum_{j=k+1}^{m} [\tilde{b}_{k-1,j}\tilde{x}_j, \tilde{b}_{kj}\tilde{x}_j] \equiv [\tilde{d}_1, \tilde{d}_2] \qquad (4.19)$$

This is a $2n \times 2n$ linear system for $\tilde{x}_k$ and $\tilde{x}_{k-1}$. By suitably ordering the variables, the coefficient matrix is obtained as upper triangular with two nonzero subdiagonals. Using Gaussian elimination with partial pivoting, this system can be solved in $6n^2$ flops once the right-hand side is computed. A $2n^2$ workspace is required to carry out the calculation.

The increase in the necessary storage for this algorithm is partly compensated by the fact that the orthogonal matrix U can be stored in a factored form below the diagonal of $\tilde{A}$. This implies that we do not need an $n \times n$ array for U as in the Bartels–Stewart algorithm.

Overall we have the following algorithm.

Algorithm 4.6 Solution of the Sylvester equation $AX + XB = C$ by the Hessenberg–Schur algorithm

Reduce A to upper Hessenberg form

$\tilde{A} = U^T A U$

Reduce B^T to upper real Schur form

$\tilde{B} = V^T B^T V$

Compute $\tilde{C} = U^T C V$

For $k = m, m - 1, \dots, 1$

　　If $\tilde{b}_{k,k-1} = 0$, compute $\tilde{x}_k$ from (4.18)
　　If $\tilde{b}_{k,k-1} \neq 0$, compute $\tilde{x}_{k-1}, \tilde{x}_k$ from (4.19)

Compute $X = U\tilde{X}V^T$

The operational count for Algorithm 4.6 is as shown in Table 4.1.

In counting the operations associated with the determination of $\tilde{X}$, it is assumed that the matrix $\tilde{B}$ has $m/2$ blocks of dimension 2×2 along its diagonal, which is the 'worst' case.

The work count for the Hessenberg–Schur algorithm shows that it

Table 4.1

Reduction of A to upper Hessenberg form and storing U in factored form	$\frac{5}{3}n^3$
Reduction of B^T to upper real Schur form	$13m^3$
Updating the right-hand side	$n^2m + nm^2$
Back substitution for $\bar{X}$	$3n^2m + \frac{1}{2}nm^2$
Obtaining the solution	$n^2m + nm^2$
Total	$\frac{5}{3}n^3 + 13m^3 + 5n^2m + \frac{5}{2}nm^2$

has an advantage over the Bartels–Stewart algorithm when $n \gg m$. If, for instance, $n = 4m$, then the Hessenberg–Schur algorithm requires 0.25 times the work associated with the Bartels–Stewart algorithm. If $n < m$ it is possible to apply the Hessenberg–Schur algorithm to the transposed problem

$$B^T X^T + X^T A^T = C^T$$

The necessary storage for the Hessenberg–Schur algorithm is proportional to $3n^2 + 2m^2 + nm$ which is slightly greater than that for the Bartels–Stewart algorithm.

The Hessenberg–Schur algorithm has the same satisfactory numerical properties as the Bartels–Stewart algorithm. The roundoff error analysis done for the Bartels–Stewart algorithm holds entirely for this case also. Comparison of the accuracy of both methods shows that they are practically indistinguishable.

The Hessenberg–Schur algorithm offers no advantage over the Bartels–Stewart method for the Lyapunov equation. This is because the latter algorithm requires only one Schur decomposition to solve $A^T X + XA = C$.

4.2.4 Solving the Lyapunov Equation for the Cholesky Factor

In some applications it is necessary to solve the Lyapunov equation

$$A^T X + XA = -D^T D \tag{4.20}$$

where A is an $n \times n$ stability matrix and D is an $l \times n$ matrix. Since the matrix $D^T D$ is non-negative definite, Equation (4.20) has a unique non-negative solution for X. Using the Cholesky decomposition (see Section

1.6), X can be factorized as

$$X = Y^T Y$$

where Y is an upper triangular matrix.

In solving (4.20), it is preferable to find the Cholesky factor Y via D rather than $D^T D$ for the following reasons. First, when X is positive definite (and hence nonsingular), its condition number with respect to inversion is

$$\text{cond}_2(X) = \text{cond}_2^2(Y)$$

That is why X may be much more ill conditioned than Y. Also, if we can find Y directly from D we will avoid the loss of accuracy associated with the computation of $D^T D$.

Example 4.13

Let $A = -I_2$ and

$$D = \begin{bmatrix} 1 & 0 \\ 0 & \mu \end{bmatrix}$$

From (4.20) we obtain that

$$X = \frac{1}{2} D^T D = \frac{1}{2} \begin{bmatrix} 1 & 0 \\ 0 & \mu^2 \end{bmatrix}$$

Since

$$2 Y^T Y = D^T D$$

and D is upper triangular, the Cholesky factor Y is given by

$$Y = \frac{1}{\sqrt{2}} D = \frac{1}{\sqrt{2}} \begin{bmatrix} 1 & 0 \\ 0 & \mu \end{bmatrix}$$

Now for $\mu < 1$ we have that

$$\text{cond}_2(X) = \frac{1}{\mu^2}$$

while

$$\text{cond}_2(Y) = \frac{1}{\mu}$$

It can be seen that perturbations of order μ^2 may cause the matrix $D^T D$ to lose its full rank, while Y becomes of rank 1 under perturbations of order μ. If μ is less than the square root of the machine precision, i.e.

$$\mu < \varepsilon^{1/2}$$

then it is not even possible to compute X as a full rank matrix.

The equation

$$A^T(Y^TY) + (Y^TY)A = -D^TD \tag{4.21}$$

can be solved directly for Y by a method due to Hammarling (1982a). This method is based on the Bartels–Stewart algorithm and involves reduction of the matrix A into real Schur form

$$T = U^TAU \tag{4.22}$$

where U is orthogonal and T is upper quasitriangular.
Let the QR decomposition of the matrix D be

$$D = Q\begin{bmatrix} R \\ 0 \end{bmatrix} \tag{4.23}$$

where Q is orthogonal and R is upper triangular. According to (4.23)

$$D^TD = R^TR$$

so that R is a Cholesky factor of D^TD. Using (4.22), Equation (4.21) is transformed into

$$T^T(\tilde{Y}^T\tilde{Y}) + (\tilde{Y}^T\tilde{Y})T = -\tilde{R}^T\tilde{R} \tag{4.24}$$

where $\tilde{Y}$ and $\tilde{R}$ are upper triangular matrices which satisfy

$$\tilde{Y}^T\tilde{Y} = U^T(Y^TY)U$$
$$\tilde{R}^T\tilde{R} = U^T(R^TR)U$$

The matrices $\tilde{Y}$ and $\tilde{R}$ can be found as the Cholesky factors of $U^T(Y^TY)U$ and $U^T(R^TR)U$, respectively. Notice that it is not necessary to form the matrix $U^T(R^TR)U$ explicitly since we can form RU and then find $\tilde{R}$ as the upper triangular matrix of the QR decomposition of RU. After $\tilde{Y}$ is found, the matrix Y is obtained in a similar manner from $\tilde{Y}U^T$.

Equation (4.24) can be solved by a forward substitution for the elements of $\tilde{Y}$ without forming $\tilde{R}^T\tilde{R}$. We refer the reader to Hammarling (1982a), where this procedure is described in detail.

To summarize, the Hammarling method allows solving Equation (4.21) directly for the Cholesky factor, without the loss of accuracy associated with the forming of D^TD.

EXERCISES

4.7   Perform a detailed roundoff error analysis of Algorithms 4.5 and 4.6.

4.8   Write computer programs implementing Algorithms 4.3–4.6.

4.9   Modify Algorithm 4.3 to balance the matrices A and B before solving the Sylvester equation.

4.10 Modify the Bartels–Stewart algorithm to solve efficiently the equations $AX + XA^T = C$ and $AXA^T - X = C$, where C is symmetric.

4.11 Develop an extension of the Bartels–Stewart algorithm to solve the equation $AXB - X = C$, where A and B are square nonsymmetric matrices.

4.12 Modify the Hessenberg–Schur algorithm to solve the equations $AXB - X = C$ and $AXM + LXB = C$, where L and M are square matrices.

4.13 Develop an algorithm to find the Cholesky factor of the Lyapunov equation $A^T X + XA = -D^T D$ using the Hammarling method.

4.3 CONTROLLABILITY/OBSERVABILITY ANALYSIS

In Section 2.4 we presented several criteria for controllability and observability of the linear system

$$\dot{x}(t) = Ax(t) + Bu(t) \tag{4.25}$$
$$y(t) = Cx(t)$$

Although mathematically equivalent, these criteria may produce different results in the presence of rounding errors. That is why, in this section we shall discuss the numerical behaviour of the various methods for testing the controllability of (4.25), the same conclusions being valid by duality for the case of testing observability. Most of the considerations will hold also for discrete-time systems.

When considering the controllability of a linear system in the presence of errors in the data, it should be borne in mind that controllability is *generic*, i.,e. 'almost every' pair (A, B) is completely controllable. Strictly speaking, if a pair (A, B) is considered as a point in a finite-dimensional parameter space, then it may be shown that the set of controllable pairs is open and dense. This means that in a sufficiently small neighbourhood of each controllable pair (A, B) there are no uncontrollable pairs, and every uncontrollable pair (F, G) can be made controllable by arbitrarily small changes in F, G. On the other hand, a controllable pair (A, B) may or may not be close to some uncontrollable pair (F, G).

The most widely used criterion for controllability of (4.25) is that the controllability matrix

$$P = [B, AB, ..., A^{n-1}B]$$

be of full rank, i.e.

$$\text{rank}(P) = n \qquad (4.26)$$

This criterion is easily implemented by the following algorithm:

Compute P
Determine the singular values of P
Check if $\sigma_{min} > \varepsilon\sigma_{max}$, where σ_{min}, σ_{max} are the minimum and maximum singular values of P

Unfortunately, this algorithm may give a wrong answer, since the computations of the controllability matrix may be related to great errors, because of subtractive cancellations in evaluating the powers of A.

Example 4.14

Consider a system with matrices

$$A = \begin{bmatrix} 32.1 & -29.6 & -65.4 & 27.4 \\ 18.0 & -16.2 & -37.2 & 15.0 \\ 9.0 & -8.4 & -18.2 & 7.8 \\ 4.0 & -3.5 & -8.4 & 3.3 \end{bmatrix}$$

$$B = \begin{bmatrix} -5.2 & -11.5 \\ 7.0 & -2.0 \\ -3.0 & -4.5 \\ 6.5 & 0.5 \end{bmatrix}$$

If the computations are carried out in floating-point arithmetic with $\varepsilon = 2^{-19} \approx 1.91 \times 10^{-6}$ we obtain that the singular values of the computed matrix P are

$$\sigma_1 = 14.572\ 8$$
$$\sigma_2 = 9.227\ 16$$
$$\sigma_3 = 0.339\ 661$$
$$\sigma_4 = 0.012\ 181\ 3$$

Since the smallest singular values σ_4 is appreciably larger than $\varepsilon\sigma_1$, we should conclude that the corresponding system is completely controllable. In fact, this system is uncontrollable, the dimension of the controllable part being equal to 3. In this example the rounding errors in computing the terms $A^i B$ cause the controllability matrix to become of numerical rank 4, so that the uncontrollable system is considered as completely controllable. If the matrix P is computed in arithmetic with $\varepsilon = 2^{-56} \approx 1.39 \times 10^{-17}$, then we obtain that $\sigma_4 = 0$ to full precision.

A decrease of the effect of rounding errors on the computation of P can be achieved if this matrix is formed as

$$P = [P_1, P_2, ..., P_n]$$

where

$$\begin{aligned} P_1 &= B \\ P_j &= AP_{j-1} \qquad j = 2, ..., n \end{aligned} \qquad (4.27)$$

This avoids explicit computation of the powers of A and thus avoids subtractive cancellations when $\| A \|$ is large, but $\| A^{n-1} \|$ is small. If we use (4.27) for the system of Example 4.14 we obtain

$$\begin{aligned} \sigma_1 &= 14.557\ 2 \\ \sigma_2 &= 9.173\ 80 \\ \sigma_3 &= 0.306\ 656 \\ \sigma_4 &= 1.497\ 24 \times 10^{-5} \end{aligned}$$

which is a considerably better result.

The errors in computing the controllability matrix P are not the only difficulties which arise in testing controllability by using (4.26). The matrix P can be computed exactly but nevertheless, the dimension of the controllable part of the system may not be determined accurately. This danger is illustrated by the following example.

Example 4.15

Consider a system of tenth-order with matrices

$$A = \begin{bmatrix} 1 & & & & \\ & 2^{-1} & & \text{\Large 0} & \\ & & 2^{-2} & & \\ & & & \ddots & \\ & \text{\Large 0} & & & 2^{-9} \end{bmatrix} \quad \text{and} \quad B = \begin{bmatrix} 1 \\ 1 \\ 1 \\ \vdots \\ 1 \end{bmatrix}$$

which is apparently controllable. The matrix P can be computed and stored exactly on computers with a base β which is a power of two, since $p_{ij} = 2^{(i-1)(1-j)}$. If, however, we compute the singular values of P, we obtain (up to three digits)

$$\sigma_1 = 3.63 \times 10^0 \qquad \sigma_6 = 2.45 \times 10^{-4}$$
$$\sigma_2 = 2.46 \times 10^0 \qquad \sigma_7 = 6.01 \times 10^{-6}$$
$$\sigma_3 = 4.50 \times 10^{-1} \qquad \sigma_8 = 7.12 \times 10^{-8}$$
$$\sigma_4 = 6.22 \times 10^{-2} \qquad \sigma_9 = 3.64 \times 10^{-10}$$
$$\sigma_5 = 5.21 \times 10^{-3} \qquad \sigma_{10} = 6.13 \times 10^{-13}$$

Thus, depending on the choice of the relative zero threshold, the dimension of the controllable part could have any value in the range from 6 to 10. It can be seen from the singular values that a perturbation in P of order 10^{-12} may cause the controllability matrix to become of rank less than 10. It may also be shown that a perturbation with a norm of order 10^{-3} should be made in A to obtain an uncontrollable system. This shows that the original problem is transformed to an intermediate problem (rank determination of the controllability matrix) which is far more sensitive than the original one.

The errors in forming the controllability matrix can be avoided using the criterion proposed by Rosenbrock (see Theorem 2.8). In this case the matrix

$$R = \begin{bmatrix} I_n & & & & & & B \\ -A & I_n & & 0 & & B & \\ & \ddots & \ddots & & & \ddots & \\ 0 & & I_n & B & & 0 \\ & & -A & B & & \end{bmatrix}$$

is obtained without computation and its rank can be determined again by checking the singular values (R is of full rank if and only if the pair (A, B) is controllable).

Example 4.16

Let us determine the singular values of the 16×20 matrix R for the pair (A, B) considered in Example 4.14. Using the same precision $(\varepsilon = 2^{-19} \approx 1.19 \times 10^{-6})$, we obtain

$$\sigma_1 = 99.608\ 5$$
$$\vdots$$
$$\sigma_{15} = 0.001\ 942\ 44$$
$$\sigma_{16} = 0.000\ 023\ 927\ 7$$

Thus the ratio

$$\frac{\sigma_{min}}{\sigma_{max}} = 2.402\ 17 \times 10^{-7}$$

is smaller than ε and the matrix R is considered of rank 15, i.e. the pair is uncontrollable. Hence the Rosenbrock criterion allows us to test reliably the controllability of this system. A disadvantage of the Rosenbrock criterion is the large amount (proportional to n^4) of space and computational work required for determining the rank of R.

The controllability of (4.25) can also be analyzed by using the Grammian

$$W_t = \int_0^t e^{As} BB^T e^{A^T s}\ ds \tag{4.28}$$

If the matrix W_Δ is positive definite for some $t = \Delta$ then by Theorem 2.7 the system is controllable.

The Grammian (4.28) can be evaluated in different ways. If for instance, A is a stability matrix, then for $t \to \infty$ the matrix W_∞ is the solution of the Lyapunov equation

$$A W_\infty + W_\infty A^T + BB^T = 0 \tag{4.29}$$

It is possible, however, for sep$(A, -A^T)$ to be small, which may cause inaccurate determination of W_∞ from (4.29) and a wrong conclusion about the controllability of the system.

In the general case, the Grammian (4.28) may be determined in the following way. Let

$$\begin{bmatrix} E_1(\Delta) & G_1(\Delta) \\ 0 & E_2(\Delta) \end{bmatrix} = \exp\left(\begin{bmatrix} -A & BB^T \\ 0 & A^T \end{bmatrix} \Delta \right) \tag{4.30}$$

where the matrices $E_1(\Delta)$, $E_2(\Delta)$ and $G_1(\Delta)$ have dimension $n \times n$. Then it may be shown (see Van Loan (1978)) that

$$W_\Delta = E_2^T(\Delta) G_1(\Delta) \tag{4.31}$$

The matrix exponential in (4.30) may be computed by some of the methods presented in Chapter 3. The positive definiteness of the matrix W_Δ can be checked by determining its eigenvalues (they must be positive). This is done by the QR algorithm for symmetric matrices.

Example 4.17

Consider the same system as in Example 4.14. Determining the exponential of

$$\begin{bmatrix} -A & BB^{\mathrm{T}} \\ 0 & A^{\mathrm{T}} \end{bmatrix} \Delta$$

for $\Delta = 1$ by diagonalization (the eigenvalues of this matrix are simple) in arithmetic with precision $\varepsilon = 2^{-19} \approx 1.91 \times 10^{-6}$, we obtain (up to three digits)

$$\overline{W}_1 = 10^2 \begin{bmatrix} 1.84 & -0.191 & 0.788 & -0.465 \\ -0.186 & 0.702 & -0.168 & 0.579 \\ 0.787 & -0.170 & 0.349 & -0.268 \\ -0.462 & 0.580 & -0.267 & 0.531 \end{bmatrix}$$

We see that the computed Grammian $\overline{W}_1$ is even nonsymmetric. In the given case this is a result of the ill conditioning of the eigenvector matrix in computing the exponential. The eigenvalues of $\overline{W}_1$, determined in the same arithmetic, are

$\lambda_1 = 2.43 \times 10^2$
$\lambda_2 = 9.89 \times 10^1$
$\lambda_3 = 7.47 \times 10^{-3}$
$\lambda_4 = -8.92 \times 19^{-4}$

The last eigenvalue gives some indication that the matrix W_1 can actually have a zero eigenvalue. Carrying out the computations with precision $\varepsilon = 2^{-56} \approx 1.39 \times 10^{-17}$, we find that the matrix W_1 has an eigenvalue which is zero to full precision.

Consider now the implementation of the eigenvalues of A in the controllability analysis of (4.25). As it was shown in Section 2.4, if the system is controllable, then the matrix $\tilde{B} = X^{-1}B$ will not have a zero row, where X is the matrix of the eigenvectors of A. Unfortunately, as is already known, the matrix X may be ill conditioned, even if the eigenvalues of A are well separated.

Example 4.18

For the system from Example 4.14, let the eigenvalues and eigenvectors of the matrix A be determined by the QR algorithm in arithmetic with precision $\varepsilon = 2^{-19} \approx 1.19 \times 10^{-6}$. As a result one obtains

$\overline{\lambda}_1 = 0.345\ 064$
$\overline{\lambda}_2 = 0.327\ 090$
$\overline{\lambda}_3 = 0.223\ 999$
$\overline{\lambda}_4 = 0.104\ 391$

$\tilde{A} = \text{diag}(\bar{\lambda}_1, \bar{\lambda}_2, \bar{\lambda}_3, \bar{\lambda}_4)$

$$\tilde{B} = \bar{X}^{-1}B = \begin{bmatrix} -1.788\ 10 & -0.140\ 949 \\ -0.012\ 358\ 0 & -0.005\ 263\ 47 \\ 2.506\ 34 & 14.223\ 8 \\ -9.053\ 89 & 0.739\ 779 \end{bmatrix}$$

Since the elements of $\tilde{B}$ are sufficiently larger than ε, we could conclude that the system is completely controllable. In fact, the exact eigenvalues of A are 0.4, 0.3, 0.2, 0.1 and the eigenvalue 0.4 is uncontrollable. In the given case, the matrix $\bar{X}$ is ill conditioned (cond$(\bar{X}) = 8.66 \times 10^5$) so that the errors in computing $\bar{X}^{-1}B$ are large. If the eigenvalues and eigenvectors are computed in arithmetic with $\varepsilon = 2^{-56} \approx 1.39 \times 10^{-17}$, then the eigenvalues of A are obtained to full precision and the elements in the row of $X^{-1}B$, corresponding to $\lambda = 0.4$, are zeros to ten digits.

Better results in using eigenvalues $\lambda_1, \lambda_2, ..., \lambda_n$ of A for controllability analysis are obtained by the criterion of Hautus (see Theorem 2.9). According to this criterion, the system is controllable if and only if

$$\text{rank}\,[A - \lambda_i I, B] = n \tag{4.32}$$

for each $i = 1, 2, ..., n$. In this case, the eigenvectors of A are not used, which avoids large errors in the inversion of an ill-condition eigenvector matrix. The computed eigenvalues, however, can be inaccurate because of their sensitivity, which may lead to a wrong decision in determining the rank in (4.32).

Example 4.19

For the system from Example 4.14, let the singular values of $[A - \lambda_i I, B]$, $i = 1, 2, 3, 4$, be computed using the eigenvalues of A, found in the previous example. The ratio $\sigma_{min}/\sigma_{max}$ for λ_i, $i = 1, 2, 3, 4$, is shown in Table 4.2.

Table 4.2

λ_i	$\sigma_{min}/\sigma_{max}$
0.345 064	$6.568\ 16 \times 10^{-6}$
0.327 090	$7.511\ 83 \times 10^{-6}$
0.223 999	$3.898\ 16 \times 10^{-6}$
0.104 391	$3.125\ 15 \times 10^{-5}$

It can be seen from Table 4.2 that all computed ratios are of order ε which makes it difficult to recover the actual rank.

There is another possibility for testing controllability based on the considerations given in Section 2.6. If there is an uncontrollable pole in the system, it cannot be removed by state feedback. Thus if we apply a state feedback $u(t) = -Kx(t)$ with a random matrix K to a controllable system

$$\dot{x}(t) = Ax(t) + Bu(t)$$

then with 'probability 1' the eigenvalues of $A - BK$ will differ from those of A. If there are common eigenvalues of A and $A - BK$, then these eigenvalues are the uncontrollable poles of the system.

Difficulties with this approach can arise when A or $A - BK$ has ill conditioned eigenvalues. In such cases, it is possible to make a wrong decision as to what eigenvalues of A are common with the eigenvalues of $A - BK$.

Example 4.20

For the system considered in the previous examples, we generate a feedback matrix K whose elements are random numbers from a uniform distribution on $(-1, 1)$. In the given case

$$K = \begin{bmatrix} 0.211\,325 & 0.000\,221\,135 & 0.665\,381 & 0.849\,745 \\ 0.756\,044 & 0.330\,327 & 0.628\,392 & 0.685\,731 \end{bmatrix}$$

Computing the eigenvalues of $A - BK$ in arithmetic with precision $\varepsilon = 2^{-19} \approx 1.91 \times 10^{-6}$, we obtain

$\bar{\lambda}_1 = 11.589\,2$
$\bar{\lambda}_2 = -1.889\,07$
$\bar{\lambda}_3 = 0.307\,471$
$\bar{\lambda}_4 = 0.402\,799$

Since these eigenvalues are sufficiently separated from those of A, we again come to the wrong conclusion that the system is controllable. If the eigenvalues of $A - BK$ are computed in arithmetic with $\varepsilon = 2^{-56}$, we find that $A - BK$ has an eigenvalue which is equal to 0.4 to twelve digits.

In this way we ascertain that almost all of the methods for controllability/observability analysis considered up to now cannot yield a reliable answer in the presence of rounding errors. This is a consequence of the fact that in these methods the original problem is transformed into a problem which may be very sensitive to rounding errors. As we know from Chapter 1, the sensitivity of a problem can be preserved if we use orthogonal transformation A and B. A method, suitable for controllability/observability analysis and based on the orthogonal reduction of the matrices A and B, is considered in the next section.

In analyzing the controllability of a pair (A, B) by some computational method, we can at best test the controllability of a nearby pair $(A + \Delta \bar{A}, B + \Delta \bar{B})$, where the perturbations $\Delta \bar{A}$, $\Delta \bar{B}$ reflect the effect of rounding errors. If, however, the pair (A, B) is uncontrollable, then the perturbations $\Delta \bar{A}$ and $\Delta \bar{B}$ may transform it to a completely controllable pair and vice versa. Thus it is not sufficient to establish that a given system is completely controllable; it is more important to know how far this system is from the nearest uncontrollable system.

Example 4.21

Let us assume a controllable nth order system with matrices

$$
A = \begin{bmatrix} -1 & -1 & \cdots & -1 & -1 \\ 1 & -1 & \cdots & -1 & -1 \\ & \ddots & \ddots & \vdots & \vdots \\ \mathbf{0} & & 1 & -1 & -1 \\ & & & 1 & 1 \end{bmatrix} \quad \text{and} \quad b = \begin{bmatrix} 1 \\ 0 \\ \vdots \\ 0 \\ 0 \end{bmatrix}
$$

Adding the vector $-2^{1-n}[1, 1, ..., 1]$ to the last row of $[A, b]$ we obtain a system that is uncontrollable. This is verified by multiplying the resulting matrix on the left with the vector $[1, 2, 2^2, ..., 2^{n-1}]$. The product is the zero vector which means that the matrix is of rank less than n.

For large n the perturbation introduced in system matrices is very small, so that the original system is close to an uncontrollable one.

As a measure of controllability, we can take the quantity

$$
\mu(A, B) = \min_{\Delta A, \Delta B} \| [\Delta A, \Delta B] \|_2
$$

such that $[(A + \Delta A), (B + \Delta B)]$ is uncontrollable. This quantity is referred to as the *distance of the controllable pair to the nearest uncontrollable pair*. If this distance is small, then small perturbations in A and B may cause the system to become uncontrollable. Hence the controllability test will produce a meaningful answer if the equivalent perturbations $\Delta \bar{A}$ and $\Delta \bar{B}$, due to the effect of rounding errors, satisfy

$$
\| [\Delta \bar{A}, \Delta \bar{B}] \|_2 \ll \mu(A, B)
$$

The distance from an uncontrollable system is invariant to orthogonal transformations of the state and input. In fact, if U and V are orthogonal matrices of dimension $n \times n$ and $m \times m$ respectively, then

$$\mu(U^T A U, U^T B V)$$

$$= \min_{\Delta A, \Delta B} \| [U^T \Delta A U, U^T \Delta B V] \|_2$$

$$= \mu(A, B)$$

Thus the orthogonal transformations do not increase the sensitivity of the original problem to the perturbations ΔA and ΔB and can be used for reliable analysis of controllability and observability.

The determination of the distance to an uncontrollable system may present a difficult problem. It may be shown (see, for instance, Eising (1984b)) that

$$\mu(A, B) = \min_{\lambda} \sigma_n [A - \lambda I, B] \tag{4.33}$$

where $\sigma_n [A - \lambda I, B]$ is the smallest singular value of $[A - \lambda I, B]$ and λ is a complex number. Since the minimum which yields $\mu(A, B)$ is generally not achieved when λ is an eigenvalue of A, the minimization in (4.33) must be carried over the whole complex plane. This is prohibitively expensive, since at each step of the minimization procedure, it is necessary to compute the singular values of a complex $n \times (n + m)$ matrix.

An estimate of $\mu(A, B)$ can be obtained using the following result, proved in Boley and Lu (1986).

Theorem 4.4

For a given completely controllable pair (A, B)

$$\mu(A, B) \leq \left(1 + \frac{\| - A_c \|_2}{\sigma_{n-1}} \right) \sigma_n \tag{4.34}$$

where A_c is the companion matrix for A and

$$\sigma_1 \geq \sigma_2 \geq \cdots \geq \sigma_{n-1} \geq \sigma_n$$

are the singular values of the controllability matrix

$$P = [B, AB, ..., A^{n-1} B]$$

Theorem 4.4 shows that the size of allowable perturbations which preserve controllability is not bounded by the smallest singular value of

P, as might be assumed on the basis of the condition

$$\text{rank}(P) = n$$

In fact, the size of these perturbations is bounded by the ratio between the two smallest singular values of P. To find the smallest perturbation which leads to an uncontrollable system, we must find the gap among these singular values.

Example 4.22

Let

$$A = \begin{bmatrix} -11 & 10 & -17 & 38 \\ 8 & -6 & 12 & -26 \\ -2 & 0 & 1 & 3 \\ -6 & 4 & -7 & 18 \end{bmatrix} \quad \text{and} \quad B = \begin{bmatrix} -1 & 4 \\ 1 & -2 \\ 1 & 1 \\ 0 & 2 \end{bmatrix}$$

The characteristic polynomial of A is

$$\lambda^4 - 2\lambda^3 - \lambda + 2$$

so that the companion matrix for A is

$$A_c = \begin{bmatrix} 0 & 1 & 0 & 0 \\ 0 & 0 & 1 & 0 \\ 0 & 0 & 0 & 1 \\ -2 & 1 & 0 & 2 \end{bmatrix}$$

and $\| A_c \|_2 = 3.095\,57$. Since the two smallest singular values of P are (to six digits)

$$\sigma_3 = 0.982\,404 \quad \text{and} \quad \sigma_4 = 0.143\,501$$

it follows that

$$\mu(A, B) \leq 0.595\,674$$

   A disadvantage of using (4.34) to estimate the distance to an uncontrollable system is that it is necessary to find the companion matrix for A which is equivalent to computing the characteristic polynomial of A. As previously noted, this may be related to large errors because of the numerical instability of the corresponding methods. Also, it is required to compute the singular values of the controllability matrix, which, as shown in Example 4.14, may present a difficult task due to the rounding errors. Finally, we note that (4.34) is an upper estimate for $\mu(A, B)$, while a lower estimate would be more useful in general. In fact, a reasonable use of (4.34) is possible only if its right-hand side is small, which indicates that the system is close to an uncontrollable system. Other ways to estimate $\mu(A, B)$ are presented in the next section.

The computational analysis of controllability/observability may be affected significantly by the choice of variables in the state–space model (4.25). The state variables are usually chosen from physical considerations and the resulting matrices A, B and C may have elements whose magnitudes are spread over a large interval. These elements are contaminated by errors due to the uncertainty in the data, and rounding errors introduced in computation and storing matrix elements. Hence the individual elements of system matrices will be characterized by absolute errors, which could be very different in size. In such a case, the final result may be obtained with a large error. To minimize the solution error, the model should be properly *scaled*, so that the absolute errors in the elements of system matrices are nearly the same in magnitude. Scaling is done by using the nonsingular transformations

$$\tilde{x}(t) = D_x x(t) \qquad \tilde{u}(t) = D_u u(t) \qquad \tilde{y}(t) = D_y y(t)$$

where D_x, D_u and D_y are diagonal matrices with positive elements. The scaled model is then given by

$$\dot{\tilde{x}}(t) = \tilde{A}\tilde{x}(t) + \tilde{B}\tilde{u}(t)$$
$$\tilde{y}(t) = \tilde{C}\tilde{x}(t)$$

where

$$\tilde{A} = D_x A D_x^{-1} \qquad \tilde{B} = D_x B D_u^{-1} \qquad \tilde{C} = D_y C D_x^{-1}$$

The matrices D_x, D_u and D_y are chosen so that the uncertainties in the individual elements of $\tilde{A}$, $\tilde{B}$ and $\tilde{C}$ become nearly the same. This means that if the relative errors in the elements of A, B and C are of the same order, then D_x, D_u and D_y must be determined so that the elements of $\tilde{A}$, $\tilde{B}$ and $\tilde{C}$ have approximately equal magnitudes. In such a case we must use some kind of equilibration of system matrices.

The scaling of the model should be done so that the sensitivity of the original problem does not deteriorate. In the case of the controllability analysis, this means that if the original system is far from an uncontrollable system, then the model should be far from an uncontrollable model and vice versa. Otherwise, the scaling should not significantly alter the distance of the given system from an uncontrollable system.

The above considerations show that the proper choice of a state–space model for a given system is not a trivial problem. This choice is hard to formalize and depends very much on the experience and intuition of the engineer who builds the system model.

EXERCISES

4.14 Perform a rounding error analysis for the computation of the controllability matrix.

4.15 Write computer programs implementing the controllability criteria of Rosenbrock and Hautus.

4.16 Check the controllability of the system considered in Example 4.15 by the Hautus criterion. Comment on the result.

4.17 Give an error bound for the eigenvalues of the matrix W_∞ in (4.29) utilizing the results from Sections 4.2 and 1.9.

4.18 Derive a minimization procedure for determining the distance from an uncontrollable system, based on the expression (4.33).

4.19 Find an estimation of the distance from an uncontrollable system for the system considered in Example 4.15.

4.20 Find estimations of the distance from an uncontrollable system for the system considered in Example 4.22, for different scaling matrices D_x and D_u. Comment on the results.

4.4 ORTHOGONAL REDUCTION TO CANONICAL FORM

The controllability/observability analysis of a linear system

$$\dot{x}(t) = Ax(t) + Bu(t)$$
$$y(t) = Cx(t)$$

is closely related to the problem of finding a suitable canonical form of the system. As shown in Section 2.5, the matrices of a controllable system can be reduced into simple form, implementing similarity transformations which involve columns of the controllability matrix. Unfortunately, for several reasons, the canonical forms considered in Section 2.5 (the controllable canonical form and the Luenberger canonical form) cannot be obtained by numerically stable methods. The first reason is that the columns of the controllability matrix may be computed with large errors because of subtractive cancellations in computing the powers of the matrix A (see Example 4.14). Next, it is possible that the matrix S of the similarity transformation $x(t) = S\tilde{x}(t)$, which reduces the system into a canonical form, is ill conditioned so that the products $S^{-1}AS$ and $S^{-1}B$ are contaminated by large errors.

Example 4.23

Consider the reduction of a single-input system with matrices

$$A = \begin{bmatrix} 47.8 & 85.5 & -16.0 & -412.8 \\ 36.3 & 61.5 & -14.25 & -306.05 \\ -8.6 & -11.5 & 6.5 & 67.1 \\ 13.6 & 23.5 & -5.0 & -115.6 \end{bmatrix} \quad \text{and} \quad b = \begin{bmatrix} 2.5 \\ -6.1 \\ 4.7 \\ 9.2 \end{bmatrix}$$

into controllable canonical form. The matrix S of the corresponding similarity transformation can be determined from

$$S^{-1} = \begin{bmatrix} s_1 \\ s_1 A \\ \vdots \\ s_1 A^{n-1} \end{bmatrix}$$

where s_1 is the last row of the inverse of the controllability matrix

$$P = [b, Ab, ..., A^{n-1}b]$$

Performing the computations in floating-point arithmetic with relative precision $\varepsilon = 2^{-19} \approx 1.91 \times 10^{-6}$ we obtain (up to four decimal digits)

$$s_1 = 10^{-2}[-0.3453, \quad 0.4200, \quad 0.2289, \quad 0.2554]$$

and

$$S^{-1} = 10^{-2} \begin{bmatrix} -0.3453 & 0.4200 & 0.2289 & 0.2554 \\ 0.2442 & -0.3258 & -0.2488 & -0.1553 \\ -0.1236 & 0.05865 & -0.1061 & 0.1413 \\ -0.9454 & -2.421 & -0.2542 & 9.622 \end{bmatrix}$$

As a result we find that the matrices of the canonical form are

$$\bar{A} = \overline{S^{-1}AS} = \begin{bmatrix} 0 & 1 & 0 & 0 \\ 0 & 0 & 1 & 0 \\ 0 & 0 & 0 & 1 \\ -0.09811 & 0.4011 & -0.2994 & 0.1963 \end{bmatrix}$$

$$\bar{b} = \overline{S^{-1}b} = \begin{bmatrix} 0.0000 \\ 0.0000 \\ 0.001346 \\ 0.9974 \end{bmatrix}$$

The exact answer is

$$\tilde{A} = \begin{bmatrix} 0 & 1 & 0 & 0 \\ 0 & 0 & 1 & 0 \\ 0 & 0 & 0 & 1 \\ -0.1 & 0.4 & -0.3 & 0.2 \end{bmatrix} \quad \text{and} \quad \tilde{b} = \begin{bmatrix} 0 \\ 0 \\ 0 \\ 1 \end{bmatrix}$$

so that the elements of $\bar{A}$ and $\bar{b}$ contain errors of order 10^{-2}.

In the given case, the controllability matrix (and hence the vector s_1) is obtained with large errors because of cancellations in computing the terms Ab, A^2b and A^3b and these errors are then magnified in finding the matrix S^{-1}. Also, the matrix S^{-1} is relatively ill conditioned $(\text{cond}_2(S^{-1}) = 1.61 \times 10^3)$ which introduces additional errors in computing $\tilde{A}, \tilde{b}$.

The same difficulties arise in the computation of the Luenberger canonical form in the multi-input case. These difficulties can be removed if we apply orthogonal transformations on the system matrices, obtaining the so-called *orthogonal canonical forms*. In such a case, it is possible to keep the rounding errors small in the computation and implementation of the transformations, so that the final result is exact for slightly perturbed system matrices. In other words, using orthogonal tranformations it is possible to reduce the system into a canonical form in a numerically stable way.

In the single-input case the pair (A, b) may be reduced to the orthogonal canonical form $(\tilde{A}, \tilde{b})$,

$$\tilde{A} = U^T A U = \begin{bmatrix} a_{11} & a_{12} & \cdots & a_{1n} \\ a_{21} & a_{22} & \cdots & a_{2n} \\ & a_{32} & \cdots & a_{3n} \\ \mathbf{0} & & \ddots & \vdots \\ & & a_{n,n-1} & a_{nn} \end{bmatrix} \qquad \tilde{b} = U^T b = \begin{bmatrix} b_1 \\ 0 \\ 0 \\ \vdots \\ 0 \end{bmatrix} \qquad (4.35)$$

where U is an orthogonal matrix. If the system is controllable, then

$$b_1 \neq 0 \qquad a_{i,i-1} \neq 0 \qquad i = 2, \ldots, n$$

This form is also referred to as the *Hessenberg form* of the pair (A, b).

In the multi-input case a controllable pair (A, B) can be reduced to the orthogonal canonical form $(\tilde{A}, \tilde{B})$

$$\tilde{A} = U^T A U = \begin{bmatrix} A_{11} & A_{12} & \cdots & A_{1p} \\ A_{21} & A_{22} & \cdots & A_{2p} \\ & A_{32} & \cdots & A_{3p} \\ \mathbf{0} & & \ddots & \vdots \\ & & A_{p,p-1} & A_{pp} \end{bmatrix} \qquad \tilde{B} = U^T B = \begin{bmatrix} B_1 \\ 0 \\ 0 \\ \vdots \\ 0 \end{bmatrix} \qquad (4.36)$$

where p is the controllability index of the system, B_1 is $m_1 \times m$ and $A_{i,i-1}$, $i = 2, \ldots, p$, are $m_i \times m_{i-1}$ matrices,

$$\text{rank}(B_1) = m_1$$
$$\text{rank}(A_{i,i-1}) = m_i \qquad i = 2, \ldots, p$$

and the numbers

$$m_1 \geq \cdots \geq m_p \qquad m_1 + \cdots + m_p = n$$

are the *conjugate Kronecker indices* of the pair (A, B). The form (4.36) is also said to be the *block-Hessenberg form* of the pair (A, B). Similarly, a completely observable pair (C, A) can be reduced in the single-output case to the form

$$\tilde{A} = U^{\mathrm{T}} A U = \begin{bmatrix} a_{11} & a_{12} & & & & \\ a_{21} & a_{22} & a_{23} & & \mathbf{0} & \\ \vdots & \vdots & \vdots & \ddots & & \\ a_{n-1,1} & a_{n-1,2} & a_{n-1,3} & \cdots & a_{n-1,n} \\ a_{n1} & a_{n2} & a_{n3} & \cdots & a_{nn} \end{bmatrix} \qquad (4.37)$$

$$\tilde{c} = cU = [c_1, \quad 0, \quad 0, \quad \ldots, \quad 0]$$

and in the multi-output case into

$$\tilde{A} = U^{\mathrm{T}} A U = \begin{bmatrix} A_{11} & A_{12} & & & & \\ A_{21} & A_{22} & A_{23} & & \mathbf{0} & \\ \vdots & \vdots & \vdots & \ddots & & \\ A_{q-1,1} & A_{q-1,2} & A_{q-1,3} & \cdots & A_{q-1,q} \\ A_{q1} & A_{q2} & A_{q3} & \cdots & A_{qq} \end{bmatrix} \qquad (4.38)$$

$$\tilde{C} = CU = [C_1, \quad 0, \quad 0, \quad \ldots, \quad 0]$$

where q is the observability index of the system, C_1 is $r \times r_1$ and $A_{i,i+1}$, $i = 1, \ldots, q-1$ are $r_i \times r_{i+1}$ matrices,

$$\mathrm{rank}(C_1) = r_1$$
$$\mathrm{rank}(A_{i,i+1}) = r_{i+1} \qquad i = 1, \ldots, q-1$$

and

$$r_1 \geq \cdots \geq r_q \qquad r_1 + \cdots + r_q = n$$

are the conjugate Kronecker indices of the pair (C, A).

Although having a greater number of nonzero elements than the canonical forms presented in Section 2.5, the orthogonal canonical forms have the same advantages in the synthesis of linear systems.

Using orthogonal transformations also allows us to determine reliably the dimension of the controllable/observable part of the system in the case of uncontrollable/unobservable system.

Later we present computational algorithms for finding the forms (4.35) and (4.36). The forms (4.37) and (4.38) can be obtained in a similar way, working with the pair $(A^{\mathrm{T}}, C^{\mathrm{T}})$ instead of (A, B), and then transposing the result. The canonical forms of a discrete-time system are obtained by the same algorithms as in the continuous-time case.

4.4.1 Single-input Case

Consider for definiteness a fifth-order system whose matrices are initially of the form

$$
A = \begin{bmatrix} x & x & x & x & x \\ x & x & x & x & x \\ x & x & x & x & x \\ x & x & x & x & x \\ x & x & x & x & x \end{bmatrix} \quad \text{and} \quad b = \begin{bmatrix} x \\ \widehat{x} \\ \widehat{x} \\ \widehat{x} \\ \widehat{x} \end{bmatrix} \tag{4.39}
$$

Let an elementary reflection U_1 be constructed so that $\tilde{b} = U_1 b$ has zeros in the positions encircled in (4.39). (The construction and implementation of elementary reflections is discussed in Section 1.7.) This reflection is also applied to the matrix A, yielding a matrix

$$
A_2 = U_1 A U_1 = \begin{bmatrix} x & x & x & x & x \\ x & x & x & x & x \\ \widehat{x} & x & x & x & x \\ \widehat{x} & x & x & x & x \\ \widehat{x} & x & x & x & x \end{bmatrix} \tag{4.40}
$$

which is not of a special form (note that the elementary reflection is symmetric, i.e. $U_1^T = U_1$).

At the second step a reflection

$$
U_2 = \begin{bmatrix} 1 & 0 \\ 0 & U_2' \end{bmatrix}
$$

is determined, such that $U_2 A_2$ has zeros in the positions distinguished in (4.40). The postmultiplication of A_2 by U_2 leaves the first column of $U_2 A_2$ unchanged. The matrix $\tilde{b}$ is also unaffected at this step. The transformed matrix $A_3 = U_2 A_2 U_2$ appears thus

$$
A_2 = \begin{bmatrix} x & x & x & x & x \\ x & x & x & x & x \\ 0 & x & x & x & x \\ 0 & \widehat{x} & x & x & x \\ 0 & \widehat{x} & x & x & x \end{bmatrix} \tag{4.41}
$$

Next, a reflection

$$
U_3 = \begin{bmatrix} I_2 & 0 \\ 0 & U_3' \end{bmatrix}
$$

is found so that $U_3 A_3$ acquires two zeros in its second column as shown in (4.41). This tranformation does not change the zeros in the first column of A_3 and in $\tilde{b}$. Also, the right multiplication by U_3 leaves the first two columns of $U_3 A_3$ unaffected. Hence the matrix $A_4 = U_3 A_3 U_3$ will have the form

$$A_4 = \begin{bmatrix} x & x & x & x & x \\ x & x & x & x & x \\ 0 & x & x & x & x \\ 0 & 0 & x & x & x \\ 0 & 0 & \widetilde{x} & x & x \end{bmatrix} \tag{4.42}$$

A transformation with

$$U_4 = \begin{bmatrix} I_3 & 0 \\ 0 & U_4' \end{bmatrix}$$

zeros the single element below the subdiagonal in (4.42).

In this way, after the first step, the described procedure is a reduction of the matrix A_2 into upper Hessenberg form $A_5 = U_4 A_4 U_4$. In the actual algorithm, presented later, an additional tranformation with

$$U_5 = \begin{bmatrix} I_4 & 0 \\ 0 & U_5' \end{bmatrix}$$

is applied for generality, which does not change the form of A_5 and $\tilde{b}$. Thus

$$\tilde{A} = U^T A U \qquad \tilde{b} = U^T b$$

where $U = U_1 U_2 \dots U_5$, are the matrices of the orthogonal canonical form of (A, b).

In the above presentation, it was implicitly assumed that at each step the part of A (or b), which is to be transformed, is not a zero vector. If it is a zero vector, then the system is not completely controllable and the next transformations may be skipped. This is illustrated by the situation when

$$A_3 = \begin{bmatrix} x & x & x & x & x \\ x & x & x & x & x \\ 0 & 0 & x & x & x \\ 0 & 0 & x & x & x \\ 0 & 0 & x & x & x \end{bmatrix} \qquad \text{and} \qquad \tilde{b} = \begin{bmatrix} x \\ 0 \\ 0 \\ 0 \\ 0 \end{bmatrix}$$

In this case, after the second step, the last three elements in the second column of A_3 become equal to zero. The remaining 3×3 sub-matrix in the bottom-right corner of A_3 corresponds to the uncontrollable part of the system. Thus, the dimension of the controllable part of the system is equal to 2. The stabilizability of the system may be checked by determining the eigenvalues of the uncontrollable part.

In practice, the computations described are contaminated by rounding errors which may turn the zero element into nonzero and vice versa. That is why the system may be considered as uncontrollable if the norm of the corresponding part of A (or b), which is to be transformed, is smaller than some tolerance, i.e. if the numerical rank of this part is zero. The tolerance should reflect the effect of rounding errors during the transformations and it is reasonable to take it as $c\varepsilon$ max-$(\| A \|_F, \| b \|_F)$ for some positive constant c. If this tolerance is increased, the determined dimension of the controllable part becomes less sensitive to perturbations in A and b, but at the same time the ability of the algorithm to recognize completely controllable systems will decrease.

The above considerations are summarized in Algorithm 4.7 which overwrites A and b with $\tilde{A}$ and $\tilde{b}$, respectively. The algorithm yields an integer l, which is equal to the dimension of the controllable part of the system. An auxiliary n-vector v is used as a working array during the reduction.

Algorithm 4.7 Reduction of a single-input system into orthogonal canonical form

Set $k = 0$, $l = 0$

Set tol $= n^2 \max \{\| A \|_F, \| b \|_F\}$

$v \leftarrow b$, $b \leftarrow 0$
$k \leftarrow k + 1$

Construct an elementary reflection U_k' of order $n - k + 1$ such that

$$
U_k' \begin{bmatrix} v_k \\ v_{k+1} \\ \vdots \\ v_n \end{bmatrix} = \begin{bmatrix} -g_k \\ 0 \\ \vdots \\ 0 \end{bmatrix}
$$

If fl(tol $+ | g_k |) =$ tol, quit
$l \leftarrow l + 1$
For $j = k, k + 1, ..., n$

$$\begin{bmatrix} a_{kj} \\ a_{k+1,j} \\ \vdots \\ a_{nj} \end{bmatrix} \leftarrow U_k{}' \begin{bmatrix} a_{kj} \\ a_{k+1,j} \\ \vdots \\ a_{nj} \end{bmatrix}$$

For $i = 1, 2, \ldots, n$

$\quad [a_{ik}, a_{i,k+1}, \ldots, a_{in}] \leftarrow [a_{ik}, a_{i,k+1}, \ldots, a_{in}] U_k{}'$

If $k = 1$, $b_1 \leftarrow -g_1$

If $k > 1$, $a_{k,k-1} \leftarrow -g_k$

If $k = n$, quit

For $i = k + 1, k + 2, \ldots, n$

$\quad v_i \leftarrow a_{ik}, \ a_{ik} \leftarrow 0$

The algorithm works as follows. At step $k = 1$, the matrix b is reduced to $\tilde{b}$ by single elementary reflection, which is implemented on A. At step $k \geq 2$ the $(k-1)$th column of A is transformed. The elementary reflections are constructed by Algorithm 1.16 and applied on A by Algorithms 1.17 and 1.18 (if required, the transformations at each step can be accumulated). If the first element of $\tilde{b}$ or a subdiagonal element of A is negligible (i.e. the numerical rank of the corresponding transformed part is zero), the reduction quits. The test for negligibility of an element e is of the form

$\quad$ if $\mathrm{fl}(\mathrm{tol} + |e|) = \mathrm{tol}$, quit

which is equivalent to

$\quad$ if $|e| < \varepsilon\, \mathrm{tol}$, quit

where

$\quad \mathrm{tol} = n^2 \max\{\| A \|_{\mathrm{F}}, \| b \|_{\mathrm{F}}\}$

The use of the factor n^2 is justified by the numerical analysis of the algorithm presented later.

In the case of a completely controllable system, Algorithm 4.7 required about $\frac{5}{3}n^3 + \frac{9}{2}n^2$ flops. The accumulation of the transformations requires additional $n^3 + 2n^2$ flops. The necessary storage is proportional to $n^2 + 2n$ ($2n^2 + 2n$, if the transformation matrix is desired).

The numerical properties of Algorithm 4.7 are very favorable. It will be shown that the computed pair $(\tilde{A}, \tilde{b})$ is the exact canonical form of a pair $(A + \Delta A, b + \Delta b)$ where $\| \Delta A \|$ and $\| \Delta b \|$ are small compared to $\| A \|$ and $\| b \|$. The error analysis we shall present is for controllable systems, which corresponds to largest error bounds.

The true desired operations on A and b are

$$\tilde{b} = \hat{U}_1 b \qquad A_1 = A$$
$$A_{k+1} = \hat{U}_k A_k \hat{U}_k \qquad k = 1, 2, ..., n$$

where $\hat{U}_k$ is the elementary reflection constructed to introduce zeros in positions $k + 1$ through n of b (if $k = 1$) or of the $(k + 1)$th column of $\hat{U}_k A_k$ (if $k > 1$).

The computed quantities are

$$\bar{b} = \text{fl}(\bar{U}_1 b) \qquad \bar{A}_1 = A$$
$$\bar{A}_{k+1} = \text{fl}(\bar{U}_k \bar{A}_k \bar{U}_k) \qquad k = 1, 2, ..., n$$

where $\bar{U}_k$ is the computed approximation to the true matrix U_k, which would introduce zeros in the positions $k + 1$ through n of the corresponding column of $U_k \bar{A}_k$ or of $U_1 b$ (notice the difference between U_k and $\hat{U}_k$).

It is convenient to define the error matrices

$$\begin{aligned}
F_1 &= \text{fl}(\bar{U}_1 b) - U_1 b \\
F_{k+1} &= \text{fl}(\bar{U}_k \bar{A}_k \bar{U}_k) - U_k \bar{A}_k U_k \\
&= \bar{A}_{k+1} - U_k \bar{A}_k U_k \qquad k = 1, 2, ..., n
\end{aligned} \qquad (4.43)$$

Using the error bounds for the computed elementary reflections given in Section 1.7, it may be shown that

$$\| \text{fl}(\bar{U}_1 b) - U_1 b \|_F \leq f_n \| b \|_F$$
$$\| \text{fl}(\bar{U}_k \bar{A}_k) - U_k \bar{A}_k \|_F \leq f_{n-k+1} \| \bar{A}_k \|_F$$
$$\| \text{fl}(\bar{A}_k \bar{U}_k) - \bar{A}_k U_k \|_F \leq f_{n-k+1} \| \bar{A}_k \|_F$$

where $f_i = (6i + 27)\varepsilon$. Hence

$$\| F_1 \|_F \leq f_n \| b \|_F$$
$$\| F_{k+1} \|_F \leq 2 f_{n-k+1} \| \bar{A}_k \|_F \qquad k = 1, 2, ..., n$$

Combining equations (4.43) for $k = 1, 2, ..., n$, gives

$$\begin{aligned}
\bar{A} &= \bar{A}_{n+1} \\
&= U_n ... U_1 A U_1 ... U_n \\
&\quad + U_{n-1} ... U_1 F_2 U_1 ... U_{n-1} \\
&\quad + U_{n-2} ... U_1 F_3 U_1 ... U_{n-2} + \cdots + F_n
\end{aligned}$$

Since the matrices U_k are orthogonal, it follows that

$$\|\bar{A} - U^{\mathrm{T}}AU\|_{\mathrm{F}} \leq 2 \sum_{k=1}^{n} f_{n-k+1} \|\bar{A}_k\|_{\mathrm{F}}$$

$$\leq 2\|A\|_{\mathrm{F}} \sum_{k=1}^{n} f_{n-k+1} \tag{4.44}$$

$$\leq (6n^2 + 60n)\varepsilon\|A\|_{\mathrm{F}}$$

$U = U_1 U_2 \ldots U_n$
Finally, we obtain

$$\begin{aligned} \bar{A} &= U^{\mathrm{T}}(A + \Delta A)U & \Delta A &= U\bar{A}U^{\mathrm{T}} - A \\ \bar{b} &= U^{\mathrm{T}}(b + \Delta b) & \Delta b &= U\bar{b} - b \end{aligned}$$

where according to (4.44)

$$\begin{aligned} \|\Delta A\|_{\mathrm{F}} &\leq (6n^2 + 60n)\varepsilon\|A\|_{\mathrm{F}} \\ \|b\|_{\mathrm{F}} &\leq (6n + 27)\varepsilon\|b\|_{\mathrm{F}} \end{aligned} \tag{4.45}$$

The computed product of n approximated elementary reflections satisfies

$$\|\mathrm{fl}(\bar{U}_1 \ldots \bar{U}_n) - U_1 \ldots U_n\|_{\mathrm{F}} \leq \sum_{k=1}^{n} f_{n-k+1} \|I_n\|_{\mathrm{F}} \leq (3n^2 + 30n)\sqrt{n}\varepsilon$$

(the factor $\sqrt{n}$ is the Frobenius norm of the unit matrix I_n).

These results show that the computed transformation matrix is near to the exact orthogonal matrix which reduces $(A + \Delta A, b + \Delta b)$ into canonical form.

Example 4.24

Consider the reduction into orthogonal canonical form of the fourth-order pair (A, B)

$$A = \begin{bmatrix} -1/2 & 0 & 5/2 & 0 \\ -\sqrt{2} & -1 & 8/\sqrt{2} & 0 \\ -3/2 & 0 & 7/2 & 0 \\ 1/\sqrt{2} & -1 & 3\sqrt{2} & -2 \end{bmatrix} \quad \text{and} \quad b = \begin{bmatrix} 0 \\ 1 \\ 0 \\ 1 \end{bmatrix}$$

This pair is uncontrollable, the dimension of the controllable part being equal to 2.

Using Algorithm 4.7 in floating-point arithmetic with $\varepsilon = 2^{-24} \approx 5.96 \times 10^{-8}$ we obtain after the first stage of the reduction (the results are rounded up to four decimal digits)

$$\bar{A}_2 = \begin{bmatrix} -0.2000 \times 10^1 & -0.3536 \times 10^0 \\ -0.7071 \times 10^0 & -0.2980 \times 10^{-7} \\ -0.1192 \times 10^{-6} & 0.1061 \times 10^1 \\ 0.7071 \times 10^0 & -0.5000 \times 10^0 \end{bmatrix}$$

$$\begin{bmatrix} -0.5500 \times 10^1 & -0.3536 \times 10^0 \\ -0.4768 \times 10^{-6} & 0.1000 \times 10^1 \\ 0.3500 \times 10^1 & 0.1061 \times 10^1 \\ -0.3536 \times 10^1 & -0.1500 \times 10^1 \end{bmatrix}$$

$$\bar{b} = \begin{bmatrix} -0.1414 \times 10^1 \\ 0 \\ 0 \\ 0 \end{bmatrix}$$

After the second step the matrix A is reduced to

$$\bar{A}_3 = \begin{bmatrix} -0.2000 \times 10^1 & 0.7153 \times 10^{-6} \\ 0.1000 \times 10^1 & -0.1000 \times 10^1 \\ 0 & -0.4768 \times 10^{-6} \\ 0 & 0.3576 \times 10^{-6} \end{bmatrix}$$

$$\begin{bmatrix} -0.5500 \times 10^1 & -0.5000 \times 10^0 \\ -0.2500 \times 10^1 & -0.1500 \times 10^1 \\ 0.3500 \times 10^1 & 0.1500 \times 10^1 \\ -0.2500 \times 10^1 & -0.5000 \times 10^0 \end{bmatrix}$$

The nonzero elements in positions $(3, 2)$ and $(4, 2)$ result from the rounding errors during the reduction. Since

$$\| A \|_F = 8.124 \qquad \| b \|_F = 1.414$$
$$\text{tol} = 129.985 \qquad \varepsilon\text{tol} = 7.75 \times 10^{-6}$$

these elements are neglected at the next step and we finally get

$$\bar{A} = \begin{bmatrix} -2 & 0 & -5.5 & -0.5 \\ 1 & -1 & -2.5 & -1.5 \\ 0 & 0 & 3.5 & 1.5 \\ 0 & 0 & -2.5 & -0.5 \end{bmatrix}$$

The computed quantities satisfy

$$\frac{\| \bar{U} \bar{A} \bar{U}^{\mathrm{T}} - A \|_{\mathrm{F}}}{\| A \|_{\mathrm{F}}} = 0.1788 \times 10^{-6}$$

$$\frac{\| \bar{U} \bar{b} - b \|_{\mathrm{F}}}{\| b \|_{\mathrm{F}}} = 0.1067 \times 10^{-6}$$

which agrees with (4.45).

The system is not stabilizable since the eigenvalues of the uncontrollable part are 1 and 2.

4.4.2 Multi-input Case

The reduction of a multi-input system into orthogonal canonical form can be accomplished using either QR or singular value decomposition. In the algorithm presented below, we implement QR decomposition with column pivoting (see Section 1.8), since it is less expensive. It should be noted, however, that generally the singular value decomposition yields more reliable results in determining the dimension of the controllable part of the system.

Let the matrix B be decomposed as (it is assumed that $m < n$)

$$B = Q_1 \begin{bmatrix} R_1 \\ 0 \end{bmatrix} P_1$$

where Q_1 is orthogonal, R_1 is $m_1 \times m$ upper trapezoidal and P_1 is a permutation matrix ($P_1^{-1} = P_1^{\mathrm{T}}$) which reflects the moving of the columns during the decomposition. The diagonal elements of R_1 satisfy

$$|r_{11}| \geq |r_{22}| \geq \cdots \geq |r_{m_1, m_1}| > 0$$

and the first conjugate Kronecker index m_1 is determined as the numerical rank of R_1 (and hence of B). Performing an orthogonal similarity transformation with the matrix Q_1 one obtains

$$A_2 = Q_1^{\mathrm{T}} A Q_1 = \begin{bmatrix} A_{11}^{(2)} & A_{12}^{(2)} \\ A_{21}^{(2)} & A_{22}^{(2)} \end{bmatrix} \quad \text{and} \quad \tilde{B} = Q_1^{\mathrm{T}} B = \begin{bmatrix} B_1 \\ 0 \end{bmatrix}$$

where $A_{11}^{(2)}$ is an $m_1 \times m_1$ and $A_{21}^{(2)}$ is an $(n - m_1) \times m_1$ matrix, and $B_1 = R_1 P_1$.

At the second stage of the reduction, the matrix $A_{21}^{(2)}$ is decomposed as

$$A_{21}^{(2)} = Q_2{}' \begin{bmatrix} R_2 \\ 0 \end{bmatrix} P_2$$

where R_2 is an $m_2 \times m_1$ upper trapezoidal matrix and m_2 is determined from

$$m_2 = \operatorname{rank}(R_2) = \operatorname{rank}(A_{21}^{(2)}) \leqq m_1$$

Since

$$\operatorname{rank}[\tilde{B}, A_2\tilde{B}] = \operatorname{rank}\begin{bmatrix} B_1 & A_{11}^{(2)} B_1 \\ 0 & A_{21}^{(2)} B_1 \end{bmatrix} = m_1 + m_2$$

it follows that

$$m_2 = \operatorname{rank}[\tilde{B}, A_2\tilde{B}] - \operatorname{rank}(\tilde{B}) = \operatorname{rank}[B, AB] - \operatorname{rank}(B)$$

An orthogonal similarity transformation with the matrix

$$Q_2 = \begin{bmatrix} I_{m_1} & 0 \\ 0 & Q_2' \end{bmatrix}$$

gives

$$A_3 = Q_2^{\mathrm{T}} A_2 Q_2 = \begin{bmatrix} & A_1^{(3)} & \\ \hline A_{21} & A_{22}^{(3)} & A_{23}^{(3)} \\ \hline 0 & A_{32}^{(3)} & A_{33}^{(3)} \end{bmatrix} \qquad A_{21} = R_2 P_2$$

where $A_{22}^{(3)}$ is an $m_2 \times m_2$ and $A_{32}^{(3)}$ is an $(n - m_1 - m_2) \times m_2$ matrix. This transformation does not affect the matrix $\tilde{B}$.

At the next stage, we decompose the matrix $A_{32}^{(3)}$ which allows us to determine

$$m_3 = \operatorname{rank} A_{32}^{(3)} = \operatorname{rank}[B, AB, A^2B] - \operatorname{rank}[B, AB] \leqq m_2$$

This process continues until one obtains

$$\tilde{A} = A_{p+1} = \begin{bmatrix} & & A_1 & & \\ \hline A_{21} & & A_2 & & \\ \hline & A_{32} & & A_3 & \\ & & \ddots & & \\ 0 & & & A_{p,p-1} & A_p \end{bmatrix}$$

Since

$$[\tilde{B}, \tilde{A}\tilde{B}, ..., \tilde{A}^{p-1}\tilde{B}] = Q^{\mathrm{T}}[B, AB, ..., A^{p-1}B]$$

$$= \begin{bmatrix} B_1 & & & & \\ \hline A_{21}B_1 & & & \sim & \\ \hline & A_{32}A_{21}B_1 & & & \\ & & \ddots & & \\ 0 & & & A_{p,p-1} ... A_{21}B_1 \end{bmatrix}$$

it follows that the matrix $A_{p,p-1}$ is of full rank if the system is completely controllable, or is a zero matrix in the opposite case.

In the actual computations, rounding errors may cause a zero diagonal element of the matrix R_k to turn into a nonzero one and vice versa. That is why the numerical rank of R_k is determined by the test

$$|r_{m_k,m_k}| > c_1 \varepsilon \max\{\|A\|_F, \|B\|_F\}$$

for some positive c_1, analogically to the single-input case. It should be noted, however, that this test may also yield an inaccurate result for the rank in certain cases. If, for instance,

$$R_{k+1} = \begin{bmatrix} 1 & -c & -c & \cdots & -c \\ & s & -cs & \cdots & -cs \\ & & s^2 & \cdots & -cs^2 \\ & 0 & & \ddots & \vdots \\ & & & & s^{m_k-1} \end{bmatrix}$$

for some positive c and s satisfying $c^2 + s^2 = 1$, then the columns of this matrix are of unit length, the inequalities

$$r_{qq}^2 \geq r_{qj}^2 + \cdots + r_{jj}^2 \qquad \begin{cases} q = 1, \ldots, m_k \\ j = q+1, \ldots, m_k \end{cases}$$

used in the column pivoting, are satisfied and the matrix is not altered (up to the signs of the columns) by the QR decomposition. That is why R_{k+1} will be considered as a matrix of rank m_k, using the above test. At the same time this matrix may have a very small singular value, i.e. it may be near to the loss of rank, the diagonal elements being non-negligible. Fortunately, matrices of such a special structure arise very rarely in practice and using a singular value decomposition in the rank determination is not justified.

The reduction of a multi-input system to an orthogonal canonical form is carried out by Algorithm 4.8, which overwrites A and B with $\tilde{A}$ and $\tilde{B}$ respectively. The algorithm produces an integer l, equal to the dimension of the controllable part of the system, an integer p, equal to the controllability index of the system and the conjugate Kronecker indices $m_1, \ldots, m_p$. An auxiliary $n \times m$ array W is used to hold the matrices which are to be decomposed.

Algorithm 4.8 Reduction of a multi-input system into orthogonal canonical form

Set $k = 0$, $l = 0$, $nb = n$, $mb = m$, $p = 0$
Set $tol = n^2 \max\{\|A\|_F, \|B\|_F\}$
$W \leftarrow B$, $B \leftarrow 0$
$k \leftarrow k + 1$

Compute the QR decomposition with column pivoting of the $nb \times mb$ matrix W,

$$Q_{k'} \begin{bmatrix} R_k \\ 0 \end{bmatrix} = W P_k^{\mathsf{T}}$$

Set $irnk = 0$
For $i = 1, 2, \ldots, \min\{nb, mb\}$

    if $\mathrm{fl}(\mathrm{tol} + |r_{ii}|) = \mathrm{tol}$, step i
    $irnk \leftarrow i$

If $irnk = 0$, quit
Set $ni = l$
$l \leftarrow l + irnk$
$p \leftarrow p + 1$
$m_p = irnk$
$A \leftarrow Q_k^{\mathsf{T}} A Q_k$, $Q_k = \mathrm{diag}(I_{n-nb}, Q_{k'})$

If $k = 1$, $B \leftarrow \begin{bmatrix} R_1 P_1 \\ 0 \end{bmatrix}$

If $irnk = nb$, quit
$mb \leftarrow irnk$
$nb \leftarrow nb - irnk$
For $i = 1, 2, \ldots, nb$

    $w_{ij} \leftarrow a_{l+i, ni+j}$ $j = 1, 2, \ldots, mb$

Algorithm 4.8 is also known under the name of the *staircase algorithm*.

    In the case of a completely controllable system, Algorithm 4.8 requires approximately

$$\tfrac{5}{3} n^3 + n^2 m + \tfrac{7}{2} n^2$$

flops (it is assumed for simplicity that $m_1 = m_2 = \cdots = m_p = m$). If the transformation matrix is desired, an additional $n^2 + 2n$ flops are required. The necessary storage is proportional to $2n^2 + 2nm + n$.

    The numerical properties of Algorithm 4.8 are similar to those of Algorithm 4.7. Using the error bounds for the QR decomposition (see Section 1.8) and proceeding in the same way as for the single-input case, it can be shown that (again for the case $m_1 = m_2 = \cdots = m_p = m$)

$$\| \bar{B} - Q^{\mathsf{T}} B \|_F \leq (6nm - 3m^2 + 30m)\varepsilon \| B \|_F$$

$$\| \bar{A} - Q^{\mathsf{T}} A Q \|_F \leq 2 \sum_{k=1}^{p} f_{p-k+1} \| A_k \|_F$$

where

$$f_{p-k+1} = \{6[n - (k-1)m]m - 3m^2 + 30m\}\varepsilon$$

This yields that the computed pair $(\bar{A}, \bar{B})$ is the exact orthogonal canonical form of a pair $(A + \Delta A, B + \Delta B)$ with

$$\| \Delta A \|_F \le (6n^2 + 60n)\varepsilon\| A \|_F$$
$$\| \Delta B \|_F \le (6nm - 3m^2 + 30m)\varepsilon\| B \|_F$$

which is again a very satisfactory result.

Example 4.25

Let

$$A = \begin{bmatrix} 0 & -3 & 0 & 0 & 0 \\ 1 & -4 & 0 & 0 & 0 \\ 0 & -2 & 0 & 0 & -8 \\ 0 & -3 & 1 & 0 & -14 \\ 0 & -1 & 0 & 1 & -7 \end{bmatrix} \quad \text{and} \quad B = \begin{bmatrix} 3 & 1 \\ 1 & 1 \\ -2 & 0 \\ 0 & 0 \\ 0 & 0 \end{bmatrix}$$

The first column of B is of a larger norm than the second one and there is no interchanging during the QR decomposition of B. The matrices obtained after the first stage of the reduction in arithmetic with precision $\varepsilon = 2^{-24} \approx 5.96 \times 10^{-8}$ are (up to three decimal digits)

$$A_2 = \begin{bmatrix} -0.429 \times 10^0 & -0.181 \times 10^1 & -0.154 \times 10^1 & 0 & -0.428 \times 10^1 \\ -0.660 \times 10^0 & -0.357 \times 10^1 & -0.321 \times 10^1 & 0 & 0.494 \times 10^1 \\ 0.617 \times 10^0 & 0.535 \times 10^0 & -0.149 \times 10^{-6} & 0 & -0.462 \times 10^1 \\ 0.134 \times 10^1 & 0.170 \times 10^1 & 0.231 \times 10^1 & 0 & -0.140 \times 10^2 \\ 0.267 \times 10^0 & 0.772 \times 10^0 & 0.577 \times 10^0 & 1 & -0.700 \times 10^1 \end{bmatrix}$$

$$\tilde{B} = \begin{bmatrix} -0.374 \times 10^1 & -0.107 \times 10^1 \\ 0 & -0.926 \times 10^0 \\ 0 & 0 \\ 0 & 0 \\ 0 & 0 \end{bmatrix}$$

The second QR decomposition, performed at the next stage, requires interchanging of the columns of the corresponding part of A_2.

The result of this transformation is

$$A_3 = \begin{bmatrix} -0.429 \times 10^0 & -0.181 \times 10^1 & 0.213 \times 10^1 \\ -0.660 \times 10^0 & -0.357 \times 10^1 & -0.108 \times 10^1 \\ \hline -0.145 \times 10^1 & -0.194 \times 10^1 & -0.552 \times 10^1 \\ 0.384 \times 10^0 & 0 & 0.556 \times 10^0 \\ \hline 0 & 0 & 0.179 \times 10^{-6} \end{bmatrix}$$

$$\begin{bmatrix} 0.255 \times 10^1 & -0.311 \times 10^1 \\ -0.578 \times 10^1 & -0.276 \times 10^0 \\ \hline -0.144 \times 10^2 & 0.573 \times 10^1 \\ -0.481 \times 10^0 & 0.505 \times 10^0 \\ \hline 0.113 \times 10^{-5} & -0.100 \times 10^1 \end{bmatrix}$$

In the given case

$$\| A \|_F = 18.7 \qquad \| B \|_F = 4 \qquad \text{tol} = 468 \qquad \varepsilon \text{tol} = 2.79 \times 10^{-5}$$

and the elements in positions $(5, 3)$ and $(5, 4)$ are neglected, which produces the final solution. Thus

$$m_1 = m_2 = 2$$

and the dimension of the controllable part is 4. The system is clearly stabilizable.

The orthogonal canonical form may be used to determine the distance to an uncontrollable system (see the previous section). In the single-input case we have the following result.

Theorem 4.5

Let the pair (A, b) be in orthogonal canonical form (4.35). If

$$\| A \|_2 + \| b \|_2 \leq \tfrac{1}{4}$$

and

$$| b_1 a_{21} a_{32} \dots a_{n, n-1} | > 4\delta$$

for $\delta < \tfrac{1}{4}$, then this pair is completely controllable and there is no pair of the form $(A + \delta E, b + \delta f)$ with $\| E \|_2 = 1$ and $\| f \|_2 = 1$ which yields a controllable part of smaller dimension.

The proof of the theorem can be found in Boley (1985).

The application of Theorem 4.5 requires the matrices A and b to be scaled so that

$$s = \| A \|_2 + \| b \|_2 < \tfrac{1}{4} \tag{4.46}$$

In such a case the quantity

$$\mu_1(A, b) = | b_1 a_{21} \ldots a_{n,n-1} |$$

gives a lower bound on the size of the perturbations needed to obtain an uncontrollable system.

    If the condition (4.46) is not fulfilled, then we must scale the pair (A, b) by a factor of $4s$ in order to apply the theorem, so that the measure becomes

$$\mu_1(A, b) = \frac{1}{(4s)^n} | b_1 a_{21} \ldots a_{n,n-1} |$$

Example 4.26

If

$$A = \begin{bmatrix} -1 & 0.3 & -0.4 & -2 \\ 0.8 & -0.5 & 0.9 & 0.5 \\ 0 & 0.001 & 0.3 & 0.6 \\ 0 & 0 & 1 & -0.2 \end{bmatrix} \quad \text{and} \quad b = \begin{bmatrix} -0.85 \\ 0 \\ 0 \\ 0 \end{bmatrix}$$

then

$$s = \| A \|_2 + \| b \|_2 = 3.456$$

and

$$\mu_1(A, b) = 1.862 \times 10^{-8}$$

Thus if $\delta < \tfrac{1}{4}\mu_1(A, b)$, then the pair $(A + \delta E, b + \delta f)$, $\| E \|_2 = 1$, $\| f \|_2 = 1$, remains completely controllable.

    Theorem 4.5 can give pessimistic results since with the required scaling, it is possible to obtain small values of $\mu_1(A, b)$ for large n even for well-controlled systems.

    In the multi-input case the quantity $\mu_1(A, B)$ is defined to be the product of the smallest singular values of the blocks B_1, A_{21}, ..., $A_{p,p-1}$ in (4.36). This definition is not based on a specific proven result, but can be used as an experimental extension of the definition for the single-input case.

EXERCISES

4.21 Develop an analogue of Algorithm 4.7 using plane rotations instead of elementary reflections. Compare both versions with respect to the computational work and accuracy.

4.22 Show that the conjugate Kronecker indices $m_1, m_2, ..., m_p$ of a multi-input system can be found from the Kronecker indices $p_1, p_2, ..., p_m$ by the rule '$m_i =$ the number of the integers in the set $\{p_1, p_2, ..., p_m\}$, which are larger or equal to i, $i = 1, 2, ..., p$.'

4.23 Write computer programs implementing Algorithms 4.7 and 4.8.

4.24 Compute the orthogonal canonical form $(\bar{A}, \bar{b})$ of the pair (A, b) considered in Example 4.22. Check the errors $\| \bar{A} - U^T A U \|_F$, $\| \bar{b} - U^T b \|_F$.

4.25 Develop an analogue of Algorithm 4.8, using the singular value decomposition instead of QR decomposition. Compare both algorithms.

4.26 Write an algorithm for finding the orthogonal canonical form of the pair (C, A) without transposition of A and C and without using Algorithm 4.8.

4.27 Show that every controllable pair (A, B), $B \equiv [b_1, b_2, ..., b_m]$ may be reduced by orthogonal transformations into the *Hermite–Hessenberg form*

$$A = U^T A U = \begin{bmatrix} A_{11} & A_{12} & ... & A_{1m} \\ & A_{22} & ... & A_{2m} \\ \mathbf{0} & & \ddots & \vdots \\ & & & A_{mm} \end{bmatrix} \qquad B = U^T B = \begin{bmatrix} B_1 \\ B_2 \\ \vdots \\ B_m \end{bmatrix}$$

where each A_{jj}, $j = 1, 2, ..., m$ is a $\delta_j \times \delta_j$ upper Hessenberg matrix,

$$B_j = \begin{bmatrix} x & x & ... & x \\ 0 & x & ... & x \\ \vdots & \vdots & & \vdots \\ 0 & x & ... & x \\ \uparrow & & & \\ j & & & \end{bmatrix}$$

and

$$\delta_j = \text{rank}\,[b_1, Ab_1, ..., A^{n-1}b_1, ..., b_j, Ab_j, ..., A^{n-1}b_j]$$
$$- \text{rank}\,[b_1, Ab_1, ..., A^{n-1}b_1, ..., b_{j-1}, Ab_{j-1}, ..., A^{n-1}b_{j-1}]$$
$$j = 1, 2, ..., m$$

4.5 SYSTEM BALANCING AND MODEL REDUCTION

Consider a stable time-invariant linear system

$$\begin{aligned} \dot{x}(t) &= Ax(t) + Bu(t) \\ y(t) &= Cx(t) \end{aligned} \tag{4.47}$$

where the pair (A, B) is assumed controllable and the pair (C, A) is assumed observable. The associated controllability and observability Grammians of the system are, respectively,

$$W_c = \int_0^\infty e^{As} BB^T e^{A^T s}\, ds$$

$$W_o = \int_0^\infty e^{A^T s} C^T C e^{As}\, ds$$

The system balancing is based upon the simultaneous diagonalization of the positive definite Grammians which is done by using a suitably chosen state similarity transformation.

Applying a nonsingular transformation

$$x(t) = S\hat{x}(t) \tag{4.48}$$

the system (4.47) is reduced to

$$\dot{\hat{x}}(t) = \hat{A}\hat{x}(t) + \hat{B}u(t)$$
$$y(t) = \hat{C}\hat{x}(t)$$

where

$$\hat{A} = S^{-1}AS \qquad \hat{B} = S^{-1}B \qquad \hat{C} = CS$$

The corresponding Grammians of the transformed system are expressed as

$$\hat{W}_c = S^{-1} W_c S^{-T} \qquad \hat{W}_o = S^T W_o S$$

It is necessary to point out that while the eigenvalues of the matrix A (the system poles) are invariant under similarity transformation, the eigenvalues of the Grammians are not.

The transformation (4.48) for which the matrices $\hat{W}_c$ and $\hat{W}_o$ are both diagonal is called a *contragradient transformation*. Several contragradient transformations may be found in the following way.

Since the matrix W_c is symmetric and positive definite, it may be reduced to

$$V_c^T W_c V_c = \Sigma_c^2$$

where V_c is an orthogonal and Σ_c is a diagonal matrix with positive entries. Further on, because of the positive definiteness of the matrix $(V_c \Sigma_c)^T W_o (V_c \Sigma_c)$ it is possible to find another orthogonal matrix U such that

$$U^T (V_c \Sigma_c)^T W_o (V_c \Sigma_c) U = \Sigma^2$$

where Σ is a diagonal matrix with positive entries. Now it is easy to check that the expression

$$S_k = V_c \Sigma_c U \Sigma^{-k}$$

defines a family of contragradient transformations with

$$S_k^{-1} W_c S_k^{-T} = \Sigma^{2k} \qquad S_k^T W_o S_k = \Sigma^{2-2k} \qquad (4.49)$$

The diagonal elements σ_i, $i = 1, \ldots, n$ of the matrix Σ are the positive square roots of the eigenvalues of $W_c W_o$ since

$$S_k^{-1} W_c W_o S_k = \Sigma^2$$

These elements are referred to as *second-order modes*. Note that the second-order modes are invariant under state similarity transformations.

The following three values of k, namely $k = 0$, $\frac{1}{2}$ and 1 are of special interest and correspond to a specific terminology:

$$
\begin{array}{llll}
k = 0 & \hat{W}_c = I & \hat{W}_o = \Sigma^2 \rightarrow \text{'input-normal coordinates'} \\
k = \frac{1}{2} & \hat{W}_c = \Sigma & \hat{W}_o = \Sigma \rightarrow \text{'internally balanced coordinates'} \\
k = 1 & \hat{W}_c = \Sigma^2 & \hat{W}_0 = I \rightarrow \text{'output-normal coordinates'}
\end{array}
$$

The most interesting of the applications is the contragradient transformation for $k = \frac{1}{2}$ which we will refer to further by the generic name of *balancing transformation*.

Consider now methods for computing the balancing transformation. One of the possible techniques is the algorithm of Laub (1980) which is described as follows;

Since the Grammians satisfy the Lyapunov equations

$$
\begin{aligned}
A W_c + W_c A^T + B B^T = 0 \\
A^T W_o + W_o A + C^T C = 0
\end{aligned}
\qquad (4.50)
$$

it is appropriate to compute them by the Bartels–Stewart algorithm (see Section 4.2). A little trick allows us to solve both equations using only the Schur decomposition of A.

Applying the Cholesky decomposition, the controllability Grammian is represented as

$$W_c = L_c L_c^T$$

where L_c is a lower triangular matrix. The observability Grammian is then transformed to the matrix $L_c^T W_o L_c$ which is reduced to the diagonal form

$$V^T (L_c^T W_o L_c) V = \Sigma^2$$

This reduction is done by the QR method for symmetric matrices. Now it is easy to verify that

$$S = L_c V \Sigma^{-1/2}$$

is a balancing transformation such that

$$S^{-1} W_c S^{-T} = S^T W_o S = \Sigma$$

The matrices of the balanced system are given by

$$\hat{A} = S^{-1} A S = \Sigma^{1/2} V^T L_c^{-1} A L_c V \Sigma^{-1/2}$$
$$\hat{B} = S^{-1} B = \Sigma^{1/2} V^T L_c^{-1} B$$
$$\hat{C} = C S = C L_c V \Sigma^{-1/2}$$

Example 4.27

If

$$A = \begin{bmatrix} 0 & 1 & 0 & 0 \\ 0 & 0 & 1 & 0 \\ 0 & 0 & 0 & 1 \\ -50 & -79 & -33 & -5 \end{bmatrix} \qquad B = \begin{bmatrix} 0 \\ 0 \\ 0 \\ 1 \end{bmatrix}$$

$$C = [50, \quad 15, \quad 1, \quad 0]$$

then using an arithmetic with $\varepsilon = 2^{-56} \approx 1.39 \times 10^{-17}$, we obtain (up to three decimal digits)

$$W_c = 10^{-3} \begin{bmatrix} 0.155 & 0 & -0.451 & 0 \\ 0 & 0.451 & 0 & -7.12 \\ -0.451 & 0 & 7.12 & 0 \\ 0 & -7.12 & 0 & 2.13 \end{bmatrix}$$

$$W_o = 10^{3} \begin{bmatrix} 1.86 & 0.888 & 0.148 & 0.025\,0 \\ 0.888 & 0.434 & 0.074\,0 & 0.012\,7 \\ 0.148 & 0.0740 & 0.013\,5 & 0.002\,26 \\ 0.0250 & 0.0127 & 0.002\,26 & 0.000\,451 \end{bmatrix}$$

$$L_c = 10^{-2} \begin{bmatrix} 1.25 & 0 & 0 & 0 \\ 0 & 2.12 & 0 & 0 \\ -3.62 & 0 & 7.62 & 0 \\ 0 & -33.6 & 0 & 31.6 \end{bmatrix}$$

$$\Sigma^2 = \operatorname{diag}(0.332, 0.0217, 0.008\,17, 0.000\,369)$$

$$S = \begin{bmatrix} 0.0117 & -0.0110 & 0.0216 & 0.0295 \\ 0.0118 & -0.0255 & -0.0351 & -0.0922 \\ 0.0123 & 0.128 & -0.214 & 0.160 \\ -0.0523 & 0.949 & 0.929 & 0.143 \end{bmatrix}$$

$$\hat{A} = \begin{bmatrix} -0.518 & 1.45 & -0.391 & -0.350 \\ -1.45 & -2.20 & 4.75 & 1.22 \\ -0.391 & -4.75 & -0.630 & -1.20 \\ 0.350 & 1.22 & 1.20 & -1.66 \end{bmatrix}$$

$$\hat{B} = [0.773, \quad 0.805, \quad 0.337, \quad -0.252]^T$$
$$\hat{C} = [0.773, \quad -0.805, \quad 0.337, \quad 0.252]$$
$$\hat{W}_c = \hat{W}_o = \text{diag}(0.576, 0.147, 0.0904, 0.0192)$$

A disadvantage of the balancing algorithm considered is that the computation of the Grammians through (4.49) requires evaluation of the matrices BB^T and C^TC which may be associated with large rounding errors in some cases. However, as was shown in Section 4.2, it is possible to find the Cholesky factor of the solution of a Lyapunov equation by the Hammarling algorithm without explicit determination of the solution. This possibility is exploited in the algorithm of Laub *et al.* (1987) which proceeds in the following way. The decompositions

$$W_c = L_c L_c^T \qquad W_o = L_o L_o^T$$

are found by the Hammarling algorithm, where the matrices W_c and W_o are not actually formed. It may be shown that the singular values of $L_o^T L_c$ are the positive square roots of the eigenvalues of $W_c W_o$, i.e. the second-order modes of the system. Using the singular value decomposition

$$L_o^T L_c = U\Sigma V^T$$

of $L_o^T L_c$, the balancing transformation matrix is determined as

$$S = L_c V \Sigma^{-1/2} \qquad S^{-1} = \Sigma^{-1/2} U^T L_o^T$$

In this way we have the following algorithm.

Algorithm 4.9 Internal balancing of a state–space representation

Compute the Cholesky factor L_c, L_o of the corresponding Grammians W_c, W_o from

$$AW_c + W_c A^T + BB^T = 0$$
$$A^T W_o + W_o A + C^T C = 0$$

   using the Hammarling algorithm
Compute the singular value decomposition

$$L_o^T L_c = U\Sigma V^T$$

Form the balancing transformation matrix

$$S = L_c V \Sigma^{-1/2}$$

Form the matrices of the balanced system

$$\hat{A} = S^{-1}AS = \Sigma^{-1/2}U^T L_o^T A L_c V \Sigma^{-1/2}$$
$$\hat{B} = S^{-1}B = \Sigma^{-1/2}U^T L_o^T B$$
$$\hat{C} = CS = CL_c V \Sigma^{-1/2}$$

Algorithm 4.9 requires approximately $40 \div 50 n^3$ flops. Within this figure, approximately $16n^3$ flops are spent in the computation of the Cholesky factors of the Grammians.

Example 4.28

For the system from Example 4.27, Algorithm 4.9 produces (again up to three digits)

$$L_o = \begin{bmatrix} 43.1 & 0 & 0 & 0 \\ 20.6 & 3.06 & 0 & 0 \\ 3.42 & 1.13 & 0.718 & 0 \\ 0.580 & 0.235 & 0.008\ 13 & 0.245 \end{bmatrix}$$

$$U = \begin{bmatrix} 0.999 & -0.0302 & 0.0400 & -0.0193 \\ 0.0497 & 0.753 & -0.438 & 0.489 \\ 0.111 & 0.259 & -0.486 & -0.835 \\ -0.169 & 0.604 & 0.756 & -0.253 \end{bmatrix}$$

$$\Sigma = \text{diag}(0.567, 0.147, 0.0904, 0.0192)$$

$$V = \begin{bmatrix} 0.711 & -0.339 & 0.521 & -0.328 \\ 0.422 & -0.461 & -0.497 & 0.602 \\ 0.461 & 0.483 & -0.596 & -0.446 \\ 0.322 & 0.663 & 0.355 & 0.576 \end{bmatrix}$$

The matrices S, $\hat{A}$, $\hat{B}$ and $\hat{C}$ obtained differ from the corresponding matrices in Example 4.27 in the last significant digits.

An operation which may be a source of large errors in Algorithm 4.9 is the formation of the product $L_o^T L_c$. If this product has widespread singular values then the smallest singular values may be affected significantly by the rounding errors in computing $L_o^T L_c$. To avoid this danger, it is recommended that the singular values of a product of two matrices be computed by the algorithm of Heath *et al.* (1986), which avoids the explicit calculation of the product.

Numerical difficulties with Algorithm 4.9 may arise in the case when the Lyapunov equations (4.50) for the Grammians are ill conditioned, i.e. when $\text{sep}(A^T, -A)$ is small. In such a case the errors in the matrices L_c and L_o may become large, which will cause the accuracy of the matrices U, V, Σ and S and thus the accuracy of $\hat{A}$, $\hat{B}$ and $\hat{C}$ to deteriorate. For the system from Example 4.27

$$\text{sep}(A^T, -A) = 0.0111$$

so that at most two digits may be lost in the computation of W_c, W_o. If $\text{sep}(A^T, -A)$ is very small, then an iterative refinement of the factors of the solution of Lyapunov equation may be necessary.

The accuracy of Algorithm 4.9 may also deteriorate by bad scaling of the system model when matrices B and C have entries differing by many orders of magnitude.

Algorithm 4.9 may yield very inaccurate results if the given system is near to an uncontrollable or unobservable system. The following example, presented in Safonov and Chiang (1989), illustrates the difficulties associated with small distances to uncontrollable and unobservable systems.

Example 4.29

If

$$A = \begin{bmatrix} -\frac{1}{2} & -\delta \\ 0 & -\frac{1}{2} \end{bmatrix} \qquad B = \begin{bmatrix} \delta \\ 1 \end{bmatrix} \qquad C = [1, \ \delta]$$

then for $\delta \to 0$ the first state of the system becomes uncontrollable and the second state becomes unobservable. Note that for this system $\text{sep}(A^T, -A) \to 1$ as $\delta \to 0$ so that the Lyapunov equations for the Grammians are perfectly conditioned. The exact solutions of (4.50) are

$$W_c = \begin{bmatrix} \delta^2 & 0 \\ 0 & 1 \end{bmatrix} \qquad W_o = \begin{bmatrix} 1 & 0 \\ 0 & \delta^2 \end{bmatrix}$$

and for $\delta > 0$ we obtain that

$$L_c = \begin{bmatrix} \delta & 0 \\ 0 & 1 \end{bmatrix} \qquad L_o = \begin{bmatrix} 1 & 0 \\ 0 & \delta \end{bmatrix}$$

$$U = V = I_2 \qquad \Sigma = \text{diag}(\delta, \delta)$$

Thus

$$S = \begin{bmatrix} \sqrt{\delta} & 0 \\ 0 & \sqrt{(1/\delta)} \end{bmatrix}$$

cond$(S) = 1/\delta$, and the balancing transformation becomes infinitely ill conditioned as $\delta \to 0$. It should be pointed out that this is not a disadvantage particular to Algorithm 4.9, but a disadvantage of balancing in general, which is inherently badly conditioned for systems with nearly uncontrollable or unobservable modes.

Algorithm 4.9 is obviously adapted for computing input-normal and output-normal balancing transformations. It is easily modified, also, for balancing a discrete-time system

$$x_{k+1} = Ax_k + Bu_k$$
$$y_k = Cx_k$$

In this case the controllability and observability Grammians (see Section 4.2) satisfy the discrete Lyapunov equations

$$A W_c A^T - W_c + BB^T = 0$$
$$A^T W_o A - W_o + C^T C = 0$$

The determination of the Cholesky factors of the solutions is described in Hammarling (1982a).

Consider now the implementation of system balancing to find a reduced-order model of a stable, completely controllable, completely observable system. We note first that if the system is not controllable/observable one can extract the controllable/observable part by using the algorithms from Section 4.4. This part of the system may be balanced then by Algorithm 4.9.

The internal balancing is implemented to obtain a reduced-order model of a given system by deleting the 'least controllable/least observable' states. Specifically, if the internally balanced system is

$$\dot{x}(t) = \hat{A}\hat{x}(t) + \hat{B}u(t)$$
$$y(t) = \hat{C}\hat{x}(t)$$
(4.51)

and

$$\sigma_q \gg \sigma_{q+1}$$
(4.52)

for some $1 \leq q < n$, then partitioning

$$\hat{A} = \begin{bmatrix} F & F_1 \\ F_2 & F^* \end{bmatrix} \begin{matrix} \}q \\ \}n-q \end{matrix} \qquad \hat{B} = \begin{bmatrix} G \\ G^* \end{bmatrix} \begin{matrix} \}q \\ \}n-q \end{matrix}$$
$$\underbrace{}_{q} \underbrace{}_{n-q}$$

$$\hat{C} = [H, \quad H^*]$$
$$\underbrace{}_{q} \underbrace{}_{n-q}$$

we obtain an *internally dominant reduced-order model*

$$\dot{w}(t) = Fw(t) + Gu(t)$$
$$z(t) = Hw(t)$$
(4.53)

In this model the vector w is an approximation of the first q components of $\hat{x}$ and the vector z is an approximation of y.

It may be proved that the reduced-order model (4.53) is also balanced with controllability and observability Grammian

$$\overline{W}_c = \overline{W}_o = \text{diag}(\sigma_1, \sigma_2, \ldots, \sigma_q)$$

Thus, the reduced-order model (4.53) is obtained from (4.51) by deleting the least controllable/least observable states. It is possible to show that this model is stable provided the diagonal elements of $\hat{W}_c$ are distinct.

To distinguish a qth-order internally dominant subsystem of (2.51), i.e. to determine the model order q, one may use different criteria. For example, instead of the condition (4.52) it is possible to use the condition

$$\left(\sum_{i=1}^{q} \sigma_i^2 \right)^{1/2} \gg \left(\sum_{i=q+1}^{n} \sigma_i^2 \right)^{1/2}$$

Note that both conditions do not guarantee closeness between the model output $z(t)$ and the system output $y(t)$. A condition based on analysis of the output error of the reduced order model is proposed, for instance, in Lastman and Sinha (1985).

It should be stressed that the determination of a suitable reduced-order model may require comparison of several models of different order for given initial conditions and given input.

Example 4.30

For the system considered in Example 4.27 we obtain the following reduced-order models.

$q = 3$

$$\frac{\sigma_4}{\sqrt{(\sigma_1^2 + \sigma_2^2 + \sigma_3^2)}} = 0.0319$$

$$F = \begin{bmatrix} -0.518 & 1.45 & -0.391 \\ -1.45 & -2.20 & 4.75 \\ -0.391 & -4.75 & -0.630 \end{bmatrix} \qquad G = \begin{bmatrix} 0.773 \\ 0.805 \\ 0.337 \end{bmatrix}$$

$$H = [0.773, \ -0.805, \ 0.337]$$

The poles of the model are -0.738, $-1.30 \pm 4.88i$ while the poles of the original system are -1, $-1 \pm 4.90i$, -2.

$q = 2$

$$\frac{\sqrt{(\sigma_3^2 + \sigma_4^2)}}{\sqrt{(\sigma_1^2 + \sigma_2^2)}} = 0.155$$

$$F = \begin{bmatrix} -0.518 & 1.45 \\ -1.45 & -2.20 \end{bmatrix} \qquad G = \begin{bmatrix} 0.773 \\ 0.805 \end{bmatrix}$$

$H = [0.773, \ -0.805]$

The poles of this model are $-1.36 \pm 1.18i$.

$q = 1$

$$\frac{\sqrt{(\sigma_2^2 + \sigma_3^2 + \sigma_4^2)}}{\sigma_1} = 0.302$$

$F = -0.518 \qquad G = 0.773 \qquad$ and $\qquad H = 0.773$.

Table 4.3

t	Original system	3rd-order model	2nd-order model	1st-order model
0	0	0	0	0
0.5	0.216	0.219	0.206	0.263
1.0	0.553	0.558	0.530	0.466
1.5	0.671	0.666	0.749	0.623
2.0	0.809	0.790	0.851	0.744
2.5	0.884	0.870	0.880	0.837
3.0	0.922	0.917	0.878	0.909
3.5	0.958	0.958	0.869	0.965
4.0	0.972	0.981	0.862	1.01
4.5	0.983	0.999	0.858	1.04
5.0	0.990	1.01	0.857	1.07
5.5	0.993	1.02	0.857	1.09
6.0	0.996	1.03	0.857	1.10
6.5	0.998	1.03	0.858	1.11
7.0	0.999	1.03	0.858	1.12
7.5	0.999	1.05	0.858	1.13
8.0	0.999	1.04	0.858	1.13
8.5	1.000	1.04	0.858	1.14
9.0	1.000	1.04	0.858	1.14
9.5	1.000	1.04	0.858	1.14
10.0	1.000	1.04	0.858	1.15

The outputs of the original system and the reduced-order models for $u(t) = 1$ and zero initial conditions are shown for different t in Table 4.3.

As already mentioned, Algorithm 4.9 can introduce large errors in the case of a nearly uncontrollable/nearly unobservable system and so it is desirable to avoid its implementation in finding a reduced-order model. In what follows, we describe a Schur method for model reduction, proposed by Safonov and Chiang (1989) which circumvents the computation of balancing transformation. In this method one uses orthonormal bases for the left and right invariant subspaces (the spaces spanned by the left and right eigenvectors) associated with the 'large' eigenvalues of the matrix $W_c W_o$. This makes it possible to compute directly without balancing a reduced-order model.

Let the matrix $W_c W_o$ be reduced into upper triangular (Schur) form using orthogonal similarity transformations. (Note that $W_c W_o$ has real non-negative eigenvalues.) Using plane rotations it is possible to reorder the Schur form (see Section 1.10) to obtain the eigenvalues on its diagonal in ascending and descending order, respectively,

$$T_a = U_a^T W_c W_o U_a = \begin{bmatrix} \lambda_{a_n} & x & \cdots & x \\ & \lambda_{a_{n-1}} & \cdots & x \\ & & \ddots & \vdots \\ 0 & & & \lambda_{a_1} \end{bmatrix}$$

$$T_d = U_d^T W_c W_o U_d = \begin{bmatrix} \lambda_{d_1} & x & \cdots & x \\ & \lambda_{d_2} & \cdots & x \\ & & \ddots & \vdots \\ 0 & & & \lambda_{d_n} \end{bmatrix}$$

where

$$\{\lambda_{a_i}, i = 1, \ldots, q\} = \{\lambda_{d_i}, i = 1, \ldots, q\}$$
$$= \{\sigma_i^2, i = 1, \ldots, q\}$$

and U_a, U_d are the matrices of accumulated orthogonal transformations. Partitioning U_a and U_d as

$$U_a = [\; U_{a_1} \;,\; U_{a_2} \;] \quad \text{and} \quad U_d = [\; U_{d_1} \;,\; U_{d_2} \;]$$
$$\underbrace{\qquad}_{n-q} \underbrace{\quad}_{q} \qquad\qquad \underbrace{\quad}_{q} \underbrace{\qquad}_{n-q}$$

we have that the matrices U_{d_1} and U_{a_1} form, respectively, an orthonormal basis for the right invariant subspace of $W_c W_o$ associated with the large eigenvalues $\{\sigma_1^2, \ldots, \sigma_q^2\}$ and small eigenvalues $\{\sigma_{q+1}^2, \ldots, \sigma_n^2\}$. (We stress that the reasonable separation of the small eigenvalues from the large may be a nontrivial task.) The columns of U_{a_2} and U_{d_2} provide in turn an analogous decomposition of the left invariant subspace.

Assuming that the matrix $U_{a_2}^T U_{d_1}$ is decomposed by SVD as

$$U\Sigma V^T = U_{a_2}^T U_{d_1}$$

and computing the matrices

$$P = U_{a_2} U\Sigma^{-1/2}$$
$$Q = U_{d_1} V\Sigma^{-1/2}$$

then it may be proved that the matrices

$$F = P^T A Q \qquad G = P^T B \qquad H = CQ \tag{4.54}$$

correspond to a reduced-order model with controllability and observability Grammians

$$\overline{W}_c = P^T W_c P \quad \text{and} \quad \overline{W}_o = Q^T W_o Q$$

The model obtained is generally not balanced but shares many of the properties owned by the internally balanced model of the same order. For instance, both models have the same poles.

The Shur method for model reduction is implemented by the following algorithm.

Algorithm 4.10 Model reduction by Schur decomposition of the Grammians product

Compute the controllability and observability Grammians W_c, W_o by
  Algorithm 4.4

Reduce the matrix $W_c W_o$ into upper Schur form by the QR algorithm

Set appropriate value $1 \leq q < n$ for the model order

Reorder the Schur form into two upper triangular forms T_a and T_d with
  diagonal elements in ascending and descending order, respectively.
  Accumulate the corresponding orthogonal transformations in the
  matrices U_a and U_d

Compute the singular value decomposition

$$U\Sigma V^T = U_{a_2}^T U_{d_1}$$

where

$$U_a \equiv [\underbrace{U_{a_1},}_{n-q} \quad \underbrace{U_{a_2}}_{q}] \qquad U_d \equiv [\underbrace{U_{d_1},}_{q} \quad \underbrace{U_{d_2}}_{n-q}]$$

Find the reduced-order model matrices

$$F = P^T A Q \qquad G = P^T B \qquad H = CQ$$

where

$$P = U_{a_2} U \Sigma^{-1/2} \qquad Q = U_{d_1} V \Sigma^{-1/2}$$

Algorithm 4.10 requires about $50n^3$ flops.

Example 4.31

Consider the fourth-order system examined in Examples 4.27 and 4.30. After reordering the Schur form of $W_c W_o$ one obtains (to three decimal digits)

$$U_a = \begin{bmatrix} 0.125 & -0.0391 & -0.419 & 0.899 \\ -0.392 & 0.160 & 0.796 & 0.433 \\ 0.678 & -0.601 & 0.220 & -0.759 \\ 0.609 & 0.782 & 0.136 & 0.0122 \end{bmatrix}$$

$$T_a = \begin{bmatrix} 0.000\,369 & 0.004\,01 & 0.0706 & -0.557 \\ 0.0 & 0.008\,17 & 0.0541 & -0.967 \\ 0.0 & 0.0 & 0.0217 & 0.0498 \\ 0.0 & 0.0 & 0.0 & 0.332 \end{bmatrix}$$

$$U_d = \begin{bmatrix} 0.207 & -0.402 & 0.658 & -0.603 \\ 0.210 & -0.372 & 0.434 & 0.793 \\ 0.220 & -0.759 & -0.608 & -0.0813 \\ -0.930 & -0.353 & 0.101 & -0.0253 \end{bmatrix}$$

$$T_d = \begin{bmatrix} 0.332 & -0.640 & 0.866 & -0.306 \\ 0.0 & 0.0217 & 0.009\,79 & -0.041\,9 \\ 0.0 & 0.0 & 0.008\,17 & -0.008\,01 \\ 0.0 & 0.0 & 0.0 & 0.000\,369 \end{bmatrix}$$

If, for instance, $q = 2$, then

$$P = \begin{bmatrix} 0.523 & -2.34 \\ 0.920 & 1.01 \\ 0.344 & 0.772 \\ 0.102 & 0.270 \end{bmatrix} \qquad Q = \begin{bmatrix} 0.499 & -0.187 \\ 0.468 & -0.219 \\ 0.882 & 0.0833 \\ 0.0461 & 2.66 \end{bmatrix}$$

$$F = \begin{bmatrix} -8.23 & 1.96 \\ -24.9 & 5.52 \end{bmatrix} \qquad G = \begin{bmatrix} 0.102 \\ 0.270 \end{bmatrix}$$

$$H = \begin{bmatrix} 32.9 & -12.6 \end{bmatrix}$$

The model obtained and the internally balanced model of second order, found in Example 4.30, have the same poles and the same output for $u(t) = 1$ (Table 4.3).

The main advantage of Algorithm 4.10 is that it circumvents the implementation of ill-conditioned balancing transformations when some modes of the system are much more observable than controllable and vice versa. The algorithm relies on the use of orthogonal transformations and performs well even for systems which are not completely controllable and completely observable. A drawback of this algorithm is the necessity to explicitly form the product $W_c W_o$ which may be associated with large rounding errors in some cases. Also this algorithm does not avoid the numerical difficulties related to small $sep(A^T, -A)$ in solving the Lyapunov equations for the Grammians.

We conclude with the following remarks. The model reduction involves a compromise between model order and the adequacy of model to the system characteristics. As pointed out Moore (1981) there can be no universal model reduction algorithm since the relative importance of various system characteristics is highly dependent upon the application. It should be borne in mind also, that there exist several other techniques for model reduction, which are not considered here.

EXERCISES

4.28 Verify (4.49).
4.29 Give an algorithm for solving the Lyapunov equations (4.50) using only the Schur decomposition of A.
4.30 Sketch a rounding error analysis of Algorithm 4.9.
4.31 Write computer programs for input, internal and output balancing of continuous-time and discrete-time systems, based on Algorithm 4.9.
4.32 Prove that the internally dominant reduced-order model is balanced.
4.33 Study the output error of an internally dominant reduced-order model for a given class of input functions.
4.34 Verify (4.54).
4.35 Write a computer program for finding reduced-order models by Algorithm 4.10.

NOTES AND REFERENCES

The stability analysis of a linear system in the presence of small perturbations is reduced to the problem of determining the sensitivity of the state matrix eigenvalues. A profound consideration of matrix eigenvalues sensitivity can be found in Wilkinson (1965). In his treatment of the multiple eigenvalues sensitivity Wilkinson made use of the Gershgorin theorem. Further results in this topic are presented in Stewart (1973a) and Van Loan (1987).

Stability analysis in the case of large perturbations is related to the problem of finding the distance to an unstable system. Several approaches to the solution of this problem are discussed in Van Loan (1985a) and Demmel (1987a). Implementation of the Gershgorin theorem, as done by Algorithm 4.2, is inspired by the algorithm of Ruhe (1970) for determining the Jordan structure of a matrix. An algorithm which estimates the distance to an order of magnitude is proposed in Byers (1988). Other bounds for this distance are proposed by Hinrichsen and Pritchard (1986), Lewkowicz and Sivan (1988), Martin and Hewer (1987) and Zhou and Khargonekar (1987).

The perturbation analysis of the Sylvester equation and the roundoff error analysis of the Bartels–Stewart algorithm are done by Golub, Nash and Van Loan (1979). The sensitivity of the Lyapunov equation and its relation to the properties of the corresponding linear system are studied in Hewer and Kenney (1988). Algorithm 4.5 for solving the discrete Lyapunov equation is due to Barraud (1977).

The numerical aspects of the controllability/observability analysis of a linear system are discussed in Paige (1981). Methods for determining the distance to an uncontrollable system are presented in Boley (1981), Eising (1984b) and Wicks and Decarlo (1987).

The properties of the orthogonal canonical forms of linear systems are studied in Konstantinov, Petkov and Christov (1982) and Laub and Linnemann (1986). Staircase algorithms for orthogonal reduction into canonical form are proposed by several authors: Van Dooren, Emami-Naeini and Silverman (1979), Varga (1979), Konstantinov, Petkov and Christov (1980), Paige (1981) and others. The error analysis of Algorithms 4.7 and 4.8 is due to Petkov, Christov and Konstantinov (1986a).

The idea of system balancing and its implementation in reduction of the model order is given by Moore (1981). Further developments and extensions may be found in Pernebo and Silverman (1982), Shokoohi, Silverman and Van Dooren (1983), Therapos (1989), Young (1985) and the references therein. The algorithms for system balancing presented in Section 4.5, are given by Laub (1980) and Laub, *et al* (1987). The Schur method for model reduction is proposed in Safonov and Chiang (1989). Several other techniques for model reduction of linear systems are described in Mahmoud and Singh (1981).

5

Pole Assignment

In this chapter we present computational methods for the pole assignment of linear systems. The pole assignment problem is related to some numerical difficulties, and most of the methods considered in Section 2.6 do not perform satisfactorily in floating-point arithmetic. That is why we describe methods based on the Schur decomposition of the closed-loop system matrix, which allow us to solve the problem in a reliable way. Another possible approach that may produce acceptable results is to minimize the condition number of the eigenvector matrix in the multi-input case. We consider also algorithms for the synthesis of stabilizing controls and methods for the synthesis of state observers.

5.1 COMPUTATIONAL CONSIDERATIONS

The pole assignment problem for the linear system

$$\dot{x}(t) = Ax(t) + Bu(t)$$

may be considered as an inverse eigenvalue problem which requires the determination of a matrix with given eigenvalues. In fact, the eigenvalues of the closed-loop system matrix $A - BK$ are known and we seek the unknown feedback matrix K. The solution of this problem is direct rather than iterative, in contrast to the determination of the eigenvalues of a matrix.

Although the pole assignment problem is relatively simple from a computational point of view, its solution in the presence of rounding errors may be a difficult task. To illustrate this, consider the following simple example.

Example 5.1

Consider a first-order system with $a = 999.9$, $b = 0.9$ and a desired pole $\lambda = -0.1$. In this case the feedback coefficient is computed as

$$\bar{k} = \mathrm{fl}((a - \lambda)/b)$$

Assuming that a, b and λ are floating-point numbers, we obtain

$$\bar{k} = (a - \lambda)(1 - e)/b$$
$$= a(1 - e)/b - \lambda(1 - e)/b$$

where, using the error bounds on the basic arithmetic operations (see Section 1.4), we have that $|e| \leq 2\varepsilon$. This is a very satisfactory result since the computed gain element $\bar{k}$ is exact for $a^* = a(1 - e)$ and $b^* = b/(1 - e)$ which are close to a and b, respectively, i.e. the computations are numerically stable from a backward error analysis point of view.

For the given data

$$\bar{k} = \frac{1000}{0.9}(1 - e)$$

and the closed-loop system pole corresponding to $\bar{k}$ is

$$a - b\bar{k} = -0.1 + 1000e$$

If we perform the computation of k in floating-point arithmetic with $\varepsilon = 2^{-11} \approx 4.88 \times 10^{-4}$ then $\bar{k} = 1111$ and $a - b\bar{k} = 0$, while the exact gain coefficient is $k = 1111.11 \ldots$ and $a - bk = -0.1$.

This elementary example shows that although the gain matrix is obtained by stable computations, there is no guarantee of the closeness of the resulting poles to the desired closed-loop poles. In the given case, $|k|$ is very large compared to $|\lambda| = |a - bk|$ and hence a small relative error in the gain element causes a large error in the pole.

Consider the sources of errors in the closed-loop system poles resulting from the implementation of the computed gain matrix. As shown by Example 5.1, in the best case we can expect that the computed feedback matrix $\bar{K}$ is exact for a system with matrices $A^* = A + \Delta A$, $B^* = B + \Delta B$, where $\| \Delta A \|$, $\| \Delta B \|$ are small compared to $\| A \|$, $\| B \|$, respectively. Hence, instead of the closed-loop system matrix $F = A - BK$ which has the prescribed set

$$\Lambda = \{\lambda_1, \lambda_2, \ldots, \lambda_n\}$$

of eigenvalues, we will have the matrix

$$\bar{F} = A - B\bar{K} = F - B\Delta K$$

where $\Delta K = \bar{K} - K$ is the absolute error in $\bar{K}$.

The distance between the eigenvalues of $\bar{F}$ and the desired poles will depend on the size of $\| B\Delta K \|$ compared to $\| F \|$. If $\| B\Delta K \|$ is large relative to $\| F \|$ then the eigenvalues of the resulting closed-loop system matrix $\bar{F}$ may differ greatly from the desired poles. Specifically, we can distinguish the following three factors affecting the accuracy of the closed-loop system poles.

First, the problem of computing the feedback matrix from the given data A, B and Λ may be ill conditioned itself, i.e. small perturbations ΔA and ΔB in A and B may cause a large perturbation ΔK in K (we do not consider perturbations in the desired spectrum since they may be reduced to equivalent perturbations in A). The situation when K is very sensitive to perturbations in A and B is demonstrated by the following example.

Example 5.2

Let

$$A = \begin{bmatrix} 0.1 & 0 & 0 & 0 \\ 0.01 & 0.1 & 0 & 0 \\ 0 & 0.01 & 0.1 & 0 \\ 0 & 0 & 1 & 0.1 \end{bmatrix} \qquad b = \begin{bmatrix} 100 \\ 0 \\ 0 \\ 0 \end{bmatrix}$$

and

$$\Lambda = \{-0.15, \ -0.2, \ -0.25, \ -0.3\}$$

Then the feedback matrix is

$$k = [0.013, \ 0.6275, \ 13.325, \ 1.05]$$

If the element a_{31} is changed to -0.001 then the new gain matrix is

$$k^* = [0.013, \ 3.01, \ 23.825, \ 1.05]$$

Thus a relative perturbation $\delta A = \|\Delta A\|/\|A\|$ in A of order 10^{-3} is causing a relative change $\delta k = \|\Delta k\|/\|k\|$ in k of order 1 and this leads to a large absolute change $\|\Delta k\| = \|k\|\delta k$. Note that high sensitivity of the gain matrix may be manifested even when the desired poles are well separated.

Second, the exact solution for K may be too large so that $\|\Delta K\|$ is very large with respect to $|\lambda|$, even when the relative error δK is small. Indeed, since the norm of the error in $\bar{F}$ is bounded by

$$\|B\Delta K\| \le \|B\|\|K\|\delta K$$

it follows that the errors in the eigenvalues of $\bar{F}$ may be large if $\|B\|\|K\|$ is large compared to $\|A - BK\|$. A feedback matrix with large elements is obtained when the pair (A, B) is close to an uncontrollable pair or when the desired poles are far from the eigenvalues of A.

Example 5.3

Consider a system with matrices

$$A = \begin{bmatrix} A_{11} & a_{12} \\ 0 & a_{22} \end{bmatrix} \quad \text{and} \quad b = \begin{bmatrix} b_1 \\ \mu \end{bmatrix}$$

where the pair (A_{11}, b_1) is controllable and μ is small. Partitioning the feedback matrix k as

$$k = [k_1, \ k_2]$$

where k_2 is a scalar, we obtain that the closed-loop system matrix is expressed as

$$A - bk = \begin{bmatrix} A_{11} - b_1 k_1 & a_{12} - b_1 k_2 \\ -\mu k_1 & a_{22} - \mu k_2 \end{bmatrix}$$

With $\mu \to 0$, which corresponds to decreasing of the distance to an uncontrollable system, the quantity $a_{22} - \mu k_2$ will approach a pole of the closed-loop system, but for $\lambda \neq a_{22}$ the element

$$k_2 = \frac{\lambda - a_{22}}{\mu}$$

tends to infinity. Hence with the decreasing of the distance to an uncontrollable system, the norm of the gain matrix will increase since larger control action is needed.

The third factor for large errors in the eigenvalues of $A - B\overline{K}$ may be the high sensitivity of the closed-loop system poles to perturbations in $A - BK$. As we know from Section 1.9, the errors in the poles depend on the eigenvalue condition numbers

$$\text{cond}(\lambda) = \frac{1}{w^T v}$$

of $A - BK$, where v and w are the normed eigenvectors of $A - BK$ and $(A - BK)^T$, respectively, corresponding to the simple eigenvalue λ. If these numbers are large, then small errors in $A - BK$ may lead to large errors in the eigenvalues (see, for instance, Example 1.27). However, matrix $A - BK$ from Example 5.1 fulfils $\text{cond}(\lambda) = 1$ which is the lowest possible value, so that the eigenvalues of $A - BK$ may be perfectly conditioned, but nevertheless the errors in the closed-loop system poles may be very large. This phenomenon is explained as follows. Let ΔK be a perturbation in K such that the relative perturbation $\delta K = \| \Delta K \| / \| K \|$ is small. Then the perturbation $\Delta \lambda$ in λ can be evaluated as

$$| \Delta \lambda | \leq \text{cond}(\lambda) \| B \| \| K \| \delta K \tag{5.1}$$

Thus if $\| B \| \| K \|$ is large, then $| \Delta \lambda |$ will also be large even if cond(λ) is near to 1, exactly as in Example 5.1. Note also that for a single-input system the eigenvalue condition numbers are uniquely determined by the desired poles since in this case the solution for the feedback matrix is unique.

Hence, we may require that a computational method for pole assignment yields a matrix $\bar{K}$ which is exact for slightly perturbed matrices A and B and a set Λ of given poles. Such methods will be referred to as *numerically stable methods for pole assignments*. The closeness of the resulting closed-loop system poles to the desired locations will depend on the factors described above.

With respect to the first factor affecting the accuracy of the poles, it should be pointed out that the sensitivity of the feedback matrix to perturbations in the system matrices depends neither on the pole assignment algorithm used, nor on the precision implemented. Hence this factor is practically out of our control (at least for a given set Λ of desired poles) and we can only ascertain its effect. Unfortunately, an estimate for the sensitivity of K is known only in the single-input case and for simple desired poles.

Concerning the second factor, the following analysis based on the computed $\bar{K}$ can be made. Since the matrix

$$
\begin{aligned}
F^* &= A + \Delta A - (B + \Delta B)\bar{K} \\
&= \bar{F} + \Delta A - \Delta B\bar{K}
\end{aligned}
$$

has the desired spectrum Λ, the closeness of the eigenvalues of $\bar{F} = A - B\bar{K}$ to Λ will depend on the size of $\| \Delta A - \Delta B\bar{K} \|$ compared to $\| F^* \|$. Thus if $\| \bar{K} \|$ is very large in comparison to $\| \bar{F} \|$, the eigenvalues of $\bar{F}$ may be far from the desired poles.

In connection with the third factor it should be noted that for a multi-input system, the spectral condition numbers of the poles may be changed in certain bounds utilizing the freedom in the eigenvectors of the closed-loop system matrix. We consider a method for such a pole assignment in Section 5.3. In the light of the second and third factors, it is obvious that in the multi-input case one should minimize not only the condition number cond(λ) but also the product cond(λ)$\| K \|$ which determines the error in λ.

An important remark concerning the numerical determination of the resulting poles of $A - B\bar{K}$ must be made. If the numbers cond(λ) are large, then the errors in computing the poles may be also large in spite of the fact that the computed gain matrix $\bar{K}$ ensures exactly the desired poles. In such a case, even high-quality programs for eigenvalue computation, such as the programs from EISPACK, may yield false results.

Example 5.4

Let

$$A = \begin{bmatrix} 4 & -1 & 2 & 1 \\ -2 & 5 & 8 & -3 \\ 0 & -1 & -6 & -1 \\ 0 & 0 & 1 & 2 \end{bmatrix} \qquad b = \begin{bmatrix} 10 \\ 0 \\ 0 \\ 0 \end{bmatrix}$$

and

$$\lambda = \{-1, -1, -1, -1\}$$

The exact gain matrix which puts all poles at -1 is

$$k = [0.9, \ -1.8, \ -2.45, \ 4.55]$$

The eigenvalues of $A - bk$ determined by the subroutine COMQR2 from EISPACK using floating-point arithmetic with $\varepsilon = 2^{-56} \approx 1.39 \times 10^{-17}$ are (up to six digits)

$$-0.999\,783 + 0.000\,227\,075i$$
$$-0.999\,773 - 0.000\,216\,776i$$
$$-1.000\,23 + 0.000\,216\,792i$$
$$-1.000\,22 - 0.000\,227\,091i$$

In the given case the eigenvalue condition numbers are theoretically infinite which leads to large errors (of order $\sqrt[4]{\varepsilon}$) in the computed poles. If, however, the subroutine JNF from Kågström and Ruhe (1980) is used, then one obtains all poles equal to -1 up to 15 decimal digits. As described in Section 1.10, in this program the decision about the grouping of eigenvalues in Jordan block is taken using the Gershgorin circles, which allows accurate determination of multiple eigenvalues.

Consider now in brief the numerical properties of the available methods for pole assignment. The methods described in Section 2.6, which are based on reduction of the open-loop system into controllable canonical form or into Luenberger canonical form in the multi-input case, are generally numerically unstable. Their weakness comes from the fact that the canonical form utilized cannot be obtained by stable similarity transformations, as demonstrated by Example 4.22. The methods based on the characteristic polynomial of the open-loop or closed-loop system matrix, such as Ackermann's formula (see Exercise 2.41) are also numerically unstable, since these methods involve computation of the powers of a matrix, which is usually accompanied by large rounding errors.

Example 5.5

Let

$$A = \begin{bmatrix} -6.360 & 8.000 & 7.450 & 20.900 \\ -0.720 & 0.670 & 1.340 & 1.980 \\ -1.664 & 2.108 & 1.950 & 5.426 \\ -1.242 & 1.684 & 1.200 & 4.278 \end{bmatrix} \qquad b = \begin{bmatrix} 13.6 \\ -25.4 \\ 8.4 \\ 16.2 \end{bmatrix}$$

and the desired poles be

$$\Lambda = \{-0.5 \pm i, \ -0.6, \ -0.7\}$$

The characteristic polynomial which corresponds to the prescribed poles location is

$$p(\lambda) = \lambda^4 + 2.3\lambda^3 + 2.97\lambda^2 + 2.045\lambda + 0.525$$

Using Ackermann's formula in floating-point arithmetic with precision $\varepsilon = 2^{-19} \approx 1.91 \times 10^{-6}$ we obtain the gain matrix (up to six digits)

$$\bar{k} = [-234.954, \ 347.862, \ 323.785, \ 574.965]$$

Since the exact gain matrix is

$$k = [-234.128, \ 348.579, \ 317.942, \ 578.405]$$

the relative error in the computed result is

$$\delta k = \frac{\|\bar{k} - k\|_2}{\|k\|_2} = 0.0088$$

The eigenvalues of the matrix $A - b\bar{k}$ are

$$0.7493 \pm 0.6393i, \ -0.1310, \ -3.9796$$

so that even the closed-loop system is unstable.

The numerical instability is inherent also in the methods utilizing the eigenvalues and eigenvectors of the state matrix, such as the Mayne–Murdoch formula, presented in Exercise 2.42. This formula produces inaccurate results when the matrix A has close eigenvalues.

Attempts to solve the pole assignment problem in a numerically stable way are related to the use of the Schur form of the closed-loop system matrix $A - BK$. As is already known from the presentation in Section 1.9, there exist an orthogonal matrix Q and an upper quasi-triangular matrix T such that

$$Q^{\mathrm{T}}(A - BK)Q = T$$

This decomposition shows that the feedback matrix K could be obtained by orthogonal transformations chosen so that the matrix

$Q^T(A - BK)Q$ is in upper triangular form with the desired eigenvalues on the diagonal. The implementation of orthogonal transformations allows construction of pole assignment methods which are numerically stable in the sense of the definition given earlier in this section, i.e. the computed gain matrix is exact for a pair (A^*, B^*) close to (A, B) and for the desired set of eigenvalues.

Methods for pole assignment utilizing the Schur form of the closed-loop system matrix are referred to as *Schur methods*. We consider several Schur methods in the next section.

Most of the widely used pole assignment algorithms involve preliminary reduction of the pair (A, B) into orthogonal canonical form, which considerably simplifies the solution. As shown in Section 4.4, the orthogonal canonical form can be obtained in a numerically stable way both in the single-input and multi-input case. To demonstrate the usefulness of this form in the solution of the pole assignment problem, consider a completely controllable single-input system whose pair (A, b) is transformed into orthogonal canonical form

$$\tilde{A} = U^T A U = \begin{bmatrix} a_{11} & a_{12} & \cdots & a_{1n} \\ a_{21} & a_{22} & \cdots & a_{2n} \\ & a_{32} & \cdots & a_{3n} \\ \mathbf{0} & & \ddots & \vdots \\ & & a_{n,n-1} & a_{nn} \end{bmatrix} \quad \text{and} \quad \tilde{b} = U^T b = \begin{bmatrix} b_1 \\ 0 \\ 0 \\ \vdots \\ 0 \end{bmatrix}$$

The transformed closed-loop system matrix is

$$\tilde{A} - \tilde{b}\tilde{k} = \begin{bmatrix} a_{11} - b_1 k_1 & a_{12} - b_1 k_2 & \cdots & a_{1n} - b_1 k_n \\ a_{21} & a_{22} & \cdots & a_{2n} \\ & \mathbf{0} & \ddots & \vdots \\ & & a_{n,n-1} & a_{nn} \end{bmatrix}$$

where

$$\tilde{k} = kU \equiv [k_1, k_2, ..., k_n]$$

is the transformed feedback matrix. Now it may be shown that the eigenvector

$$\tilde{v} = [v_1, \ v_2, \ ..., \ v_n]^T$$

of $\tilde{A} - \tilde{b}\tilde{k}$, corresponding to a desired eigenvalue λ, can be easily computed from the equation

$$(\tilde{A} - \tilde{b}\tilde{k})\tilde{v} = \tilde{v}\lambda$$

or

$$
\begin{bmatrix}
a_{11} - b_1 k_1 & \cdots & a_{1n} - b_1 k_n \\
a_{21} & \cdots & a_{2n} \\
0 & \ddots & \vdots \\
& a_{n,n-1} & a_{nn}
\end{bmatrix}
\begin{bmatrix}
v_1 \\ v_2 \\ \vdots \\ v_n
\end{bmatrix}
=
\begin{bmatrix}
v_1 \\ v_2 \\ \vdots \\ v_n
\end{bmatrix} \lambda
\qquad (5.2)
$$

before the gain matrix is known. In fact, if we set an arbitrary nonzero v_n we obtain from (5.2) that

$$
v_i = \frac{v_{i+1} - \displaystyle\sum_{j=i+1}^{n} a_{i+1,j} v_j}{a_{i+1,i}} \qquad i = n-1, n-2, \ldots, 1 \qquad (5.3)
$$

where $a_{21}, a_{32}, \ldots, a_{n,n-1}$ are nonzero due to the controllability of the system (the Hessenberg matrix $\tilde{A}$ is unreduced). Thus for prescribed closed-loop poles the eigenvectors of $\tilde{A} - \tilde{b}\tilde{k}$ can be obtained with high accuracy by the backsubstitution process (5.3). For simple desired poles $\lambda_1, \lambda_2, \ldots, \lambda_n$ the feedback matrix may be obtained then as

$$
\tilde{k} = \frac{1}{b_1} (a_1 - v^{(1)} \Lambda V^{-1})
$$

$$
k = \tilde{k} U^{\mathrm{T}}
$$

where a_1 is the first row of $\tilde{A}$, $\Lambda = \mathrm{diag}(\lambda_1, \lambda_2, \ldots, \lambda_n)$, V is the matrix of corresponding eigenvectors and $v^{(1)}$ is the first row of V. Of course, such a method fails in the case when the eigenvector matrix V is ill conditioned. More sophisticated methods utilizing the eigenvectors of the closed-loop system matrix are presented in the next sections.

EXERCISES

5.1 Find an estimate of the relative perturbation δk in the feedback matrix k due to perturbations δA and δb in the matrices A and b of a single-input system for a given set of simple desired poles.

5.2 Find a bound on the allowable perturbations in the gain matrix which do not affect the stability of the closed-loop system, utilizing the distance to an unstable system (Section 4.1).

5.3 Analyze the accuracy of the resulting closed-loop poles of a high-order system, using the subroutines for pole assignment from your local program library.

5.4 Perform an analysis of the rounding errors in the eigenvectors of closed-loop system matrix, computed by the recurrence relationship (5.3).

5.2 SCHUR METHODS

In this section we present two methods for pole assignment which utilize the Schur form of the closed-loop system matrix. These methods may be implemented with continuous-time as well as with discrete-time systems.

5.2.1 Orthogonal Transformation of the Eigenvectors

As was shown in Section 5.1 for a completely controllable single-input system with prescribed eigenvalue λ of the closed-loop system matrix, we can find the corresponding eigenvector v before the feedback matrix k is found. The eigenvector elements are determined simply by the expression (5.3), if the pair (A, b) is first reduced to orthogonal canonical form. Assume now that using an orthogonal transformation all eigenvector elements, except the first one, are annulled. As a result we obtain from the equation

$$(\tilde{A} - \tilde{b}\tilde{k})\tilde{v} = \lambda\tilde{v}$$

that

$$Q^{\mathrm{T}}(\tilde{A} - \tilde{b}\tilde{k})Qv^* = \lambda v^*$$

where

$$v^* = Q^{\mathrm{T}}v = [x, \ 0, \ ..., \ 0]^{\mathrm{T}}$$

and Q is the matrix of the orthogonal transformation. Hence by necessity, the transformed closed-loop system matrix will have the form

$$Q^{\mathrm{T}}(\tilde{A} - \tilde{b}\tilde{k})Q = \begin{bmatrix} \lambda & x & ... & x \\ 0 & x & ... & x \\ \vdots & \vdots & & \vdots \\ 0 & x & ... & x \end{bmatrix} \tag{5.4}$$

The relation (5.4) yields an equation for the first element of the transformed gain matrix $\tilde{k}Q$. This element can be found by the corresponding elements of $Q^{\mathrm{T}}\tilde{A}Q$ and $Q^{\mathrm{T}}\tilde{b}$. Later we will show that the annulment of eigenvector elements may be done by using plane rotations (Section 1.7) in such a way that the open-loop and closed-loop state matrices remain in Hessenberg form. This allows us to proceed further in the same way, finding at each stage one element of the feedback matrix. The closed-loop poles may be prescribed independently, and after each stage the algorithm is applied on a subsystem of lower order, which reduces the volume of computations. Since every subsystem matrix is in Hessenberg form, its eigenvector may be computed by using an expression similar to (5.3), i.e. by solving a triangular system of linear equations. However, we will show that it is more efficient to compute the eigenvector elements simultaneously with the transformation of the eigenvector, exploiting the fact that some of the previous elements are already annulled. This may reduce significantly the number of computational operations.

To describe the actual algorithm we assume that the desired closed-loop system pole is real. The corresponding eigenvector obeys equation (5.2). Since the eigenvector is determined up to an arbitrary nonzero scalar, we may choose the element v_n equal to 1. Now according to (5.3) the element v_{n-1} is computed by

$$v_{n-1} = \frac{(\lambda - a_{nn})v_n}{a_{n,n-1}}$$

and if $n \geq 3$ we can find

$$v_{n-2} = \frac{(\lambda - a_{n-1,n-1})v_{n-1} - a_{n-1,n}v_n}{a_{n-1,n-2}}$$

(Recall that the subdiagonal entries of $\tilde{A}$ are nonzero because of the controllability of the system.)

Let a plane rotation R_1 in the $(n-1, n)$-plane be implemented, which zeros the element v_n, i.e.

$$R_1 \tilde{v} = v^{(1)} \qquad v^{(1)} = [x, \ldots, x, v_{n-2}, v_{n-1}^*, 0]^T$$

Since the multiplication by plane rotation preserves the 2-norm of the transformed eigenvector part, it follows that $v_{n-1}^* \neq 0$.

Equation (5.2) is rewritten as

$$R_1(\tilde{A} - \tilde{b}\tilde{k})R_1^T v^{(1)} = v^{(1)}\lambda \tag{5.5}$$

The transformed closed-loop system matrix $R_1(\tilde{A} - \tilde{b}\tilde{k})R_1^T$ has the form $(n = 5)$

$$\begin{bmatrix} x & x & x & x & x \\ x & x & x & x & x \\ & x & x & x & x \\ & \quad & x & x & x \\ \mathbf{0} & \quad & + & x & x \end{bmatrix}$$

where the element denoted by $+$ is introduced in computing $R_1\tilde{A}R_1^T$. If $n > 2$, the matrix $\tilde{b}$ is not affected at this step. If $n > 3$, we may now compute the element v_{n-3} from

$$v_{n-3} = \frac{(\lambda - a_{n-2,n-2})v_{n-2} - a_{n-2,n-1}v_{n-1}^*}{a_{n-2,n-3}}$$

where a_{ij} are the corresponding elements of $R_1 \tilde{A} R_1^{\mathsf{T}}$. In the above expression we exploited the fact that the element v_n is annulled.

At the next step we construct a plane rotation R_2 which zeros the element v_{n-1}^*, i.e.

$$R_2 v^{(1)} = v^{(2)} \qquad v^{(2)} = [x, \ \ldots, \ x, \ v_{n-3}, \ v_{n-2}^*, \ 0, \ 0]^{\mathsf{T}}$$

Clearly,

$$v_{n-2}^* \neq 0$$

From (5.5) we have that

$$R_2 R_1 (\tilde{A} - \tilde{b}\tilde{k}) R_1^{\mathsf{T}} R_2^{\mathsf{T}} v^{(2)} = v^{(2)} \lambda \tag{5.6}$$

and the transformed closed-loop state matrix is of the form

$$\begin{bmatrix} x & x & x & x & x \\ x & x & x & x & x \\ & x & x & x & x \\ & + & x & x & x \\ 0 & & \oplus & x & x \end{bmatrix}$$

It may be shown that the encircled element in the $(n, n-2)$-position, introduced at the previous step, now becomes zero. In fact, the equation for the last row in (5.6) yields

$$a_{n,n-2} v_{n-2}^* = 0$$

where $a_{n,n-2}$ is the corresponding element of $R_2 R_1 (\tilde{A} - \tilde{b}\tilde{k}) R_1^{\mathsf{T}} R_2^{\mathsf{T}}$ and since $v_{n-2}^* \neq 0$ it follows that $a_{n,n-2} = 0$. However, the transformation with the matrix R_2 introduces a nonzero element in position $(n-1, n-3)$. Thus the nonzero element below the subdiagonal is shifted one position up and left. This process continues at the next steps, 'chasing' the nonzero element outside the Hessenberg form toward the upper left-hand corner, as shown below

$$\begin{bmatrix} x & \ldots & & x & x \\ x & \ldots & & x & x \\ \oplus & & & & \\ & \ddots & & \vdots & \vdots \\ 0 & \searrow & \oplus & x & x \end{bmatrix}$$

After $n-1$ transformations we obtain the equation

$$Q^{\mathsf{T}}(\tilde{A} - \tilde{b}\tilde{k}) v^{(n-1)} = v^{(n-1)} \lambda$$
$$Q = R_1^{\mathsf{T}} R_2^{\mathsf{T}} \ldots R_{n-1}^{\mathsf{T}} \tag{5.7}$$
$$v^{(n-1)} = [v_1^*, \ 0, \ \ldots, \ 0]^{\mathsf{T}} \qquad v_1^* \neq 0$$

The transformed open-loop state matrix $Q^T \tilde{A} Q$ is represented as

$$Q^T \tilde{A} Q = \begin{bmatrix} a_{11} & x & x & \dots & x \\ \hline a_{21} & & & & \\ 0 & & A^{(2)} & & \\ \end{bmatrix}$$

where $A^{(2)}$ is an $(n-1) \times (n-1)$ Hessenberg matrix. Hence, the implemented orthogonal transformation retains the Hessenberg form of $\tilde{A}$.

At this step the vector $\tilde{b}$ is reduced to

$$Q^T \tilde{b} = \begin{bmatrix} b_1^* \\ \hline b_2^* \\ 0 \\ \vdots \\ 0 \end{bmatrix}$$

According to (5.7) the transformed closed-loop system matrix must be in the form

$$Q^T(\tilde{A} - \tilde{b}\tilde{k})Q = \begin{bmatrix} \lambda & x & x & \dots & x \\ \hline 0 & & A_c^{(2)} & & \end{bmatrix} \tag{5.8}$$

where $A_c^{(2)}$ is an $(n-1) \times (n-1)$ Hessenberg matrix (this follows from the form of $Q^T \tilde{A} Q$ and $Q^T \tilde{b}$). Since the closed-loop system is also controllable, we have that $b_2^* \neq 0$. The relation (5.8) yields

$$b_1^* k_1 = a_{11} - \lambda$$
$$b_2^* k_1 = a_{21}$$

where k_1 is the first element of $\tilde{k}Q$. Hence k_1 can be determined from

$$k_1 = \frac{a_{21}}{b_2^*}$$

which gives

$$a_{11} - b_1^* k_1 \equiv \lambda_1$$

In this way we obtain one element of the gain matrix and the dimension of the problem is reduced to $n-1$. Since the matrices of the reduced dimension problem are also in orthogonal canonical form, it is possible to continue in the same way. Note that the poles of each subsystem are prescribed independently and the corresponding eigenvectors are used separately. Hence the eigenvectors may even be linearly dependent when there are multiple eigenvalues. That is why the method works equally well with distinct and multiple poles. The multiple poles will appear in a single Jordan block, since the closed-loop system is controllable. It can be observed that the final transformed closed-loop system matrix, which is in Schur form, will have a nonsingular upper triangular eigenvector matrix. In this way the method described produces the QR decomposition of the eigenvectors (and principal eigenvectors) of the closed-loop system matrix.

The method described may also be applied to determine elements of the gain matrix in the case of complex conjugate poles. If the desired pole is complex, then the corresponding eigenvector is also complex and to zero a specified eigenvector element it is necessary to use complex (unitary) rotation in the form

$$
R =
\begin{array}{c}
\\ \\ i \\ \\ j \\ \\
\end{array}
\overset{\begin{array}{ccc} & i & \quad\quad j \end{array}}{
\begin{bmatrix}
1 & & & & & & \\
& \ddots & & & & & \\
& & 1 & & & & \\
& & p_1 & & q & & \\
& & & \ddots & & & \\
& & -\bar{q} & & 1 & & \\
& & & & p_1 & & \\
& & & & & \ddots & \\
& & & & & & 1
\end{bmatrix}
}
$$

where $\bar{q}$ is the complex conjugate of q. The elements p and q may be computed by an appropriate modification of Algorithm 1.14.

Thus we have the following algorithm, which for a given pair (A, b) in orthogonal canonical form and a set of desired poles $\lambda_1, \ldots, \lambda_n$ computes the feedback matrix k. The Schur form of $A - bk$ overwrites A and the corresponding unitary transformation matrix is obtained in the array Q.

Algorithm 5.1 Pole assignment of a single-input system by orthogonal transformation of the eigenvectors

Set $Q = I_n$
For $l = 1, 2, \ldots, n - 1$
    Set $v_n = 1$
    For $i = n - 1, n - 2, \ldots, l$
        Compute v_i and (possibly) v_{i-1}

Determine a (complex) plane rotation $R_{i,i+1}$ such that

$$R_{i,i+1} \begin{bmatrix} x \\ \vdots \\ x \\ v_i \\ v_{i+1} \\ x \\ \vdots \\ x \end{bmatrix} = \begin{bmatrix} x \\ \vdots \\ x \\ g_i \\ 0 \\ 0 \\ \vdots \\ 0 \end{bmatrix}$$

$A \leftarrow R_{i,i+1} A R_{i,i+1}^H$

If $i = l$, $b \leftarrow R_{i,i+1} b$

$Q \leftarrow Q R_{i,i+1}^H$

$k_l = a_{l+1,l} / b_{l+1}$

For $i = 1, 2, \ldots, l+1$

$\quad a_{il} \leftarrow a_{il} - b_i k_l$

$k_n = (a_{nn} - \lambda_n) / b_n$

$a_{nn} \leftarrow \lambda_n$

$k \leftarrow k Q^H$

Algorithm 5.1 requires approximately $4n^3 + 9n^2$ flops in the case of real poles. Adding the number of necessary operations for reducing the pair (A, b) into orthogonal canonical form (see Section 4.4) one obtains

$$\tfrac{20}{3} n^3 + \tfrac{31}{2} n^2$$

flops, i.e. the full implementation of the algorithm requires less than $7n^3$ flops. With respect to storage, the algorithm requires $2n^2 + 4n$ working precision words.

Algorithm 5.1 may be modified to treat the case of complex conjugate poles by a slightly complicated technique using real arithmetic only. As a result, two elements of the transformed gain matrix may be determined simultaneously and the transformed closed-loop matrix will have 2×2 blocks on its diagonal. Thus at the final stage this matrix will be in real Schur form. The details of this technique may be found in Petkov, Christov and Konstantinov (1984a).

It is possible to show that the algorithm presented is numerically stable, i.e. it produces a feedback matrix $\bar{k}$, which is exact for matrices $A + \Delta A$ and $b + \Delta b$ and for the desired poles, where $\| \Delta A \|$ and $\| \Delta b \|$ are small multiples of $\varepsilon \| A \|$ and $\varepsilon \| b \|$, respectively (Cox and Moss, 1989). Thus the accuracy of the computed gain matrix relies on the conditioning of the problem. The case of ill conditioning manifests itself by small values of the elements b_1^* and b_2^* which cause the errors in the determination of k to be magnified. The closeness of the eigenvalues of

$A - b\bar{k}$ to the desired poles depends on the factors described in Section 5.1. In some cases, the eigenvalue condition numbers may be large and the errors in the resulting closed-loop system poles will also be large, independent of the numerical stability of the method. Thus it is necessary to perform an *a posteriori* check of the eigenvalues of $A - b\bar{k}$.

Example 5.6

Consider a fifth-order system with matrices

$$A = \begin{bmatrix} 0.1 & 0 & 0 & 0 & 0 \\ 0.01 & 0.1 & 0 & 0 & 0 \\ 0 & 0.01 & 0.1 & 0 & 0 \\ 0 & 0 & 1. & 0.1 & 0 \\ 0 & 0 & 0 & 1. & 0.1 \end{bmatrix} \quad \text{and} \quad b = \begin{bmatrix} 100 \\ 0 \\ 0 \\ 0 \\ 0 \end{bmatrix}$$

which is already in orthogonal canonical form, and let the desired poles be

$$-0.1, -0.15, -0.2, -0.25, -0.3.$$

Using the algorithm described above in floating-point arithmetic with $\varepsilon = 2^{-24} \approx 5.96 \times 10^{-8}$ we obtain (up to seven significant digits)

$$\bar{k} = [0.015\,000\,00, \quad 0.887\,500\,3, \quad 25.875\,02, \quad 3.715\,004, \quad 0.210\,000\,4]$$

The exact gain matrix is

$$k = [0.015, \quad 0.088\,75, \quad 25.875, \quad 3.715, \quad 0.21]$$

so that the relative error in $\bar{k}$ is

$$\frac{\|\bar{k} - k\|_2}{\|k\|_2} \approx 6.716 \times 10^{-7}$$

This is a satisfactory result since the pole assignment problem for the given system is ill conditioned. In fact, a perturbation of -0.001 in the element a_{31} leads to a gain matrix

$$k^* = [0.015\,000\,00, \quad 9.290\,000, \quad 84.025\,00, \quad 5.815\,000, \quad 0.210\,000\,0]$$

so that a relative perturbation in A,

$$\frac{\|A^* - A\|_2}{\|A\|_2} \approx 9.472 \times 10^{-4}$$

causes a relative change in the gain matrix,

$$\frac{\|k^* - k\|_2}{\|k\|_2} \approx 2.248$$

This shows that the effect of the errors introduced by the algorithm is equivalent to perturbations in the data which are much smaller than 10^{-3}.

The eigenvalues of $A - b\bar{k}$ are

$-0.099\ 995\ 73,\ -0.150\ 045\ 1,\ -0.199\ 866\ 2,\ -0.250\ 160\ 1,$
$-0.299\ 932\ 7$

The difference between the desired and actual poles is due to the difference $b(\bar{k} - k)$ and the ill conditioning of the eigensystem of $A - bk$ (the eigenvalues condition numbers of this matrix are of order $10^4 \div 10^7$).

The method presented may be extended to the multi-input case by reducing the controllable pair (A, B) into orthogonal canonical form

$$\tilde{A} = U^{\mathrm{T}} A U = \begin{bmatrix} A_1 \\ A_{21} & A_2 \\ & \ddots \\ 0 & & A_{p,p-1} & A_p \end{bmatrix} \quad \text{and} \quad \tilde{B} = U^{\mathrm{T}} B = \begin{bmatrix} B_1 \\ 0 \end{bmatrix}$$

where B_1 is an $m_1 \times m$ matrix and $A_{i,i-1}$ is an $m_i \times m_{i-1}$ matrix

$$\text{rank } B_1 = \text{rank } B = m_1 \qquad \text{rank } A_{i,i-1} = m_i \qquad m_1, ..., m_p$$

$(m \geq m_1 \geq \cdots \geq m_p \geq 1,\ m_1 + \cdots + m_p = n)$ are the conjugate Kronecker indices, and p is the controllability index of (A, B). Since the closed-loop system matrix $\tilde{A} - \tilde{B}\tilde{K}$, $\tilde{K} = KU$, is in block-Hessenberg form as $\tilde{A}$, it is possible to find an eigenvector $\tilde{v}$ of this matrix, knowing only $\tilde{A}$ and the desired eigenvalue λ. Indeed, from the equation $(\tilde{A} - \tilde{B}\tilde{K})\tilde{v} = \tilde{v}\lambda$, or

$$\begin{bmatrix} A_1 - B_1\tilde{K} \\ A_{21} & A_2 \\ & \ddots \\ & & A_{p-1,p-2} & A_{p-1} \\ 0 & & & A_{p,p-1} & A_p \end{bmatrix} \begin{bmatrix} v_1 \\ v_2 \\ \vdots \\ v_{p-1} \\ v_p \end{bmatrix} = \begin{bmatrix} v_1 \\ v_2 \\ \vdots \\ v_{p-1} \\ v_p \end{bmatrix} \lambda$$

where v_i is an m_i-vector, it follows that

$$A_{i,i-1}v_{i-1} = v_i\lambda - A_i[v_i^{\mathrm{T}}, ..., v_p^{\mathrm{T}}]^{\mathrm{T}} \qquad i = p, p-1, ..., 2 \qquad (5.9)$$

so that by setting v_p it is possible to compute recursively $v_{p-1}, ..., v_1$. The vector v_p must be nonzero and different choices of this vector will lead to different solutions for $\tilde{K}$.

Equation (5.9) is underdetermined when $m_i < m_{i-1}$ and may be solved using QR decomposition with column pivoting.

Consider now a sequence of plane rotations

$$R_{n-1,n}, \ldots, R_{n-m_p+1,n-m_p+2}$$

in the corresponding planes chosen so as to annul successively all elements of v_p except the first one moving from bottom to the top. It is then possible to compute v_{p-2} by exploiting the fact that the vector v_p has only one nonzero element. Zeroing the next m_{p-1} elements of the eigenvector by implementing plane rotations, we obtain

where the element $v_{p-1,1}$ is the norm of $[v_{p-1}^T, v_p^T]^T$ and hence is nonzero. The elements denoted by $+$ are introduced during the annulment of the first element of v_p. The encircled elements in the first column of $A_{p,p-1}^*$ must be zeros since $v_{p-1,1} \neq 0$.

At this step we can compute the eigenvector part v_{p-3} (if $p > 3$) using only v_{p-2} and the single nonzero element of v_{p-1}.

This process continues until we obtain

$$Q^T(\tilde{A} - \tilde{B}\tilde{K})Qv^* = v^*\lambda$$

$$v^* = Q^T v = [v_{11}, 0, \ldots, 0]^T \qquad v_{11} \neq 0$$

where $A_{i,i-1}^0$ is an $m_i \times m_{i-1}$ matrix, $i = 2, \ldots, p-1$; $A_{p,p-1}^0$ is an $(m_p - 1) \times m_p$ matrix and B_1^0 is an $m_1 \times m$ matrix.

Denoting the first column of the transformed gain matrix $\tilde{K}Q$ by k_1 we obtain the linear, consistent system of equations

$$\left[\begin{array}{cccc} x & x & \dots & x \\ \hline & & B_1^0 & \end{array}\right] k_1 = a_1 - [\lambda, \ 0, \ \dots, \ 0]^{\mathrm{T}}$$

which yields a solution for k_1.

Since the closed-loop system with state matrix

$$Q^{\mathrm{T}}(\tilde{A} - \tilde{B}\tilde{K})Q = \left[\begin{array}{c|ccccc} \lambda & x & x & \dots & x \\ \hline & A_1^0 - B_1^0 K_1^0 \\ \hline & A_{21}^0 & A_2^0 \\ \hline 0 & 0 & & \ddots \\ & & & A_{p,p-1}^0 & A_p^0 \end{array}\right]$$

where K_1^0 is an $m \times (n-1)$ matrix, is controllable then the pair

$$A^0 = \left[\begin{array}{c|ccc} & & A_1^0 \\ \hline A_{21}^0 & & A_2^0 \\ \hline 0 & & \ddots \\ & A_{p,p-1}^0 & A_p^0 \end{array}\right] \qquad B^0 = \left[\begin{array}{c} B_1^0 \\ \hline 0 \end{array}\right]$$

must be also controllable. This requires

rank $B_1^0 = m_1,$ rank $A_{21}^0 = m_2, \dots$
$$\text{rank } A_{p-1,p-2}^0 = m_{p-1}, \ \text{rank } A_{p,p-1}^0 = m_p - 1$$

In this way the pair (A^0, B^0) retains the structure of the pair $(\tilde{A}, \tilde{B})$, only the last conjugate Kronecker index being decreased by 1. This makes it possible to proceed in the same way at the next stage working with a system of order $n - 1$. Thus each pole is prescribed independently. Notice that different orderings of the desired poles will lead to different gain matrices. In contrast to the single-input case, here it is possible to assign multiple poles which do not correspond to a single block in the Jordan form of the closed-loop system matrix.

When complex conjugate poles are prescribed one may use the real and imaginary parts of the corresponding complex eigenvector, thus avoiding implementation of complex arithmetic. This leads to the appearance of a 2×2 block on the diagonal of the transformed closed-loop system matrix. In such a case, we obtain two columns of the gain matrix simultaneously and the block $A_{p,p-1}^0$ that results will be of rank $m_p - 2$. We refer the reader to Petkov, Christov and Konstantinov (1986b) where further details of the algorithm are presented.

The numerical properties of the algorithm are favourable, as in the single-input case.

Example 5.7

Consider a fifth-order, two-input system with matrices

$$A = \begin{bmatrix} -0.1094 & 0.0628 & 0 & 0 & 0 \\ 1.306 & -2.132 & 0.9807 & 0 & 0 \\ 0 & 1.595 & -3.149 & 1.547 & 0 \\ 0 & 0.0355 & 2.632 & -4.257 & 1.855 \\ 0 & 0.002\,27 & 0 & 0.1636 & -0.1625 \end{bmatrix}$$

$$B = \begin{bmatrix} 0 & 0.0638 & 0.0838 & 0.1004 & 0.0063 \\ 0 & 0 & -0.1396 & -0.206 & -0.0128 \end{bmatrix}^T$$

and let

$$\Lambda = \{-0.2, -0.5, -1, -1 \pm i\}$$

Using the described algorithm in floating-point arithmetic with $\varepsilon = 2^{-24} \approx 5.96 \times 10^{-8}$, we obtain the following five solutions for K, for different ordering of the desired poles (the results are rounded to five digits):

1. $\Lambda = \{-0.2, -0.5, -1, -1 \pm i\}$

$$\bar{K}_1 = \begin{bmatrix} 32.3475 & -90.7662 & 238.756 & -325.572 & 32.5829 \\ 14.6555 & -42.7159 & 94.3053 & -123.444 & 8.520\,55 \end{bmatrix}$$

2. $\Lambda = \{-1 \pm i, -0.5, -0.2, -1\}$

$$\bar{K}_2 = \begin{bmatrix} 31.1730 & -121.674 & 256.187 & -226.680 & 36.6433 \\ 18.1663 & -69.3181 & 122.000 & -98.4440 & 41.8742 \end{bmatrix}$$

3. $\Lambda = \{-1, -1 \pm i, -0.2, -0.5\}$

$$\bar{K}_3 = \begin{bmatrix} 55.6569 & -118.964 & 249.921 & -214.853 & 53.0666 \\ -13.8077 & -58.3225 & 99.5142 & -78.1973 & 34.6007 \end{bmatrix}$$

4. $\Lambda = \{-0.5, -1 \pm i, -0.2, -0.1\}$

$$\bar{K}_4 = \begin{bmatrix} 67.9485 & -83.4495 & 174.716 & -150.637 & 21.9620 \\ -12.3529 & -39.9713 & 49.0104 & -32.4347 & 21.9613 \end{bmatrix}$$

5. $\Lambda = \{-0.2, -1 \pm i, -0.5, -0.1\}$

$$\bar{K}_5 = \begin{bmatrix} 42.6271 & -62.2297 & 123.327 & -104.265 & 7.061\,98 \\ -9.098\,05 & -34.2915 & 13.5014 & -0.859\,191 & 26.7902 \end{bmatrix}$$

All solutions are computed with relative error

$$\frac{\| \bar{K}_i - K_i \|_2}{\| K_i \|_2}$$

less than 10^{-6}.

The absolute errors in the closed-loop poles, found in arithmetic with $\varepsilon = 2^{-56} \approx 1.39 \times 10^{-17}$, are less than 10^{-4}.

5.2.2 Modified QR Algorithm

The matrices K and Q in the Schur decomposition of the closed-loop system matrix

$$Q^{\mathrm{T}}(A - BK)Q = T$$

may also be determined by an algorithm which resembles the QR algorithm for finding the eigenvalues of a matrix (see Section 1.10). Since in the given case the desired eigenvalues are known, the QR algorithm becomes a direct, rather than an iterative, method. In what follows, we describe an algorithm for the pole assignment of single-input systems which may be considered as a modification of the QR method with explicit shifts. For simplicity of presentation, we shall assume the use of complex arithmetic for allocation of a complex eigenvalue. As in the previous Schur algorithm, the pair (A, b) is preliminarily reduced to orthogonal form $(\tilde{A}, \tilde{b}) = (U^{\mathrm{T}} A U, U^{\mathrm{T}} b)$ and we seek for the transformed gain matrix $\tilde{k} = kU$ which assigns the eigenvalues of $\tilde{A} - \tilde{b}\tilde{k}$ at desired locations $\lambda_1, \lambda_2, \ldots, \lambda_n$.

Let $A^{(1)} = \tilde{A}$, $b^{(1)} = \tilde{b}$ and $k^{(1)} = \tilde{k}$. For a given eigenvalue λ_1, the matrix $A^{(1)} - b^{(1)}k^{(1)} - \lambda_1 I$ must be singular. Hence if we apply (unitary) rotations R_{ij} from the right to reduce this matrix to upper triangular form, then one of its diagonal elements must be zero. For $n = 5$ the matrix $A^{(1)} - \lambda_n I$ will take the form

$$(A^{(1)} - \lambda_1 I)\underbrace{R_{11} \ldots R_{1,n-1}}_{Q_1}$$

$$= \begin{bmatrix} x & x & x & x & x \\ \overset{(x)}{\underset{4}{\longrightarrow}} & x & x & x & x \\ & \overset{(x)}{\underset{3}{\longrightarrow}} & x & x & x \\ 0 & & \overset{(x)}{\underset{2}{\longrightarrow}} & x & x \\ & & & \overset{(x)}{\underset{1}{\longrightarrow}} & x \end{bmatrix} \equiv \begin{bmatrix} t_{11} & x & x & x & x \\ & x & x & x & x \\ & & x & x & x \\ 0 & & & x & x \\ & & & & x \end{bmatrix}$$

where the arrows numbered by $1, 2$, etc., indicate the action of the rotations R_{11}, R_{12}, etc.

Since $A^{(1)}$ is an unreduced Hessenberg matrix each rotation is nontrivial and the resulting diagonal elements in positions from $(2, 2)$ to (n, n) are nonzero. Because $k^{(1)}$ affects only the first row of $A^{(1)}$, these diagonal elements are the same as in $(A^{(1)} - b^{(1)}k^{(1)} - \lambda_1 I)Q_1$ and the $(1, 1)$-element of this matrix must be zero. That is why if we write

$$k^{(1)}Q_1 = [k_1, k^{(2)}]$$

where k_1 is a scalar and $k^{(2)}$ is an $(n - 1) \times 1$ row vector, then we have that

$$t_{11} - b_1 k_1 = 0$$

and so $k_1 = t_{11}/b_1$ which will zero the first column of $(A^{(1)} - b^{(1)}k^{(1)} - \lambda_1 I)Q_1$. The left multiplication by the matrix Q_1^H gives

$$Q_1^H(A^{(1)} - \lambda_1 I)Q_1 + \lambda_1 I = Q_1^H A^{(1)} Q_1$$

$$= \begin{bmatrix} x & x & x & x & x \\ + & x & x & x & x \\ & + & x & x & x \\ & & + & x & x \\ & 0 & & + & x \\ & & & & + \end{bmatrix} = \begin{bmatrix} x & x & x & x & x \\ \hline x & & & & \\ 0 & & A^{(2)} & & \\ & & & & \end{bmatrix}$$

The elements denoted by $+$ are introduced during the multiplication by Q_1^H. The matrix $A^{(2)}$ is necessarily in unreduced Hessenberg form since the rotations implemented are nontrivial and all but the first diagonal elements are nonzero.

Transformation of the vector $b^{(1)}$ yields

$$Q_1^H b^{(1)} = \begin{bmatrix} x \\ \hline b_2 \\ \hline 0 \end{bmatrix} = \begin{bmatrix} x \\ \hline b^{(2)} \end{bmatrix} \qquad b_2 \neq 0$$

Now we have

$$Q_1^H(A^{(1)} - b^{(1)}k^{(1)})Q_1 = \begin{bmatrix} \lambda_2 & x & x & x & x \\ \hline 0 & & A^{(2)} - b^{(2)}k^{(2)} \end{bmatrix}$$

where the pair $(A^{(2)}, b^{(2)})$ is controllable.

In this way the problem is reduced to the allocation of the remaining $n - 1$ eigenvalues $\lambda_2, \ldots, \lambda_n$ to $A^{(2)} - b^{(2)}k^{(2)}$ by choosing $k^{(2)}$. The form of the matrices $A^{(2)}$ and $b^{(2)}$ prompts that we may continue in the same way and the choice of $k^{(2)}$ will not alter the eigenvalue λ_1. After the $(n-1)$th step we will obtain $A^{(n)}$ as a 1×1 matrix and the unknown scalar $k^{(n)} = k_n$ is determined from

$$A^{(n)} - b^{(n)}k^{(n)} = \lambda_n$$

so that

$$k_n = (A^{(n)} - \lambda_n)/b_n$$

To obtain the gain matrix $k = k^{(1)}$ it is necessary to perform a backward sweep. Since in the forward sweep

$$k^{(j)}Q_j = [k_j, k^{(j+1)}] \qquad j = 1, 2, ..., n-1$$

where $k_1, ..., k_n$ and $Q_1, ..., Q_{n-1}$ are already determined, the backward sweep is

$$k^{(n)} = k_n$$
$$k^{(j)} = [k_j, k^{(j+1)}] Q_j^{\mathrm{H}} \qquad j = n-1, n-2, ..., 1$$

Overall we have the following algorithm.

Algorithm 5.2 Pole assignment of a single-input system by modified QR algorithm

Set $A^{(1)} = \tilde{A}$
For $j = 1, 2, ..., n-1$
    Choose $Q_j = R_{j1} ... R_{j,n-j}$ as a product of (complex) rotations so
        that $(A^{(j)} - \lambda_j I)Q_j = T^{(j)}$ is upper triangular
    Set $t_{jj} = [T^{(j)}]_{11}$
    $k_j = t_{jj}/b_j$
    Set

$$\left[\begin{array}{c} x \\ \hline b^{(j+1)} \end{array} \right] = R_{j,n-j}^{\mathrm{H}} \left[\begin{array}{c} b_j \\ 0 \end{array} \right]$$

    Set

$$\left[\begin{array}{c|ccc} x & x & ... & x \\ \hline x & & & \\ 0 & & A^{(j+1)} & \end{array} \right] = Q_j^{\mathrm{H}} T^{(j)} + \lambda_j I$$

Set
    $k^{(n)} = k_n = (A^{(n)} - \lambda_n)/b_n$
For $j = n-1, n-2, ..., 1$
    $k^{(j)} = [k_j, k^{(j+1)}] Q_j^{\mathrm{H}}$
$\tilde{k} = k^{(1)}$

Algorithm 5.2 requires about $\frac{23}{11}n^3$ flops as well as $4n^2 + 2n$ working precision words for its execution. It should be pointed out that explicit forming of the matrices Q_j is unnecessary since it is time- and space-consuming. Instead, it is only necessary to save the elements of the rotations R_{ij}.

This algorithm is numerically stable and has the same numerical behaviour as Algorithm 5.1.

Algorithm 5.2 has much in common with the QR algorithm with explicit shifts. The difference is that in the QR algorithm the shifts converge to the eigenvalues to be computed, whereas in the given algorithm the shifts are the desired eigenvalues, which are known a priori. As in the QR algorithm, it is possible to implement implicit shifts which yield more accurate results and enable us to assign complex conjugate pairs of eigenvalues using real arithmetic only.

The modified QR method for pole assignment may be extended to the multi-input case, exploiting the existing freedom in the solution in an appropriate way. The corresponding algorithm, along with a proof of the numerical stability, can be found in Miminis and Paige (1988).

EXERCISES

5.5 Consider in detail the procedure for finding two elements of the feedback matrix by orthogonal transformation of the real and imaginary part of the eigenvector, corresponding to complex conjugate pair of eigenvalues. Use only real arithmetic.

5.6 Apply the algorithm for the pole assignment of multi-input systems using orthogonal transformation of the eigenvectors to a third-order system with two inputs. Try to exploit the freedom in eigenvectors to obtain different gain matrices.

5.7 Write a computer program for the pole assignment of single-input systems by orthogonal transformation of the eigenvectors.

5.8 Derive an algorithm for pole assignment by computing successively the columns of the orthogonal transformation matrix Q and the superdiagonal elements of the Schur form T in the equation

$$(\tilde{A} - \tilde{B}\tilde{K})Q = QT$$

where $\tilde{A}$ and $\tilde{B}$ are in orthogonal canonical form.

5.9 Write a computer program for the pole assignment of single-input systems by the modified QR algorithm.

5.10 Compare the accuracy of computer programs implementing Algorithms 5.1 and 5.2.

5.3 ROBUST POLE ASSIGNMENT

In the case of a multi-input system, the pole assignment problem has several solutions for a given set of desired poles. These solutions correspond to closed-loop systems with different properties. Thus the freedom in the feedback matrix may be used to satisfy additional requirements for the closed-loop system. In particular, this freedom can be utilized to minimize the sensitivity of the closed-loop poles to perturbations in the matrices A, B and K.

Since the poles' sensitivity depends on the corresponding eigenvectors, it may be controlled by suitable choice of the eigenvectors. A pole assignment, accompanied by setting the eigenvectors, is referred to as *eigenstructure assignment*.

As is already known, if the closed-loop system matrix $A - BK$ is nondefective, the sensitivity of the eigenvalue λ_j to perturbations in $A - BK$ depends on the spectral condition number

$$\text{cond}(\lambda_j) = \frac{1}{w_j^T v_j}$$

where v_j and w_j are the normalized ($\| v_j \|_2 = \| w_j \|_2 = 1$) right and left eigenvectors of $A - BK$. For sufficiently small perturbation E in the matrix $A - BK$ the perturbation $\Delta\lambda_j$ in λ_j satisfies

$$| \Delta\lambda_j | \leq \text{cond}(\lambda_j) \| E \|_2$$

If the matrix $A - BK$ is defective, then the perturbation in λ_j is proportional to $\| E \|_2^{1/k}$ where k is the size of the maximal Jordan block containing λ_j. Thus to reduce the sensitivity of the poles it is preferable to choose the feedback matrix K so that $A - BK$ is nondefective. This, in particular, will be fulfilled when the closed-loop system poles are distinct.

Since

$$\max \text{cond}(\lambda_j) \leq \text{cond}_2(V) = \| V \|_2 \| V^{-1} \|_2$$

where $V = [v_1, v_2, ..., v_n]$, a possible way to minimize the sensitivity of the poles is to minimize the condition number $\text{cond}_2(V)$ of the eigenvector matrix. Hence we can formulate the following problem. Given a completely controllable pair (A, B) and a set of desired poles $\Lambda = \{\lambda_1, \lambda_2, ..., \lambda_n\}$, find a matrix K and a non-singular matrix V satisfying

$$\begin{aligned} (A - BK)V &= VD \\ D &= \text{diag}(\lambda_1, \lambda_2, ..., \lambda_n) \end{aligned} \tag{5.10}$$

such that some measure ν of the conditioning of the eigenproblem for

$A - BK$ is minimized. This problem is referred to as the *robust pole assignment problem*. The measure ν could be taken, for instance, as

$$\nu_1 = \sum_{j=1}^{n} \text{cond}(\lambda_j)$$

or

$$\nu_2 = \text{cond}_2(V)$$

The conditions for the existence of a solution to the above problem are given by the following theorem.

Theorem 5.1

Given A, B, Λ and V, with rank $B = m$, and V nonsingular, there exists a matrix K, which is a solution to (5.10) if and only if

$$U_2^T(AV - VD) = 0 \tag{5.11}$$

where

$$B = [U_1, \ U_2]\begin{bmatrix} Z \\ 0 \end{bmatrix} \tag{5.12}$$

with $U = [U_1, \ U_2]$ orthogonal and Z nonsingular. Moreover, K is found explicitly by

$$K = Z^{-1}U_1^T(A - VDV^{-1}) \tag{5.13}$$

The proof of the theorem is easily carried out taking into account that K must satisfy

$$BK = A - VDV^{-1} \tag{5.14}$$

Since B is of full rank it may be decomposed as shown in (5.12) and the multiplication of (5.14) by U^T then gives

$$ZK = U_1^T(A - VDV^{-1})$$
$$0 = U_2^T(A - VDV^{-1})$$

from which (5.11) and (5.13) follow directly, since V is invertible.

The decomposition (5.12) can be done by using QR decomposition or singular value decomposition.

In this way the robust pole assignment problem reduces to the problem of selecting n linearly independent vectors v_i, $i = 1, 2, ..., n$ such that the eigenproblem (5.10) is as well conditioned as possible.

It may be shown that the multiplicity of the prescribed eigenvalue λ_j must be less than or equal to the number m of the inputs. This is because the maximum number of linearly independent eigenvectors, which can be chosen to correspond to λ_j, is equal to m. Indeed, from (5.11) and (5.12) we have that

$$U^T[B, \; A - \lambda_j I] = \begin{bmatrix} Z & U_1^T(A - \lambda_j I) \\ 0 & U_2^T(A - \lambda_j I) \end{bmatrix}$$

and since

rank $Z = m$
rank $[B, \; A - \lambda_j I] = n$

it follows that

rank $U_2^T(A - \lambda_j I) = n - m$
  Hence the equation

$$U_2^T(A - \lambda_j I)v_j = 0 \tag{5.15}$$

has at most m linearly independent solutions.
  From expression (5.13) we obtain that

$$\| K \|_2 \leq \| Z^{-1} \|_2 \| U_1^T \|_2 [\| A \|_2 + \| V \|_2 \| V^{-1} \|_2 \| D \|_2]$$

$$\leq \frac{\| A \|_2 + \max_j \{| \lambda_j |\} \mathrm{cond}_2(V)}{\sigma_m(B)} \tag{5.16}$$

where $\sigma_m(B)$ is the smallest singular value of B. Hence minimizing the conditioning of the assigned eigensystem also minimizes a bound on the norm of the feedback matrix. Note, however, that this does not imply that the gain matrix obtained will be of minimal norm. Also, the bound (5.16) may considerably overestimate $\| K \|_2$ when the eigenvector matrix is ill conditioned. Usually an estimate of the form

$$\| K \|_2 \leq \frac{\| A \|_2}{\sigma_m(B)}$$

gives a more reliable result in practice.
  In this way the feedback matrix which produces minimum sensitivity of the closed-loop poles may be computed by an algorithm which is sketched as follows.

Compute the decomposition (5.12) of the matrix B
Select vectors v_j with $\| v_j \|_2 = 1$, $j = 1, 2, \dots, n$ which satisfy (5.15), such
  that $V = [v_1, v_2, \dots, v_n]$ is as well conditioned as possible
Find the matrix $F = A - BK$ by solving

$$FV = VD \tag{5.17}$$

and compute K explicitly from

$$K = Z^{-1} U_1^T (A - F) \tag{5.18}$$

The first and third steps of this algorithm are performed easily by standard subroutines. The implementation of QR decomposition to find (5.12) is computationally less expensive, but the singular value decomposition gives more information about the problem. The key step of the algorithm is the second one. In this step the vectors v_j, $j = 1, 2, \ldots, n$, are chosen by an iterative process to minimize the measure of conditioning. Every eigenvector v_j, $j = 1, 2, \ldots, n$, which is a solution of (5.15) may be expressed as

$$v_j = Q_j w_j$$

where the $n \times m$ matrix Q_j satisfies

$$U_2^T (A - \lambda_j I) Q_j = 0$$

and w_j is an unknown m-vector. The matrix Q_j may be found from the QR decomposition of the matrix $(A - \lambda_j I)^T U_2$. Indeed, from

$$(A - \lambda_j I)^T U_2 = [P_j, Q_j] \begin{bmatrix} R_j \\ 0 \end{bmatrix}$$

where the matrices P_j and Q_j have orthogonal columns and R_j is of rank $n - m$, it follows that

$$U_2^T (A - \lambda_j I) Q_j = R_j^T P_j^T Q_j = 0$$

Another way is to use the singular value decomposition

$$U_2^T (A - \lambda_j I) = T_j [S_j, 0] [P_j, Q_j]^T$$

where S_j is the diagonal matrix of singular values. In this way the problem reduces to the computation of the vector w_j such that the corresponding measure ν is minimal. This is done by an iterative procedure, since the change in the sensitivity of the jth pole alters the sensitivity of the previous poles. Several versions of such a procedure are described in detail in Kautsky, Nichols and Van Dooren (1985). From a computational point of view it is convenient also to use the measure

$$\nu_3 = \frac{1}{\sqrt{n}} \| V^{-1} \|_F$$

If $\| v_j \|_2 = 1$ then

$$\| V \|_F = \sqrt{n} \qquad \| V^{-1} \|_F = \sqrt{\sum_j \text{cond}^2(\lambda_j)}$$

so that

$$\nu_3 = \frac{1}{\sqrt{n}} \sqrt{\sum_j \text{cond}^2(\lambda_j)} = \frac{1}{n} \text{cond}_F(V)$$

and

$$1 \leqq \nu_3 \leqq \nu_1 \leqq \nu_2 \leqq n\nu_3$$

Hence the measures ν_1, ν_2 and ν_3 are mathematically equivalent and obtain their minimums $\nu_1 = 1$ simultaneously when the eigensystem is perfectly conditioned and V is unitary.

The following example taken from Kautsky, Nichols and Van Dooren (1985) illustrates the robust pole assignment.

Example 5.8

Let

$$A = \begin{bmatrix} -0.1094 & 0.0628 & 0 & 0 & 0 \\ 1.306 & -2.132 & 0.9807 & 0 & 0 \\ 0 & 1.595 & -3.149 & 1.547 & 0 \\ 0 & 0.0355 & 2.632 & -4.257 & 1.855 \\ 0 & 0.00227 & 0 & 0.1636 & -0.1625 \end{bmatrix}$$

$$B = \begin{bmatrix} 0 & 0.0638 & 0.0838 & 0.1004 & 0.0063 \\ 0 & 0 & -0.1396 & -0.206 & -0.0128 \end{bmatrix}^{\mathrm{T}}$$

and

$$\Lambda = \{-0.2, -0.5, -1.0, -1.0 \pm 1.0\mathrm{i}\}$$

The computed feedback matrix, which minimizes the measure ν_3, is

$$K = \begin{bmatrix} -47.690 & 102.01 & -213.07 & 179.86 & -42.552 \\ -22.596 & 30.633 & -48.077 & 33.799 & 2.2776 \end{bmatrix}$$

For this solution

$$\mathrm{cond}_2(V) = 39.4$$

and

$$\sqrt{\sum_j \mathrm{cond}^2(\lambda_j)} = 22.4$$

The gain elements are large in the given case, $\| K \|_2 = 311.5$, since $\| A \|_2 = 6.246$, $\sigma_m(B) = 0.057$ and the ratio $\| A \|_2 / \sigma_m(B)$ is of order 100. To demonstrate the robustness of the solution, the elements of the computed gain matrix K are rounded to three significant digits and the eigenvalues of the resulting closed-loop system matrix are calculated. Rounding the elements of K corresponds to introducing maximum absolute errors of order

$$\tfrac{1}{2} \times 10^{-3} \times \max_{i,j} \{| B || K |\}_{ij}$$

Table 5.1

λ_j	-0.2	-0.5	-1	$-1 \pm i$
Percentage error	1.8	0.1	0.2	2.4

into the matrix $A - BK$. If the solution is robust, such perturbations should cause errors of the same order of magnitude in the closed-loop system poles. In the given case the percentage errors in the poles are shown in Table 5.1.

The robust pole assignment will produce acceptable numerical results only when it is possible to find an eigenvector matrix V which is well conditioned. If such a matrix does not exist, the algorithm described will produce an inaccurate result for the feedback matrix and therefore inaccurate closed-loop poles, since the errors in the computed feedback depend on the condition number of the eigenvector matrix. In such a case the condition numbers $\mathrm{cond}(\lambda_j)$, $j = 1, 2, ..., n$, will be large and the perturbations in the poles will also be large. More precisely, it follows from (5.17) and (5.18) that the relative error in the computed gain matrix

$$\delta K = \frac{\| \bar{K} - K \|_2}{\| K \|_2}$$

will be bounded by $\mathrm{cond}_2(V)\varepsilon$. Hence, according to (5.1) the errors in the resulting closed-loop system poles in the worst case will be of order $\mathrm{cond}_2^2(V)\varepsilon$.

Note that $\mathrm{cond}_2(V)$ increases together with the system order.

When $\mathrm{cond}(V)$ is large, it may be possible to obtain a more accurate solution for the gain matrix by some of the Schur methods for pole assignment.

EXERCISES

5.11 Show that the robust pole assignment minimizes a bound on the norm of the closed-loop system transient response.

5.12 Develop a detailed iterative procedure for choosing the eigenvectors which minimize $\mathrm{cond}_2(V)$.

5.13 Implement the orthogonal canonical form of (A, B) in solving (5.15).

5.14 Develop a version of the robust pole assignment algorithm intended for the case of complex conjugate desired poles, which uses only real arithmetic.

5.15 Is it possible to control the conditioning of uncontrollable modes? Develop a method for robust pole assignment of uncontrollable systems.

5.4 DEADBEAT SYNTHESIS

The problem for finding a deadbeat control law

$$u_k = -Kx_k$$

for the discrete-time system

$$x_{k+1} = Ax_k + Bu_k$$

may be considered as a special pole assignment problem. In fact, if we set all desired closed-loop system poles equal to zero and use some of the pole assignment methods presented in Section 5.2, we will obtain a matrix K such that

$$(A - BK)^n = 0$$

and

$$x_n = (A - BK)^n x_0 = 0$$

However, the synthesis of deadbeat systems may require specific methods, since in the general case it is possible to find a feedback matrix K such that

$$(A - BK)^p = 0$$

where $p \leq n$ is the controllability index of the system (see Section 2.6). Thus in the multi-input case, when $p < n$ for rank$(B) > 1$, we may determine K such that the closed-loop transient response reaches the zero state after less than n steps. Unfortunately, this may require larger gains in comparison with the case when the response must 'die' after n steps, as shown by the following example.

Example 5.9

If

$$A = \begin{bmatrix} 0 & 1 \\ a & 0 \end{bmatrix} \quad \text{and} \quad B = \begin{bmatrix} \mu & 0 \\ 0 & 1 \end{bmatrix}$$

then the controllability index of the corresponding system is $p = 1$ and we may find K such that $A - BK = 0$

$$K = B^{-1}A = \begin{bmatrix} 0 & \dfrac{1}{\mu} \\ a & 0 \end{bmatrix}$$

If we want to find a feedback matrix K such that

$$(A - BK)^2 = 0$$

then a possible solution is

$$K = \begin{bmatrix} 0 & 0 \\ a & 0 \end{bmatrix}$$

and

$$(A - BK)^2 = \begin{bmatrix} 0 & 1 \\ 0 & 0 \end{bmatrix}^2 = 0$$

In this way for small μ the feedback yielding the fastest transient response will have large gains.

In this section we will consider a method for deadbeat synthesis of completely controllable multi-input discrete-time systems which produces a feedback matrix K such that $(A - BK)^p = 0$. The method is based on successive implementation of orthogonal transformations and has favourable numerical properties. It starts with reduction of the pair (A, B) to orthogonal canonical form

$$\tilde{A} = U^T A U = \begin{bmatrix} A_{11} & A_{12} & \cdots & & A_{1p} \\ A_{21} & A_{22} & \cdots & & A_{2p} \\ & A_{32} & \cdots & & A_{3p} \\ & & \ddots & & \vdots \\ \mathbf{0} & & & A_{p,p-1} & A_{pp} \end{bmatrix} \qquad \tilde{B} = U^T B = \begin{bmatrix} B_1 \\ 0 \\ 0 \\ \vdots \\ 0 \end{bmatrix}$$

where each block A_{ij} is of dimension $m_i \times m_j$ and B_1 is an $m_1 \times m$ matrix. Our aim is to find a feedback matrix K such that

$$V^T(A - BK)V = \begin{matrix} m_1 \quad m_2 \quad m_3 \qquad\quad m_p \\ \begin{bmatrix} 0 & A_{12} & A_{13} & \cdots & A_{1p} \\ & 0 & A_{23} & \cdots & A_{2p} \\ & & & \ddots & \vdots \\ & & & & A_{p-1,p} \\ \mathbf{0} & & & & 0 \end{bmatrix} \begin{matrix} \}m_1 \\ \}m_2 \\ \\ \}m_{p-1} \\ \}m_p \end{matrix} \end{matrix} \qquad (5.19)$$

for some orthogonal matrix V, since in this case

$$(A - BK)^p = 0$$

The desired feedback matrix K may be found recursively in p steps. After the ith step we obtain the matrices

$$V_i^T(A - BK_i)V_i = \begin{bmatrix} A_d^i & X \\ 0 & A_h^i \end{bmatrix} \begin{matrix} \}d_i \\ \}n - d_i \end{matrix} \tag{5.20}$$

$$V_i^T B = \begin{bmatrix} B_d^i \\ B_h^i \end{bmatrix} \begin{matrix} \}d_i \\ \}n - d_i \end{matrix} \qquad d_i = \sum_{j=1}^{i} m_j \tag{5.21}$$

Here the pair (A_d^i, B_d^i) corresponds to a closed-loop subsystem which is already 'beaten to death',

$$A_d^i = \begin{bmatrix} 0 & A_{12}^i & \cdots & A_{1i}^i \\ & 0 & \cdots & A_{2i}^i \\ & & \ddots & \vdots \\ & \mathbf{0} & & A_{i-1,i}^i \\ & & & 0 \end{bmatrix} \begin{matrix} \}m_1 \\ \}m_2 \\ \\ \}m_{i-1} \\ \}m_i \end{matrix} \tag{5.22}$$

$$B_d^i = \begin{bmatrix} B_1^i \\ B_2^i \\ \vdots \\ B_i^i \end{bmatrix} \begin{matrix} \}m_1 \\ \}m_2 \\ \\ \}m_i \end{matrix} \tag{5.23}$$

and the pair (A_h^i, B_h^i) is still in block-Hessenberg form

$$A_h^i = \begin{bmatrix} A_{i+1,i+1}^i & \cdots & & A_{i+1,p}^i \\ A_{i+2,i+1}^i & \cdots & & A_{i+2,p}^i \\ & \mathbf{0} & \ddots & \vdots \\ & & A_{p,p-1}^i & A_{pp}^i \end{bmatrix} \begin{matrix} \}m_{i+1} \\ \}m_{i+2} \\ \\ \}m_p \end{matrix}$$

$$B_h^i = \begin{bmatrix} B_{i+1}^i \\ 0 \\ \vdots \\ 0 \end{bmatrix} \begin{matrix} \}m_{i+1} \\ \}m_{i+2} \\ \\ \}m_p \end{matrix}$$

The blocks $A_{j,j-1}^i$, $j = i + 2, \ldots, p$, and B_{i+1}^i are of full row rank due to the controllability of the closed-loop system.

In the next step we construct an orthogonal transformation matrix

$$\hat{V}_{i+1} = \begin{bmatrix} I_{d_i} & 0 \\ 0 & V_h^{i+1} \end{bmatrix}$$

and a feedback matrix

$$\hat{K}_{i+1} = [0, K_h^{i+1}]$$

which affect only the subsystem (A_h^i, B_h^i). We want the matrices V_h^{i+1} and K_h^{i+1} to satisfy

$$(V_h^{i+1})^{\mathrm{T}}(A_h^i - B_h^i K_h^{i+1})V_h^{i+1} = \left[\begin{array}{cc} 0 & X \\ 0 & X \end{array}\right] \begin{array}{l} \}m_{i+1} \\ \}n - d_i + 1 \end{array}$$
$$\underbrace{}_{m_{i+1}} \quad \underbrace{}_{n - d_i + 1}$$

or, equivalently

$$(A_h^i - B_h^i K_h^{i+1})V_h^{i+1} = [\ \underbrace{0}_{m_{i+1}}\ ,\ \underbrace{X}_{n - d_i + 1}\ ] \tag{5.24}$$

These matrices are found in the following way. Construct an orthogonal transformation with matrix V_h^{i+1} which reduces A_h^i into upper triangular form

$$A_h^i V_h^{i+1} = R^{i+1}$$

$$= \left[\begin{array}{ccc} R_{i+1,i+1}^{i+1} & \cdots & R_{i+1,p}^{i+1} \\ 0 & \ddots & \vdots \\ & & R_{p,p}^{i+1} \end{array}\right] \begin{array}{l} \}m_{i+1} \\ \\ \}m_p \end{array} \tag{5.25}$$

If we solve the equation

$$B_{i+1}^i G_{i+1} = R_{i+1,i+1}^{i+1} \tag{5.26}$$

with respect to G_{i+1} and implement the feedback

$$K_h^{i+1} = [G_{i+1}, 0, ..., 0]\, V_h^{i+1}$$

then the matrix

$$(A_h^i - B_h^i K_h^{i+1})V_h^{i+1}$$

will take the form

$$(A_h^i - B_h^i K_h^{i+1})V_h^{i+1} = \left[\begin{array}{cccc} 0 & R_{i+1,i+2}^{i+1} & \cdots & R_{i+1,p}^{i+1} \\ & R_{i+2,i+2}^{i+1} & \cdots & R_{i+2,p}^{i+1} \\ & 0 & \ddots & \vdots \\ & & & R_{p,p}^{i+1} \end{array}\right] \begin{array}{l} \}m_{i+1} \\ \}m_{i+2} \\ \\ \}m_p \end{array}$$

which satisfies (5.24).

Equation (5.26) has a solution since B_{i+1}^i has full row rank. A minimum norm solution is found as

$$G_{i+1} = (B_{i+1}^i)^+ R_{i+1,i+1}^{i+1}$$

where $(+)$ denotes the pseudoinverse (see Section 1.8). The representation (5.25) may be obtained by using the QR decomposition of $(A_h^i)^{\mathrm{T}}$.

The same decomposition may be implemented in solving (5.26) as a least squares problem.

It is possible to show that the matrix $(V_h^{i+1})^T$ has the same block structure as A_h^i, i.e.

$$(V_h^{i+1})^T = \begin{bmatrix} V_{i+1,i+1} & V_{i+1,i+2} & \cdots & V_{i+1,p} \\ V_{i+2,i+1} & V_{i+2,i+2} & \cdots & V_{i+2,p} \\ & 0 & \ddots & & \vdots \\ & & & V_{p,p-1} & V_{pp} \end{bmatrix} \begin{matrix} \}m_{i+1} \\ \}m_{i+2} \\ \\ \}m_p \end{matrix}$$

(the proof of this fact is left as an exercise) and hence

$$(V_h^{i+1})^T (A_h^i - B_h^i K_h^{i+1}) V_h^{i+1}$$

$$= \begin{bmatrix} 0 & A_{i+1,i+2}^{i+1} & \cdots & & & A_{i+1,p}^{i+1} \\ & A_{i+2,i+2}^{i+1} & \cdots & & & A_{i+2,p}^{i+1} \\ 0 & A_{i+3,i+2}^{i+1} & \cdots & & & A_{i+3,p}^{i+1} \\ & 0 & \ddots & & & \vdots \\ & & & A_{p,p-1}^i & & A_{pp}^i \end{bmatrix} \qquad (5.27)$$

$$(V_h^{i+1})^T B_h^i = \begin{bmatrix} B_{i+1}^{i+1} \\ B_{i+2}^{i+1} \\ 0 \end{bmatrix} \qquad (5.28)$$

$$K_h^{i+i} V_h^{i+1} = [G_{i+1}, 0]$$

Since

$$\begin{aligned} A_{j+1,j}^{i+1} &= V_{j+1,j} R_{jj}^{i+1} & j = i+1, \ldots, p-1 \\ B_{i+2}^{i+1} &= V_{i+2,i+1} B_{i+1}^i \end{aligned} \qquad (5.29)$$

and since the blocks R_{jj}^{i+1} are invertible, it follows that the blocks in (5.29) are of full column rank. In this way, at the end of the $(i+1)$th step we have again the form shown in (5.20)–(5.23) with i replaced by $i+1$. This allows us to continue using the same procedure. The orthogonal transformation matrix and the feedback matrix are updated at each step as

$$\begin{aligned} V_{i+1} &= V_i \hat{V}_{i+1} \\ K_{i+1} V_{i+1} &= K_i V_{i+1} + \hat{K}_{i+1} \hat{V}_{i+1} = K_i V_i + \hat{K}_{i+1} \hat{V}_{i+1} \end{aligned}$$

Finally we obtain

$$V_p^T(A - BK_p)V_p = \begin{bmatrix} 0 & A_{12}^p & \cdots & A_{1p}^p \\ & 0 & \cdots & A_{2p}^p \\ & & \ddots & \vdots \\ & \mathbf{0} & & A_{p-1,p}^p \\ & & & 0 \end{bmatrix}$$

$$V_p^T B = \begin{bmatrix} B_1^p \\ B_2^p \\ \vdots \\ B_p^p \end{bmatrix}$$

where $V_p = U\hat{V}_1\hat{V}_2 \dots \hat{V}_p$
$\quad\ \, K_p = \hat{K}_1 + \hat{K}_2 + \cdots + \hat{K}_p$
$\quad\ \, \hat{K}_i = [0, ..., 0, G_i, 0, ..., 0] V_i$

Hence we obtain the form (5.19) as requested.

The method presented is realized simply, the basic computations at each step being the decomposition (5.25), the solution of Equation (5.26) and the orthogonal transformations (5.27) and (5.28). An important property of this method is that it produces a feedback matrix K which is the unique minimum Frobenius norm solution to the problem, i.e.

$$\| K \|_F^2 = \sum_i \| G_i \|_F^2 \to \min$$

If the pair (A, B) is already reduced into orthogonal canonical form, the solution requires $2n^3 + 2mn^2 + m^2n$ flops, so that the overall implementation requires $\frac{14}{3}n^3 + 3mn^2 + m^2n$ flops.

It may be shown (see for details Van Dooren (1984a)) that the computed gain matrix $\bar{K}$ satisfies

$$[(A + \Delta A) + (B + \Delta B)\bar{K}]^p = 0$$

where

$$\| \Delta A \| = \eta_1 \| A \| + \eta_2 \| B \| \| K \|$$
$$\| \Delta B \| = \eta_3 \| B \|$$

and η_1, η_2 and η_3 are of the order of ε. Hence for a moderate size of $\| K \|$ the method is numerically stable.

EXERCISES

5.16 Develop a computer program for deadbeat synthesis which implements the method described in this section.

5.17 What is the influence of errors in the computed feedback matrix on the transient response of a deadbeat system? Perform an analysis.

5.18 Write an algorithm for deadbeat synthesis of multi-input discrete systems utilizing the Hermite–Hessenberg form (Exercise 4.27).

5.19 Derive conditions for the existence of deadbeat control for uncontrollable systems. Develop an algorithm for finding the feedback matrix.

5.5 STABILIZING ALGORITHMS

In some cases it is desirable to find a control law which stabilizes the system without setting the closed-loop poles explicitly. Methods for determining a feedback matrix which stabilize the closed-loop system are referred to as *stabilizing algorithms*. In this section we consider several stabilizing algorithms for continuous-time and discrete-time systems.

5.5.1 Using the Controllability Grammian

The following theorem due to Kleinman (1970) gives a simple way to stabilize a continuous-time system.

Theorem 5.2

If the system

$$\dot{x}(t) = Ax(t) + Bu(t) \tag{5.30}$$

is completely controllable, then $u(t) = -Kx(t)$ is a stabilizing control law with

$$K = B^{\mathrm{T}} W^{-1}(\Delta)$$

$$W(\Delta) = \int_0^\Delta e^{-A\tau} BB^{\mathrm{T}} e^{-A^{\mathrm{T}}\tau} \, d\tau$$

Δ is arbitrary nonzero scalar and

$$V(x) = x^{\mathrm{T}} W^{-1} x$$

is a suitable Lyapunov function for the closed-loop system.

The pre- and postmultiplication of W by A and A^T, respectively, yields

$$AW + WA^T = - \int_0^\Delta \frac{d}{d\tau} (e^{-A\tau}BB^Te^{-A^T\tau})\, d\tau = - e^{-A\Delta}BB^Te^{-A^T\Delta} + BB^T$$

$$(5.31)$$

Since the system is completely controllable, the Grammian W is positive definite, so that adding

$$-2BB^T = - WW^{-1}BB^T - BB^TW^{-1}W$$

to both sides of Equation (5.31) gives

$$FW + WF^T = -e^{-A\Delta}BB^Te^{-A^T\Delta} - BB^T \equiv -Q \qquad (5.32)$$

where $F = A - BB^TW^{-1}$ and $Q > 0$.

Since $W > 0$ and $Q > 0$ the second method of Lyapunov guarantees that the matrix F^T (and hence F) will have eigenvalues with negative real parts. Pre- and postmultiplication of equation (5.32) by W^{-1} shows that $V(x)$ is a suitable Lyapunov function which completes the proof.

To determine the matrix $W(\Delta)$ we may use the identity

$$\exp\left(\begin{bmatrix} A & BB^T \\ 0 & -A^T \end{bmatrix}\Delta\right) = \begin{bmatrix} E_1(\Delta) & G_1(\Delta) \\ 0 & E_2(\Delta) \end{bmatrix} \qquad (5.33)$$

where

$$E_1(\Delta) = e^{A\Delta}$$
$$E_2(\Delta) = e^{-A^T\Delta}$$
$$G_1(\Delta) = \int_0^\Delta e^{A(\Delta-\tau)}BB^Te^{-A^T\tau}\, d\tau$$

and hence

$$W(\Delta) = E_2^T(\Delta)G_1(\Delta)$$

The matrix exponential in (5.33) can be computed efficiently by using Padé approximations (see Chapter 3). The value of Δ should be chosen so that $W(\Delta)$ is well conditioned.

If the system (5.30) is uncontrollable but stabilizable, the feedback matrix may be determined as

$$K = B^TW^+(\Delta)$$

where $W^+(\Delta)$ is the generalized inverse of $W(\Delta)$ (in the given case the matrix $W(\Delta)$ is singular). To prove this result we reduce the system

(5.30) into the form

$$\dot{\tilde{x}}(t) = \tilde{A}\tilde{x}(t) + \tilde{B}u(t)$$

$$\tilde{A} = S^{-1}AS = \begin{bmatrix} A_{11} & A_{12} \\ 0 & A_{22} \end{bmatrix} \qquad \tilde{B} = S^{-1}B = \begin{bmatrix} B_1 \\ 0 \end{bmatrix}$$

where the pair (A_{11}, B_1) is controllable and A_{22} is a stability matrix. Then it is easy to find that

$$W^+(\Delta) = S^{-T}\begin{bmatrix} W_{11}^{-1}(\Delta) & 0 \\ 0 & 0 \end{bmatrix} S^{-1}$$

where

$$W_{11}(\Delta) = \int_0^\Delta e^{-A_{11}\tau}B_1 B_1^T e^{-A_{11}^T\tau}\, d\tau$$

Hence the closed-loop system with state matrix

$$A - BB^TW^+(\Delta) = S\begin{bmatrix} A_{11} - B_1 B_1^T W_{11}^{-1}(\Delta) & A_{12} \\ 0 & A_{22} \end{bmatrix} S^{-1}$$

is asymptotically stable according to Theorem 5.2 and taking into account the stability of A_{22}.

The pseudo-inverse $W^+(\Delta)$ can be computed reliably by the expression

$$W^+(\Delta) = U\begin{bmatrix} D^{-1} & 0 \\ 0 & 0 \end{bmatrix} U^T$$

where

$$W(\Delta) = U\begin{bmatrix} D & 0 \\ 0 & 0 \end{bmatrix} U^T$$

is the eigenvalue–eigenvector decomposition of the symmetric positive semidefinite matrix $W(\Delta)$, D is a positive diagonal matrix and U is an orthogonal matrix.

In this way we obtain the following algorithm.

Algorithm 5.3 Determining a stabilizing feedback matrix for a continuous-time system using the controllability Grammian

Set

$$H = \begin{bmatrix} A & BB^T \\ 0 & -A^T \end{bmatrix}$$

For $\Delta > 0$ compute

$$\exp(H) \equiv \begin{bmatrix} E_1(\Delta) & G_1(\Delta) \\ 0 & E_2(\Delta) \end{bmatrix}$$

Find

$$W(\Delta) = E_2^{\mathrm{T}}(\Delta)G_1(\Delta)$$

Compute K from

$$K = B^{\mathrm{T}}U\begin{bmatrix} D^{-1} & 0 \\ 0 & 0 \end{bmatrix}U^{\mathrm{T}}$$

where

$$W(\Delta) = U\begin{bmatrix} D & 0 \\ 0 & 0 \end{bmatrix}U^{\mathrm{T}}$$

The decomposition of $W(\Delta)$ may be found by the QR algorithm. For this purpose one may use the EISPACK subroutines TRED2 and TQL2 (or IMTQL2).

Example 5.10

Let

$$A = \begin{bmatrix} -11 & -2 & 14 & 0 \\ -28 & -11 & 27 & -13 \\ -20 & -6 & 22 & -5 \\ 5 & 1 & -7 & -1 \end{bmatrix} \quad \text{and} \quad B = \begin{bmatrix} -2 & 4 \\ 5 & 0 \\ 1 & 3 \\ 1 & -2 \end{bmatrix}$$

The pair (A, B) is stabilizable, the uncontrollable modes being -1 and -2. For $\Delta = 1$ we have (up to four digits)

$$W(1) = \begin{bmatrix} 49.58 & 5.027 & 39.70 & -24.79 \\ 5.027 & 6.185 & 6.863 & -2.514 \\ 39.70 & 6.863 & 33.21 & -19.85 \\ -24.79 & -2.514 & -19.85 & 12.40 \end{bmatrix}$$

$$U = \begin{bmatrix} -0.7189 & 0.2674 & -0.6064 & 0.2098 \\ -0.096\,15 & -0.9185 & -0.3473 & -0.1626 \\ -0.5872 & -0.2587 & 0.6946 & 0.3253 \\ 0.3594 & -0.1337 & -0.1709 & 0.9076 \end{bmatrix}$$

$$D = \mathrm{diag}(95.08,\ 6.289)$$

and

$$K = \begin{bmatrix} -0.240\,2 & 0.805\,5 & 0.222\,6 & -0.120\,1 \\ 0.064\,33 & -0.076\,47 & 0.010\,01 & -0.032\,16 \end{bmatrix}$$

The eigenvalues of the closed-loop system matrix $A - BK$ are $-1.601 \pm 1.980\text{i}, -2, -1$.

For $\varDelta = 0.1$ we obtain the gain matrix

$$K = \begin{bmatrix} -0.1385 & 2.606 & 1.199 & 0.069\,26 \\ 0.9981 & 0.8457 & 1.171 & -0.4991 \end{bmatrix}$$

which results in closed-loop system poles

$$-10.54 \pm 4.339\text{i}, -2, -1.$$

A stabilizing feedback matrix for the discrete-time system

$$x_{k+1} = Ax_k + Bu_k$$

may be found by the formula (see Kleinman (1974))

$$K = B^{\mathrm{T}}[(A^{\mathrm{T}})^N W_N^+ A^N] A = (I + B^{\mathrm{T}} V_N B)^{-1} B^{\mathrm{T}} V_N A \qquad (5.34)$$

for $N \geqq n$, where W_N and V_N are given by

$$W_N = \sum_{i=0}^{N} A^i B B^{\mathrm{T}} (A^{\mathrm{T}})^i$$
$$V_N = (A^{\mathrm{T}})^N W_N^+ + A^N \qquad (5.35)$$

In the case when the pair (A, B) is controllable, the pseudo-inverses in (5.34) and (5.35) can be replaced by the inverses.

The implementation of (5.34) and (5.35) requires a nonconvergent matrix A to be raised to the power N which may be accompanied by large rounding errors, especially for high-order systems. That is why this method is not recommended as a numerical stabilizing algorithm.

5.5.2 Using the Second Method of Lyapunov

A stabilizing gain matrix for a continuous-time system may be determined by means of the following theorem.

Theorem 5.3 (Bass algorithm)

Let the pair (A, B) be controllable. Then

$$K = B^{\mathrm{T}} Z^{-1}$$

stabilizes the system (5.30) where $Z = Z^\mathrm{T} > 0$ satisfies the Lyapunov equation

$$-(A + \beta I_n)Z + Z[-(A + \beta I_n)]^\mathrm{T} = -2BB^\mathrm{T} \tag{5.36}$$

with $\beta > |\lambda_{\max}(A)|$

The pair $[-(A + \beta I_n), B]$ is controllable due to the controllability of (A, B). Since $-(A + \beta I_n)$ is a stability matrix (this is easily verified by using the Schur or Jordan form of A) it follows that the equation (5.36) has a unique positive definite solution for Z. Writing (5.36) as

$$(A - BB^\mathrm{T} Z^{-1})Z + Z(A - BB^\mathrm{T} Z^{-1})^\mathrm{T} = -2\beta Z$$

and applying again the second method of Lyapunov, we obtain that $(A - BB^\mathrm{T} Z^{-1})$ is a stability matrix.

Since $|\lambda_{\max}(A)| < \|A\|$ for any matrix norm for which there is a consistent vector norm, the parameter β in (5.36) may be taken also as $\beta > \|A\|$.

In the general case, when the pair (A, B) is uncontrollable, but stabilizable, the stabilizing gain matrix may be found from Armstrong (1975)

$$K = B^\mathrm{T} Z^+ \tag{5.37}$$

where $Z = Z^T \geqq 0$ satisfies (5.36) and Z^+ is the pseudo-inverse of Z. This result is proved by reducing the pair (A, B) into the form

$$\tilde{A} = U^\mathrm{T} A U = \begin{bmatrix} A_{11} & A_{12} \\ 0 & A_{22} \end{bmatrix} \qquad \tilde{B} = U^\mathrm{T} B = \begin{bmatrix} B_1 \\ 0 \end{bmatrix} \tag{5.38}$$

where (A_{11}, B_1) is controllable and A_{22} is stable. (The representation (5.38) can be obtained by using the orthogonal canonical form of (A, B).) Equation (5.36) can now be transformed into

$$-(\tilde{A} + \beta I_n)\tilde{Z} + \tilde{Z}[-(\tilde{A} + \beta I_n)]^\mathrm{T} = -2\tilde{B}\tilde{B}^\mathrm{T} \tag{5.39}$$

where

$$-(\tilde{A} + \beta I_n) = -U(A + \beta I_n)U^\mathrm{T} = \begin{bmatrix} -(A_{11} + \beta I) & -A_{12} \\ 0 & -(A_{22} + \beta I) \end{bmatrix}$$

and

$$\tilde{Z} = U^\mathrm{T} Z U$$

Since the matrices $-(A_{11} + \beta I)$ and $-(A_{22} + \beta I)$ are stable it follows that the equation (5.39) has a unique positive definite solution for $\tilde{Z}$. It is easy to show that this solution has the form

$$\tilde{Z} = \begin{bmatrix} Z_{11} & 0 \\ 0 & 0 \end{bmatrix}$$

where $Z_{11} = Z_{11}^T > 0$ satisfies

$$-(A_{11} + \beta I)Z_{11} + Z_{11}[-(A_{11} + \beta I)]^T = -2B_1 B_1^T$$

Applying Theorem 5.3 to the pair (A_{11}, B_1) and noting that

$$\tilde{Z}^+ = \begin{bmatrix} Z_{11}^{-1} & 0 \\ 0 & 0 \end{bmatrix}$$

we obtain that the system is stabilized by the gain matrix

$$K = \tilde{B}^T \tilde{Z}^+ V^T = B^T Z^+$$

The above considerations are summarized in the following algorithm.

Algorithm 5.4 Determining a stabilizing feedback matrix for a continuous-time system using the second method of Lyapunov

Select $\beta \geq \| A \|$
Compute $Q = 2BB^T$
Solve the Lyapunov equation

$$-(A + \beta I_n)Z + Z[-(A + \beta I_n)]^T = -Q$$

Find

$$K = B^T V \begin{bmatrix} D^{-1} & 0 \\ 0 & 0 \end{bmatrix} V^T$$

where

$$Z = V \begin{bmatrix} D & 0 \\ 0 & 0 \end{bmatrix} V^T$$

is the eigenvalue–eigenvector decomposition of the symmetric positive semidefinite matrix Z.

The Lyapunov equation may be solved by Algorithm 4.4 (the Bartels–Stewart algorithm) or by determining the Cholesky factor of Z (the Hammarling method).

A modest saving of computational operations in Algorithm 5.4 may be realized by preliminary transformation of the matrix A into real Schur form

$$\hat{A} = U^T A U$$

Denoting

$$\hat{Z} = U^{T}ZU \qquad \text{and} \qquad \hat{B} = U^{T}B$$

we obtain

$$-(\hat{A} + \beta I_n)\hat{Z} + \hat{Z}[-(\hat{A} + \beta I_n)]^{T} = -2\hat{B}\hat{B}^{T} \tag{5.40}$$

Then from (5.37) we have that

$$K = \hat{B}^{T}U^{T}(U\hat{Z}U^{T})^{+} = \hat{B}^{T}\hat{Z}^{+}U^{T} \tag{5.41}$$

The use of (5.40) and (5.41) allows us to omit the unnecessary computations $\tilde{Q} = U^{T}QU$ and $Z = U\hat{Z}U^{T}$ which are implied in solving the Lyapunov equation by the Bartels–Stewart algorithm. A saving of computational operations of about $3n^{3} - 2n^{2}m$ flops is thus obtained.

Algorithm 5.4 has some advantage over Algorithm 5.3 owing to the potential difficulties related to the computation of the Grammian by matrix exponential in Algorithm 5.3.

Example 5.11

Consider the stabilizable pair (A, B) from Example 5.10. Since the eigenvalues of A are $1 \pm 2i$, -1, -2 it is necessary to choose $\beta > \sqrt{5}$. For $\beta = 3$, Algorithm 5.4 yields (up to four digits)

$$Z = \begin{bmatrix} 13.62 & -2.812 & 8.812 & -6.812 \\ -2.812 & 3.906 & -0.1562 & 1.406 \\ 8.812 & -0.1562 & 6.531 & -4.406 \\ -6.812 & 1.406 & -4.406 & 3.406 \end{bmatrix}$$

$$U = \begin{bmatrix} -0.7633 & 0.074\,52 & 0.3403 & 0.5440 \\ 0.1421 & -0.912\,6 & -0.080\,75 & 0.3749 \\ -0.5015 & -0.400\,4 & 0.1615 & -0.7498 \\ 0.3817 & -0.037\,26 & 0.9228 & -0.036\,59 \end{bmatrix}$$

$$D = \text{diag}(23.34, 4.125)$$

and

$$K = \begin{bmatrix} -0.1623 & 1.152 & 0.4544 & 0.081\,14 \\ 0.1590 & 0.1509 & 0.1947 & -0.079\,51 \end{bmatrix}$$

The resulting closed-loop system poles are

$$-3 \pm 3.343i, -2, -1$$

For $\beta = 5$ we obtain the gain matrix

$$K = \begin{bmatrix} -0.1506 & 1.581 & 0.6776 & 0.075\ 29 \\ 0.3482 & 0.3435 & 0.4329 & -0.1741 \end{bmatrix}$$

which results in closed-loop system poles

$$-5 \pm 3.795\mathrm{i},\ -2,\ -1$$

In the discrete-time case we have the following result (see for a proof Armstrong and Rublein (1976)).

Theorem 5.4

Let the pair (A, B) be stabilizable. Then

$$K = B^{\mathrm{T}}(Z + BB^{\mathrm{T}})^{+} A$$

is a stabilizing gain matrix, where $Z = Z^{\mathrm{T}}$ satisfies

$$AZA^{\mathrm{T}} = \alpha^2 Z + 2BB^{\mathrm{T}}$$

with

$$0 < \alpha < \min\left(1, \min_{i}\ \{|\ \lambda_i(A)\ |\}\right)$$

Theorem 5.4 is the basis of the following algorithm.

Algorithm 5.5 Determining a stabilizing feedback matrix for a discrete-time system by the second method of Lyapunov

Select α

Compute $\hat{A} = \dfrac{A}{\alpha}$, $\hat{Q} = \dfrac{2}{\alpha^2} BB^{\mathrm{T}}$

Solve the Lyapunov equation

$$\hat{A}Z\hat{A}^{\mathrm{T}} - Z = \hat{Q}$$

Determine

$$K = B^{\mathrm{T}}(Z + BB^{\mathrm{T}})^{+} A$$

The discrete Lyapunov equation can be solved efficiently by Algorithm 4.5. As in the continuous-time case, a saving in operations can be realized if the matrix A is transformed into real Schur form

$$\tilde{A} = U^{\mathrm{T}} A U$$

In this case we obtain the equation

$$\tilde{A} \tilde{Z} \tilde{A}^{\mathrm{T}} = \alpha^2 \tilde{Z} + \tilde{Q}$$
$$\tilde{Q} = 2 \tilde{B} \tilde{B}^{\mathrm{T}} \qquad \tilde{B} = U^{\mathrm{T}} B$$

where α can be easily selected by inspection of the diagonal of A. Then

$$K = \tilde{B}^{\mathrm{T}} (\tilde{Z} + \tilde{Q})^+ \tilde{A} U^{\mathrm{T}}$$

Example 5.12

Let

$$A = \begin{bmatrix} 7.5 & 3.2 & -8.6 & -3.4 \\ 0.5 & 1.0 & 0.5 & 2.0 \\ 5.5 & 3.2 & -7.1 & -4.4 \\ -3.5 & -1.6 & 4.3 & 2.2 \end{bmatrix} \quad \text{and} \quad B = \begin{bmatrix} -8 & 4 \\ 2 & 1 \\ -7 & 2 \\ 4 & -2 \end{bmatrix}$$

The eigenvalues of A are 1.5, 1.0, 0.6 (controllable) and 0.5 (uncontrollable). Choosing $\alpha = 0.1$ we obtain by Algorithm 5.5

$$Z = 10^3 \begin{bmatrix} 1.315 & -0.116\,6 & 1.356 & -0.657\,7 \\ -0.1166 & 0.011\,24 & -0.1188 & 0.058\,28 \\ 1.356 & -0.118\,8 & 1.402 & -0.678\,2 \\ -0.6577 & 0.058\,28 & -0.6782 & 0.328\,9 \end{bmatrix}$$

rank $Z = 3$ and

$$K = \begin{bmatrix} 0.040\,91 & 0.3447 & 0.1728 & 0.4054 \\ 0.8613 & 0.4282 & -0.4984 & 0.6827 \end{bmatrix}$$

The closed-loop system poles corresponding to this gain matrix are

$$0.053\,83 \pm 0.1682i, \ 0.7072, \ 0.5.$$

Taking α as 0.2 we find

$$K = \begin{bmatrix} -0.003\,920 & 0.3042 & 0.1943 & 0.3634 \\ 0.7530 & 0.3371 & -0.4430 & 0.5880 \end{bmatrix}$$

which results in closed-loop system poles

$$0.1848 \pm 0.2726i, \ 0.7095, \ 0.5.$$

EXERCISES

5.20 Write computer programs implementing Algorithms 5.3–5.5.

5.21 For a particular example, analyze the influence of the free parameter Δ in Algorithm 5.3 on the location of the closed-loop system poles. What about the norm of the feedback matrix?

5.22 Compare the numerical behaviour of Algorithm 5.3 and 5.4 on high-order system examples.

5.23 Analyze the influence of the parameter β on the numerical properties of Algorithm 5.4 using the results from Section 4.2.

5.6 DESIGN OF STATE OBSERVERS

It was shown in Section 2.8 that the matrices of the state observer

$$\dot{z}(t) = Fz(t) + Gy(t) + Bu(t)$$

of a linear continuous-time system

$$\dot{x}(t) = Ax(t) + Bu(t)$$
$$y(t) = Cx(t)$$

must satisfy the equations

$$TA - FT = GC \tag{5.42}$$

$$H = TB \tag{5.43}$$

(The same relations also hold in the discrete-time case.)

In the case of a full-order observer the matrix T may be chosen as unit matrix so that

$$F = A - GC$$
$$H = B$$

and the problem of observer design reduces to the determination of a matrix G which assigns the eigenvalues of F at desired locations. As we already know, this is always possible for an observable pair (C, A). Hence, the design of a full-order observer can be done by some of the pole assignment methods considered in Sections 5.2 and 5.3 applied to the controllable pair (A^T, C^T).

The design of reduced-order observers may be performed using several methods. In this section, we compare some of the available methods from the computational point of view and present an algorithm which has favourable numerical properties.

As mentioned in Section 2.8, one of the possible ways for synthesis of a reduced-order observer is to choose a matrix F with the desired spectrum, set an arbitrary nonzero matrix G and then solve the Sylvester equation (5.42) with respect to T. For this purpose it is appropriate to use the Bartels–Stewart or Hessenberg–Schur algorithm (see Section 4.2). Equation (5.42) will have a unique solution for T if and only if the matrices A and F have no eigenvalues in common. Moreover, the analysis presented in Section 4.2 in connection with the Bartels–Stewart algorithm shows that the solution error is inversely proportional to the separation between the matrices A and F. Hence, the eigenvalues of F should be chosen so as to be well separated from those of A. This is usually the case, since the eigenvalues of F are placed sufficiently far left (in the continuous-time case) from the eigenvalues of A to guarantee fast convergence of the observation error. A potential disadvantage of this approach is that the matrix

$$\begin{bmatrix} T \\ C \end{bmatrix}$$

which is used in the reconstruction of the state vector, may be obtained singularly or badly conditioned. (Hereafter the matrix C is assumed to be of full rank.) This requires a trial-and-error procedure for finding the appropriate matrix G.

The second method for the design of reduced-order observers, presented in Section 2.8, allows us to find a matrix T which guarantees nonsingularity of the matrix

$$\begin{bmatrix} T \\ C \end{bmatrix}$$

For this purpose the system is transformed by using a nonsingular transformation, which involves the rows of the matrix C. The matrix F is determined in the given case by a pole assignment method, which is applied to a problem of order $(n - r)$. Using this approach, the matrix G is found directly. Note also, that the observer poles in this case are not necessarily different from the eigenvalues of A. The only disadvantage of the method under consideration is that there is some uncertainty in the determination of the nonsingular transformation, which should be chosen to be as well conditioned as possible to reduce the effect of rounding errors. In this way we reach the idea for using orthogonal transformations in the design of reduced-order observers, which are best conditioned.

Most of the computational methods for the synthesis of state observers involve, as a preliminary step, reduction of the pair (C, A)

into observable orthogonal canonical form $(\tilde{C}, \tilde{A})$

$$\tilde{A} = U^{T}AU = \begin{bmatrix} A_{11} & A_{12} & & & \\ A_{21} & A_{22} & A_{23} & & \mathbf{0} \\ \vdots & \vdots & \vdots & \ddots & \\ A_{q-1,1} & A_{q-1,2} & A_{q-1,3} & \cdots & A_{q-1,q} \\ A_{q1} & A_{q2} & A_{q3} & \cdots & A_{qq} \end{bmatrix}$$

$$\tilde{C} = CU = [C_1, \ 0, \ 0, \ ..., \ 0]$$

where U is an orthogonal matrix, q is the observability index of the system and the matrices $C_1, A_{12}, A_{23}, ..., A_{q-1,q}$ are of full column rank (see Section 4.4). Equation (5.42) is now transformed to

$$\tilde{T}\tilde{A} - F\tilde{T} = G\tilde{C} \tag{5.44}$$

where $\tilde{T} = TU$.

Let the matrix $\tilde{T}$ be represented by

$$\tilde{T} = [T_1, \ T_2, \ ..., \ T_q]$$

where the partition is done according to that of $\tilde{A}$. It follows then from (5.44) that

$$\tilde{T}\begin{bmatrix} A_{11} \\ \vdots \\ A_{q1} \end{bmatrix} - FT_1 = GC_1$$

$$T_{11}A_{12} + [T_2, \ ..., \ T_q]\begin{bmatrix} A_{22} \\ \vdots \\ A_{q2} \end{bmatrix} - FT_2 = 0$$

$$\vdots$$

$$T_{q-1}A_{q-1,q} + T_q A_{qq} - FT_q = 0$$

The equations

$$T_1 A_{12} = FT_2 - T_2 A_{22} - \cdots - T_q A_{q2}$$

$$\vdots$$

$$T_i A_{i,i+1} = FT_{i+1} - T_{i+1} A_{i+1,i+1} - \cdots - T_q A_{q,i+1} \tag{5.45}$$

$$\vdots$$

$$T_{q-1} A_{q-1,q} = FT_q - T_q A_{qq}$$

are underdetermined with respect to $T_1, T_2, ..., T_{q-1}$ and may be solved recursively for $i = q-1, q-2, ..., 1$ provided the matrix T_q is known. The matrix G is then found as a solution of

$$GC_1 = \tilde{T}\begin{bmatrix} A_{11} \\ \vdots \\ A_{q1} \end{bmatrix} - FT_1$$

The matrix T_q should be chosen so that the matrix $[T_2, T_3, \ldots, T_q]$ is of full rank. This requirement follows from the expression

$$\begin{bmatrix} T \\ C \end{bmatrix}^{-1} = U \begin{bmatrix} \tilde{T} \\ \tilde{C} \end{bmatrix}^{-1} = U \begin{bmatrix} T_1 & T_2 & \ldots & T_q \\ C_1 & 0 & \ldots & 0 \end{bmatrix}^{-1}$$

Unfortunately, with this method it is not clear how to set the matrix T_q so that the matrix

$$\begin{bmatrix} T \\ C \end{bmatrix}$$

has full rank. Also, it is possible that a magnification of the rounding errors will occur in the recursive solution of (5.45).

A better approach to solving (5.44) is to assume that the matrix F is in lower Schur form. In this case, the solution of (5.42) resembles the Hessenberg–Schur method for solving the Sylvester equation. Such an approach makes it possible to obtain a recursive solution for the rows of F, $\tilde{T}$ and G. An important property of this solution is that it guarantees nonsingularity of the matrix

$$\begin{bmatrix} T \\ C \end{bmatrix}$$

We refer the reader to Van Dooren (1984b) where the corresponding algorithm is described in detail.

Further improvement of the computational solution of (5.42) can be achieved if we pose the restriction that the matrix T has orthogonal rows. This allows us to solve (5.42) in a numerically stable way and makes it possible to decrease the condition number of the matrix

$$\begin{bmatrix} T \\ C \end{bmatrix}$$

We shall now consider a method which reduces the observer design to a pole assignment problem and produces a matrix T with orthogonal rows.

Let the matrix G in (5.42) be chosen as

$$G = TK \tag{5.46}$$

for some $n \times r$ matrix K, so that the equation (5.42) becomes

$$T(A - KC) = FT \tag{5.47}$$

Using some of the Schur methods for pole assignment presented in Section 5.2, we can find a matrix K^T and an orthogonal matrix Q such that

$$(A^{\mathrm{T}} - C^{\mathrm{T}}K^{\mathrm{T}})Q = Q \begin{bmatrix} \overset{n-r}{F^{\mathrm{T}}} & \overset{r}{S} \\ 0 & R \end{bmatrix} \begin{matrix} \}n-r \\ \}r \end{matrix} \tag{5.48}$$

where the spectrum of the upper (quasi) triangular matrix F^{T} consists of the desired observer poles and the spectrum of the matrix R is set arbitrarily. Partitioning the matrix Q conformably with the Schur form on the right-hand side of (5.48) as

$$Q = [Q_1, Q_2]$$

we obtain

$$Q_1^{\mathrm{T}}(A - KC) = FQ_1^{\mathrm{T}}$$

which, according to (5.47), (5.46) and (5.43), yields

$$T = Q_1^{\mathrm{T}} \qquad G = Q_1^{\mathrm{T}}K \qquad H = Q_1^{\mathrm{T}}B \tag{5.49}$$

In this way the matrix F is obtained directly in a lower Schur form and the matrices G and H are computed by multiplication with a matrix having orthonormal columns.

The representation (5.48) shows that in the given case the problem of reduced-order observer design is 'imbedded' in an nth-order pole assignment problem. This problem may be solved by a numerically stable method.

It is possible to show that using the multi-input version of the pole assignment algorithm, which involves orthogonal triangularization of the eigenvectors (see Section 5.2), the matrix T obtained will always fulfil the condition

$$\mathrm{rank} \begin{bmatrix} T \\ C \end{bmatrix} = n$$

To prove this, suppose that the corresponding algorithm is applied on the pair $(A^{\mathrm{T}}, C^{\mathrm{T}})$ and the first $(n - r)$ prescribed poles are the desired observer poles. As a result we obtain

$$Q^{\mathrm{T}}(A^{\mathrm{T}} - C^{\mathrm{T}}K^{\mathrm{T}})Q = \begin{bmatrix} F^{\mathrm{T}} & S \\ 0 & R \end{bmatrix}$$

$$Q^{\mathrm{T}}C^{\mathrm{T}} = \begin{bmatrix} C_1^{\mathrm{T}} \\ C_2^{\mathrm{T}} \end{bmatrix} \begin{matrix} \}n-r \\ \}n \end{matrix}$$

where the matrix C_2 is of rank r (we leave this assertion as an exercise). According to (5.49) $G = \tilde{K}_1$, where $\tilde{K}_1^{\mathrm{T}} = K^{\mathrm{T}}Q_1$ is the part of the gain matrix which assigns the first $(n - r)$ poles at the desired places. We then have

$$\begin{bmatrix} T \\ C \end{bmatrix}^{-1} = \begin{bmatrix} Q_1^T \\ [C_1, C_2]Q^T \end{bmatrix}^{-1} = Q\begin{bmatrix} Q_1^TQ \\ C_1 & C_2 \end{bmatrix}^{-1}$$

$$= Q\begin{bmatrix} I_{n-r} & 0 \\ C_1 & C_2 \end{bmatrix}^{-1} = Q\begin{bmatrix} I_{n-r} & 0 \\ -C_2^{-1}C_1 & C_2^{-1} \end{bmatrix}$$

Thus the matrix

$$\begin{bmatrix} T \\ C \end{bmatrix}$$

is nonsingular and its inversion is reduced to the inversion of the $r \times r$ matrix C_2.

Example 5.13

Consider an observable system with matrices

$$A = \begin{bmatrix} 4 & -7 & -1 & 6 & -2 \\ -8 & -5 & 0 & -5 & 1 \\ 9 & 3 & 7 & -6 & 4 \\ 1 & 3 & 2 & -2 & 8 \\ -4 & -9 & -12 & 4 & -7 \end{bmatrix} \qquad B = \begin{bmatrix} -7 & 2 & 4 \\ 5 & -11 & 3 \\ -6 & -9 & 1 \\ -4 & 0 & 7 \\ 1 & -5 & 3 \end{bmatrix}$$

$$C = \begin{bmatrix} -2 & -5 & 3 & -7 & 6 \\ 1 & -4 & 0 & 8 & 2 \end{bmatrix}$$

The matrix C is of rank 2 so that the reduced-order observer is of third-order. Let the desired observer poles be

$$\lambda_1 = -2 + i, \ -2 - i, \ -3$$

and the other two poles (the eigenvalues of R) are set to zero.

Using the algorithm for pole assignment by orthogonal tri-angularization of the eigenvectors in arithmetic with precision $\varepsilon = 2^{-56} \approx 1.4 \times 10^{-17}$ we obtain (up to three digits)

$$Q^T(A^T - C^TK^T)Q = \begin{bmatrix} -2.68 & 5.19 & 8.51 & 33.4 & 6.49 \\ -0.283 & -1.32 & 18.1 & 5.36 & 0.439 \\ 0 & 0 & -3.00 & 14.5 & -17.3 \\ \hline 0 & 0 & 0 & 0 & -2.13 \\ 0 & 0 & 0 & 0 & 0 \end{bmatrix}$$

$$
Q^{\mathrm{T}}C^{\mathrm{T}} = \left[\begin{array}{rr}
8.76 & -1.70 \\
-0.132 & -2.73 \\
6.05 & -1.50 \\
\hline
-3.09 & 0.768 \\
0 & -8.48
\end{array}\right]
$$

$$
Q = [Q_1, Q_2] = \left[\begin{array}{rrr|rr}
-0.713 & 0.0165 & 0.421 & -0.550 & -0.105 \\
-0.431 & -0.367 & 0.000\,954 & 0.413 & 0.714 \\
0.283 & -0.699 & -0.232 & -0.595 & 0.156 \\
-0.402 & -0.413 & -0.431 & 0.299 & -0.627 \\
0.253 & -0.454 & 0.764 & 0.290 & -0.250
\end{array}\right]
$$

$$
\tilde{K}_1^{\mathrm{T}} = K^{\mathrm{T}}Q_1 = \left[\begin{array}{rrr}
-0.195 & -0.890 & -0.847 \\
-0.570 & 0.0559 & 1.44
\end{array}\right]
$$

Hence

$$
T = \left[\begin{array}{rrrrr}
-0.713 & -0.431 & 0.283 & -0.402 & 0.253 \\
0.0165 & -0.367 & -0.699 & -0.413 & -0.454 \\
0.421 & 0.000\,954 & -0.232 & -0.431 & 0.764
\end{array}\right]
$$

$$
F = \left[\begin{array}{rrr}
-2.68 & -0.283 & 0 \\
5.19 & -1.32 & 0 \\
8.51 & 18.1 & -3.00
\end{array}\right]
$$

$$
G = \left[\begin{array}{rr}
-0.195 & -0.570 \\
-0.890 & 0.0559 \\
-0.847 & 1.44
\end{array}\right]
$$

$$
H = \left[\begin{array}{rrr}
3.00 & -0.490 & -5.92 \\
3.44 & 12.6 & -5.99 \\
0.941 & -0.895 & 0.728
\end{array}\right]
$$

$$
\left[\begin{array}{c} T \\ C \end{array}\right]^{-1} = \left[\begin{array}{rrrrr}
-2.28 & 0.0742 & -0.657 & 0.181 & 0.0124 \\
0.777 & -0.617 & 0.808 & -0.154 & -0.0842 \\
-1.39 & -0.724 & -1.40 & 0.188 & -0.0184 \\
0.408 & -0.221 & 0.153 & -0.0781 & 0.0740 \\
1.06 & -0.385 & 1.33 & -0.0865 & 0.0295
\end{array}\right]
$$

It should be pointed out that in the method described it is not necessary to set the spectrum of the matrix R. The pole assignment algorithm can be carried out until the desired $(n - r)$ observer poles are assigned and the matrix $\tilde{K}_1$ is found. In this way the observer design may be reduced to an $(n - r)$th order pole assignment problem which decreases the volume of computations.

EXERCISES

5.24 Write a computer program for the synthesis of full-order observers utilizing some algorithm for pole assignment.

5.25 Develop an algorithm for determining the matrices G and F of a full-order observer avoiding transposition of the matrices A and C.

5.26 Develop an algorithm for the synthesis of a full-order deadbeat observer of a discrete-time system which reconstructs the state vector in q steps, where q is the observability index of the system.

5.27 Consider in detail the algorithm for synthesis of reduced-order observers which utilizes the orthogonal canonical form of the pair (C, A) and assumes the matrix F is in lower Schur form.

5.28 Develop a computer program for the design of reduced-order observers implementing the algorithm which produces a matrix T with orthogonal rows.

NOTES AND REFERENCES

A large number of pole assignment algorithms have been proposed in the control literature during the last two decades. Few of these algorithms, however, are suitable for implementation on a computer.

Schur algorithms for pole assignment are proposed in Miminis and Paige (1982, 1988), Patel (1985), Petkov, Christov and Konstantinov (1984a, 1986b), Varga (1981) and Lee and Liaw (1986). Other computational algorithms for pole assignment are described in Datta (1987), Saad (1988), Shafai and Bhattacharyya (1988) and Tsui (1986).

The possibility of simultaneous assignment of eigenvalues and eigenvectors in the multi-input case was first pointed out in Moore (1976). Algorithms for robust eigenvalue design are presented in Klein and Moore (1977), Kautsky and Nichols (1984), Kautsky, Nichols and Van Dooren (1985) and Byers and Nash (1989). The advantages and disadvantages of the Schur methods and robust pole assignment methods are discussed in Nichols (1986) and Petkov, Christov and Konstantinov (1988a).

Algorithms for deadbeat synthesis of discrete-time systems are described in Eising (1984a) and Van Dooren (1984a).

The stabilization of linear control systems without setting poles is considered in Kleinman (1970, 1974), Armstrong (1975), Armstrong and Rublein (1976), Farias and Bingulac (1978) and Sima (1981).

Computational algorithms for synthesis of state observers are presented in Petkov, Christov and Konstantinov (1984c), Tsui (1985, 1987) and Van Dooren (1984b).

The computational aspects of eigenvalue allocation by output feedback or dynamic compensator are discussed in Fletcher *et al.* (1985), Ho and Fletcher (1988) and Misra and Patel (1989).

6

Solution of Matrix Riccati Equations

The main computational problem arising in the quadratic optimization of linear control systems (see Section 2.7) is the solution of the corresponding matrix Riccati equation. Since the analytical solution of this equation is possible only in restricted cases, it is necessary in general to implement some numerical method. In this chapter we present several methods for solving Riccati equations and discuss their numerical properties. We consider first the conditioning of Riccati equations, which reflects the sensitivity of the solution to perturbations in the data.

6.1 CONDITIONING OF MATRIX RICCATI EQUATIONS

Consider the continuous matrix algebraic Riccati equation

$$A^T P + PA + Q - PBR^{-1}B^T P = 0$$

arising in the quadratic optimization of linear control systems (Section 2.7).

For convenience we shall write this equation in the form

$$A^T P + PA + Q - PSP = 0 \tag{6.1}$$

where $S = BR^{-1}B^T$.

When Equation (6.1) is solved on a computer, the computed result $\bar{P}$ will satisfy the perturbed equation

$$(A + E)^T \bar{P} + \bar{P}(A + E) + Q + F - \bar{P}(S + G)\bar{P} = 0 \tag{6.2}$$

where E, F and G are the perturbation matrices. If we use a numerically stable method for solving (6.1) we can expect at best that these matrices will fulfil the conditions

$$\|E\| \leq \varepsilon \|A\| \qquad \|F\| \leq \varepsilon \|Q\| \qquad \|G\| \leq \varepsilon \|S\| \tag{6.3}$$

where ε is the relative machine precision of the computer used. In some cases small perturbations may cause a large change in the solution, i.e. the Riccati equation may be very sensitive to perturbations in the data. In such cases the solution is obtained with large errors, even if a numerically stable method is implemented. Notice that the sensitivity is a property inherent to the given problem and it depends neither on the solution method nor on the arithmetic used.

Example 6.1

Let

$$A = \begin{bmatrix} -6 & -2 & 1 \\ 5 & 1 & -1 \\ -4 & -2 & -1 \end{bmatrix}$$

$$Q = \begin{bmatrix} 305.63 & 300.21 & -5.21 \\ 300.21 & 300.07 & -0.07 \\ -5.21 & -0.07 & 5.07 \end{bmatrix}$$

$$S = \begin{bmatrix} 101.01 & -101.02 & 102.01 \\ -101.02 & 101.04 & -102.02 \\ 102.01 & -102.02 & 104.01 \end{bmatrix}$$

The exact solution of (6.1) for these matrices is

$$P = \begin{bmatrix} 101.09 & 100.03 & -1.03 \\ 100.03 & 100.01 & -0.01 \\ -1.03 & -0.01 & 1.01 \end{bmatrix}$$

Let the matrix S be perturbed by

$$G = \begin{bmatrix} 0.01 & -0.02 & 0.01 \\ -0.02 & 0.04 & -0.02 \\ 0.01 & -0.02 & 0.01 \end{bmatrix}$$

The solution of the perturbed equation (6.2) is (up to five digits)

$$\bar{P} = \begin{bmatrix} 83.378 & 82.318 & -1.0300 \\ 82.318 & 82.298 & -0.0100 \\ -1.0300 & -0.0100 & 1.0100 \end{bmatrix}$$

Thus a relative perturbation

$$\frac{\|G\|_F}{\|S\|_F} = 1.9646 \times 10^{-4}$$

in the matrix S leads to a relative perturbation in the solution equal to

$$\frac{\|\bar{P} - P\|_F}{\|P\|_F} = 0.176\ 60$$

i.e. the change in the solution is more than 10^3 times larger than the change in the data.

The sensitivity of the Riccati equation (6.1) may be estimated as follows. Let $\Delta P = \bar{P} - P$. Subtracting (6.1) from (6.2) and neglecting the higher order terms in ΔP we obtain

$$(A - SP)^T \Delta P + \Delta P(A - SP) + E^T P + PE + F - PGP = 0 \qquad (6.4)$$

Equation (6.4) is a Lyapunov equation with respect to the perturbation ΔP. Notice that the matrix $A - SP$ is the closed-loop state matrix $A - BK$, $K = R^{-1}B^T P$. Implementing the same technique as in the perturbation analysis of the Sylvester equation (see Section 4.2) we get

$$\frac{\|\Delta P\|_F}{\|P\|_F} \leqq \frac{2\|E\|_F + \dfrac{\|F\|_F}{\|P\|_F} + \|G\|_F \|P\|_F}{\text{sep}[(A - SP)^T, -(A - SP)]} \qquad (6.5)$$

where

$$\text{sep}[(A - SP)^T, -(A - SP)] = \min_{X \neq 0} \frac{\|(A - SP)^T X + X(A - SP)\|_F}{\|X\|_F}$$

is the separation between the matrices $(A - SP)^T$ and $-(A - SP)$.

The bound (6.5) suggests that the Riccati equation may be very sensitive if the norm of the solution is very small or very large and/or the separation between the matrices $(A - SP)^T$ and $-(A - SP)$ is very small.

The quantity $\text{sep}[(A - SP)^T, -(A - SP)]$ can be computed as the minimal singular value of the matrix

$$I_n \otimes (A - SP)^T + (A - SP)^T \otimes I_n$$

which requires a great deal of space for large n. A LINPACK-style estimator of the separation between $(A - SP)^T$ and $-(A - SP)$ is described in Byers (1983).

If the perturbations E, F and G obey (6.3) then we have that

$$\frac{\|\Delta P\|_F}{\|P\|_F} \leqq \varepsilon \frac{2\|A\|_F + \dfrac{\|Q\|_F}{\|P\|_F} + \|S\|_F \|P\|_F}{\text{sep}[(A - SP)^T, -(A - SP)]} \qquad (6.6)$$

This inequality reveals that the relative perturbation in the solution may be c_R times larger than the perturbations in the data, where

$$c_R = \frac{2\|A\|_F + \dfrac{\|Q\|_F}{\|P\|_F} + \|S\|_F \|P\|_F}{\text{sep}[(A - SP)^T, -(A - SP)]}$$

will be referred to as the *condition number of the Riccati equation* (6.1). For the equation from Example 6.1, $c_R = 5.5351 \times 10^4$ which shows that the solution is very sensitive to perturbations. In the given case this is because of the large norms of the matrices P and S while

$$\text{sep}[(A - SP)^T, -(A - SP)] = 1.1072$$

is relatively large.

It should be stressed that in some cases

$$\text{sep}[(A - SP)^T, -(A - SP)]$$

can be extremely small so that the corresponding Riccati equation is very ill conditioned.

Consider now the discrete Riccati equation

$$A^T PA - P + Q - A^T PB(R + B^T PB)^{-1}B^T PA = 0$$

For simplicity, we shall present this equation as (see Exercise 2.54)

$$Q - P + A^T P(I_n + SP)^{-1}A = 0 \tag{6.7}$$

where $S = BR^{-1}B^T$. The perturbed equation is then

$$Q + F - (P + \Delta P) \\ + (A + E)^T(P + \Delta P)[I_n + (S + G)(P + \Delta P)]^{-1}(A + E) = 0$$

An analysis, similar to that in the continuous-time case, yields

$$\frac{\|\Delta P\|_F}{\|P\|_F} \leq \frac{2\|A\|_F \|E\|_F + \dfrac{\|F\|_F}{\|P\|_F} + \|A\|_F^2 \|G\|_F \|P\|_F}{\text{sep}_d(A_c^T, A_c)} \tag{6.8}$$

where

$$\text{sep}_d(A_c^T, A_c) = \min_{X \neq 0} \frac{\|A_c^T X A_c - X\|_F}{\|X\|_F}$$

and

$$A_c = (I_n + SP)^{-1}A = A - B(R + B^T PB)^{-1}B^T PA$$

is the closed-loop state matrix.

The quantity $\text{sep}_d(A_c^T, A_c)$ can be computed as the minimal singular value of the matrix

$$A_c^T \otimes A_c^T - I_{n^2}$$

For ε perturbations in the matrices A, Q and S we have that

$$\frac{\|\Delta P\|_F}{\|P\|_F} \leq \varepsilon \; \frac{2\|A\|_F^2 + \dfrac{\|Q\|_F}{\|P\|_F} + \|A\|_F^2 \|S\|_F \|P\|_F}{\mathrm{sep}_d(A_c^T, A_c)}$$

The quantity

$$d_R = \frac{2\|A\|_F^2 + \dfrac{\|Q\|_F}{\|P\|_F} + \|A\|_F^2 \|S\|_F \|P\|_F}{\mathrm{sep}_d(A_c^T, A_c)}$$

is said to be a *condition number of the discrete Riccati equation* (6.7)

Example 6.2

Let

$$A = \begin{bmatrix} 0.76 & 0.14 & -0.38 \\ 0.42 & 0.12 & 0.46 \\ 0.06 & 0.34 & 0.72 \end{bmatrix} \qquad Q = \begin{bmatrix} 0.11 & -0.10 & 0.11 \\ -0.10 & 0.15 & -0.15 \\ 0.11 & -0.15 & 0.16 \end{bmatrix}$$

$$S = \begin{bmatrix} 1.401 & -2.201 & 1.199 \\ -2.201 & 4.101 & -2.099 \\ 1.199 & -2.099 & 1.101 \end{bmatrix}$$

The solution of (6.7) is (up to five digits)

$$P = \begin{bmatrix} 0.439\,25 & -0.325\,17 & -0.651\,95 \\ -0.325\,17 & 0.402\,19 & 0.585\,27 \\ -0.651\,95 & 0.585\,27 & 2.392\,7 \end{bmatrix}$$

If the matrix S is perturbed by

$$G = \begin{bmatrix} -0.0005 & 0.0005 & 0.0005 \\ 0.0005 & -0.0005 & -0.0005 \\ 0.0005 & -0.0005 & -0.0005 \end{bmatrix}$$

then the solution becomes

$$\bar{P} = \begin{bmatrix} 0.543\,27 & -0.429\,11 & -0.963\,85 \\ -0.429\,11 & 0.506\,06 & 0.896\,95 \\ -0.963\,85 & 0.896\,95 & 3.327\,9 \end{bmatrix}$$

In this way a relative perturbation

$$\frac{\|G\|_F}{\|S\|_F} = 2.3321 \times 10^{-4}$$

in the matrix S causes a change in the solution equal to

$$\frac{\|\Delta P\|_F}{\|P\|_F} = 0.408\ 61$$

i.e. the changes in the data are magnified about 1750 times in the solution. In the given case

$$\text{sep}_d(A_c^T, A_c) = 1.1265 \times 10^{-2}$$

and

$$d_R = 3.1743 \times 10^3$$

EXERCISES

6.1   Verify (6.5).

6.2   Investigate the conditioning of (6.1) for

$$A = \begin{bmatrix} -30.0 & 56.0 & -122.0 \\ 59.8 & -104.0 & 227.8 \\ 29.2 & -56.0 & 111.9 \end{bmatrix}$$

$$Q = \begin{bmatrix} 0.0703 & -0.1400 & 0.2803 \\ -0.1400 & 0.2850 & -0.5700 \\ 0.2803 & -0.5700 & 1.1403 \end{bmatrix}$$

$$S = \begin{bmatrix} 14.0 & -26.0 & -14.0 \\ -26.0 & 49.4 & 26.2 \\ -14.0 & 26.2 & 14.1 \end{bmatrix}$$

6.3   Verify (6.8). *Hint:* use the relationships
$P(I + SP)^{-1} = (I + PS)^{-1}P$
$\|P(I + SP)^{-1}\|_F \leq \|P\|_F$
$(Y + Z)^{-1} \approx Y^{-1} - Y^{-1}ZY^{-1}$ for any matrices Y and Z such that Y^{-1} and $(Y + Z)^{-1}$ exist and $\|Z\| \ll \|Y\|$

6.2 NEWTON'S METHOD

Newton's method is an iterative method for finding a root of the nonlinear equation $f(x) = 0$. Applied to the matrix Riccati equation this method yields a sequence of matrices, which under certain conditions approach the positive (semi-) definite solution of the equation. In this section we consider the implementation of Newton's method for solving continuous and discrete Riccati equations.

Consider first the continuous matrix Riccati equation

$$A^T P + PA + Q - PBR^{-1}B^T P = 0 \tag{6.9}$$

where the pair (A, B) is stabilizable and the pair (C, A) ($C^T C = Q$, rank(C) = rank(Q)) is detectable. The unique positive semidefinite solution P of (6.9) may be found using the following theorem.

Theorem 6.1

Let P_k, $k = 0, 1, \ldots$, be the (unique) positive semidefinite solution of the linear equation

$$A_k^T P_k + P_k A_k + Q + K_k^T R K_k = 0 \tag{6.10}$$

where, recursively,

$$K_k = R^{-1} B^T P_{k-1} \qquad k = 1, 2, \ldots$$
$$A_k = A - BK_k \qquad k = 0, 1, \ldots$$

and where K_0 is chosen such that the matrix $A - BK_0$ has eigenvalues with negative real parts. Then

1. $0 \leq P \leq P_{k+1} \leq P_k \leq \cdots \leq P_0 \qquad k = 0, 1, 2, \ldots$
2. $\lim_{k \to \infty} P_k = P$

The proof of this theorem may be found in Kleinman (1968) and Sandell (1974), and see also Wonham (1986, ch. 12).

It may be shown that under the conditions of Theorem 6.1 the convergence of the sequence $P_0, P_1, \ldots$ to P is quadratic, i.e.

$$\| P_{k+1} - P \| \leq c_1 \| P_k - P \|^2$$

for some constant c_1 and for all k. This guarantees fast convergence of P_k to P in the vicinity of P.

Since the pair (A, B) is stabilizable, it is always possible to find a matrix K_0, for which $A - BK_0$ is a stability matrix. This requirement is essential, for if the matrix $A - BK_0$ is not stable, then the iteration may converge to a nonstabilizing solution of (6.9) or may diverge. The determination of K_0 can be done by some of the stabilizing algorithms presented in Section 5.5. The use of a pole assignment method is also possible. It is important to note that different choices of K_0 may require substantially different volumes of computation for a prescribed accuracy of the solution. If the initial gain matrix is chosen so that P_0 is far from P ($\| P_0 - P \| / \| P \|$ is large) then the convergence may be very slow.

This method requires at each step the solution of the matrix Lyapunov equation (6.10). This should be done by some of the methods considered in Section 4.2. It is efficient to implement the Bartels–Stewart algorithm, which allows us to find, in addition, the eigenvalues of the closed-loop system matrix $A - BR^{-1}B^{\mathrm{T}}P$.

Let us consider the numerical properties of Newton's method. Let $\bar{P}_k$ be the approximation computed at the kth step of the iteration by a numerically stable method for solving the Lyapunov equation. According to the results from Section 4.2 we have that

$$\frac{\| \bar{P}_k - P_k \|_{\mathrm{F}}}{\| P_k \|_{\mathrm{F}}} \leq \varepsilon \, \frac{2 \| A_k \|_{\mathrm{F}} + \dfrac{\| Q_k \|_{\mathrm{F}}}{\| P_k \|_{\mathrm{F}}}}{\mathrm{sep}(A_k^{\mathrm{T}}, -A_k)}$$

In the vicinity of the solution it is fulfilled that

$$\frac{\| \bar{P}_k - P_k \|_{\mathrm{F}}}{\| P_k \|_{\mathrm{F}}} \leq \varepsilon c_{\mathrm{R}}$$

where

$$c_{\mathrm{R}} = \frac{2 \| A \|_{\mathrm{F}} + \dfrac{\| Q \|_{\mathrm{F}}}{\| P \|_{\mathrm{F}}} + \| BR^{-1}B^{\mathrm{T}} \|_{\mathrm{F}} \| P \|_{\mathrm{F}}}{\mathrm{sep}\,[(A - BR^{-1}B^{\mathrm{T}}P)^{\mathrm{T}}, -(A - BR^{-1}B^{\mathrm{T}}P)]}$$

is the condition number of the continuous Riccati equation. Hence Newton's method will continue to approach the solution until the size of $\| P_k - P_{k-1} \|_{\mathrm{F}} / \| P_{k-1} \|_{\mathrm{F}}$ decreases to the magnitude of $\varepsilon c_{\mathrm{R}}$. Further implementation of the iteration is useless since rounding errors will prevent improvement of the accuracy of the solution. Thus the approximate solution $\bar{P}$ of the Riccati equation will satisfy

$$\frac{\| \bar{P} - P \|_{\mathrm{F}}}{\| P \|_{\mathrm{F}}} \leq \varepsilon c_{\mathrm{R}}$$

Comparison of this result with the bound (6.6) shows that Newton's method is a numerically stable method for solving the Riccati equation.

Example 6.3

Let

$$A = \begin{bmatrix} 3 & 1 & -3 \\ 2 & -5 & 3 \\ -1 & -2 & 1 \end{bmatrix} \quad B = \begin{bmatrix} 2 & -1 \\ -3 & -5 \\ 8 & 1 \end{bmatrix}$$

$$Q = \begin{bmatrix} 95 & 106 & -54 \\ 106 & 258 & 44 \\ -54 & 44 & 109 \end{bmatrix} \quad R = I_2$$

The exact solution of (6.9) is

$$P = \begin{bmatrix} 2 & 1 & -1 \\ 1 & 3 & 1 \\ -1 & 1 & 2 \end{bmatrix}$$

Newton's method is implemented by using the Bartels–Stewart algorithm in arithmetic with precision $\varepsilon = 2^{-56} \approx 1.39 \times 10^{-17}$. The behaviour of the relative error

$$\frac{\| P_k - P \|_F}{\| P \|_F}$$

at each step for three different initial matrices K_0:

$$1. \ K_0 = \begin{bmatrix} -5 & 0 & 5 \\ -5 & -5 & 0 \end{bmatrix}$$

$$2. \ K_0 = \begin{bmatrix} 3 & 12 & 26 \\ -8 & -9 & -6 \end{bmatrix}$$

$$3. \ K_0 = \begin{bmatrix} -107 & -17 & -10 \\ -198 & -31 & -19 \end{bmatrix}$$

Table 6.1

		$\| P_k - P \|_F / \| P \|_F$	
k	*1st guess*	*2nd guess*	*3rd guess*
0	0.957×10^0	0.604×10^1	0.190×10^3
1	0.155×10^0	0.337×10^1	0.103×10^3
2	0.751×10^{-2}	0.115×10^1	0.619×10^2
3	0.466×10^{-4}	0.215×10^0	0.422×10^2
4	0.187×10^{-8}	0.111×10^{-1}	0.329×10^2
5	0.127×10^{-15}	0.320×10^{-4}	0.280×10^2
6	0.288×10^{-15}	0.267×10^{-9}	0.239×10^2
7	0.647×10^{-15}	0.243×10^{-15}	0.176×10^2
8	0.363×10^{-15}	0.112×10^{-15}	0.845×10^1
9	0.212×10^{-15}	0.324×10^{-15}	0.279×10^1
10	0.235×10^{-15}	0.473×10^{-15}	0.804×10^0
11	0.232×10^{-15}	0.278×10^{-15}	0.119×10^0
12	0.150×10^{-15}	0.164×10^{-15}	0.349×10^{-2}
13	0.828×10^{-16}	0.107×10^{-15}	0.317×10^{-5}
14	0.421×10^{-15}	0.212×10^{-15}	0.264×10^{-11}
15	0.252×10^{-15}	0.196×10^{-15}	0.274×10^{-15}

is shown in Table 6.1. In the given case $c_R \approx 356.7$ and $\varepsilon c_R \approx 4.95 \times 10^{-15}$.

It is seen from the table that while six iterations are required to reach relative accuracy of order εc_R for the first initial gain matrix, sixteen iterations are required to reach the same accuracy for the third initial gain matrix.

The relation

$$\frac{\| P_k - P_{k-1} \|_F}{\| P_{k-1} \|_F} \leqq \varepsilon c_R$$

may serve as a stopping criterion in Newton's method. As an approximation of the exact solution P in the computation of c_R we may use the matrix P_k. The calculation of an approximation of

$$\text{sep}\,[(A - BR^{-1}B^T P)^T, \, -(A - BR^{-1}B^T P)]$$

may be done by the Byers algorithm (1984) exploiting the fact that the matrix A_k is reduced to Schur form in the Bartels–Stewart algorithm. In this way we obtain the following algorithm.

Algorithm 6.1 Solution of the continuous matrix Riccati equation by Newton's method

Choose K_0, such that $A - BK_0$ is a stability matrix
For $k = 0, 1, 2, \ldots$
    Set $A_k = A - BK_k$
    Set $Q_k = Q + K_k^T R K_k$
    Solve $A_k^T P_k + P_k A_k + Q_k = 0$ for P_k
    Compute $K_{k+1} = R^{-1} B^T P_k$

$$\text{Determine } c_R = \frac{2 \| A \|_F + \dfrac{\| Q \|}{\| P_k \|_F} + \| BR^{-1}B^T \|_F \| P_k \|_F}{\text{sep}\,[(A - BK_{k+1})^T, \, -(A - BK_{k+1})]}$$

If $\dfrac{\| P_k - P_{k-1} \|_F}{\| P_{k-1} \|_F} \leq \varepsilon c_R$, quit

If the solution of the Lyapunov equation $A_k^T P_k + P_k A_k + Q_k = 0$ is done by the Bartels–Stewart algorithm then Algorithm 6.1 requires about $20n^3$ flops at each step. Hence in the typical case (10 iterations) this algorithm will require approximately $200n^3$ flops, which is very expensive. The necessary storage is proportional to $10n^2$.

Newton's method can also be applied to find the positive semi-definite solution P of the discrete matrix Riccati equation

$$A^T PA - P + Q - A^T PB(R + B^T PB)^{-1} B^T PA = 0 \qquad (6.11)$$

This can be done by using the following theorem.

Theorem 6.2

Let P_k, $k = 0, 1, 2, \ldots$, be the solution of the Lyapunov equation

$$A_k^T P_k A_k - P_k + Q + K_k^T R K_k = 0 \qquad (6.12)$$

where

$$
\begin{aligned}
K_k &= (R + B^T P_{k-1} B)^{-1} B^T P_{k-1} A & k &= 1, 2, \ldots \\
A_k &= A - B K_k & k &= 0, 1, \ldots
\end{aligned}
$$

and K_0 is chosen such that A_0 is a convergence matrix. Then

$$P \leqq P_{k+1} \leqq P_k \cdots \leqq P_0 \qquad k = 0, 1, 2, \ldots$$

and

$$\lim_{k \to \infty} P_k = P$$

The proof of this theorem is given in Hewer (1971).

The comments pertaining to Theorem 6.1 are also valid in the given case. The iterations defined by Theorem 6.2 are quadratically convergent, i.e.

$$\| P_{k+1} - P \| \leqq c_2 \| P_k - P \|^2$$

where c_2 is a constant, independent of the iteration index k. The initial gain matrix K_0 may be computed by Algorithm 5.5 and the discrete Lyapunov equation (6.12) is efficiently solved by Algorithm 4.5. The iterations are performed until the following is satisfied

$$\frac{\| P_k - P_{k-1} \|_F}{\| P_{k-1} \|_F} \leqq \varepsilon d_R$$

where

$$d_R = \frac{2 \| A \|_F^2 + \dfrac{\| Q \|_F}{\| P \|_F} + \| A \|_F^2 \| BR^{-1} B^T \|_F \| P \|_F}{\mathrm{sep}_d(A_c^T, A_c)}$$

$$A_c = A - B(R + B^T PB)^{-1} B^T PA$$

is the condition number of the discrete Riccati equation (6.11) and $\text{sep}_d(A_c^T, A_c)$ is the minimum singular value of the matrix $A_c^T \otimes A_c^T - I_{n^2}$.

These considerations are summarized in the following algorithm.

Algorithm 6.2 Solution of the discrete matrix Riccati equation by Newton's method

Choose K_0, such that $A - BK_0$ is a convergence matrix
For $k = 0, 1, 2, \ldots$
  Set $A_k = A - BK_k$
  Set $Q_k = Q + K_k^T R K_k$
  Solve $A_k^T P_k A_k - P_k + Q_k = 0$ for P_k
  Compute $K_{k+1} = (R + B^T P_k B)^{-1} B^T P_k A$
  Determine

$$d_R = \frac{2\|A\|_F^2 + \dfrac{\|Q\|_F}{\|P_k\|_F} + \|A\|_F^2 \|BR^{-1}B^T\|_F \|P_k\|_F}{\text{sep}_d[(A - BK_{k+1})^T, (A - BK_{k+1})]}$$

  If $\dfrac{\|P_k - P_{k-1}\|_F}{\|P_{k-1}\|_F} \leqq \varepsilon d_R$, quit

Algorithm 6.2 requires about $20n^3$ flops per iteration provided that Algorithm 4.5 is used to solve (6.12). It produces an approximate solution of the discrete Riccati equation which satisfies

$$\frac{\|\bar{P} - P\|_F}{\|P\|_F} \leqq \varepsilon d_R$$

Newton's method for solving the matrix Riccati equation is also implemented in the form of an *incremental iteration*. In this case the new iteration is determined by

$$P_k = P_{k-1} + \Delta P_k \tag{6.13}$$

Substituting (6.13) into the continuous Riccati equation (6.9) and neglecting the second-order term in ΔP_k yields

$$(A - BR^{-1}B^T P_{k-1})^T \Delta P_k + \Delta P_k (A - BR^{-1}B^T P_{k-1})$$
$$+ A^T P_{k-1} + P_{k-1} A + Q - P_{k-1} BR^{-1}B^T P_{k-1} = 0 \tag{6.14}$$

Note that this equation may be written in the form

$$A_k^T \Delta P_k + \Delta P_k A_k + Y_k = 0$$

where

$$Y_k = A^T P_{k-1} + P_{k-1} A + Q - P_{k-1} B R^{-1} B^T P_{k-1}$$

is the residual matrix corresponding to the approximation of P found at the $(k-1)$th step.

It may be proved (see Merriam (1974)) that for P_k determined by (6.13) and (6.14) the following is satisfied:

$$\lim_{k \to \infty} P_k = P$$

and

$$\lim_{k \to \infty} \Delta P_k = 0$$

provided the matrix P_0 is chosen so that $A - BR^{-1}B^T P_0$ is a stability matrix.

The incremental iteration (6.13) and (6.14) is mathematically equivalent to the direct iteration for P_k as specified by (6.10). However, in using the incremental iteration we solve the Lyapunov equation (6.14) for the increment ΔP_k, not for P_k directly, which leads to some numerical differences. The incremental method is more accurate, since a given relative error in solving the Lyapunov equation for ΔP_k has smaller effect on P_k in the vicinity of the solution than the same relative error in the matrix P_k, computed directly. The incremental iteration, however, requires more memory and computations. To use this method it is necessary to provide an initial approximation P_0 of the solution. To achieve this, it is necessary to implement some other method for solving the Riccati equation or to perform an initial step according to (6.10).

The incremental iteration for the discrete Riccati equation has the form $P_k = P_{k-1} + \Delta P$

$$[A - B(R + B^T P_{k-1} B)^{-1} B^T P_{k-1} A]^T \Delta P_k$$
$$A - B(R + B^T P_{k-1} B)^{-1} B^T P_{k-1} A] - \Delta P_k + A^T P_{k-1} A \quad (6.15)$$
$$- P_{k-1} + Q - A^T P_{k-1} B(R + B^T P_{k-1} B)^{-1} B^T P_{k-1} A = 0$$

Algorithms 6.1 and 6.2 are easily modified to realize the incremental iteration version of Newton's method.

The Lyapunov equations (6.10) and (6.14) may be solved for the Cholesky factor of the solution using the Hammarling method (see Section 4.2). This allows us to improve the accuracy and to save computations in the iterative solution of (6.9). Details of the corresponding algorithm may be found in Hammarling (1982b).

EXERCISES

6.4 Perform a detailed roundoff error analysis of Algorithms 6.1 and 6.2.

6.5 Write computer programs implementing Algorithms 6.1 and 6.2.

6.6 Verify (6.14) and (6.15).

6.7 Modify Algorithms 6.1 and 6.2 to realize Newton's method in the version of incremental iterations.

6.8 Develop an algorithm for an iterative solution of the continuous Riccati equation using the Hammarling method for solving the Lyapunov equation.

6.3 MATRIX SIGN FUNCTION METHOD

Let A be an $n \times n$ matrix with Jordan canonical form

$$J = X^{-1}AX \equiv D + N$$

where X is a matrix composed of the eigenvectors and principal eigenvectors of A, $D = \mathrm{diag}(d_1, d_2, ..., d_n)$ is a diagonal matrix whose entries are the eigenvalues of A, and N is a nilpotent matrix. The *matrix sign function* of A is defined as

$$\mathrm{Sign}(A) = XYX^{-1}$$

where

$$Y = \mathrm{diag}(y_1, y_2, ..., y_n) \qquad y_i = \begin{cases} 1 & \text{if } \mathrm{Re}(d_i) > 0 \\ -1 & \text{if } \mathrm{Re}(d_i) < 0 \end{cases}$$

If A has eigenvalues on the imaginary axis then $\mathrm{Sign}(A)$ is not defined.

The matrix sign function is a generalization of the scalar sign function

$$\mathrm{sign}(a) = \begin{cases} 1 & \text{if } \mathrm{Re}(a) > 0 \\ -1 & \text{if } \mathrm{Re}(a) < 0 \end{cases}$$

It is easy to establish that $\mathrm{Sign}(A)$ has the following properties:

$$\left.\begin{array}{l} 1. \ [\mathrm{Sign}(A)]^2 = I_n \\ 2. \ \mathrm{Sign}(cA) = \mathrm{Sign}(c)\mathrm{Sign}(A) \\ 3. \ \mathrm{Sign}(TAT^{-1}) = T\mathrm{Sign}(A)T^{-1} \\ 4. \ A\mathrm{Sign}(A) = \mathrm{Sign}(A)A \\ 5. \ \text{The eigenvalues of } \mathrm{Sign}(A) \text{ are } \pm 1. \end{array}\right\} \qquad (6.16)$$

Consider the application of the matrix sign function to the solution of the continuous Riccati equation

$$A^T P + PA + Q - PSP = 0 \qquad (6.17)$$

Defining the Hamiltonian matrix corresponding to (6.17) as

$$H = \begin{bmatrix} A & -S \\ -Q & -A^T \end{bmatrix} \qquad S = BR^{-1}B^T$$

the Hamiltonian matrix is decomposed as (see (2.53))

$$H = \begin{bmatrix} I_n & 0 \\ P & I_n \end{bmatrix} \begin{bmatrix} A - SP & -S \\ 0 & -(A - SP)^T \end{bmatrix} \begin{bmatrix} I_n & 0 \\ -P & I_n \end{bmatrix} \tag{6.18}$$

Since the eigenvalues of $A - SP$ have negative real parts, the matrix sign function of H is defined. Applying this function to (6.18) and using the third property in (6.16) we obtain

$$W \equiv \begin{bmatrix} W_{11} & W_{12} \\ W_{21} & W_{22} \end{bmatrix} = \text{Sign}(H) = \begin{bmatrix} I_n & 0 \\ P & I_n \end{bmatrix} \begin{bmatrix} -I_n & Z \\ 0 & I_n \end{bmatrix} \begin{bmatrix} I_n & 0 \\ -P & I_n \end{bmatrix} \tag{6.19}$$

where W_{ij}, $i, j = 1, 2$, and Z are $n \times n$ matrices. Exploiting the fourth identity of (6.16) we find that Z is the solution of the Lyapunov equation

$$(A - SP)Z + Z(A - SP)^T = 2S$$

It follows from equation (6.19) that

$$\begin{bmatrix} W_{11} & W_{12} \\ W_{21} & W_{22} \end{bmatrix} \begin{bmatrix} I_n \\ P \end{bmatrix} = - \begin{bmatrix} I_n \\ P \end{bmatrix}$$

or

$$MP = -N \tag{6.20}$$

where

$$M = \begin{bmatrix} W_{12} \\ W_{22} + I_n \end{bmatrix} \qquad \text{and} \qquad N = \begin{bmatrix} W_{11} + I_n \\ W_{21} \end{bmatrix}$$

Equation (6.20) is a consistent system of $2n^2$ equations in the n^2 unknown entries of P. From (6.19) we have

$$\begin{bmatrix} W_{12} \\ W_{22} \end{bmatrix} = \begin{bmatrix} Z \\ PZ + I_n \end{bmatrix}$$

so that

$$M = \begin{bmatrix} Z \\ PZ + 2I_n \end{bmatrix}$$

and

$$[-P, \; I_n] M = [-P, \; I_n] \begin{bmatrix} Z \\ PZ + 2I_n \end{bmatrix} = 2I_n$$

The last equation shows that

$$\text{rank}(M) = n.$$

Equation (6.20) is reliably solved by using the QR decomposition of the matrix M (see Section 1.8).

The matrix sign function of the Hamiltonian matrix may be computed in the following way. The first identity of (6.16) shows that $\text{Sign}(H)$ is a square root of the identity matrix. Roberts (1980) has shown that applying Newton's method to the square root of the identity matrix one obtains the sequence

$$W_0 = H$$
$$W_{k+1} = \tfrac{1}{2}(W_k + W_k^{-1}) \qquad\qquad (6.21)$$

which converges to $\text{Sign}(H)$ quadratically. If $\text{Sign}(H)$ is undefined, then either the iteration (6.21) does not converge, or some of the matrices W_k are singular. It is possible to show that if H is a Hamiltonian matrix, so is H^{-1} and hence $\tfrac{1}{2}(H + H^{-1})$. In this way the Hamiltonian property of W_0 propagates in $W_{k+1} = \tfrac{1}{2}(W_k + W_k^{-1})$.

More accurate results in computing $\text{Sign}(H)$ may be obtained if, instead of (6.21), one uses the expression

$$W_0 = H$$
$$W_{k+1} = W_k - \tfrac{1}{2}(W_k - W_k^{-1}) \qquad\qquad (6.22)$$

In this case the rounding errors are confined to the eventually small correction $\tfrac{1}{2}(W_k - W_k^{-1})$.

The above considerations are summarized in the following algorithm.

Set $H = \begin{bmatrix} A & -S \\ -Q & -A^{\text{T}} \end{bmatrix}$

Compute $W = \text{Sign}(H)$ through (6.22)
Solve for P in (6.20)

This algorithm is one of the simplest methods for solving the Riccati equation. It requires only inversion of matrices and solution of an overdetermined but consistent system of equations. In contrast to Newton's method it does not require an initial approximation to the solution.

Example 6.4

Let

$$A = \begin{bmatrix} 3 & 1 & -3 \\ 2 & -5 & 3 \\ -1 & -2 & 1 \end{bmatrix} \qquad Q = \begin{bmatrix} 95 & 106 & -54 \\ 106 & 258 & 44 \\ -54 & 44 & 109 \end{bmatrix}$$

$$S = \begin{bmatrix} 5 & -1 & 15 \\ -1 & 34 & -29 \\ 15 & -29 & 65 \end{bmatrix}$$

The exact solution of (6.17) is

$$P = \begin{bmatrix} 2 & 1 & -1 \\ 1 & 3 & 1 \\ -1 & 1 & 2 \end{bmatrix}$$

An approximate solution P_k is obtained by using (6.22) and (6.20) in arithmetic with $\varepsilon = 2^{-56} \approx 1.39 \times 10^{-17}$. The relative error for each k, along with the condition number of the matrix W_k, is shown in Table 6.2.

The recursion (6.22) has the disadvantage that its initial convergence may be very slow. The convergence can be accelerated by scaling W_k at each iteration so that its eigenvalues are as close to ± 1 as possible. Byers (1987) has shown that maximal acceleration is achieved if W_k is scaled by $|\det(W_k)|^{-1/(2n)}$, i.e. by the reciprocal of the geometric

Table 6.2

k	$cond(W_k)$	$\|P_k - P\|_F / \|P\|_F$
1	377.4	0.100×10^2
2	307.3	0.500×10^1
3	192.4	0.146×10^1
4	101.1	0.115×10^1
5	53.2	0.791×10^0
6	30.1	0.328×10^0
7	19.9	0.701×10^{-1}
8	15.6	0.424×10^{-2}
9	14.7	0.168×10^{-4}
10	14.7	0.266×10^{-9}
11	14.7	0.647×10^{-16}
12	14.7	0.329×10^{-16}

Table 6.3

k	cond(Z_k)	$\|P_k - P\|_F / \|P\|_F$
1	377.4	0.100×10^1
2	35.0	0.995×10^{-1}
3	16.5	0.448×10^{-2}
4	14.7	0.516×10^{-5}
5	14.7	0.589×10^{-11}
6	14.7	0.203×10^{-15}
7	14.7	0.200×10^{-15}

mean of the eigenvalues. Incorporating this scaling into (6.22) gives

$$W_0 = H$$
$$Z_k = W_k |\det(W_k)|^{-1/(2n)} \qquad (6.23)$$
$$W_{k+1} = Z_k - \tfrac{1}{2}(Z_k - Z_k^{-1})$$

It is interesting to apply (6.23) to some trivial cases. If, for instance, H is a real 1×1 matrix then

$$W_0/\det(W_0) = W_1 = \text{Sign}(H)$$

so that (6.23) produces Sign(H) in one iteration. If H is a 2×2 matrix with real eigenvalues, then W_1 has two eigenvalues with equal modules and $W_2 = \text{Sign}(H)$.

The results of implementing (6.23) to the Riccati equation considered in Example 6.4 are shown in Table 6.3. In the given case a solution accurate to the working precision is obtained only in six iterations.

The inversion of the unsymmetric matrix in (6.23) can be replaced by symmetric inversion as follows. Taking into account that for

$$J = \begin{bmatrix} 0 & I_n \\ -I_n & 0 \end{bmatrix}$$

the matrix JZ is symmetric, Z_k^{-1} can be computed as $Z_k^{-1} = (JZ_k)^{-1} J$. This allows us to reduce the necessary computations and storage requirements twice. The iteration (6.23) is thus reorganized as

$$W_0 = H$$
$$Z_k = W_k |\det(W_k)|^{-1/(2n)} \qquad (6.24)$$
$$W_{k+1} = Z_k - \tfrac{1}{2}(Z_k - (JZ_k)^{-1} J)$$

Notice that it is not necessary to perform the multiplication of JZ_k explicitly; it suffices to rearrange the components of Z_k, changing signs where necessary.

Consider the numerical properties of the iteration (6.24). The most significant rounding errors are introduced during the inversion of JZ_k. According to the results from Section 1.6, the relative error in the computed inverse is bounded by

$c\varepsilon \ \text{cond}(JZ_k)$

where c is a low-degree polynomial in n and $\text{cond}(JZ_k)$ is the condition number of the matrix JZ_k. This suggests that the iteration should stop when

$$\frac{\| Z_k - (JZ_k)^{-1}J \|}{\|(JZ_k)^{-1}\|} \leq \ n\varepsilon \ \text{cond}(JZ_k) \qquad (6.25)$$

Another possibility is to stop the iteration when W_k no longer changes significantly, i.e. when

$$\| Z_k - (JZ_k)^{-1}J \| \leq \varepsilon \| Z_k \| \qquad (6.26)$$

Experience with these criteria shows that (6.25) sometimes stops the iteration too early while (6.26) stops it too late. In practice, a compromise between these criteria is reached.

Overall we have the following algorithm.

Algorithm 6.3 Solution of the continuous matrix Riccati equation by using the matrix sign function

Set $W = \begin{bmatrix} A & -S \\ -Q & -A^T \end{bmatrix}$

For $k = 1, 2, \ldots$
  $c \ = \text{cond}(JW)$
  $d \ = | \det(JW)|^{1/(2n)}$
  $Y = d(JW)^{-1}$
  $Z = W/d$
  $s \ = \| Z - YJ \|$
  $W = Z - (Z - YJ)/2$
  If $s \leq \varepsilon \| Z \|$, terminate the loop
Partition

$$W = \begin{bmatrix} W_{11} & W_{12} \\ W_{21} & W_{22} \end{bmatrix}$$

Solve for P in

$$\begin{bmatrix} W_{12} \\ W_{22} + I_n \end{bmatrix} P = - \begin{bmatrix} W_{11} + I_n \\ W_{21} \end{bmatrix}$$

The inversion of the matrix JW and the computation of $\det(JW)$ can be done efficiently by the subroutines DSICO, DSIFA and DSIDI from LINPACK. Equation (6.20) is solved by the subroutines DQRDC and DQRSL from the same package. The algorithm incorporates the stopping criterion (6.26).

Algorithm 6.3 requires about $4n^3$ flops per iteration. The necessary storage is proportional to $6n^2$, provided that only the triangle parts of the symmetric matrices Q, S and JW are stored.

The solution obtained by Algorithm 6.3 can be further refined by using Newton's method.

The numerical analysis done in Byers (1986b) shows that the matrix sign function method itself is not a numerically stable method. However, combined with iterative refinement it produces an efficient, numerically stable procedure for solving the Riccati equation.

The matrix sign function method can be applied to the solution of the discrete Riccati equation

$$A^T PA - P + Q - A^T PB(R + B^T PB)^{-1} B^T PA = 0 \qquad (6.27)$$

in the following way. If the matrix A is nonsingular, the discrete Riccati equation is associated with the symplectic matrix (see Section 2.7)

$$Z = \begin{bmatrix} A + BR^{-1}B^T A^{-T} Q & -BR^{-1}B^T A^{-T} \\ -A^{-T} Q & A^{-T} \end{bmatrix}$$

whose eigenvalues are

$$\lambda_i \qquad i = 1, 2, \ldots, n$$

and

$$\frac{1}{\lambda_i} \qquad i = 1, 2, \ldots, n$$

Using the bilinear transformation

$$H = (Z - I)(Z + I)^{-1} \qquad (6.28)$$

we obtain a matrix H which is Hamiltonian, since its eigenvalues are

$$\frac{\lambda_i - 1}{\lambda_i + 1} \qquad i = 1, 2, \ldots, n$$

and

$$-\frac{\lambda_i - 1}{\lambda_i + 1} \qquad i = 1, 2, \ldots, n$$

(The matrix $Z + I$ in (6.28) is nonsingular since under the conditions of Theorem 2.15 it is fulfilled that $\lambda_i \neq -1$.)

The matrix H is associated with a continuous Riccati equation which has the same solution as (6.27). Hence the solution of the discrete Riccati equation can be found using Algorithm 6.3 with H defined by (6.28). This method, however, may produce an acceptable result only when A is a matrix well conditioned with respect to inversion.

EXERCISES

6.9 Verify (6.16).

6.10 Prove that the scaling implemented in (6.23) causes maximal acceleration of the convergence.

6.11 Perform an error analysis of Algorithm 6.3.

6.12 Write a computer program implementing Algorithm 6.3.

6.13 Develop a computer program for solving the discrete Riccati equation (6.27) by using the bilinear transformation (6.28).

6.4 THE EIGENSYSTEM APPROACH

In this section we present two methods for solving the matrix Riccati equations which utilize the eigensystem of the Hamiltonian (symplectic) matrix.

6.4.1 The Eigenvector Method

Consider the continuous Riccati equation

$$A^T P + PA + Q - PSP = 0 \tag{6.29}$$

where the pair (A, B) $(BB^T = S$, $\text{rank}(B) = \text{rank}(S))$ is stabilizable and the pair (C, A) $(C^T C = Q$, $\text{rank}(C) = \text{rank}(Q))$ is detectable. The Hamiltonian matrix

$$H = \begin{bmatrix} A & -S \\ -Q & -A^T \end{bmatrix}$$

associated with (6.29) has eigenvalues (see Section 2.7)

$$-\lambda_1, -\lambda_2, \ldots, -\lambda_n, \lambda_1, \lambda_2, \ldots, \lambda_n$$

where λ_i, $i = 1, 2, \ldots, n$, are the eigenvalues with positive real parts.

(Under the above conditions the matrix H has no pure imaginary eigenvalues, see Mårtensson (1971) and Kučera (1972).) If the elementary divisors of H corresponding to $\lambda_1, \lambda_2, ..., \lambda_n$ are linear, then H may be transformed into the diagonal form

$$V^{-1}HV = \begin{bmatrix} -\Lambda & 0 \\ 0 & \Lambda \end{bmatrix}$$

where $\Lambda = \text{diag}(\lambda_1, \lambda_2, ..., \lambda_n)$.

Let the eigenvector matrix V be conformably partitioned into $n \times n$ blocks,

$$V = \begin{bmatrix} V_{11} & V_{12} \\ V_{21} & V_{22} \end{bmatrix}$$

where $V_1 = [V_{11}^{\mathrm{T}}, V_{21}^{\mathrm{T}}]^{\mathrm{T}}$ is a matrix which consists of the eigenvectors corresponding to eigenvalues with negative real parts. Then it may be proved (see Potter (1966) and Mårtensson (1971)), that the matrix V_{11} is nonsingular and the unique positive semidefinite solution of (6.29) is given by

$$P = V_{21}V_{11}^{-1} \tag{6.30}$$

In addition, the eigenvalues of the closed-loop system matrix $A - SP$ (the poles of the optimal system) are $-\lambda_1, -\lambda_2, ..., -\lambda_n$ and U_{11} is the eigenvector matrix of $A - SP$. Notice that the actual ordering of the eigenvectors in the matrix V_1 is not significant.

If some of the eigenvalues of H are complex then relationship (6.30) still holds. To avoid using complex arithmetic, instead of the complex conjugate eigenvectors $v_k = y_k + iz_k$ and $v_{k+1} = y_k - iz_k$ corresponding to the complex conjugate eigenvalues $\lambda_k = \sigma_k + i\omega_k$ and $\lambda_{k+1} = \sigma_k - i\omega_k$, one can put consecutively in the matrix V_1 the real vectors y_k and z_k. Then in place of λ_k and λ_{k+1} in the diagonal form of H will appear a 2×2 block in the form

$$\begin{bmatrix} \sigma_k & \omega_k \\ -\omega_k & \sigma_k \end{bmatrix}$$

In this way the solution of the continuous Riccati equation may be found by the following algorithm.

Set

$$H = \begin{bmatrix} A & -S \\ -Q & -A^{\mathrm{T}} \end{bmatrix}$$

Compute the eigenvalues and eigenvectors of H by the QR method. Select the eigenvectors corresponding to eigenvalues with negative real parts and put them into the matrix

$$V = \begin{bmatrix} V_{11} \\ V_{21} \end{bmatrix}$$

Solve $V_{11}^T P = V_{21}^T$ for P

This algorithm requires about $120n^3$ flops, and storage proportional to $10n^2$.

    The main advantage of the eigenvector method is that it allows us to find the solution P without iterations in P. As a result of the computations, one also obtains the eigenvalues and eigenvectors of the matrix $A - SP$. This method, however, is numerically unstable when the Hamiltonian matrix is nearly defective (H has almost multiple eigenvalues) since in this case the matrix U_{11} is poorly conditioned. It must be stressed that the ill conditioning of U_{11}, caused by closeness of some eigenvalues, is independent of the conditioning of the Riccati equation.

    The difficulties arising in the case of nonlinear elementary divisors of H can be avoided if H is reduced to Jordan canonical form. The solution is then again in the form (6.30) but the matrix V must consist of the eigenvectors and principal vectors of H. However, the reduction of H into Jordan form requires about n^4 flops, making the corresponding method inefficient. Besides that, in some cases the Jordan form may be ill conditioned, which causes ill conditioning of V_{11}.

Example 6.5

Let

$$A = \begin{bmatrix} -3 & 0 & 1 \\ 2 & 1 & -1 \\ -1 & 1 & 0 \end{bmatrix} \qquad Q = \begin{bmatrix} 53 & 18 & -20 \\ 18 & 12 & -1 \\ -20 & -1 & 14 \end{bmatrix}$$

and

$$S = \begin{bmatrix} 3 & -4 & 4 \\ -4 & 6 & -5 \\ 4 & -5 & 6 \end{bmatrix}$$

The exact solution of (6.29) is

$$P = \begin{bmatrix} 11 & 4 & -4 \\ 4 & 3 & 0 \\ -4 & 0 & 3 \end{bmatrix}$$

    The eigenvalues $-\lambda_1$, $-\lambda_2$ and $-\lambda_3$ of the Hamiltonian matrix and their corresponding eigenvectors v_1, v_2 and v_3 computed by the subroutine COMQR2 from EISPACK in arithmetic with $\varepsilon = 2^{-56} \approx 1.39 \times 10^{-17}$ are (up to four digits)

$$-\lambda_1 = -3.000 \qquad -\lambda_2 = -3.000 \qquad -\lambda_3 = -3.000$$

$$
v_1 = \begin{bmatrix} -0.2887 \\ 0.5774 \\ -0.2887 \\ 0.2887 \\ 0.5774 \\ 0.2887 \end{bmatrix}
\qquad
v_2 = 10^7 \begin{bmatrix} 0.4212 \\ -0.8424 \\ 0.4212 \\ -0.4212 \\ -0.8424 \\ -0.4212 \end{bmatrix}
\qquad
v_3 = \begin{bmatrix} 0.5334 \\ -0.5409 \\ 0.5337 \\ 1.572 \\ 0.5121 \\ -0.5337 \end{bmatrix}
$$

The solution $\bar{P}$ of (6.29) computed through (6.30) has a relative error

$$\frac{\|\bar{P} - P\|_F}{\|P\|_F} = 7.657 \times 10^{-10}$$

so that approximately seven digits of accuracy have been lost in the computations. This is because the first two eigenvalues correspond to a 2×2 Jordan block in the Jordan form of H and the eigenvectors v_1 and v_2 are almost linearly dependent

$$\text{cond}_2(V_{11}) = 1.432 \times 10^{15}$$

(Owing to the large errors in computing the multiple eigenvalues $-\lambda_1$ and $-\lambda_2$ the vectors v_1 and v_2 are not strictly linearly dependent, which allows us to find an approximate solution.) Since the condition number of the Riccati equation is $c_R = 67.77$, the relative error in $\bar{P}$ is about 8×10^5 times εc_R, which indicates the numerical instability of the method. If, however, the Hamiltonian matrix is reduced into Jordan form by using the subroutine JNF from Kågström and Ruhe (1980) then the solution is obtained with relative error 1.50×10^{-16}.

The eigenvector method may also be applied to the solution of the discrete Riccati equation

$$A^T P A + Q - P - A^T P B (R + B^T P B)^{-1} B^T P A = 0 \tag{6.31}$$

provided the matrix A is nonsingular.

Suppose that the pair (A, B) is stabilizable and the pair (A, C) ($C^T C = Q$, rank(C) = rank(Q)) is detectable, which ensures the existence of a unique positive semidefinite solution of (6.31). If the symplectic matrix

$$Z = \begin{bmatrix} A + SA^{-T}Q & -SA^{-T} \\ -A^{-T}Q & A^{-T} \end{bmatrix} \qquad S = BR^{-1}B^T \tag{6.32}$$

associated with (6.31) has linear elementary divisors it may be reduced to

$$V^{-1}ZV = \begin{bmatrix} \Lambda & 0 \\ 0 & \Lambda^{-1} \end{bmatrix} \qquad V = \begin{bmatrix} V_{11} & V_{12} \\ V_{21} & V_{22} \end{bmatrix}$$

where $\Lambda = \text{diag}(\lambda_1, \lambda_2, ..., \lambda_n)$ and λ_i, $i = 1, 2, ..., n$, are the eigenvalues of Z whose moduli are less than 1. The solution of (6.31) is then found from $P = V_{21}V_{11}^{-1}$. Notice that instead of (6.32) it is possible to work with the matrix

$$Z^{-1} = \begin{bmatrix} A^{-1} & A^{-1}S \\ QA^{-1} & A^T + QA^{-1}S \end{bmatrix}$$

In such a case the matrix $[V_{11}^T, \ V_{21}^T]^T$ must consist of the eigenvectors corresponding to eigenvalues outside the unit circle.

Besides the difficulties related to defective symplectic matrices, the eigenvector method for solving the discrete Riccati equation has a drawback in that it can be implemented only when A is a well-conditioned matrix.

6.4.2 The Schur Method

The continuous Riccati equation can be solved by reducing H into quasi-triangular (real Schur) form using orthogonal similarity transformations. This makes it possible to avoid the numerical difficulties related to the use of eigenvectors of nearly defective Hamiltonian matrices.

According to the presentation in Section 1.8, there exists an orthogonal matrix U which transforms the Hamiltonian matrix H into real Schur form

$$T = U^T H U = \begin{bmatrix} T_{11} & T_{12} \\ 0 & T_{22} \end{bmatrix} \tag{6.33}$$

where T_{11} and T_{22} are $n \times n$ upper quasitriangular matrices. The blocks on the diagonals of T_{11} and T_{22} have dimensions at most 2×2.

The reduction (6.33) is non-unique and it is always possible to choose the matrix U so that the eigenvalues of T_{11} have negative real parts while those of T_{22} have positive real parts.

Let the matrix U be partitioned as

$$U = \begin{bmatrix} U_{11} & U_{12} \\ U_{21} & U_{22} \end{bmatrix}$$

where each block is of dimension $n \times n$. Then it may be proved (see Laub (1979)) that the matrix U_{11} is nonsingular and the positive semidefinite solution of (6.29) is found from

$$P = U_{21}U_{11}^{-1} \tag{6.34}$$

In addition, the eigenvalues of T_{11} are the eigenvalues of the matrix $A - SP$, i.e. the poles of the optimal closed-loop system due to the relationship $A - SP = U_{11}T_{11}U_{11}^{-1}$.

The orthogonal reduction of the Hamiltonian matrix into real Schur form is done by the QR method (see Section 1.10). Since the matrix H is nonsymmetric, it is not possible to guarantee a special ordering of the eigenvalues on the diagonal of the quasitriangular form. It is possible, however, to reorder this form arbitrarily by systematically interchanging adjacent pairs of eigenvalues implementing orthogonal similarity transformations (see Section 1.10).

In this way we obtain the following algorithm.

Algorithm 6.4 Solution of the continuous matrix Riccati equation by the Schur method

Set

$$H = \begin{bmatrix} A & -S \\ -Q & -A^{\mathrm{T}} \end{bmatrix}$$

Reduce H into real Schur form $T = U^{\mathrm{T}}HU$ by using orthogonal similarity transformations.

Reorder the real Schur form by using orthogonal transformations so that the quasitriangular block T_{11} in the upper left-hand corner of T has eigenvalues with negative real parts, and accumulate the transformations in U

Solve $U_{11}^{\mathrm{T}}P = U_{21}^{\mathrm{T}}$ for P where $[U_{11}^{\mathrm{T}}, U_{21}^{\mathrm{T}}]^{\mathrm{T}}$ are the Schur vectors corresponding to T_{11}.

The orthogonal reduction of H into real Schur form may be done by a shortened version of the subroutine HQR2 from EISPACK. The preliminary reduction of H into upper Hessenberg form and the accumulation of the orthogonal transformations is performed by the subroutines ORTHES and ORTRAN from the same package. To improve the accuracy, the Hamiltonian matrix can be preliminarily balanced. The reordering of the Schur form is carried out using subrountines EXCHNG and QRSTEP from Stewart (1976). The solution of the linear equation $U_{11}^{\mathrm{T}}P = U_{21}^{\mathrm{T}}$ with estimation of the conditioning is done by the subroutines DGECO, DGEFA and DGESL from LINPACK.

Algorithm 6.4 requires approximately $63n^3$ flops if the re-ordering of the Schur form is not taken into account. Since the reordering requires about 25% of the whole volume of computations, it follows that $75n^3$ flops are necessary to solve the continuous Riccati equation by the Schur method. The necessary storage is proportional to $10n^2$.

The Schur method has several advantages over the eigenvector method. First, the reduction into quasitriangular form is an intermediate step in the computation of the eigenvectors by the QR method so that the Schur method requires fewer computations than the eigenvector method. More important is that the Schur method does not suffer from the difficulties inherent in the use of eigenvectors of defective or nearly defective Hamiltonian matrices. This method works equally well in the presence of linear and nonlinear elementary divisors of H.

The Schur method is significantly faster than Newton's method and it is as fast as the matrix sign function method with iterative refinement.

As in the eigenvector method, as a result of the solution of the Riccati equation one also obtains the eigenvalues of the closed-loop system matrix.

Example 6.6

If

$$A = \begin{bmatrix} -3 & 0 & 1 \\ 2 & -1 & -1 \\ -1 & 1 & 0 \end{bmatrix} \qquad Q = \begin{bmatrix} 53 & 18 & -20 \\ 18 & 12 & -1 \\ -20 & -1 & 14 \end{bmatrix}$$

$$S = \begin{bmatrix} 3 & -4 & 4 \\ -4 & 6 & -5 \\ 4 & -5 & 6 \end{bmatrix}$$

then Algorithm 6.4 produces, in arithmetic with $\varepsilon = 2^{-56} \approx 1.39 \times 10^{-17}$, the matrices (up to four digits)

$$T_{11} = \begin{bmatrix} -3.000 & 0.5148 & 0.8738 \\ & -3.000 & 0.000 \\ 0 & & -3.000 \end{bmatrix}$$

$$
\begin{bmatrix} U_{11} \\ U_{21} \end{bmatrix} =
\left[
\begin{array}{rrr}
-0.2887 & -0.3541 & -0.07453 \\
0.5774 & 0.3293 & 0.03249 \\
-0.2887 & -0.5027 & -0.3268 \\
\hline
0.2887 & -0.5669 & 0.6172 \\
0.5774 & -0.4284 & -0.2006 \\
0.2887 & -0.09176 & -0.6822
\end{array}
\right]
$$

$$
\bar{P} =
\begin{bmatrix}
11.000 & 4.000 & -4.000 \\
4.000 & 3.000 & 0.000 \\
-4.000 & 0.000 & 3.000
\end{bmatrix}
$$

the relative error in the solution being

$$
\| \bar{P} - P \|_F / \| P \|_F = 1.307 \times 10^{-16}
$$

The condition number of the matrix U_{11} is $\mathrm{cond}_2(U_{11}) = 13.93$. Hence the accuracy of the method is not diminished by the presence of a 2×2 Jordan block in the Jordan form of H (cf. Example 6.5).

Consider the numerical properties of the Schur method. The computed Schur form $\bar{T}$ of H is orthogonally similar to a matrix $H + E$ where

$$
\| E \|_2 \le c_1 \varepsilon \| H \|_2
$$

with c_1 a low-degree polynomial in n. In this way

$$
\bar{U}^T (H + E) \bar{U} = \begin{bmatrix} \bar{T}_{11} & \bar{T}_{12} \\ 0 & \bar{T}_{22} \end{bmatrix}
$$

where $\bar{U}$ is the computed orthogonal matrix.

To determine the error in $\bar{P} = \mathrm{fl}(\bar{U}_{21}\bar{U}_{11}^{-1})$ we need the relationship between the matrices $\bar{U}$ and U. If the conditions of Theorem 1.2 are satisfied, then we have that

$$
\bar{U}_1 = (U_1 + U_2 X)(I_n + X^T X)^{-1/2}
$$
$$
\bar{U} = [\bar{U}_1, \ \bar{U}_2]
$$

and

$$
\| X \|_2 \le 2 c_1 \varepsilon \| H \|_2 / \delta
$$

where

$$
\delta = \mathrm{sep}(T_{11}, T_{22}) - 2 c_1 \varepsilon \| H \|_2
$$

Partitioning the matrix $\bar{U}_1$ as $[\bar{U}_{11}^T, \bar{U}_{21}^T]^T$ we get

$$
\bar{U}_{11} = (U_{11} + U_{12} X)(I_n + X^T X)^{-1/2}
$$
$$
\bar{U}_{21} = (U_{21} + U_{22} X)(I_n + X^T X)^{-1/2} \tag{6.35}
$$

The solution of the Riccati equation is computed from

$$P\bar{U}_{11} = \bar{U}_{21}$$

If this equation is solved by Gaussian elimination with pivoting then we have that (see Section 1.6)

$$\|\bar{P} - P^0\|_2 \leqq c_2 \, \text{cond}_2(\bar{U}_{11})\varepsilon \|\bar{P}\|_2$$

where $P^0 = \bar{U}_{21}\bar{U}_{11}^{-1}$ and c_2 is a coefficient characterizing the elements growth during the elimination.

To find an estimate of $\|\bar{P} - P\|_2$ we shall find first an estimate of $\|P^0 - P\|_2$ where P is the exact solution of (6.29). Utilizing (6.35) we obtain

$$P^0 = (P + U_{22}XU_{11}^{-1})(I_n + U_{12}XU_{11}^{-1})^{-1} \tag{6.36}$$

It follows from (6.36) that

$$P^0 - P = (U_{22} - PU_{12})X(I_n + X^{\mathrm{T}}X)^{-1/2}\bar{U}_{11}^{-1} \tag{6.37}$$

Since

$$\|(I_n + X^{\mathrm{T}}X)^{-1/2}\|_2 = \sigma_{\min}^{-1/2}$$

where $\sigma_{\min} \geqq 1$ is the minimal singular value of $I_n + X^{\mathrm{T}}X$, it is fulfilled that

$$\|(I_n + X^{\mathrm{T}}X)^{-1/2}\|_2 \leqq 1$$

Taking the norm of both sides of (6.37) and bearing in mind that

$$\|U_{12}\|_2 \leqq 1 \qquad \text{and} \qquad \|U_{22}\|_2 \leqq 1$$

we obtain

$$\|P^0 - P\|_2 \leqq (1 + \|P\|_2)\|X\|_2 \|\bar{U}_{11}^{-1}\|_2$$

In this way

$$\|\bar{P} - P\|_2 \leqq \|\bar{P} - P^0\|_2 + \|P^0 - P\|_2$$
$$\leqq c_2 \, \text{cond}_2(\bar{U}_{11})\varepsilon \|\bar{P}\|_2 + (1 + \|P\|_2)\|X\|_2 \, \bar{U}_{11}^{-1}\|_2$$

The final result is stated as follows.

Theorem 6.3

If

$$\delta = \text{sep}(T_{11}, T_{22}) - c_1\varepsilon \|H\|_2 > 0$$

and

$$c_1\varepsilon \| H \|_2^2 (1 + c_1\varepsilon) \leq \tfrac{1}{4}\delta^2$$

then the solution of the continuous Riccati equation (6.29) computed by the Schur method satisfies

$$\frac{\| \bar{P} - P \|_2}{\| P \|_2} \leq 2c_1\varepsilon(1 + 1/\| P \|_2)(2\| A \|_2 + \| Q \|_2 + \| S \|_2)\| \bar{U}_{11}^{-1} \|_2/\delta$$

$$+ c_2 \, \mathrm{cond}_2(\bar{U}_{11})\varepsilon \| \bar{P} \|_2/\| P \|_2 \tag{6.38}$$

The bound (6.38) can be used to obtain an a posteriori estimate of the relative error in the computed solution. The quantities $\| \bar{U}_{11}^{-1} \|_2$ and $\mathrm{cond}_2(\bar{U}_{11})$ can be computed during the solution. The separation between matrices T_{11} and T_{22}

$$\mathrm{sep}(T_{11}, T_{22}) \approx \mathrm{sep}(\bar{T}_{11}, \bar{T}_{22}) = \sigma_{\min}(I_n \otimes \bar{T}_{11} - \bar{T}_{22}^{\mathrm{T}} \otimes I_n)$$

can be estimated efficiently by the algorithm of Byers (1984). The quantity $1/\| P \|_2$ may be approximated by $1/\| \bar{P} \|_2$ and $\| \bar{P} \|_2/\| P \|_2$ is of order 1.

If the bound (6.38) is applied to the Riccati equation from Example 6.6 we obtain ($c_1 = 1, c_2 = 1$)

$$\frac{\| \bar{P} - P \|_F}{\| P \|_F} \leq \sqrt{n} \, \frac{\| \bar{P} - P \|_2}{\| P \|_2} \leq 9.905 \times 10^{-15}$$

which is slightly larger than the actual relative error in $\bar{P}$.

The bound (6.38) shows that the error in $\bar{P}$ will be large if $\| \bar{U}_{11}^{-1} \|_2$ is large, $\bar{U}_{11}$ is poorly conditioned or $\mathrm{sep}(T_{11}, T_{22})$ is small. Notice that $\| \bar{U}_{11}^{-1} \|_2$ may be large even if $\mathrm{cond}_2(\bar{U}_{11})$ is low.

The numerical behaviour of the Schur method can be revealed by comparing the estimate (6.38) with the bound

$$\frac{\| P^* - P \|_2}{\| P \|_2} \leq \frac{2\| \Delta A \|_2 + \| \Delta Q \|_2/\| P \|_2 + \| \Delta S \|_2\| P \|_2}{\mathrm{sep}\,[(A - SP)^{\mathrm{T}}, -(A - SP)]} \tag{6.39}$$

characterizing the sensitivity of the solution to small perturbations ΔA, ΔQ and ΔS in the matrices A, Q and S. (Notice that $\mathrm{sep}\,[(A - SP)^{\mathrm{T}}, -(A - SP)]$ is in general different from $\mathrm{sep}(T_{11}, T_{22})$.) It is possible to show that there are cases when the bound (6.38) is very large in comparison with (6.39) i.e. the Schur method is numerically unstable. The cause for instability is the following. If the matrix $\bar{U}_{11}$ is well conditioned, the computed solution $\bar{P}$ is exact for a slightly perturbed Hamiltonian matrix. However, the absolute perturbations ΔA, ΔQ and ΔS in the individual matrices A, Q and S which have small norms compared to $\|H\|_2$

$$\|E\|_2 \le c_1\varepsilon\|H\|_2 \le c_1\varepsilon(2\|A\|_2 + \|Q\|_2 + \|S\|_2) \tag{6.40}$$

$$E = \Delta A, \Delta Q, \Delta S$$

may correspond to relative perturbations

$$\frac{\|\Delta A\|_2}{\|A\|_2} \qquad \frac{\|\Delta Q\|_2}{\|Q\|_2} \qquad \frac{\|\Delta S\|_2}{\|S\|_2}$$

which are not small. Specifically, if the norm of some matrix involved in the Riccati equation is much smaller compared with the norms of the other matrices, then the equivalent relative perturbation in this matrix is very large. This leads to numerical instability from the point of view of backward error analysis.

If

$$\|A\| \approx \|Q\| \approx \|S\|$$

then according to (6.40) the relative perturbations in A, Q and S will be small.

Example 6.7

Consider Equation (6.29) with $n = 3$,

$$A = VA_0V \qquad Q = VQ_0V$$
$$A_0 = \mathrm{diag}(10^k, 2 \times 10^k, 3 \times 10^k)$$
$$Q_0 = \mathrm{diag}(10^{-k}, 1, 10^k)$$
$$S = \mathrm{diag}(10^{-k}, 10^{-k}, 10^{-k})$$

where V is an elementary reflection of third order

$$V = I_3 - 2vv^T/3 \qquad v = [1, 1, 1]^T$$

and $k \ge 0$ is an integer. The exact positive definite solution of (6.29) is

$$P = VP_0V$$

where

$$P_0 = \mathrm{diag}(p_1, p_2, p_3)$$

and

$$p_1 = 10^k(10^k + (10^{2k} + 10^{-2k})^{1/2})$$
$$p_2 = 10^k(2 \times 10^k + (4 \times 10^{2k} + 10^{-k})^{1/2})$$
$$p_3 = 10^k(3 \times 10^k + (9 \times 10^{2k} + 1)^{1/2})$$

For this example the condition number of the matrix U_{11} is of order 1 for every k and

$$\text{sep}(T_{11}, T_{22}) = 2(10^{2k} + 10^{-2k})^{1/2}$$

is increasing with k. The results obtained for different k by using Algorithm 6.4 in arithmetic with $\varepsilon = 2^{-56} \approx 1.39 \times 10^{-17}$ are shown in Table 6.4.

The estimate

$$\frac{\| \bar{P} - P \|_2}{\| P \|_2} \leq \text{est}$$

is obtained from (6.38),

$$\text{est} = 2(1 + 1/\| P \|_2)\| \bar{U}_{11}^{-1} \|_2 \varepsilon \| H \|_2 / \text{sep}(T_{11}, T_{22}) + \varepsilon \, \text{cond}_2(\bar{U}_{11})$$

where c_1 and c_2 are set equal to 1.

It is seen from the table that the solution error is due to the large norms of $\bar{U}_{11}^{-1}$ and H but not to $\text{cond}(\bar{U}_{11})$ which is very low. For this example the error estimate est and the actual error are of the same order.

On the other hand the Riccati equation in the given case is very well conditioned. Indeed, using (6.39) it may be shown that

$$\frac{\| P^* - P \|_2}{\| P \|_2} \leq \text{est}_1 = 3\left(\frac{\| \Delta A \|_2}{\| A \|_2} + \frac{\| \Delta S \|_2}{\| S \|_2}\right) \tag{6.41}$$

for sufficiently large k (the perturbations in Q are negligible for this example). Setting in (6.41) the right-hand side equal to $\| \bar{P} - P \|_2/\| P \|_2$ for $k = 6$, we find that the norms of the equivalent relative perturbations in A and S are of order 10^{-5} which is 10^{12} times the machine precision. Hence the Schur method is numerically unstable in the case under consideration. For our example the rounding errors in reducing H into Schur form cause a large relative error in S, which in accordance with (6.41) leads to a large relative error in the solution. This is confirmed by the results presented in Table 6.5, where we give the estimate est_1 of the relative perturbation in P, found from (6.41) for absolute errors in A and S

$$\| \Delta A \|_2 = \varepsilon \| H \|_2$$
$$\| \Delta S \|_2 = \varepsilon \| H \|_2$$

The computed estimate is nearly the same as the actual relative error.

A scaling procedure, enhancing the numerical stability of the Schur method, is proposed in Kenney, Laub and Wette (1989).

The Schur method may be used to solve the discrete Riccati

Table 6.4

k	$cond_2(\bar{U}_{11})$	$\|\bar{U}_{11}^{-1}\|_2$	$\|H\|_2$	$sep(T_{11}, T_{22})$	est	$\|\bar{P} - P\|_2/\|P\|_2$
0	2	6	3	3	3×10^{-16}	2×10^{-16}
1	3	6×10^2	4×10	2×10	3×10^{-14}	2×10^{-14}
2	3	6×10^4	4×10^2	2×10^2	3×10^{-12}	2×10^{-12}
3	3	6×10^6	4×10^3	2×10^3	3×10^{-10}	6×10^{-11}
4	3	6×10^8	4×10^4	2×10^4	3×10^{-8}	1×10^{-8}
5	3	6×10^{10}	4×10^5	2×10^5	3×10^{-6}	2×10^{-6}
6	3	6×10^{12}	4×10^6	2×10^6	3×10^{-4}	1×10^{-4}

Table 6.5

k	$\|\Delta A^{+}\|_2, \|\Delta S\|_2$	$\|A\|_2$	$\|\Delta A\|_2/\|A\|_2$	$\|S\|_2$	$\|\Delta S\|_2/\|S\|_2$	est_1
0	4×10^{-17}	3	1×10^{-17}	1	4×10^{-17}	2×10^{-16}
1	6×10^{-16}	3×10	2×10^{-17}	10^{-1}	6×10^{-15}	2×10^{-14}
2	6×10^{-15}	3×10^2	2×10^{-17}	10^{-2}	6×10^{-13}	2×10^{-12}
3	6×10^{-14}	3×10^3	2×10^{-17}	10^{-3}	6×10^{-11}	2×10^{-10}
4	6×10^{-13}	3×10^4	2×10^{-17}	10^{-4}	6×10^{-9}	2×10^{-8}
5	6×10^{-12}	3×10^5	2×10^{-17}	10^{-5}	$6 \times 10^{-7'}$	2×10^{-6}
6	6×10^{-11}	3×10^6	2×10^{-17}	10^{-6}	6×10^{-5}	2×10^{-4}

equation (6.31) if the matrix A is nonsingular. For this aim the symplectic matrix Z defined by (6.32) is reduced using orthogonal similarity transformations into the real Schur form

$$T = U^T Z U = \begin{bmatrix} T_{11} & T_{12} \\ 0 & T_{22} \end{bmatrix} \qquad U = \begin{bmatrix} U_{11} & U_{12} \\ U_{21} & U_{22} \end{bmatrix}$$

where each block is of dimension $n \times n$. If this form is ordered so that the eigenvalues of T_{11} are inside the unit circle, then the positive semi-definite solution of (6.31) is given again by (6.34). The advantages of this method over the corresponding eigenvector method are the same as in the continuous-time case. Unfortunately, its application is restricted only to the cases when A is a well-conditioned matrix, as in the eigenvector method.

EXERCISES

6.14 Verify (6.30) and (6.34).
6.15 Write a computer program for solving the continuous Riccati equation implementing Algorithm 6.4.
6.16 Verify (6.37).
6.17 Write a program which estimates the relative error in the solution of the Riccati equation obtained by the Schur method, utilizing the bound (6.38).
6.18 Sketch a rounding error analysis of the Schur method for solving the discrete Riccati equation.
6.19 Develop a Schur algorithm for solving the unsymmetric matrix Riccati equation

$$AX + XB + C + XDX = 0$$

6.5 THE GENERALIZED EIGENSYSTEM APPROACH

The solution of a matrix Riccati equation can be found by the solution of an associated generalized eigenvalue problem. This makes it possible to obtain the solution in cases when the methods described in the previous section fail.

6.5.1 Generalized Eigensystem Methods for the Discrete Riccati Equation

Consider the discrete Riccati equation

$$A^T P A + Q - P - A^T P B (R + B^T P B)^{-1} B^T P A = 0 \tag{6.42}$$

where the pair (A, B) is stabilizable and the pair (C, A) $(C^T C = Q)$ is detectable.

Equation (6.42) may be solved via the generalized eigenvalue problem

$$Mx = \lambda Nx \tag{6.43}$$

where

$$M = \begin{bmatrix} A & 0 \\ -Q & I_n \end{bmatrix} \qquad N = \begin{bmatrix} I_n & S \\ 0 & A^T \end{bmatrix}$$

$$S = BR^{-1}B^T$$

Notice that if A is nonsingular, then the generalized eigenvalue problem (6.43) reduces to the usual eigenvalue problem

$$Zx = \lambda x$$

where

$$Z = N^{-1}M = \begin{bmatrix} A + SA^{-T}Q & -SA^{-T} \\ -A^{-T}Q & A^{-T} \end{bmatrix}$$

is the symplectic matrix associated with (6.42).

The generalized eigenvalue problem (6.43) also possesses symplectic properties in the sense that if λ is an eigenvalue of the pencil $M - \lambda N$ with multiplicity r, then $1/\lambda$ is also an eigenvalue with the same multiplicity. Under the restrictions assumed above, none of the eigenvalues of (6.43) lies on the unit circle. If A is nonsingular, then all of the generalized eigenvalues are nonzero. If A is singular, then (6.43) has at least one eigenvalue $\lambda = 0$. If $\lambda = 0$ is an eigenvalue with multiplicity r, then the problem (6.43) has only $(2n - r)$ finite eigenvalues, the r missing eigenvalues being considered as 'infinite' eigenvalues (see Section 1.11). In this way when A is a singular matrix, the generalized eigenvalues of the problem (6.43) can be arranged as follows

$$\underbrace{0, \ldots, 0}_{r}, \underbrace{\lambda_{r+1}, \ldots, \lambda_n}_{n-r}, \underbrace{\frac{1}{\lambda_n}, \ldots, \frac{1}{\lambda_{r+1}}}_{n-r}, \underbrace{\infty, \ldots, \infty}_{r}$$

with

$$0 < |\lambda_i| < 1 \qquad i = r + 1, \ldots, n$$

Let V be the matrix of the generalized eigenvectors for the problem (6.43) where the first n vectors correspond to the eigenvalues inside the unit circle. (For simplicity we suppose that the eigenvalues are distinct.)

If the matrix V is partitioned into $n \times n$ blocks

$$V = \begin{bmatrix} V_{11} & V_{12} \\ V_{21} & V_{22} \end{bmatrix}$$

then it may be shown (see Pappas, Laub and Sandell (1980)) that the matrix V_{11} is nonsingular and the positive semidefinite solution of (6.42) is given by

$$P = V_{21} V_{11}^{-1}$$

In addition, the closed-loop poles, i.e. the eigenvalues of

$$A - B(R + B^T P B)^{-1} B^T P A$$

are the generalized eigenvalues of (6.43) which have moduli less than 1.

Hence the solution of the discrete Riccati equation can be found by the following algorithm.

Set

$$M = \begin{bmatrix} A & 0 \\ -Q & I_n \end{bmatrix} \qquad N = \begin{bmatrix} I_n & S \\ 0 & A^T \end{bmatrix}$$

Compute the generalized eigenvalues and eigenvectors of the pencil $M - \lambda N$ by the QZ method

Select the generalized eigenvectors corresponding to eigenvalues with moduli less than 1 and put them into the matrix $V_1 = [V_{11}^T, V_{21}^T]^T$

Solve $V_{11}^T P = V_{21}^T$ for P

This algorithm requires about $150n^3$ flops and storage proportional to $15n^3$.

The main future of the generalized eigenvector method is that it allows us to solve the discrete Riccati equation in the case of a singular or ill-conditioned matrix A, in contrast to the eigenvector method presented in Section 6.4. The price to be paid is the increased volume of computations. The method breaks down if there are multiple generalized eigenvalues to which linearly dependent generalized eigenvectors correspond. In such a case it is possible to use the generalized principal vectors, but the corresponding computational method requires a large volume of computations. (For an algorithm for finding the generalized eigenvectors and principal vectors of a pencil see Kågström (1986).)

Similar to the continuous-time case, the numerical difficulties related to the use of the generalized eigenvectors can be avoided if the pencil $M - \lambda N$ is reduced into generalized Schur form, using orthogonal equivalent transformations.

Let U and V be orthogonal matrices such that

$$\tilde{M} = UMV = \begin{bmatrix} M_{11} & M_{12} \\ 0 & M_{22} \end{bmatrix}$$

is in upper quasitriangular form and

$$\tilde{N} = UNV = \begin{bmatrix} N_{11} & N_{12} \\ 0 & N_{22} \end{bmatrix}$$

is in upper triangular form, where M_{11}, M_{22}, N_{11} and N_{22} are $n \times n$ matrices. The blocks on the diagonal of $\tilde{M}$ are of dimension at most 2×2. It is always possible to chose the matrices U and V so that the eigenvalues of the pencil $M_{11} - \lambda N_{11}$ are inside the unit circle. If the matrix V is partitioned as

$$V = \begin{bmatrix} V_{11} & V_{12} \\ V_{21} & V_{22} \end{bmatrix}$$

where V_{ij}, $i, j = 1, 2$, are $n \times n$ blocks, then it may be proved (see Pappas, Laub and Sandell (1980)) that the matrix V_{11} is nonsingular and the positive semidefinite solution of (6.42) is found from

$$P = V_{21} V_{11}^{-1}$$

In addition, the generalized eigenvalues of the pencil $M_{11} - \lambda N_{11}$ are the closed-loop system poles, i.e. the eigenvalues of

$$A_c = A - B(R + B^T PB)^{-1} B^T PA$$

The reduction of the pencil $M - \lambda N$ into generalized Schur form is done by the QZ method. Unfortunately, the order in which the generalized eigenvalues (as determined by the diagonal elements of $\tilde{M}$ and $\tilde{N}$) appear on the diagonal of the generalized Schur form can be arbitrary. That is why it is necessary to reorder the generalized Schur form so that the eigenvalues of the pencil $M_{11} - \lambda N_{11}$ are inside the unit circle. This can be done by systematically interchanging adjacent pairs of generalized eigenvalues implementing orthogonal equivalent transformations (see Section 1.11).

The preceding considerations are summarized in the following algorithm.

Algorithm 6.5   Solution of the discrete matrix Riccati equation by the generalized Schur method

Set

$$M = \begin{bmatrix} A & 0 \\ -Q & I_n \end{bmatrix} \qquad N = \begin{bmatrix} I_n & S \\ 0 & A^{\mathrm{T}} \end{bmatrix}$$

Reduce the pencil $M - \lambda N$ into generalized Schur form

$$\tilde{M} - \lambda \tilde{N} = U(M - \lambda N)V$$

by using orthogonal equivalent transformations

Reorder the generalized Schur form by using orthogonal equivalent transformations, so that the pencil $M_{11} - \lambda N_{11}$ has eigenvalues with moduli less than 1, and accumulate the transformations in U, V

Solve $V_{11}^{\mathrm{T}} P = V_{21}^{\mathrm{T}}$ for P, where $V_1 = [V_{11}^{\mathrm{T}}, V_{21}^{\mathrm{T}}]^{\mathrm{T}}$ are the generalized Schur vectors corresponding to the pencil $M_{11} - \lambda N_{11}$

The orthogonal reduction of the pencil $M - \lambda N$ into generalized Schur form is done by the subroutines QZHES, QZIT and QZVAL from EISPACK. To improve the accuracy, the pencil can be preliminarily balanced by the algorithm of Ward (1981). The reordering of the generalized Schur form is carried out by the subroutines DSUBSP and EXCHQZ developed by Van Dooren (1982) (see also the remarks in Petkov, Christov and Konstantinov (1984b)). The solution of the linear matrix equation $V_{11}^{\mathrm{T}} P = V_{21}^{\mathrm{T}}$ with estimate of the conditioning is done by the corresponding routines from LINPACK.

Algorithm 6.5 requires about $180n^3$ flops. The necessary storage is proportional to $15n^2$.

It is worthwhile making some comments concerning the efficiency of Algorithm 6.5. Experience shows that there is a tendency for the QZ algorithm to yield infinite generalized eigenvalues at the upper-left corner of the generalized Schur form. Since the large eigenvalues must be moved to the right-lower corner, this tendency causes delay in the reordering procedure. There is a possibility of avoiding the movement of large eigenvalues by working with the 'inverse' generalized eigenvalue problem $Ny = \lambda My$ (the large eigenvalues of this problem corresponding to the small eigenvalues of $Mx = \lambda Nx$). In this case, however, the initial reduction of the pencil $N - \lambda M$ into Schur form may take much more time which reduces the advantage of this formulation to zero. Also, the generalized eigenvalues of the inverse problem may be more ill conditioned than the eigenvalues of the original problem.

The main advantage of the generalized Schur method over the generalized eigenvector method is that it works well in the case of multiple generalized eigenvalues.

Example 6.8

Consider a discrete Riccati equation with matrices

$$A = \begin{bmatrix} 5 & -3 & -13 \\ -1 & 0 & 2 \\ 2 & -1 & -5 \end{bmatrix} \qquad B = \begin{bmatrix} -6 \\ 2 \\ -3 \end{bmatrix}$$

$$Q = \begin{bmatrix} 4 & -6 & -12 \\ -6 & 9 & 18 \\ -12 & 18 & 36 \end{bmatrix} \qquad R = 1$$

The matrix A is singular and the eigenvalues of the pencil $M - \lambda N$ are

$$0, 0, 0, \infty, \infty, \infty$$

Using Algorithm 6.5 in arithmetic with $\varepsilon = 2^{-56} \approx 1.39 \times 10^{-17}$ we obtain (up to four digits)

$$M_{11} = \begin{bmatrix} -0.1998 & 0.7257 & 1.244 \\ 0.2667 & -0.9685 & -2.304 \\ 0 & 0 & -0.1494 \times 10^{-4} \end{bmatrix}$$

$$N_{11} = \begin{bmatrix} 0.2240 & 0.1968 & -1.328 \\ 0 & -1.348 & 0.011\,30 \\ 0 & 0 & 2.058 \end{bmatrix}$$

$$\begin{bmatrix} V_{11} \\ V_{21} \end{bmatrix} = \left[\begin{array}{ccc} 0.035\,31 & 0.8876 & -0.2165 \\ -0.168\,4 & 0.1035 & 0.6495 \\ 0.067\,89 & 0.2614 & -0.3608 \\ \hline 0.2362 & 0.1578 & 0.2887 \\ -0.5051 & 0.3106 & 0.4330 \\ -0.8092 & -0.1086 & -0.3608 \end{array} \right]$$

and

$$\bar{P} = \begin{bmatrix} 6.000 & -7.000 & -17.00 \\ -7.000 & 10.00 & 21.00 \\ -17.00 & 21.00 & 49.00 \end{bmatrix}$$

The relative error in the solution is

$$\frac{\|\bar{P} - P\|_{\mathrm{F}}}{\|P\|_{\mathrm{F}}} = 0.2943 \times 10^{-14}$$

so that approximately 15 digits are true in the norm of the solution. The condition number of the matrix V_{11} is

$$\text{cond}_F(V_{11}) = 78.40$$

The computed generalized eigenvalues of the pencil $M_{11} - \lambda N_{11}$ are

$$\bar{\lambda}_{1,2} = 0.3628 \times 10^{-5} \pm 0.6284 \times 10^{-5}i$$
$$\bar{\lambda}_3 = -0.7256 \times 10^{-5}$$

The large difference between the computed and exact generalized eigenvalues is due to their high sensitivity as a result of the multiplicity $(r = 3)$.

The analysis of the numerical properties of Algorithm 6.5 is done similarly to the analysis of the Schur method for the continuous Riccati equation presented in Section 6.4. To do this, we shall make use of Theorem 1.3.

The computed quasitriangular matrices in the generalized Schur form of the pencil $M - \lambda N$ satisfy

$$\begin{bmatrix} \bar{M}_{11} & \bar{M}_{12} \\ 0 & \bar{M}_{22} \end{bmatrix} = \bar{U}(M + E)\bar{V}$$

$$\begin{bmatrix} \bar{N}_{11} & \bar{N}_{12} \\ 0 & \bar{N}_{22} \end{bmatrix} = \bar{U}(N + F)\bar{V}$$

where $\bar{U}$ and $\bar{V}$ are the computed matrices of the orthogonal equivalent transformation

$$\|E\|_F \leq c_1\varepsilon \|M\|_F$$
$$\|F\|_N \leq c_2\varepsilon \|N\|_F$$

and c_1 and c_2 are low-degree polynomials in n.

Assuming that the conditions of Theorem 1.3 are fulfilled, we have that

$$\bar{U}_1 = U_1 + U_2 X \qquad \bar{U} = [\bar{U}_1, \bar{U}_2]$$
$$\bar{V}_1 = V_1 + V_2 Y \qquad \bar{V} = [\bar{V}_1, \bar{V}_2]$$

where

$$\|(X, Y)\|_F \leq 2\|(E, F)\|_F/\delta$$

and

$$\delta = \text{dif}(M_{11}, N_{11}; M_{22}, N_{22}) - 2\|(E, F)\|_F$$

Here

$$\|(E, F)\|_F \leq (c_1^2 \varepsilon^2 \| M \|_F^2 + c_2^2 \varepsilon^2 \| N \|_F^2)^{1/2} \leq c\varepsilon \|(M, N)\|_F$$

where $c = \max(c_1, c_2)$ and

$$\|(M, N)\|_F \leq (2 \| A \|_F^2 + \| Q \|_F^2 + \| S \|_F^2 + 2n)^{1/2}$$

Thus

$$\bar{V}_{11} = V_{11} + V_{12} Y$$
$$\bar{V}_{21} = V_{21} + V_{22} Y \qquad\qquad (6.44)$$

where

$$\| Y \|_F \leq 2 \|(E, F)\|_F / \delta$$

The solution of the discrete Riccati equation is computed from

$$P\bar{V}_{11} = \bar{V}_{21}$$

using Gaussian elimination with pivoting. Hence

$$\| \bar{P} - P \|_F \leq c_3 \, \text{cond}_F(\bar{V}_{11})\varepsilon \| \bar{P} \|_F$$

where

$$P^0 = \bar{V}_{21} \bar{V}_{11}^{-1}$$

and c_3 is a coefficent characterizing the element's growth during the elimination.

Using (6.44) we obtain that

$$P^0 = (P + V_{22} Y V_{11}^{-1})(I_n + V_{12} Y V_{11}^{-1})^{-1} \qquad\qquad (6.45)$$

The relationship (6.45) can be expressed as

$$P^0 - P = (V_{22} - PV_{12}) Y V_{11}^{-1} (I_n + V_{12} Y V_{11}^{-1})^{-1}$$
$$= (V_{22} - PV_{12}) Y \bar{V}_{11}^{-1} \qquad\qquad (6.46)$$

Taking the norm of both sides of (6.46) and using the estimates

$$\| V_{12} \|_F \leq \sqrt{n} \qquad \| V_{22} \|_F \leq \sqrt{n}$$

we obtain that

$$\| P^0 - P \|_F \leq \sqrt{n}(1 + \| P \|_F) \| Y \|_F \| \bar{V}_{11}^{-1} \|_F$$

Thus

$$\| \bar{P} - P \|_F \leq \| \bar{P} - P^0 \|_F + \| P^0 - P \|_F \leq c_3 \, \text{cond}_F(\bar{V}_{11})\varepsilon \| \bar{P} \|_F$$
$$+ \sqrt{n}(1 + \| P \|_F) \| Y \|_F \| \bar{V}_{11}^{-1} \|_F$$

The final result can be stated in the following way.

Theorem 6.4

If

$$\delta = \mathrm{dif}(M_{11}, N_{11};\ M_{22}, N_{22}) - 2c\varepsilon\,\|(M, N)\|_F > 0$$

and

$$c\varepsilon\,\|(M, N)\|_F < \tfrac{1}{2}\delta$$

then the solution of the discrete Riccati equation (6.42) computed by the generalized Schur method satisfies

$$\frac{\|\bar{P} - P\|_F}{\|P\|_F} \leq 2c\sqrt{n}\varepsilon(1 + 1/\|P\|_F)$$

$$\times (2\|A\|_F^2 + \|Q\|_F^2 + \|S\|_F^2 + 2n)^{1/2}\|\bar{V}_{11}^{-1}\|/\delta$$

$$+ c_3\,\mathrm{cond}_F(\bar{V}_{11})\varepsilon\,\|\bar{P}\|_F/\|P\|_F \tag{6.47}$$

The bound (6.47) can be used to obtain an a posteriori estimate of the relative error in the solution, similarly to the continuous-time case. The quantity

$$\mathrm{dif}(M_{11}, N_{11};\ M_{22}, N_{22}) \approx \mathrm{dif}(\bar{M}_{11}, \bar{N}_{11};\ \bar{M}_{22}, \bar{N}_{22})$$

$$= \sigma_{\min}\begin{bmatrix} I_n \otimes \bar{M}_{11} & -\bar{M}_{22}^T \otimes I_n \\ I_n \otimes \bar{N}_{11} & -\bar{N}_{22}^T \otimes I_n \end{bmatrix}$$

can be estimated efficiently by the algorithm presented by Kågström and Westin (1989).

For the discrete Riccati equation from Example 6.8 we have

$$\|\bar{V}_{11}^{-1}\|_F = 64.02$$
$$\mathrm{dif}(\bar{M}_{11}, \bar{N}_{11};\ \bar{M}_{22}, \bar{N}_{22}) = 0.029\ 79$$

so that $(c = 1, c_3 = 1)$

$$\frac{\|\bar{P} - P\|_F}{\|P\|_F} \leq 0.4404 \times 10^{-11}$$

For the same equation we have

$$\mathrm{sep}_d(A_c^T, A_c) = 0.3348 \times 10^{-2}$$

$$d_R = \frac{2\|A\|_F^2 + \|Q\|_F/\|P\|_F + \|A\|_F^2\|P\|_F\|S\|_F}{\mathrm{sep}_d(A_c^T, A_c)} = 2.231 \times 10^8$$

In the given case the relative error in the solution is less than $\varepsilon d_R = 3.096 \times 10^{-9}$ which shows that the solution by the generalized Schur method is equivalent to introducing ε – relative perturbations in the matrices A, Q and S.

The bound (6.47) indicates that the relative error in $\bar{P}$ may be large if $\| \bar{V}_{11}^{-1} \|_F$ is large, $\bar{V}_{11}$ is ill conditioned or $\mathrm{dif}(M_{11}, N_{11}; M_{22}, N_{22})$ is small.

Experience with the generalized Schur method shows that in some cases the estimate (6.47) and the actual relative error in the computed solution may be very large in comparison to the bound

$$\frac{\| P^* - P \|_F}{\| P \|_F} \leq \varepsilon d_R$$

characterizing the sensitivity of (6.42). Thus the generalized Schur method can be numerically unstable. This happens when the norms of the matrices involved in the Riccati equation differ significantly, since in such a case the equivalent relative perturbations in the matrices of smaller norm, owing to the reduction of M and N, are large.

Example 6.9

Consider a discrete Riccati equation with $n = 3$,

$$A = V A_0 V \qquad A_0 = \mathrm{diag}(0, 1, 3)$$
$$Q = \mathrm{diag}(10^k, 10^k, 10^k)$$
$$S = \mathrm{diag}(10^{-k}, 10^{-k}, 10^{-k})$$

where $V = I_3 - 2\nu\nu^T/3$, $\nu = [1, 1, 1]^T$ and $k \geq 0$ is an integer. The stable eigenvalues of the pencil $M - \lambda N$ are

$$\lambda_1 = 0 \qquad \lambda_2 = \frac{2}{3 + \sqrt{5}} \qquad \lambda_3 = \frac{6}{11 + \sqrt{85}}$$

The exact positive definite solution of (6.42) is given by $P = V P_0 V$, where

$$P_0 = \mathrm{diag}(p_1, p_2, p_3)$$
$$p_1 = 10^k$$
$$p_2 = (1 + \sqrt{5})10^k/2$$
$$p_3 = (9 + \sqrt{85})10^k/2$$

The results obtained for different k by using Algorithm 6.5 in arithmetic with $\varepsilon = 2^{-56} \approx 1.39 \times 10^{-17}$, are shown in Table 6.6.

The estimate

$$\frac{\| \bar{P} - P \|_F}{\| P \|_F} \leq \mathrm{est}$$

is obtained from (6.47) for $c\sqrt{n} = 1$ and $c_3 = 1$,

$$\mathrm{est} = 2(1 + 1/\| P \|_F)\| \bar{V}_{11}^{-1} \|_F \varepsilon \| (M, N) \|_F/\mathrm{dif} + \varepsilon \, \mathrm{cond}_F(\bar{V}_{11})$$

Table 6.6

k	$cond_F(\bar{V}_{11})$	$\|\bar{V}_{11}^{-1}\|_F$	$\|M\|_F$	$\|N\|_F$	dif	est	$\|\bar{P}-P\|_F/\|P\|_F$
0	8	9	4	4	1	2×10^{-15}	7×10^{-17}
1	11	9×10^{1}	2×10^{1}	3	2×10^{-1}	3×10^{-13}	4×10^{-15}
2	11	9×10^{2}	2×10^{2}	3	2×10^{-2}	2×10^{-10}	3×10^{-14}
3	11	9×10^{3}	2×10^{3}	3	2×10^{-3}	2×10^{-7}	8×10^{-13}
4	11	9×10^{4}	2×10^{4}	3	2×10^{-4}	2×10^{-4}	1×10^{-11}
5	11	9×10^{5}	2×10^{5}	3	2×10^{-5}	2×10^{-1}	7×10^{-8}
6	11	9×10^{6}	2×10^{6}	3	2×10^{-6}	2×10^{2}	4×10^{-6}

The results in Table 6.6 show that the increase of the solution error with the increase of k is due, according to (6.47), to the increase of $\| \bar{V}_{11}^{-1} \|_F$, $\| M \|_F$ and the decrease of $\mathrm{dif}(M_{11}, N_{11}; M_{22}, N_{22})$. This example indicates the numerical instability of the generalized Schur method since in the given case the Riccati equation is relatively well conditioned for every k,

$$\mathrm{sep}_d(A_c^T, A_c) = 1 - \lambda_2^2 = \frac{5 + 3\sqrt{5}}{7 + 3\sqrt{5}} \approx 0.8541$$

$$d_R = 212.4$$

There are two reasons for the numerical instability. The first reason is that the equivalent relative errors in the matrices A and S, due to the reduction of $M - \lambda N$ to generalized Schur form, are increasing with k. The second reason is that the sensitivity of the corresponding generalized eigenvalue problem (characterized by dif^{-1}) is increasing with k, while the sensitivity of the Riccati equation remains constant. Thus the generalized Schur method may introduce high sensitivity, which is not inherent in the Riccati equation itself and which leads to numerical instability.

We note in passing that the generalized Schur method may be used for solving the continuous Riccati equation by setting $N = I_{2n}$, $M = H$ and reordering the generalized eigenvalues as in the Schur method (Algorithm 6.4). This approach, however, is inefficient, since

$$\mathrm{sep}(T_1, T_2) \geq \mathrm{dif}(T_1, I_n; T_2, I_n)$$

which means that the generalized eigenvalue formulation of the continuous Riccati equation is more ill conditioned than the usual eigenvalue formulation.

6.5.2 Generalized Schur Methods for Avoiding Inversion of the Control Weighting Matrix

All methods for solving the matrix Riccati equations considered up to now require inversion of the control weighting matrix R. However, in some cases this matrix may be ill conditioned, so it is desirable to avoid its inversion. In what follows, we present generalized Schur methods intended for solving the continuous and discrete Riccati equations in the case of an ill-conditioned matrix R.

Consider the generalized eigenvalue problem of order $(2n + m)$,

$$M_0 x = \lambda N_0 x \tag{6.48}$$

$$M_0 = \begin{bmatrix} A & 0 & B \\ -Q & -A^T & 0 \\ 0 & B^T & R \end{bmatrix} \qquad N_0 = \begin{bmatrix} I_n & 0 & 0 \\ 0 & I_n & 0 \\ 0 & 0 & I_m \end{bmatrix}$$

associated with the continuous Riccati equation

$$A^T P + PA + Q - PBR^{-1}B^T P = 0 \tag{6.49}$$

The problem (6.48) can be reduced to a problem of order $2n$. To achieve this we construct an $(n + m) \times (n + m)$ orthogonal matrix W such that

$$W \begin{bmatrix} B \\ R \end{bmatrix} = \begin{bmatrix} 0 \\ \hat{R} \end{bmatrix}$$

The matrix W may be taken, for instance, as a product of m elementary reflections, zeroing successively the corresponding elements in the mth, $(m-1)$th, ..., 1st column of the matrix $[B^T, R^T]^T$.

Partitioning the matrix W as

$$W = \begin{bmatrix} W_{11} & W_{12} \\ W_{21} & W_{22} \end{bmatrix}$$

where W_{11} is an $n \times n$ block, we form the orthogonal matrix

$$W_0 = \begin{bmatrix} W_{11} & 0 & W_{12} \\ 0 & I_n & 0 \\ W_{21} & 0 & W_{22} \end{bmatrix}$$

The problem (6.48) is then equivalent to the problem

$$W_0 M_0 x = \lambda W_0 N_0 x$$

We have that

$$W_0 M_0 = \begin{bmatrix} M_{11} & M_{12} & 0 \\ M_{21} & M_{22} & 0 \\ M_{31} & M_{32} & M_{33} \end{bmatrix}$$

where

$$\begin{array}{llll} M_{11} = W_{11}A & M_{12} = W_{12}B^T & M_{21} = -Q & M_{22} = -A^T \\ M_{31} = W_{21}A & M_{32} = W_{22}B^T & M_{33} = \hat{R} \end{array}$$

Also,

$$W_0 N_0 = \begin{bmatrix} N_{11} & 0 & 0 \\ 0 & I_n & 0 \\ N_{31} & 0 & 0 \end{bmatrix}$$

where

$$N_{11} = W_{11} \qquad \text{and} \qquad N_{31} = W_{21}$$

Now we construct a generalized eigenvalue problem of order $2n$,

$$Mx = \lambda Nx \qquad\qquad (6.50)$$

$$M = \begin{bmatrix} M_{11} & M_{12} \\ M_{21} & M_{22} \end{bmatrix} \qquad N = \begin{bmatrix} N_{11} & 0 \\ 0 & I_n \end{bmatrix}$$

It can be shown (see Van Dooren (1981b)) that the matrices M and N are represented as

$$M = \begin{bmatrix} W_{11} & 0 \\ 0 & I_n \end{bmatrix} H \qquad N = \begin{bmatrix} W_{11} & C \\ 0 & I_n \end{bmatrix} I_{2n}$$

where H is the Hamiltonian matrix associated with (6.49),

$$H = \begin{bmatrix} A & -S \\ -Q & -A^{\mathrm{T}} \end{bmatrix} \qquad S = BR^{-1}B^{\mathrm{T}}$$

and W_{11} is nonsingular. Hence the problem (6.50) is equivalent to the problem

$$Hx = \lambda x$$

In this way we obtain the following algorithm for solving (6.49).

Construct an orthogonal matrix

$$W = \begin{bmatrix} W_{11} & W_{12} \\ W_{21} & W_{22} \end{bmatrix}$$

which compresses the rows of $[B^{\mathrm{T}}, R^{\mathrm{T}}]^{\mathrm{T}}$

Form the matrices

$$M = \begin{bmatrix} W_{11}A & W_{12}B^{\mathrm{T}} \\ -Q & -A^{\mathrm{T}} \end{bmatrix} \qquad N = \begin{bmatrix} W_{11} & 0 \\ 0 & I_n \end{bmatrix}$$

Reduce the pencil $M - \lambda N$ into generalized Schur form

$$\tilde{M} - \lambda \tilde{N} = U(M - \lambda N)V$$

by using the QZ method

Reorder the generalized Schur form implementing orthogonal equivalent transformations, so that the eigenvalues with negative real parts appear first

Solve $V_{11}^T P = V_{21}^T$ for P

The numerical properties of this algorithm are similar to the properties of Algorithm 6.5.

The same approach may be used to solve the discrete Riccati equation

$$A^T PA + Q - P - A^T PB(R + B^T PB)^{-1} B^T PA = 0 \qquad (6.51)$$

In the given case we consider the generalized eigenvalue problem of order $(2n + m)$

$$M_0 x = \lambda N_0 x \qquad (6.52)$$

$$M_0 = \begin{bmatrix} A & 0 & -B \\ -Q & I_n & 0 \\ 0 & 0 & R \end{bmatrix} \qquad N_0 = \begin{bmatrix} I_n & 0 & 0 \\ 0 & A^T & 0 \\ 0 & B^T & 0 \end{bmatrix}$$

Let us construct an orthogonal matrix W such that

$$W \begin{bmatrix} -B \\ R \end{bmatrix} = \begin{bmatrix} 0 \\ \hat{R} \end{bmatrix} \qquad W = \begin{bmatrix} W_{11} & W_{12} \\ W_{21} & W_{22} \end{bmatrix}$$

The problem (6.52) is then equivalent to the problem

$$W_0 M_0 x = \lambda W_0 N_0 x$$

where

$$W_0 = \begin{bmatrix} W_{11} & 0 & W_{12} \\ 0 & I_n & 0 \\ W_{21} & 0 & W_{22} \end{bmatrix}$$

We have that

$$W_0 M_0 = \begin{bmatrix} M_{11} & 0 & 0 \\ M_{21} & M_{22} & 0 \\ M_{31} & 0 & M_{33} \end{bmatrix}$$

where

$$M_{11} = W_{11} A \qquad M_{21} = -Q \qquad M_{22} = I_n$$
$$M_{31} = W_{21} A \qquad M_{33} = \hat{R}$$

Also

$$W_0 N_0 = \begin{bmatrix} N_{11} & N_{12} & 0 \\ 0 & N_{22} & 0 \\ N_{31} & N_{32} & 0 \end{bmatrix}$$

where

$$N_{11} = W_{11} \qquad N_{12} = W_{12}B^T \qquad N_{22} = A^T$$
$$N_{31} = W_{21} \qquad N_{32} = W_{22}B^T$$

It is possible to show that the generalized eigenvalue problem

$$Mx = \lambda Nx$$

where

$$M = \begin{bmatrix} M_{11} & 0 \\ M_{21} & M_{22} \end{bmatrix} \quad \text{and} \quad N = \begin{bmatrix} N_{11} & N_{12} \\ 0 & N_{22} \end{bmatrix}$$

is equivalent to the problem

$$\begin{bmatrix} A & 0 \\ -Q & I_n \end{bmatrix} x = \lambda \begin{bmatrix} I_n & S \\ 0 & A^T \end{bmatrix} x$$

associated with (6.51). This leads us to the following algorithm for solving the discrete Riccati equation.

Construct an orthogonal matrix

$$W = \begin{bmatrix} W_{11} & W_{12} \\ W_{21} & W_{22} \end{bmatrix}$$

which compresses the rows of $[-B^T, R^T]^T$

Form the matrices

$$M = \begin{bmatrix} W_{11}A & 0 \\ -Q & I_n \end{bmatrix} \qquad N = \begin{bmatrix} W_{11} & W_{12}B^T \\ 0 & A^T \end{bmatrix}$$

Reduce the pencil $M - \lambda N$ into generalized Schur form $\tilde{M} - \lambda \tilde{N} = U(M - \lambda N)V$ by using the QZ method
Reorder the generalized Schur form so that the eigenvalues with moduli less than 1 appear first
Solve $V_{11}^T P = V_{21}^T$ for P

The properties of this algorithm are again covered by the numerical analysis of Algorithm 6.5.

EXERCISES

6.20 Write a computer program implementing Algorithm 6.5.
6.21 Verify (6.45) and (6.46).

6.22 Write a program estimating the accuracy in the solution of the discrete Riccati equation found by Algorithm 6.5.

6.23 Write programs for solving the continuous and discrete Riccati equations without inversion of the control weighting matrices.

6.24 Perform detailed error analysis of the algorithms for solving the Riccati equation which avoid the inversion of R.

6.6 SYMPLECTIC METHODS

The methods for solving the continuous Riccati equation presented in Section 6.4 have the disadvantage that they do not fully exploit the special structure of the Hamiltonian matrix H. In these methods the QR algorithm is applied to H as to a general $2n \times 2n$ matrix, which is inefficient in both computational work and storage. The rounding errors made in the course of the QR algorithm do not respect the Hamiltonian structure of the problem, so that the computed eigenvalues may not appear in exact plus/minus pairs.

These difficulties may be overcome if, in reducing the Hamiltonian matrix to Schur form, one uses similarity transformations which are both unitary and symplectic. The corresponding methods for solving the Riccati equation are referred to as *symplectic methods*.

The most important property of the symplectic transformations is that they preserve the Hamiltonian structure. Let the $2n \times 2n$ matrix H be Hamiltonian, i.e.

$$J^{-1}HJ = -H^{\mathrm{H}}$$

where

$$J = \begin{bmatrix} 0 & I_n \\ -I_n & 0 \end{bmatrix} \qquad J^{-1} = J^{\mathrm{H}} = J^{\mathrm{T}} = -J$$

and the $2n \times 2n$ matrix Z be symplectic, i.e.

$$J^{-1}ZJ = Z^{-\mathrm{H}}$$

Then we have

$$J^{-1}(Z^{-1}HZ)J = (J^{-1}ZJ)^{-1}(J^{-1}HJ)(J^{-1}ZJ)$$
$$= -Z^{\mathrm{H}}H^{\mathrm{H}}Z^{-\mathrm{H}} = -(Z^{-1}HZ)^{\mathrm{H}}$$

If the $2n \times 2n$ matrix U is both unitary and symplectic, then it follows from $J^{-1}UJ = U^{-\mathrm{H}}$ that U has the form

$$U = \begin{bmatrix} U_1 & U_2 \\ -U_2 & U_1 \end{bmatrix}$$

where U_1 and U_2 are $n \times n$ matrices.

From $U^H U = I_{2n}$ we obtain that

$$U_1^H U_1 + U_2^H U_2 = I_n$$
$$U_1^H U_2 - U_2^H U_1 = 0$$

Note that in using the matrix U it is necessary to store only the blocks U_1 and U_2.

The unitary symplectic transformations can be implemented to zero specific elements of vectors or matrices. For this purpose it is possible to use two types of unitary symplectic matrices. The *Householder symplectic matrices* have the form

$$U(k, v) = \begin{bmatrix} I_{k-1} & & 0 & \\ & V & & \\ & & I_{k-1} & \\ 0 & & & V \end{bmatrix} \qquad 1 \leqq k \leqq n$$

where

$$V = I_k - 2vv^H/(v^H v)$$

is a unitary Householder reflection of kth order. The *Givens symplectic matrices* are defined as

$$J(k, c, s) = \begin{bmatrix} C & S \\ -S & C \end{bmatrix}$$

where

$$C = \mathrm{diag}(\underbrace{1, 1, \ldots, 1}_{k-1}, c, 1, 1, \ldots, 1)$$

$$S = \mathrm{diag}(\underbrace{0, 0, \ldots, 0}_{k-1}, s, 0, 0, \ldots, 0)$$

and $|c|^2 + |s|^2 = 1$.

The Householder symplectic matrices can be used to zero several components of a vector while leaving others unchanged. Specifically, it is possible to construct a matrix $U(k, v)$ such that for given k and n-vectors a, b

$$U(k, v) \begin{bmatrix} a \\ b \end{bmatrix} = \begin{bmatrix} x \\ y \end{bmatrix}$$

where $y_i = 0$ for $i = k + 1, \ldots, n$.

The Givens symplectic matrices can be exploited to zero single entries of a vector or a matrix. Specifically, for given k and n-vectors

a, b it is possible to determine a matrix $J(k, c, s)$ such that

$$J(k, c, s)\begin{bmatrix} a \\ b \end{bmatrix} = \begin{bmatrix} x \\ y \end{bmatrix}$$

where $y_k = 0$.

It may be shown that every unitary symplectic matrix can be represented as a product of Householder symplectic matrices and Givens symplectic matrices.

The following theorem, due to Paige and Van Loan (1981), reveals the canonical structure of a Hamiltonian matrix under the action of unitary symplectic transformations.

Theorem 6.5 (Hamiltonian–Schur decomposition)

Let H be a Hamiltonian matrix whose eigenvalues have nonzero real parts. Then there is a unitary symplectic matrix U such that

$$U^H H U = T = \begin{bmatrix} T_1 & T_2 \\ 0 & -T_1^H \end{bmatrix} \qquad (6.53)$$

where T_1 is upper triangular and T_2 is Hermitian. The matrix T is referred to as *Hamiltonian triangular form* of H.

Using the transformation

$$H = (Z + I_{2n})(Z - I_{2n})^{-1}$$

it is easy to prove the following corollary of Theorem 6.5.

Corollary (symplectic-Schur decomposition)

If Z is symplectic and has no eigenvalues of magnitude 1 then there is a unitary symplectic matrix U such that

$$U^H Z U = R = \begin{bmatrix} R_1 & R_2 \\ 0 & R_1^{-H} \end{bmatrix}$$

where R_1 is upper triangular and $R_2 R_1^H$ is Hermitian. The matrix R is called *symplectic triangular*.

The decomposition (6.53) can be used to solve the Riccati equation

$$A^T P + PA + Q - PBR^{-1}B^T P = 0 \qquad (6.54)$$

Provided the pair (A, B) is stabilizable and the pair $(C, A)(C^T C = Q)$ is detectable, the conditions of Theorem 6.5 are satisfied. If the matrix

$$U = \begin{bmatrix} U_1 & U_2 \\ -U_2 & U_1 \end{bmatrix}$$

is chosen so that the eigenvalues of T have negative real parts, then the positive semidefinite solution of (6.54) is found from

$$P = -U_2 U_1^{-1}$$

The computation of the Hamiltonian–Schur decomposition in the general case is a difficult problem. For the case of a single-input system $(\text{rank}(B) = 1)$ there is a variant of the QR algorithm, referred to as *Hamiltonian QR algorithm*, which allows us to find the Hamiltonian triangular form of H efficiently, using a sequence of unitary symplectic transformations. As in the usual QR algorithm, the volume of computations can be decreased substantially, if the original Hamiltonian matrix is preliminarily reduced to some condensed form. In the given case $(\text{rank}(B) = 1)$ the Hamiltonian matrix can be reduced to *Hamiltonian–Hessenberg form*

$$W^H H W = \begin{bmatrix} F & Y \\ 0 & -F^H \end{bmatrix}$$

where F is upper Hessenberg and W is a unitary symplectic matrix. The corresponding algorithm requires approximately $10n^3/3$ flops.

The Hamiltonian QR algorithm itself consists of several symplectic transformations, applied iteratively to the Hamiltonian–Hessenberg form of H. The computations can be arranged so that each step of the algorithm requires only $O(n^2)$ flops. Usually $2n$ iterations of the algorithm are sufficient to transform H into Hamiltonian triangular form within the machine precision. We refer the reader to Byers (1986a) where this algorithm is fully described.

The Hamiltonian QR algorithm is numerically stable. The computed similarity transformation matrix $\bar{U}$ and the final Hamiltonian triangular matrix $\bar{T}$ satisfy

$$\| H\bar{U} - \bar{U}\bar{T} \| \le c_1 \varepsilon \| H \|$$
$$\| \bar{U}^H \bar{U} - I_{2n} \| \le c_2 \varepsilon$$

where c_1 and c_2 are low-order polynomials in n. Note, however, that this algorithm, like the algorithms described in Section 6.4, may introduce in some cases large perturbations in the individual matrices A, Q and $BR^{-1}B^T$ which leads to numerical instability of the overall method for solving the Riccati equation.

EXERCISES

6.25 Give algorithms for the construction of Householder and Givens symplectic matrices which avoid overflows and destructive underflows.

6.26 Prove that if H is a Hamiltonian matrix then there exists a unitary symplectic matrix Z such that

$$Z^H H Z = \begin{bmatrix} F & Y \\ D & -F^H \end{bmatrix}$$

where F is upper Hessenberg, Y is Hermitian and D is diagonal.

6.27 Prove that every symplectic matrix can be factored into a product of a unitary symplectic matrix and a symplectic triangular matrix.

6.28 Prove that every unitary symplectic matrix

$$Q = \begin{bmatrix} Q_1 & Q_2 \\ -Q_2 & Q_1 \end{bmatrix}$$

can be represented as

$$Q = \begin{bmatrix} U & 0 \\ 0 & U \end{bmatrix} \begin{bmatrix} \Sigma & \Delta \\ -\Delta & \Sigma \end{bmatrix} \begin{bmatrix} V & 0 \\ 0 & V \end{bmatrix}^H$$

where U and V are unitary matrices

$$\Sigma = \text{diag}(\sigma_1, \sigma_2, ..., \sigma_n)$$
$$\Delta = \text{diag}(\delta_1, \delta_2, ..., \delta_n)$$
$$0 \leq \sigma_1 \leq \cdots \leq \sigma_n \leq 1$$

and $\Sigma^2 + \Delta^2 = I$.

6.29 Give an algorithm for the reduction of a Hamiltonian matrix in the case of a single-input system into Hamiltonian–Hessenberg form using unitary symplectic transformations.

6.30 Construct a unitary symplectic similarity transformation which exchanges two eigenvalues of the Hamiltonian triangular form.

6.7 COMPARISON OF THE METHODS FOR SOLVING THE RICCATI EQUATIONS

The only computational method for solving the Riccati equations which is proved to be numerically stable is Newton's method. This method, however, has the disadvantage that it requires a large volume of computations if the initial approximation is chosen far from the exact solution. The determination of the initial guess requires implementation of some algorithm for the design of a stabilizing gain matrix. The use of

Newton's method in the version of incremental iterations may produce more accurate results but it demands more computations and storage. Also, if the initial guess is unsymmetric, then the incremental method produces at each iteration an approximation to the solution which is also unsymmetric.

The matrix sign function method is more efficient than Newton's method and it does not require computation of a stabilizing initial guess. This method is numerically unstable, but it can be combined efficiently with Newton's method to produce a numerically stable way of solving the continuous Riccati equation. Experience shows that usually one or two iterations of Newton's method are sufficient to refine the solution obtained by the matrix sign function up to the maximal allowable accuracy.

One of the reliable and efficient methods for solving the continuous Riccati equation is the Schur algorithm. This algorithm is implemented using high-quality software and it requires significantly fewer computations than Newton's method. The reduction of the Hamiltonian matrix into triangular form allows us to obtain an estimation of the conditioning of the Riccati equation easily. The Schur method may be numerically unstable but this situation is easily recognized. The solution obtained by the Schur algorithm can also be refined using Newton's method.

The generalized Schur method allows us to solve the discrete Riccati equation in case of a singular or ill-conditioned state matrix. It also gives the possibility of estimating the conditioning of the equation efficiently. Similarly to the Schur algorithm, this method is not numerically stable and the result obtained should be refined by Newton's method. The other versions of the generalized Schur method presented in Section 6.5 make it possible to solve the continuous and discrete Riccati equations when the control weight matrix is ill conditioned and possess similar numerical properties.

The symplectic methods for the continuous Riccati equation are more efficient than the Schur methods but their use up to now has been restricted to particular cases. These methods are also unstable.

To summarize, the continuous Riccati equation can be solved reliably in two ways. The first one is to use the Schur (or generalized Schur) method, refining the solution by Newton's method. The second way is to employ the matrix sign function method in conjunction with Newton's method. The two ways require nearly the same work and space.

The best way to solve the discrete Riccati equation is to implement the generalized Schur algorithm followed by Newton's method.

NOTES AND REFERENCES

There is a vast amount of literature on the computational solution of the matrix Riccati equations. Surveys and comparisons of the available methods can be found in Jamshidi (1980), Byers (1983), Farrar and DiPietro (1977) and Hewer and Nazaroff (1973).

The sensitivity of the Riccati equation is studied in Byers (1983, 1985), Arnold (1983), and Kenney and Hewer (1987). Non-local perturbation bounds for the continuous and discrete Riccati equations are presented in Konstantinov, Christov and Petkov (1987).

The solution of the Riccati equation can be obtained analytically provided the equation matrices fulfil certain conditions, see Shubert (1974), Jones (1976), Repperger (1976), Incertis (1983) and Pearce (1986). The application of the analytical methods, however, is restricted.

Iterative methods for the Riccati equations are considered in Kleinman (1968), Sandell (1974), Vit (1972), Hewer (1971), Fernando and Nicholson (1978), Anderson (1978) and Hammarling (1982b). The numerical properties of Newton's method are discussed in Byers (1983).

The matrix sign function methods for solving the Riccati equations are developed by many researchers, including Roberts (1980), Beavers and Denman (1974), Bierman (1984), Gardiner and Laub, (1986), Byers (1987), Charlier and Van Dooren (1989), and Kenney, Laub and Wette (1989). Byers (1986b) has shown that in some cases these methods can be numerically unstable.

The eigensystem methods for the continuous Riccati equation are introduced by MacFarlane (1963) and Potter (1966) and for the discrete Riccati equation by Vaughan (1970). Further development of these methods are the Schur algorithms, proposed by Laub (1978, 1979). The roundoff error properties of the Schur method for the continuous Riccati equation are discussed in Petkov, Christov and Konstantinov (1987), and Kenny, Laub and Wette (1989).

The next step in the search for reliable methods for solving the Riccati equations is the development of generalized eigensystem methods. These methods are considered in Pappas, Laub and Sandell (1980), Van Dooren (1981b), Walker, Emami-Naeini and Van Dooren (1982), Arnold (1983), Laub (1983) and Arnold and Laub (1984).

The symplectic methods for the continuous Riccati equation are based on the theoretical results obtained in Paige and Van Loan (1981). Computational methods, making use of symplectic transformations, are presented in Byers (1983, 1986a), Van Loan (1982) and Bunse-Gerstner and Mehrmann (1986).

The solution of the continuous Riccati equation can be obtained as the steady-state solution of an associated differential matrix Riccati equation, see Kalman and Englar, (1966), Hitz and Anderson (1972) and Prussing (1972). This approach, however, is time-consuming and inefficient. The solution of the differential matrix Riccati equation is of independent interest in the finite-time quadratic optimization. It is considered in Davison and Maki (1973), Laub (1982), Kenney and Leipnik (1985), Oshman and Bar-Itzhack (1985) and Kunkel and Mehrmann (1989).

Other methods for solving the matrix Riccati equations are surveyed in Jamshidi (1980) and Singer and Hammarling (1983).

7

Computations in Geometric Linear Systems Theory

In this chapter we present methods for computing subspaces and other geometric objects related to linear control systems. First we consider the computation of basic subspaces associated with matrices as well as the intersection, sum, distances and angles between subspaces. We then apply these computations to the determination of controllable/unobservable subspaces and their complements. Finally we discuss in brief the computation of controlled invariant and controllability subspaces. All these subspaces play an important role in the solution of several analysis and design problems for linear systems.

7.1 BASIC SUBSPACES COMPUTATIONS

When computing subspaces and related objects they are usually represented in terms of certain matrices. In turn, sets of matrices are often associated with a number of subspaces and other geometric objects, which are of interest both from theoretical and computational points of view.

In this section we present some basic techniques for calculating and comparing subspaces on a computer, which rely on the QR and singular value decompositions of a matrix (see Sections 1.8 and 1.12).

We shall need some notation and terminology. If L and M are subspaces of $\mathbb{R}^n$, then their *sum* $S = L + M$ is the subspace consisting of all vectors $\{z: z = x + y,\ x \in L,\ y \in M\}$. S is referred to as the *direct sum* of L and M, written $S = L \oplus M$, if each $z \in S$ has a unique representation $z = x + y$, $x \in L$, $y \in M$. The dimension of the direct sum S, $\dim(S)$, is the sum of the dimensions of L and M.

The *intersection* $L \cap M$ of two subspaces $L, M \subset \mathbb{R}^n$ is the subspace defined by

$$L \cap M = \{x: x \in L,\ x \in M\}$$

If $L \cap M = \{0\}$ then the subspaces L and M are said to be *mutually independent*. In this case $L + M = L \oplus M$.

For two arbitrary subspaces L and M it is fulfilled that

$$\dim(L + M) + \dim(L \cap M) = \dim(L) + \dim(M) \qquad (7.1)$$

The set $L^{\perp}$ of all vectors $y \in \mathbb{R}^n$, such that $y^{\mathrm{T}} x = 0$ for each $x \in L$, is said to be the *orthogonal complement* of L. Obviously, $L^{\perp}$ is a subspace of $\mathbb{R}^n$. In this way $\mathbb{R}^n$ is decomposed into the direct sum: $\mathbb{R}^n = L \oplus L^{\perp}$. It may be shown that

$$
\begin{aligned}
(L^{\perp})^{\perp} &= L \\
(L + M)^{\perp} &= L^{\perp} \cap M^{\perp} \\
(L \cap M)^{\perp} &= L^{\perp} + M^{\perp}
\end{aligned}
\qquad (7.2)
$$

Consider the vector spaces $\mathbb{R}^m$ and $\mathbb{R}^n$ with the corresponding canonical bases being the columns of the unit matrices I_m and I_n. The function $\varphi: \mathbb{R}^m \to \mathbb{R}^n$ is said to be a *linear operator* (or *linear transformation* when $m = n$) if $\varphi(c_1 x + c_2 y) = c_1 \varphi(x) + c_2 \varphi(y)$ for each $x, y \in \mathbb{R}^m$ and $c_1, c_2 \in \mathbb{R}$. Thus a real $n \times m$ matrix A ($A \in \mathbb{R}^{n \times m}$) determines a linear operator φ_A which maps $\mathbb{R}^m$ in $\mathbb{R}^n$ according to the rule $\varphi_A(x) = Ax$, $x \in \mathbb{R}^m$. Conversely, for every linear operator $\varphi: \mathbb{R}^m \to \mathbb{R}^n$ there exists a unique matrix $A_\varphi \in \mathbb{R}^{n \times m}$ such that $\varphi(x) = A_\varphi x$, $x \in \mathbb{R}^m$. That is why we shall identify linear operators with their matrices (in the canonical bases).

Let $L, M \subset \mathbb{R}^n$ and $N, P \subset \mathbb{R}^m$ be subspaces, and $A \in \mathbb{R}^{n \times m}$ be a given matrix. The set of all vectors $Ax \in \mathbb{R}^n$, $x \in N$, is said to be the *image* of N (under A) and is denoted by AN. The set of all $x \in \mathbb{R}^m$ such that $Ax \in L$ is said to be the *pre-image* of L (under A) and is denoted by $A^{-1}L$. (The symbol A^{-1} used here should not be confused with the inverse of A which may not exist.) Clearly, both $AN \subset \mathbb{R}^n$ and $A^{-1}L \subset \mathbb{R}^m$ are subspaces. It may be shown that

$$
\begin{aligned}
A(N + P) &= AN + AP \\
A(N \cap P) &\subset (AN) \cap (AP) \\
A^{-1}(L + M) &\supset A^{-1}L + A^{-1}M \\
A^{-1}(L \cap M) &= A^{-1}L \cap A^{-1}M
\end{aligned}
$$

There are four important subspaces called fundamental subspaces that arise in connection with each matrix $A \in \mathbb{R}^{n \times m}$: the range (or image, or column space) $\mathrm{Rg}(A) \subset \mathbb{R}^n$ and its orthogonal complement $\mathrm{Rg}(A)^{\perp} \subset \mathbb{R}^n$ and the kernel (or null space) $\mathrm{Ker}(A) \subset \mathbb{R}^m$ and its orthogonal complement $\mathrm{Ker}(A)^{\perp}$ (see also Section 1.1). We recall that $\dim[\mathrm{Rg}(A)] = r$ and $\dim[\mathrm{Ker}(A)] = m - r$, where $r = \mathrm{rank}(A)$. It may be shown that

$$Rg(A)^\perp = Ker(A^T) = Ker(A^+)$$
$$Ker(A)^\perp = Rg(A^T) = Rg(A^+)$$

and

$$A\ Rg(A^T) = Rg(A) \qquad A^+Rg(A) = Rg(A^T)$$
$$A\ Ker(A) = 0 \qquad A^+Ker(A^T) = 0$$

where A^+ is the pseudo-inverse of A (Section 1.8).

The relations between the subspaces are illustrated in Figure 7.1.

Each subspace may be characterized in different ways. Let $L \subset \mathbb{R}^n$ be a subspace with $r = \dim(L)$, $1 \le r \le n - 1$. Then there exist two full-rank matrices $F \in \mathbb{R}^{n \times r}$ and $G \in \mathbb{R}^{(n-r) \times n}$ such that

$$L = Rg(F) = Ker(G) \tag{7.3}$$

The matrices F and G are not unique and are determined within linear transformations: the matrices $\hat{F} = FF_1$, and $\hat{G} = G_1G$, where $F_1 \in \mathbb{R}^{r \times r}$ and $G_1 \in \mathbb{R}^{(n-r) \times (n-r)}$ are nonsingular, give rise to the same subspace L.

The columns $f_i \in \mathbb{R}^n$ of the matrix $F = [f_1, f_2, ..., f_r]$ constitute a basis for the subspace L. Similarly, there are exactly r linearly independent solutions of the linear equation $Gx = 0$, $x \in \mathbb{R}^n$, which also constitute a basis for L.

From both a theoretical and a computational point of view it is important to find an *orthonormal basis* for L, i.e. a set of r pairwise

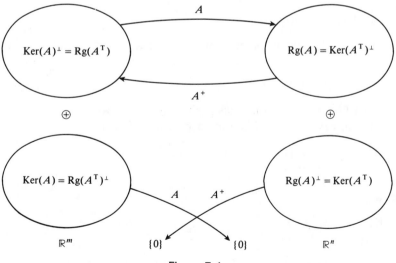

Figure 7.1

orthogonal vectors of unit length which span L. Theoretically, such a basis may be determined from the columns of the matrix

$$W = F(F^T F)^{-1/2} = [w_1, w_2, \ldots, w_r] \tag{7.4}$$

where F is any matrix for which (7.3) holds. Unfortunately, the computation of W from (7.4) may be contaminated with large errors in forming the product $F^T F$. Another way is via the classical Gramm–Schmidt procedure

$$h_1 = f_1 \qquad\qquad w_1 = h_1/\| h_1 \|_2$$

$$h_k = f_k - \sum_{j=1}^{k-1} (w_j^T f_k) w_j \qquad w_k = h_k/\| h_k \|_2 \qquad k = 2, \ldots, r$$

or its modified version (Golub and Van Loan, 1983, ch. 6), which is computationally efficient, but may produce vectors $w_1, \ldots, w_r$ which are far from orthogonal.

A reliable approach to computing an orthonormal basis of a subspace is through the QR or singular value decomposition. Consider the general case $L = \mathrm{Rg}(A)$, where the matrix $A \in \mathbb{R}^{n \times m}$, $m \leq n$ is not necessarily of full rank, i.e. $r = \dim(L) = \mathrm{rank}(A) \leq m$. Let

$$A = [Q_1, \underset{r}{\underbrace{}} \; Q_2 \; \underset{n-r}{\underbrace{}}] \begin{bmatrix} R \\ 0 \end{bmatrix} \qquad \mathrm{rank}(R) = r \tag{7.5}$$

be the QR decomposition with column pivoting of A. It follows from (7.5) that

$$L = \mathrm{Rg}(A) = \mathrm{Rg}(Q_1)$$

In addition, the QR decomposition produces an orthonormal basis for $L^\perp$

$$L^\perp = \mathrm{Rg}(A)^\perp = \mathrm{Rg}(Q_2)$$

To find an orthogonal basis for $\mathrm{Ker}(A)$ and $\mathrm{Ker}(A)^\perp$ one needs the QR decomposition with column pivoting of A^T. If

$$A^T = [\hat{Q}_1, \underset{r}{\underbrace{}} \; \hat{Q}_2 \; \underset{m-r}{\underbrace{}}] \hat{R} \qquad \mathrm{rank}(\hat{R}) = r$$

then

$$\mathrm{Ker}(A) = \mathrm{Rg}(\hat{Q}_2)$$
$$\mathrm{Ker}(A)^\perp = \mathrm{Rg}(\hat{Q}_1)$$

An orthonormal basis for all four fundamental subspaces can be found by performing only one singular value decomposition of A. If

$$A = [\underbrace{U_1,}_{r} \ \underbrace{U_2}_{n-r} \ ] \begin{bmatrix} \Sigma & 0 \\ 0 & 0 \end{bmatrix} \begin{bmatrix} V_1^T \\ V_2^T \end{bmatrix} \begin{matrix} \}r \\ \}m-r \end{matrix} \tag{7.6}$$

$$\Sigma = \text{diag}(\sigma_1, \dots, \sigma_r) \qquad \sigma_i \neq 0$$

then

$$\text{Rg}(A) = \text{Rg}(U_1)$$
$$\text{Rg}(A)^\perp = \text{Rg}(U_2)$$
$$\text{Ker}(A) = \text{Rg}(V_2)$$
$$\text{Ker}(A)^\perp = \text{Rg}(V_1)$$

The accurate determination of orthonormal bases relies on the proper determination of matrix rank. In this respect SVD is the most reliable way to compute subspaces and their combinations. However, QR decomposition with column pivoting is more efficient and is almost as reliable as SVD, which makes it preferable in practice. Note that the orthonormal bases obtained by QR decomposition are generally different from the corresponding bases obtained by SVD.

Example 7.1

If

$$A = \begin{bmatrix} -7 & 2 & 11 \\ 4 & 9 & 14 \\ -6 & 1 & 8 \\ -8 & -3 & 2 \\ 4 & -5 & -14 \end{bmatrix}$$

then the singular value decomposition, done in arithmetic with $\varepsilon \approx 1.39 \times 10^{-17}$ yields (up to three decimal digits)

$$\sigma_1 = 26.4 \qquad \sigma_2 = 13.5 \qquad \sigma_3 = 0.0$$

so that $\dim[\text{Rg}(A)] = \text{rank}(A) = 2$ and $\dim[\text{Ker}(A)] = m - \text{rank}(A) = 1$.
The four fundamental subspaces are given by

$$\text{Rg}(A) = \text{Rg}(U_1)$$

$$U_1 = \begin{bmatrix} 0.467 & 0.343 \\ 0.562 & -0.628 \\ 0.342 & 0.325 \\ 0.102 & 0.617 \\ -0.582 & -0.0322 \end{bmatrix}$$

$$\text{Rg}(A)^{\perp} = \text{Rg}(U_2)$$

$$U_2 = \begin{bmatrix} 0.638 & -0.506 & -0.0258 \\ 0.250 & 0.394 & 0.269 \\ -0.504 & -0.136 & 0.711 \\ 0.190 & 0.755 & -0.0524 \\ 0.490 & 0.0265 & 0.647 \end{bmatrix}$$

$$\text{Ker}(A) = \text{Rg}(V_2)$$

$$V_2 = \begin{bmatrix} 0.408 \\ -0.816 \\ 0.408 \end{bmatrix}$$

$$\text{Ker}(A)^{\perp} = \text{Rg}(V_1)$$

$$V_1 = \begin{bmatrix} -0.235 & -0.882 \\ 0.338 & -0.468 \\ 0.911 & -0.0538 \end{bmatrix}$$

Consider the accuracy of computed orthonormal bases. Since both QR and singular value decompositions are found by numerically stable algorithms, the computed bases are the exact bases of subspaces, corresponding to a matrix $A + \Delta A$, where $\| \Delta A \|_F$ is of order $\varepsilon \| A \|_F$. Hence the accuracy of the results obtained will depend on the sensitivity of the orthonormal bases to perturbations in the matrix A. In the case of QR decomposition, it is shown in Stewart (1977) that if $\text{rank}(A) = m$, then the changes in the matrices Q and R caused by a perturbation ΔA are bounded by $\text{cond}(A) \| \Delta A \| / \| A \|$, where $\text{cond}(A) = \| A \| \| A^+ \|$. Thus if $\text{cond}(A)$ is large, then the corresponding subspaces may be very sensitive. The same conclusion is valid for the use of SVD.

Example 7.2

If

$$A = \begin{bmatrix} 1.24 & -8.90 \\ 6.20 & -44.46 \\ -8.68 & 62.26 \\ 27.28 & -195.68 \\ -53.32 & 382.46 \end{bmatrix}$$

then using QR decomposition with column pivoting we obtain (to three decimal digits)

$$Rg(A) = Rg(Q_1)$$

$$Q_1 = \begin{bmatrix} -0.0204 & 0.409 \\ -0.102 & -0.887 \\ 0.143 & 0.0682 \\ -0.448 & 0.205 \\ 0.876 & 0.0 \end{bmatrix}$$

If the matrix A is perturbed by

$$\Delta A = 10^{-5} \begin{bmatrix} 1 & -1 \\ -2 & -4 \\ -8 & 1 \\ 4 & -7 \\ 1 & 2 \end{bmatrix}$$

then

$$Rg(A + \Delta A) = Rg(\tilde{Q}_1)$$

$$\tilde{Q}_1 = \begin{bmatrix} -0.0204 & 0.410 \\ -0.102 & -0.884 \\ 0.143 & 0.110 \\ -0.448 & 0.194 \\ 0.876 & -0.0120 \end{bmatrix}$$

so that

$$\| \tilde{Q}_1 - Q_1 \|_F = 0.0450$$

In the given case

$$cond(A) = \| A \|_F \| A^+ \|_F = 2.34 \times 10^5$$
$$cond(A) \| \Delta A \|_F / \| A \|_F = 0.0665$$

which causes a large change in the solution.

To compute the image $CL \subset \mathbb{R}^p$ of $L = Rg(A)$ under $C \in \mathbb{R}^{p \times n}$ we use the fact that

$$CL = Rg(CA) = Rg(CQ_1) = Rg(CU_1)$$

Hence an orthonormal basis for CL is formed from the first q columns ($q = rank(CA)$) for $\hat{Q}$ or $\hat{U}$, where $CQ_1 = \hat{Q}\hat{R}$ is the QR decomposition of CQ_1 and $CU_1 = \hat{U}\hat{\Sigma}\hat{V}^T$ is the SVD of CU_1. Note that to improve the accuracy we avoid the computation of CA followed by a single decomposition. Instead, we perform two decompositions in order to multiply C by a matrix with orthonormal columns, thus reducing the rounding errors (see Section 1.4).

The computation of the pre-image $A^{-1}M$, $M = \text{Rg}(B)$, is done using the equality

$$A^{-1}M = \text{Ker}(W^T A)$$

where the columns of W form a basis of the subspace $M^\perp$. Hence, to obtain an orthonormal basis of $A^{-1}M$ it is necessary to perform two decompositions: the first one of B to find an orthonormal basis W of $M^\perp$, and the second one of $W^T A$ to find a basis of $\text{Ker}(W^T A)$. To achieve this one can use QR decomposition with column pivoting or singular value decomposition.

Consider now the problem of checking the inclusion of subspaces. Given two subspaces $L, M \subset \mathbb{R}^n$ one may have to determine whether $L \supset M$, or $L = M$. Let $L = \text{Rg}(A)$ and $M = \text{Rg}(B)$, where $A \in \mathbb{R}^{n \times m}$, $B \in \mathbb{R}^{n \times p}$ (A and B may not be of full rank). Then it may be shown that $L \supset M$ if and only if $AA^+B = B$. Using the SVD (7.6) of A it is easy to show that $AA^+ = U_1 U_1^T$, i.e. the inclusion $L \supset M$ is checked by the equality $U_1 U_1^T B = B$. (Obviously, because of rounding errors, exact equality cannot be achieved.) An alternative way is to check if the number r of nonzero singular values $\sigma_1, \ldots, \sigma_r$ of A (or $\text{rank}(A)$) equals the number of nonzero singular values of $[A, B]$ (or $\text{rank}([A, B])$).

The equality $L = M$ is checked by both the inclusions $L \supset M$ and $L \subset M$.

Example 7.3

If

$$A = \begin{bmatrix} -2 & -4 \\ 3 & -7 \\ 5 & 1 \\ -1 & 6 \end{bmatrix} \quad \text{and} \quad B = \begin{bmatrix} -2 & 0 & 2 \\ -10 & -13 & -3 \\ -4 & -9 & 0 \\ 7 & 8 & 1 \end{bmatrix}$$

then (to three decimal digits)

$$A^+ = \begin{bmatrix} -0.0687 & 0.0550 & 0.139 & -0.004\,76 \\ -0.0487 & -0.0611 & 0.0288 & 0.0582 \end{bmatrix}$$

$$B^+ = \begin{bmatrix} -0.161 & -0.0701 & 0.200 & 0.111 \\ 0.0714 & 0.0311 & -0.200 & -0.0494 \\ 0.282 & -0.154 & 0.200 & -0.0247 \end{bmatrix}$$

and we obtain that $BB^+A = A$, $AA^+B \neq B$ which means $\text{Rg}(A) \subset \text{Rg}(B)$.

Consider finally the problem of computing the sum $L + M$ and the intersection $L \cap M$ of two subspaces L and M. If $L = \text{Rg}(A)$, $M = \text{Rg}(B)$, $A \in \mathbb{R}^{n \times m}$ and $B \in \mathbb{R}^{n \times p}$, then

$$L + M = \text{Rg}[A, B]$$

and an orthonormal basis for $L + M$ is found by using the QR or singular value decomposition of $[A, B]$. For instance, the QR decomposition with column pivoting yields

$$[A, B] = \underbrace{[Q_1,}_{r} \underbrace{Q_2]}_{n-r} \begin{bmatrix} R \\ 0 \end{bmatrix} \begin{matrix} \}r \\ \}n - r \end{matrix}$$

$$r = \text{rank}([A, B]) = \dim(L + M)$$

so that

$$L + M = \text{Rg}[A, B] = \text{Rg}(Q_1)$$

At the same time the columns of Q_2 span an orthonormal basis for the orthogonal complement

$$(L + M)^{\perp} = L^{\perp} \cap M^{\perp}$$

of $L + M$.

Example 7.4

Let

$$A = \begin{bmatrix} 4 & -1 & 8 \\ 3 & 6 & 1 \\ 7 & -9 & -2 \\ -5 & -4 & -6 \\ 3 & -7 & 0 \end{bmatrix} \quad \text{and} \quad B = \begin{bmatrix} -3 & 5 & 2 \\ -9 & -3 & 0 \\ 2 & 16 & -4 \\ 9 & -1 & 8 \\ 4 & 10 & -5 \end{bmatrix}$$

In this case $\dim(L + M) = \text{rank}([A, B]) = 4$ and the QR decomposition with column pivoting allows us to obtain (to three decimal digits)

$$\text{Rg}(A) + \text{Rg}(B) = \text{Rg}(Q_1)$$

$$Q_1 = \begin{bmatrix} -0.253 & -0.298 & 0.632 & 0.668 \\ 0.152 & -0.634 & 0.262 & -0.508 \\ -0.809 & -0.0804 & 0.125 & -0.435 \\ 0.0506 & 0.692 & 0.612 & -0.270 \\ -0.506 & 0.157 & -0.376 & 0.183 \end{bmatrix}$$

$$\text{Rg}(A)^{\perp} \cap \text{Rg}(B)^{\perp} = \text{Rg}(Q_2)$$
$$Q_2 = [0.0464, \quad 0.498, \quad -0.366, \quad 0.267, \quad 0.738]^{T}$$

The computation of an orthonormal basis for the intersection $L \cap M$ may be done using (7.2):

$$L \cap M = (L^{\perp} + M^{\perp})^{\perp}$$

The orthonormal bases for $L^{\perp}$ and $M^{\perp}$ are denoted by Q_2 and W_2 (obtained by QR or singular value decomposition of A and B). Then

$$L \cap M = \text{Rg}(S_2)$$

where S_2 is an orthonormal basis for the orthogonal complement of the subspace

$$L^{\perp} + M^{\perp} = \text{Rg}([Q_2, W_2])$$

This procedure requires two decompositions and may be expensive in some cases. Another way to compute $L \cap M$ uses the SVD

$$[A, B] = \hat{U}\hat{\Sigma}\hat{V}^{\text{T}} \qquad \hat{V} = [\underbrace{\hat{V}_1,}_{q} \underbrace{\hat{V}_2}_{m+p-q}] \qquad q = \text{rank}([A, B])$$

of the matrix $[A, B]$. Partitioning the matrix $\hat{V}_2$ as

$$\hat{V}_2 = \begin{bmatrix} X \\ Y \end{bmatrix} \begin{matrix} \}m \\ \}p \end{matrix}$$

and choosing

$$D = \begin{cases} AX & \text{rank}(A) < \text{rank}(B) \\ BY & \text{rank}(A) \geq \text{rank}(B) \end{cases}$$

then an orthonormal basis for

$$L \cap M = \text{Rg}(D)$$

is easily obtained by QR decomposition or SVD of the matrix D.

Example 7.5

If the subspaces L and M are spanned by the columns of the matrices A and B given in Example 7.4, then

$$L^{\perp} = \text{Rg}(A)^{\perp} = \text{Rg}(Q_2)$$

$$Q_2 = \begin{bmatrix} 0.439 & -0.235 \\ 0.569 & 0.221 \\ -0.0311 & -0.418 \\ 0.691 & -0.137 \\ 0.0698 & 0.838 \end{bmatrix}$$

$$M^\perp = \mathrm{Rg}(B)^\perp = \mathrm{Rg}(W_2)$$

$$W_2 = \begin{bmatrix} -0.635 & 0.213 \\ 0.617 & 0.355 \\ 0.329 & -0.463 \\ 0.328 & 0.190 \\ 0.0811 & 0.760 \end{bmatrix}$$

$$L \cap M = (\mathrm{Rg}(A)^\perp + \mathrm{Rg}(B)^\perp)^\perp$$
$$= \mathrm{Rg}(Q_2 + W_2)^\perp = \mathrm{Rg}(S_2)$$

$$S_2 = \begin{bmatrix} 0.173 & 0.350 \\ 0.642 & 0.112 \\ -0.246 & 0.775 \\ -0.615 & -0.320 \\ -0.344 & 0.402 \end{bmatrix}$$

Using the alternative approach based on SVD, we obtain

$$Y = \begin{bmatrix} 0.532 & 0.224 \\ -0.224 & 0.532 \\ 0.0 & 0.0 \end{bmatrix}$$

$$D = \begin{bmatrix} -2.71 & 1.99 \\ -4.12 & -3.61 \\ -2.51 & 8.96 \\ 5.01 & 1.48 \\ -0.106 & 6.22 \end{bmatrix}$$

The orthogonal basis of $L \cap M$ found by QR decomposition of D is given by

$$\begin{bmatrix} -0.169 & -0.352 \\ 0.307 & -0.575 \\ -0.763 & -0.282 \\ -0.126 & 0.682 \\ -0.529 & 0.0237 \end{bmatrix}$$

EXERCISES

7.1 Verify (7.2).

7.2 Compute the four fundamental subspaces of the matrix

$$A = \begin{bmatrix} -5 & 8 \\ 1 & 6 \\ -11 & -3 \\ 0 & 7 \\ 4 & -2 \end{bmatrix}$$

7.3 Compare the orthonormal bases of Rg(A) computed by the Gramm–Schmidt procedure and QR decomposition for the matrix A from Example 7.2.

7.4 Give an algorithm for finding an orthonormal basis for the intersection of null spaces of two matrices. *Hint*: Use (7.2).

7.5 Give an algorithm for finding an orthonormal basis of Rg(A_1) $\cap$ Rg(A_2) $\cap \cdots \cap$ Rg(A_q).

7.2 PROJECTIONS, DISTANCES AND ANGLES

In studying the geometric properties of objects in linear vector spaces, it is necessary to use quantitative measures which are generalizations of the usual notions of length and angle. These measures are closely related to the notion of projection in n-dimensional space.

7.2.1 Orthogonal Projections

If $L \subset \mathbb{R}^n$ is a subspace then each vector $x \in \mathbb{R}^n$ is represented uniquely in the form $x = y + z$ where $y \in L$ and $z \in L^\perp$. The matrix $P \in \mathbb{R}^{n \times n}$ defined by the equation $Px = y$ is referred to as the *orthogonal projection* onto L. The orthogonal projection is a linear transformation which maps each vector from $\mathbb{R}^n$ into a vector in L. If $x \in L$ then $Px = x$, if $x \in L^\perp$ then $Px = 0$.

The orthogonal projection satisfies the following conditions:

1. Rg(P) = L, Ker(P) = $L^\perp$
2. $P^2 = P$
3. $P^T = P$

It may be shown that P is a positive semidefinite matrix possessing the property $\| P \|_2 = 1$. Its eigenvalues are 0 and 1.

If P is the orthogonal projection onto L, then $P^\perp \equiv I - P$ is the orthogonal projection onto $L^\perp$ since the vectors $y = Px$ and $z = (I - P)x$ are orthogonal,

$$y^T z = x^T P(I - P)x = 0$$

Thus we have

$$x^T x = y^T y + z^T z$$

i.e.

$$\| x \|_2^2 = \| y \|_2^2 + \| z \|_2^2 \tag{7.7}$$

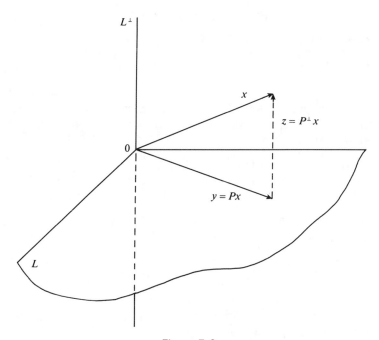

Figure 7.2

Equation (7.7) is a generalization of the Pythagorean theorem in $\mathbb{R}^n$ (Figure 7.2). It follows from (7.7) that $\| Px \|_2 \leq \| x \|_2$ for all x.

Let P_L and P_M be orthogonal projections onto the subspaces L and M. It may be shown that if $\| P_M - P_L \|_2 < 1$, then $\dim(L) = \dim(M)$. In such a case no vector in L can be orthogonal to M and vice versa. Conversely, if $\| P_M - P_L \|_2 = 1$, then there is a vector in L or M that is orthogonal to M or L. It is said that the subspaces L and M are *acute* whenever $\| P_M - P_L \|_2 < 1$. If the subspaces L and M are orthogonal, then $P_L P_M = P_M P_L = 0$.

Let the columns of $V = [v_1, \ldots, v_k]$ form an orthonormal basis for a subspace L. It may be proved that

$$P = VV^{\mathrm{T}}$$

is the unique orthogonal projection onto L. Note that any other basis VZ, where Z is orthogonal, will produce the same P.

There is a close link between the orthogonal projections onto the four fundamental subspaces of a given matrix $A \in \mathbb{R}^{n \times m}$ and its pseudo-inverse A^+. Using the properties of the pseudo-inverse (see

Exercise 1.46) it may be shown that

$$P_{\text{Rg}(A)} = AA^+$$
$$P_{\text{Ker}(A^T)} = I - AA^+$$
$$P_{\text{Rg}(A^T)} = A^+A$$
$$P_{\text{Ker}(A)} = I - A^+A$$

The computation of these projections is easily done by using the singular value decomposition of A. Assuming that rank$(A) = r$ and partitioning the SVD of A as

$$A = [U_1, \ U_2]\underbrace{}_{r \quad n-r}\begin{bmatrix} \Sigma & 0 \\ 0 & 0 \end{bmatrix}\begin{bmatrix} V_1^T \\ V_2^T \end{bmatrix}\begin{matrix} \}r \\ \}n-r \end{matrix}$$

then it is fulfilled that

$$P_{\text{Rg}(A)} = U_1 U_1^T$$
$$P_{\text{Ker}(A^T)} = U_2 U_2^T$$
$$P_{\text{Rg}(A^T)} = V_1 V_1^T$$
$$P_{\text{Ker}(A)} = V_2 V_2^T$$

The accurate computation of the projections by these formulas depends on the proper determination of the numerical rank by means of the singular values (see Section 1.12). Let

$$A = [Q_1, \ Q_2]\underbrace{}_{r \quad n-r}\begin{bmatrix} R \\ 0 \end{bmatrix}$$

be the QR decomposition with column pivoting of A. The projections $P_{\text{Rg}(A)}$ and $P_{\text{Ker}(A^T)}$ can be computed efficiently by

$$P_{\text{Rg}(A)} = Q_1 Q_1^T$$
$$P_{\text{Ker}(A^T)} = Q_2 Q_2^T$$

Similarly, $P_{\text{Rg}(A^T)}$ and $P_{\text{Ker}(A)}$ can be found by the QR decomposition of A^T.

Example 7.6

Let

$$A = \begin{bmatrix} -1 & 3 \\ 2 & -5 \\ -6 & -9 \\ 7 & 1 \end{bmatrix}$$

Using the QR decomposition of A we obtain (to three decimal digits)

$$P_{\text{Rg}(A)} = \begin{bmatrix} 0.149 & -0.259 & -0.160 & -0.185 \\ -0.259 & 0.452 & 0.256 & 0.339 \\ -0.160 & 0.256 & 0.772 & -0.291 \\ -0.185 & 0.339 & -0.291 & 0.627 \end{bmatrix}$$

$$P_{\text{Ker}(A^T)} = \begin{bmatrix} 0.851 & 0.259 & 0.160 & 0.185 \\ 0.259 & 0.548 & -0.256 & -0.339 \\ 0.160 & -0.256 & 0.228 & 0.291 \\ 0.185 & -0.339 & 0.291 & 0.373 \end{bmatrix}$$

In this case $\text{rank}(A) = 2$ and

$$P_{\text{Rg}(A^T)} = \begin{bmatrix} 1 & 0 \\ 0 & 1 \end{bmatrix}$$

$$P_{\text{Ker}(A)} = \begin{bmatrix} 0 & 0 \\ 0 & 0 \end{bmatrix}$$

If

$$x = \begin{bmatrix} -2 \\ 8 \\ 4 \\ -6 \end{bmatrix}$$

then the projections of x onto the column space $\text{Rg}(A)$ of A and its orthogonal complement are given by

$$y = P_{\text{Rg}(A)}x = \begin{bmatrix} -1.91 \\ 3.12 \\ 7.20 \\ -1.85 \end{bmatrix} \qquad z = P_{\text{Ker}(A^T)}x = \begin{bmatrix} -0.0944 \\ 4.88 \\ -3.20 \\ -4.15 \end{bmatrix}$$

and

$$\| x \|_2^2 = \| y \|_2^2 + \| z \|_2^2 = 120$$

7.2.2 Distances between Vectors and Subspaces

Let x and y be arbitrary vectors (points) in a linear space. By analogy with the notion of distance between points in $\mathbb{R}^2$, the distance $d(x, y)$ between x and y is defined by

$$d(x, y) = \| x - y \|_2 \tag{7.8}$$

The distance from x to the origin 0 is simply $\| x \|_2$ — the length of the vector x.

The distance between two vectors in a normed space possesses the properties of a *metric*:

1. $d(x, y) > 0 \quad (x \neq y), \; d(x, x) = 0$ (positivity)
2. $d(y, x) = d(x, y)$ (symmetry)
3. $d(x, y) \leq d(x, z) + d(z, y)$ (triangle inequality)

Note that the distance function (7.8) has also the additional properties:

4. $d(x + z, y + z) = d(x, y)$ (translation invariance)
5. $d(cx, cy) = |c| \, d(x, y)$ (absolute homogeneity)

Let L be a subspace of $\mathbb{R}^n$. The *distance from the vector* x *to the subspace* L is the quantity

$$d(x, L) = \min_{y \in L} \| x - y \|_2$$

The vector $y^* \in L$ is referred to as the *best approximation* to the vector x by the vectors in L if

$$\| x - y^* \|_2 = d(x, L)$$

Expanding the vector x as $x = x_L + x_L^\perp$ where x_L and $x_L^\perp$ are the projections of x onto L and $L^\perp$, respectively, we have that $x_L - y$ and $x_L^\perp$ are orthogonal (Figure 7.3). Hence by the Pythagorean theorem

$$\| x - y \|_2^2 = \| x_L - y \|_2^2 + \| x_L^\perp \|_2^2 \tag{7.9}$$

Expression (7.9) means that

$$\| x - y \|_2^2 \geq \| x_L^\perp \|_2^2$$

the equality holding only when $y = y^* = x_L$. Thus

$$d(x, L) = \| x - x_L \|_2 = \| x_L^\perp \|_2 = \| P_L^\perp x \|_2 = \| (I - P_L)x \|_2$$

In this way the problem of finding the distance between a vector and a subspace reduces to the computation of the orthogonal projection P_L or $P_L^\perp$. Clearly, x_L is the best approximation of x by vectors in L.

The determination of the distance between a vector and a subspace is related to the linear least squares problem $Ax = b$. The vector Ax^*, where $x^* = A^+b$ is the solution which minimizes $\| r \|_2 = \| b - Ax \|_2$, is the best approximation of the vector b by the vectors in the range of A (Figure 7.4). The minimum norm of the residual vector

$$\rho = \| r \|_2 = \| b - AA^+b \|_2 = \| (I - P_{\mathrm{Rg}(A)})b \|_2$$
$$= \| P_{\mathrm{Rg}(A)}^\perp b \|_2 = \| P_{\mathrm{Ker}(A^\mathsf{T})}b \|_2$$

is the distance between the vector b and the subspace $\mathrm{Rg}(A)$, spanned by the columns of A.

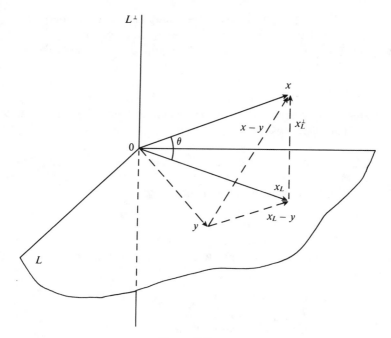

Figure 7.3

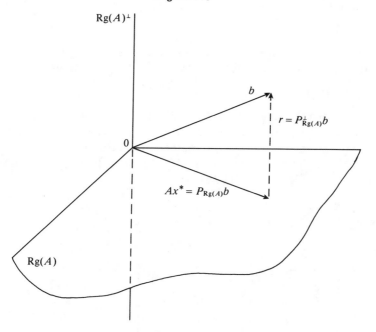

Figure 7.4

Consider now a measure of the nearness of two subspaces bearing in mind that these subspaces have at least one common vector (the zero vector). For a nontrivial subspace L denote by S_L the unit sphere

$$S_L = \{x \in L: \| x \|_2 = 1\}$$

If $L = 0$ set $S_L = 0$. Let max $d(x, L)$, $x \in S_M$, be the maximum distance between x and L as x varies on S_M and define analogically max $d(x, M)$, $x \in S_L$. The *gap* between the subspaces L and M is the quantity

$$\gamma(L, M) = \max \left\{ \max_{x \in S_M} d(x, L), \max_{x \in S_L} d(x, M) \right\}$$

The gap function satisfies the inequality

$$\gamma(L, M) \leq 1$$

since $\| x \|_2 = 1$ implies $d(x, L)$, $d(x, M) \leq 1$.

In a linear space the gap function is a metric since it possesses the properties:

1. $\gamma(L, M) > 0 \ (L \neq M) \qquad \gamma(L, L) = 0$
2. $\gamma(M, L) = \gamma(L, M)$ \hfill (7.10)
3. $\gamma(L, M) \leq \gamma(L, N) + \gamma(N, M)$

An important result concerning the gap function is that it satisfies

$$\gamma(L, M) = \| P_M - P_L \|_2 \tag{7.11}$$

where P_L and P_M are the orthogonal projections onto L and M. An obvious corollary of this result is the self-duality of the gap

$$\gamma(L^\perp, M^\perp) = \gamma(L, M)$$

To have the condition $\gamma(L, M) = 1$ satisfied in $\mathbb{R}^n$, it is necessary and sufficient for at least one of the subspaces L or M to contain a nonzero vector, orthogonal to the other subspace, i.e. $L \cap M^\perp \neq 0$ or $L^\perp \cap M \neq 0$. That is why if $\gamma(L, M) < 1$ then $\dim(L) = \dim(M)$. If $\dim(L) \neq \dim(M)$ then $\gamma(L, M) = 1$.

Example 7.7

If

$$A = \begin{bmatrix} 4 & -2 & 7 \\ -3 & -5 & 1 \\ -6 & 2 & -9 \\ -1 & 8 & -4 \\ 0 & -7 & 3 \end{bmatrix} \quad \text{and} \quad B = \begin{bmatrix} -6 & 1 & 5 \\ 4 & -3 & -2 \\ 11 & -7 & -4 \\ -3 & -6 & 9 \\ -8 & 6 & 2 \end{bmatrix}$$

then (to three decimal digits)

$$P_{\mathrm{Rg}(A)} = \begin{bmatrix} 0.405 & 0.134 & -0.470 & 0.0401 & -0.0146 \\ 0.134 & 0.944 & 0.103 & 0.0858 & 0.129 \\ -0.470 & 0.103 & 0.628 & 0.0432 & 0.003\,70 \\ 0.0401 & 0.0858 & 0.0432 & 0.644 & -0.467 \\ -0.0146 & 0.129 & 0.003\,70 & -0.467 & 0.379 \end{bmatrix}$$

$$P_{\mathrm{Rg}(B)} = \begin{bmatrix} 0.214 & 0.0 & -0.240 & 0.297 & 0.148 \\ 0.0 & 1.0 & 0.0 & 0.0 & 0.0 \\ -0.240 & 0.0 & 0.564 & 0.0359 & -0.432 \\ 0.297 & 0.0 & 0.0359 & 0.879 & -0.128 \\ 0.148 & 0.0 & -0.432 & -0.128 & 0.343 \end{bmatrix}$$

and the gap between the ranges of A and B is

$$\gamma[\mathrm{Rg}(A), \mathrm{Rg}(B)] = \| P_{\mathrm{Rg}(B)} - P_{\mathrm{Rg}(A)} \|_2 = 0.668$$

7.2.3 Angles between Vectors and Subspaces

Let x and y be vectors in $\mathbb{R}^2$. The cosine of the angle between these vectors (see Figure 7.5) is given by

$$\cos \theta = \cos(\alpha - \beta) = \cos \alpha \cos \beta + \sin \alpha \sin \beta$$

$$= \frac{x_1 y_1 + x_2 y_2}{\sqrt{(x_1^2 + x_2^2)}\sqrt{(y_1^2 + y_2^2)}} = \frac{x^{\mathrm{T}} y}{\| x \|_2 \| y \|_2}$$

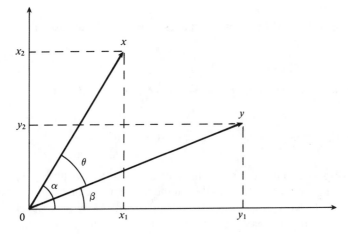

Figure 7.5

By analogy, the angle between two vectors x and y in $\mathbb{R}^n$ is defined as

$$\cos \theta = \frac{x^T y}{\| x \|_2 \| y \|_2} \qquad 0 \le \theta \le \pi$$

If x and y are orthogonal, then $\theta = \arccos(0) = \pi/2$ so that x and y lie at right angles.

Let L be a subspace of $\mathbb{R}^n$ and x be a given vector. The smallest angle between x and L is defined by (see Figure 7.3)

$$\cos \theta = \max_{y \in L} \frac{x^T y}{\| x \|_2 \| y \|_2} \qquad 0 \le \theta \le \pi/2$$

Obviously, the maximum is attained for $y = x_L = P_L x$. Thus

$$\cos \theta = \frac{x^T P_L x}{\| x \|_2 \| P_L x \|_2} = \frac{\| P_L x \|_2}{\| x \|_2}$$

and

$$\sin \theta = \frac{\| P_L^\perp x \|_2}{\| x \|_2}$$

Example 7.8

The angle between the vector x and the range of the matrix A (Example 7.6) is given by

$$\theta = \arccos \frac{\| P_{\mathrm{Rg}(A)} x \|_2}{\| x \|_2} = \arccos 0.756 = 0.713 \ (= 40.8 \text{ grad})$$

Let L and M be given subspaces of $\mathbb{R}^n$. Without loss of generality we shall assume that

$$n > p = \dim(L) \ge \dim(M) = q \ge 1$$

The smallest angle $\theta_1(L, M) \equiv \theta_1 \in [0, \pi/2]$ between L and M is defined by

$$\cos \theta_1 = \max_{u \in L} \max_{v \in M} u^T v \qquad \| u \|_2 = 1 \qquad \| v \|_2 = 1$$

Suppose that the maximum is attained for $u = u_1$ and $v = v_1$. Then $\theta_2(L, M)$ is defined as the smallest angle between the orthogonal complement of L with respect to u_1 and that of M with respect to v_1. Continuing in this way we are led to the following definition.

The *principal angles* $\theta_1, \ldots, \theta_q \in [0, \pi/2]$ between L and M are defined recursively by

$$\cos \theta_k = \max_{u \in M} \max_{v \in L} u^T v = u_k^T v_k$$

subject to the constraints

$$\| u \|_2 = 1 \qquad \| v \|_2 = 1$$
$$u^T u_i = 0 \qquad v^T v_i = 0 \qquad i = 1, \ldots, k - 1$$

Note that

$$0 \leq \theta_1 \leq \theta_2 \leq \cdots \leq \theta_q \leq \pi/2$$

The vectors $\{u_1, \ldots, u_q\}$ and $\{v_1, \ldots, v_q\}$ are called *principal vectors* of the subspace pair (L, M).

It should be pointed out that the principal vectors are not uniquely defined, but the principal angles always are. The vectors $V = [v_1, \ldots, v_q]$ form an orthonormal basis for M and the vectors $U = [u_1, \ldots, u_q]$ can be complemented with $(p - q)$ orthonormal vectors so that $u_1, \ldots, u_p$ form an orthonormal basis for L.

Suppose that the spaces L and M are defined as the ranges of given matrices $A \in \mathbb{R}^{n \times p}$ and $B \in \mathbb{R}^{n \times q}$. If the elements of A and B are subject to perturbations, then the principal angles between $\mathrm{Rg}(A)$ and $\mathrm{Rg}(B)$ change. Assume that A and B are perturbed by ΔA and ΔB, respectively, where

$$\| \Delta A \|_2 / \| A \|_2 \leq \delta A \qquad \text{and} \qquad \| \Delta B \|_2 / \| B \|_2 \leq \delta B$$

The perturbation analysis, done in Björck and Golub (1973) and Sun (1987) shows that the perturbations $\Delta \theta_k$ of the principal angles satisfy

$$| \Delta \theta_k | \leq \frac{\pi}{2} [\mathrm{cond}(A) \delta A + \mathrm{cond}(B) \delta B] \tag{7.12}$$

where

$$\mathrm{cond}(A) = \frac{\sigma_1(A)}{\sigma_p(A)} \qquad \text{and} \qquad \mathrm{cond}(B) = \frac{\sigma_1(B)}{\sigma_q(B)}$$

Thus if both $\mathrm{cond}(A)$ and $\mathrm{cond}(B)$ are small, then the angles θ_k are well determined.

There is an important decomposition of orthogonal matrices which is related to the problem for determining the angles between subspaces.

Theorem 7.1 CS decomposition

Let the orthogonal matrix $Q \in \mathbb{R}^{n \times n}$ be partitioned in the form

$$Q = \begin{bmatrix} Q_{11} & Q_{12} \\ Q_{21} & Q_{22} \end{bmatrix} \begin{matrix} \}r \\ \}n-r \end{matrix}$$
$$\underbrace{\phantom{Q_{11}}}_{r} \quad \underbrace{\phantom{Q_{12}}}_{n-r}$$

with $r \leq n/2$. Then there exist orthogonal matrices

$$U = \operatorname{diag}(\underbrace{U_1,}_{r} \underbrace{U_2}_{n-r}) \text{ and } V = \operatorname{diag}(\underbrace{V_1,}_{r} \underbrace{V_2}_{n-r})$$

such that

$$U^{\mathrm{T}} Q V = \begin{bmatrix} I & 0 & 0 \\ 0 & C & S \\ 0 & -S & C \end{bmatrix} \begin{matrix} \}n-2r \\ \}r \\ \}r \end{matrix}$$
$$\underbrace{}_{n-2r} \quad \underbrace{}_{r} \quad \underbrace{}_{r}$$

where

$$C = \operatorname{diag}(c_1, c_2, \ldots, c_r) \qquad c_i = \cos \theta_i$$
$$S = \operatorname{diag}(s_1, s_2, \ldots, s_r) \qquad s_i = \sin \theta_i$$

and

$$0 \leq \theta_1 \leq \theta_2 \leq \cdots \leq \theta_r \leq \frac{\pi}{2}$$

The proof of this theorem can be found in Stewart (1977). Algorithm for computing the CS decomposition is presented in Van Loan (1985b).

Theorem 7.1 can be used to establish the following result concerning the principal angles and vectors (Björck and Golub, 1973).

Theorem 7.2

Let the columns of Q_A and Q_B form orthonormal bases for two subspaces L and M ($\dim(L) = p \geq \dim(M) = q$) of $\mathbb{R}^n$. Put

$$Q = Q_A^{\mathrm{T}} Q_B$$

and let the singular value decomposition of this $p \times q$ matrix be

$$Q = Y \Sigma Z^{\mathrm{T}} \qquad \Sigma = \operatorname{diag}(\sigma_1, \ldots, \sigma_q)$$

where $Y^{\mathrm{T}} Y = Z Z^{\mathrm{T}} = I_q$. If the singular values are ordered as

$\sigma_1 \geqq \sigma_2 \geqq \cdots \geqq \sigma_q$, then the principal angles and principal vectors associated with this pair of subspaces are given by

$$\cos \theta_k = \sigma_k(Q) \qquad k = 1, \ldots, q$$
$$[u_1, u_2, \ldots, u_p] = Q_A Y \qquad [v_1, v_2, \ldots, v_q] = Q_B Z$$

There is a close link between the orthogonal projections onto two subspaces and the principal angles between these subspaces, which is seen by the following theorem.

Theorem 7.3

Let $L, M \subset \mathbb{R}^n$ be subspaces with $\dim(L) = \dim(M) = l, 2l \leqq n$. (If $2l > n$ we can work with $L^\perp$ and $M^\perp$.) Let P_L and P_M be the orthogonal projections onto L and M. Then the nonzero eigenvalues of $P_M - P_L$ are $\pm \sin \theta_i$, $i = 1, 2, \ldots, l$.

The proof can be found in Stewart (1973a).

Using the fact that the 2-norm of a symmetric matrix is equal to its eigenvalue of largest modulus, we obtain the following corollary.

Corollary

Let $0 \leqq \theta_1 \leqq \theta_2 \leqq \cdots \leqq \theta_l$. Then $\gamma(L, M) = \| P_M - P_L \|_2 = \sin \theta_l$.   (7.13)

Equation (7.13) shows that the gap between two equidimensional subspaces is just the sine of the largest principal angle between them. We have also by Theorem 7.2 that if Q_A and Q_B form orthonormal bases for L and M then

$$\gamma(L, M) = \sqrt{[1 - \sigma_{\min}^2(Q_A^T Q_B)]}$$

Consider now the computation of the principal angles and principal vectors. Having the orthonormal bases Q_A and Q_B of L and M, the most efficient way is to use Theorem 7.2. If L and M are the ranges of the matrices A and B, then Q_A and Q_B can be determined by QR decomposition of A and B. Theorem 7.2, however, is not well suited for computation of small principal angles, since in such cases θ_k is not well determined from $\cos \theta_k$. To circumvent this difficulty, instead of the subspace $\text{Rg}(A)$, we can use $\text{Rg}(A)^\perp$ obtaining as a result $\cos(\pi/2 - \theta_k) = \sin \theta_k$.

In this way we have the following algorithm which, for given $A \in \mathbb{R}^{n \times p}$ and $B \in \mathbb{R}^{n \times q}$ $(p \geqq q)$ with linearly independent columns, computes $\{\sin \theta_k, v_k\}_{k=1}^q$ and $\{w_k\}_{k=1}^{n-p}$, where the θ_k are the principal angles of the subspace pair $\{\mathrm{Rg}(A), \mathrm{Rg}(B)\}$, v_k are the principal vectors in $\mathrm{Rg}(B)$, and w_k are the principal vectors in the orthogonal complement $\mathrm{Rg}(A)^\perp$ of $\mathrm{Rg}(A)$.

Algorithm 7.1 Computation of the principal angles and vectors between subspaces

Use the QR decomposition to compute

$$A = [Q_A, \ Q_A^\perp] \begin{bmatrix} R_A \\ 0 \end{bmatrix} \qquad \begin{aligned} Q_A^T Q_A &= I_p \\ (Q_A^\perp)^T Q_A^\perp &= I_{n-p} \\ R_A &\in \mathbb{R}^{p \times p} \end{aligned}$$

$$B = Q_B R_B \qquad \begin{aligned} Q_B^T Q_B &= I_q \\ R_B &\in \mathbb{R}^{q \times q} \end{aligned}$$

Set

$$Q = (Q_A^\perp)^T Q_B$$

Compute the SVD

$$Y^T Q Z = \mathrm{diag}(\sin \theta_q, \ldots, \sin \theta_{l+1})$$
$$l = \max\{0, p + q - n\}$$

If $n - p < q$ set $\theta_l = \cdots = \theta_1 = 0$

$$[v_1, \ldots, v_q] = Q_B Z$$
$$[w_1, \ldots, w_{n-p}] = Q_A^\perp Y$$

The rounding error analysis, done in Björck and Golub (1973) reveals that the quantities $\bar{\theta}_k$ computed by Algorithm 7.1 satisfy

$$|\bar{\theta}_k - \theta_k| \leq c[p \ \mathrm{cond}(A) + q \ \mathrm{cond}(B)]\varepsilon$$

for some small constant c, i.e. the small principal angles will be as accurately computed as allowed by (7.12).

If the matrices A and/or B do not have full column rank, then the problem of computing principal angles and vectors is not well posed, since arbitrary small perturbations in A and B will change the rank of A and/or B. In this case, the orthonormal bases of the ranges of A and B and their orthogonal complements should be determined by QR decomposition with column pivoting or by singular value decomposition.

Example 7.9

For the subspaces $\mathrm{Rg}(A)$ and $\mathrm{Rg}(B)$ with matrices A and B from Example 7.7 we have (to three decimal digits)

$$Q_A = \begin{bmatrix} -0.508 & -0.0970 & 0.371 \\ 0.381 & -0.470 & 0.760 \\ 0.762 & 0.0620 & -0.208 \\ 0.127 & 0.651 & 0.452 \\ 0.0 & -0.585 & -0.192 \end{bmatrix}$$

and

$$Q_B = \begin{bmatrix} -0.383 & -0.249 & 0.0710 \\ 0.255 & -0.118 & 0.960 \\ 0.701 & -0.172 & -0.207 \\ -0.191 & -0.916 & -0.0616 \\ -0.510 & 0.236 & 0.164 \end{bmatrix}$$

The singular values of the matrix $Q = Q_A^T Q_B$ are

$$\cos \theta_1 = 1.0 \qquad \cos \theta_2 = 0.944 \qquad \cos \theta_3 = 0.745$$

so that

$$\theta_1 = 0 \qquad \theta_2 = 0.335 \text{ rad} \qquad \theta_3 = 0.731 \text{ rad}$$

The presence of a zero principal angle is due to the fact that

$$\dim[\mathrm{Rg}(A)] + \dim[\mathrm{Rg}(B)] = 6 > n = 5$$

which according to (7.1) implies $\dim[\mathrm{Rg}(A) \cap \mathrm{Rg}(B)] = 1$.

Using Algorithm 7.1 one obtains

$$\sin \theta_3 = 0.668 \qquad \sin \theta_2 = 0.329$$

i.e.

$$\theta_3 = 0.731 \qquad \theta_2 = 0.335$$

In this case $p = q = 3$, $l = 1$ and hence $\theta_1 = 0$.

The principal vectors in $\mathrm{Rg}(A)$ and $\mathrm{Rg}(B)$ are the columns of

$$U = \begin{bmatrix} 0.0928 & 0.358 & -0.518 \\ 0.841 & 0.445 & 0.199 \\ 0.169 & -0.386 & 0.671 \\ 0.472 & -0.537 & -0.364 \\ -0.182 & 0.486 & 0.331 \end{bmatrix}$$

and

$$V = \begin{bmatrix} 0.0928 & 0.0864 & -0.444 \\ 0.841 & 0.471 & 0.267 \\ 0.169 & -0.565 & 0.466 \\ 0.472 & -0.468 & -0.661 \\ -0.182 & 0.482 & -0.277 \end{bmatrix}$$

Notice that U and V have the same first column which is the basis of $\text{Rg}(A) \cap \text{Rg}(B)$.

In the given case the principal angles are well determined since

$$\text{cond}(A) = 9.67 \qquad \text{cond}(B) = 34.9$$

EXERCISES

7.6   Prove that in the linear least squares problem $Ax \cong b$ the minimum norm of the residual vector satisfies

$$\| r \|_2 = \| b \|_2 \sin \theta$$

where θ is the angle between b and $\text{Rg}(A)$.

7.7   Verify (7.10).

7.8   Show that the individual condition number $\text{cond}(\lambda_j)$ of an eigenvalue λ_j (Section 1.9) is inversely proportional to the cosine of the angle between the left and right eigenvectors corresponding to λ_j.

7.9   Give a bound on the angles of rotation of an invariant subspace of a matrix, due to a perturbation. Prove that in terms of Theorem 1.2 it is fulfilled that

$$\gamma [\text{Rg}(\hat{U}_1), \text{Rg}(U_1)] \leq \frac{2}{\delta} \| E \|_2$$

7.10 Determine the angles between $\text{Rg}(A)$ and $\text{Rg}(A + \Delta A)$ for the matrices A and ΔA given in Example 7.2. Comment on the results.

7.11 Give an algorithm for finding the angles between the ranges of matrices with linearly dependent columns.

7.3 COMPUTING BASIC SUBSPACES OF LINEAR MULTIVARIABLE SYSTEMS

For a linear multivariable system

$$\begin{aligned} \dot{x}(t) &= Ax(t) + Bu(t) \qquad x(0) = x_0 \\ y(t) &= Cx(t) \end{aligned} \qquad (7.14)$$

where $x(t) \in \mathbb{R}^n$, $u(t) \in \mathbb{R}^m$, $y(t) \in \mathbb{R}^r$ and $A \in \mathbb{R}^{n \times n}$, $B \in \mathbb{R}^{n \times m}$, $C \in \mathbb{R}^{r \times n}$, there are two subspaces of $\mathbb{R}^n$ which are of special interest.

The first important subspace is the *controllable subspace* defined as

$$R = \text{Rg}[B, AB, ..., A^{n-1}B] = \sum_{j=1}^{n-1} A^j \text{Rg}(B) \tag{7.15}$$

For every vector $x_T \in R$ and $T > 0$ there exists a continuous control vector $u: [0, T] \to \mathbb{R}^m$ such that the state vector $x(t)$ of (7.14) with $x_0 \in R$ satisfies $x(T) = x_T$, i.e. each $x_T \in R$ may be reached starting from each initial state $x_0 \in R$.

The controllable subspace R is A-invariant in the sense that $Ax \in R$ for each vector $x \in R$. Moreover, it may be shown that R is the smallest A-invariant subspace of $\mathbb{R}^n$ which contains $\text{Rg}(B)$. The orthogonal complement $R^\perp$ of R is said to be the *uncontrollable subspace* of (7.14).

The second important subspace associated with (7.14) is the *unobservable subspace* defined as

$$N = \text{Ker} \begin{bmatrix} C \\ CA \\ \vdots \\ CA^{n-1} \end{bmatrix} = \bigcap_{j=0}^{n-1} \text{Ker}(CA^j) \tag{7.16}$$

The unobservable subspace is also A-invariant and it is the largest A-invariant subspace contained in $\text{Ker}(C)$. The orthogonal complement $N^\perp$ of N is referred to as the *observable subspace* of (7.14). The information about the output $y: [0, T] \to \mathbb{R}^r$ allows us to determine only the observable part $x'(0)$ of the initial state $x(0)$, where $x(0) = x'(0) + x''(0)$ is a decomposition of $x(0)$ such that $x'(0) \in N^\perp$, $x''(0) \in N$. In other words, if $x(0) = x''(0) \in N$ then $y(t) = 0$, $t \geq 0$.

Using the concepts of controllable and observable subspaces, the state space $\mathbb{R}^n$ can be decomposed as a direct sum

$$R^n = L_1 \oplus L_2 \oplus L_3 \oplus L_4 \tag{7.17}$$

where the subspaces L_1, L_2 and L_3 are determined from

$L_1 = R \cap N$
$L_1 \oplus L_2 = R$
$L_1 \oplus L_3 = N$

and L_4 is chosen to satisfy (7.17).

Obviously, L_1 is the subspace of states which are controllable and unobservable. Each state $x \in L_2$ is controllable, but only its projection x' on $N^\perp$ is observable, while each state in L_3 is both uncontrollable and unobservable. Finally, each state $x \in L_4$ is uncontrollable but its projection x' on $N^\perp$ is observable.

The decomposition (7.17), known as the *Kalman decomposition* (see Kalman (1963)), is not unique. If, however, the subspaces L_i are chosen from

$$\left. \begin{aligned} L_1 &= R \cap N & L_2 &= L_1^\perp \cap R \\ L_3 &= L_1^\perp \cap N & L_4 &= (L_1 \oplus L_2 \oplus L_3)^\perp \\ & & &= (R + N)^\perp \\ & & &= L_1^\perp \cap L_2^\perp \cap L_3^\perp \end{aligned} \right\} \quad (7.18)$$

(i.e. L_2 is the orthogonal complement of L_1 in R, L_3 is the orthogonal complement of L_1 in N and L_4 is the orthogonal complement of $R + N$) then this decomposition is unique. Further on we shall assume that the subspaces L_i are defined by (7.18). In this case L_1 is orthogonal to L_2, L_3 and L_4, while L_4 is orthogonal to L_1, L_2 and L_3. Note that for a particular system some of these subspaces may be empty.

According to (7.17) and (7.18) there exists a nonsingular matrix $S \in \mathbb{R}^{n \times n}$ such that the transformation $x = S\tilde{x}$ reduces the system (7.14) into the form

$$\begin{aligned} \dot{\tilde{x}} &= \tilde{A}\tilde{x} + \tilde{B}u \\ y &= \tilde{C}\tilde{x} \end{aligned} \quad (7.19)$$

where

$$\tilde{A} = S^{-1}AS = \begin{bmatrix} A_{11} & A_{12} & A_{13} & A_{14} \\ 0 & A_{22} & 0 & A_{24} \\ 0 & 0 & A_{33} & A_{34} \\ 0 & 0 & 0 & A_{44} \end{bmatrix} \quad \tilde{B} = S^{-1}B = \begin{bmatrix} B_1 \\ B_2 \\ 0 \\ 0 \end{bmatrix} \quad (7.20)$$

$$\tilde{C} = CS = [0, \ C_2, \ 0, \ C_4]$$

and

$$A_{ij} \in \mathbb{R}^{n_i \times n_j} \qquad B_i \in \mathbb{R}^{n_i \times m} \qquad C_j \in \mathbb{R}^{r \times n_j}$$
$$n_i = \dim(L_i)$$

(see also Theorem 2.11).

The matrices A_{ij}, B_i and C_j in (7.20) are such that the pair

$$\left(\begin{bmatrix} A_{11} & A_{12} \\ 0 & A_{22} \end{bmatrix}, \begin{bmatrix} B_1 \\ B_2 \end{bmatrix} \right)$$

is controllable and the pair

$$\left(\begin{bmatrix} A_{22} & A_{24} \\ 0 & A_{44} \end{bmatrix}, [C_2, C_4] \right)$$

is observable.

Any matrix S that realizes the Kalman decomposition is partitioned as

$$S = [S_1, S_2, S_3, S_4]$$

where the matrices $S_i \in \mathbb{R}^{n \times n_i}$ form bases for L_i, respectively.

Consider the problem of computing the subspaces R, $R^\perp$, N, $N^\perp$ and L_i defined by (7.15)–(7.18). The computation of R, according to (7.15), as the span of the columns of the controllability matrix is not satisfactory from a numerical point of view because of the large rounding errors associated with the calculation of the powers A^j of A. (The numerical difficulties related to the use of the controllability matrix were discussed in Section 4.3.)

If the matrix A has distinct eigenvalues, then R is the subspace spanned by the right eigenvectors corresponding to controllable modes. This approach for determining R, however, is not to be recommended since the eigenvectors of A may be ill conditioned, while the problem of computing R may be not ill conditioned itself.

The above considerations are valid also for the computation of N via (7.16) as the kernel of the observability matrix.

As was shown in Section 4.4, the pair (A, B) can be reduced into the orthogonal canonical form

$$(\tilde{A}, \tilde{B}) = (U^T A U, U^T B)$$

where $U \in \mathbb{R}^{n \times n}$ is an orthogonal matrix. If the pair (A, B) is not controllable, i.e. $n_c = \dim(R) < n$, then the matrices $\tilde{A}, \tilde{B}$ have the form

$$\tilde{A} = \begin{bmatrix} \tilde{A}_{11} & \tilde{A}_{12} \\ 0 & \tilde{A}_{22} \end{bmatrix} \quad \text{and} \quad \tilde{B} = \begin{bmatrix} \tilde{B}_1 \\ 0 \end{bmatrix}$$

where $\tilde{A}_{11} \in \mathbb{R}^{n_c \times n_c}$ and $\tilde{B}_1 \in \mathbb{R}^{n_c \times m}$ and the pair $(\tilde{A}_{11}, \tilde{B}_1)$ is controllable. The matrix $\tilde{A}_{11}$ is upper block-Hessenberg with full row rank subdiagonal blocks and the matrix $\tilde{B}_1$ is upper trapezoidal.

If the matrix U is partitioned as

$$U = [U_1, U_2] \qquad U_1 \in \mathbb{R}^{n \times n_c} \tag{7.21}$$

then the controllable subspace R is spanned by the columns of U_1

$$R = \text{Rg}(U_1) = \text{Ker}(U_2^T) \tag{7.22}$$

while its orthogonal complement $R^\perp$ is given by

$$R^\perp = \text{Rg}(U_2) = \text{Ker}(U_1^T) \tag{7.23}$$

Example 7.10

Let

$$
A = \begin{bmatrix} 0 & -3 & 0 & 0 & 0 \\ 1 & -4 & 0 & 0 & 0 \\ 0 & -2 & 0 & 0 & -8 \\ 0 & -3 & 1 & 0 & -14 \\ 0 & -1 & 0 & 1 & -7 \end{bmatrix} \quad \text{and} \quad B = \begin{bmatrix} 3 & 1 \\ 1 & 1 \\ -2 & 0 \\ 0 & 0 \\ 0 & 0 \end{bmatrix}
$$

Using orthogonal transformation with matrix

$$
U = \begin{bmatrix} -0.802 & -0.154 & -0.159 & 0.329 & 0.447 \\ -0.267 & -0.772 & 0.159 & -0.329 & -0.447 \\ 0.535 & -0.617 & -0.159 & 0.329 & 0.447 \\ 0 & 0 & -0.875 & 0.185 & -0.447 \\ 0 & 0 & -0.398 & -0.801 & 0.447 \end{bmatrix}
$$

the pair (A, B) is reduced into the form $(\tilde{A}, \tilde{B})$, where (see Example 4.25)

$$
\tilde{A} = \left[\begin{array}{cccc:c} -0.429 & -1.81 & 2.13 & 2.55 & -3.11 \\ -0.660 & -3.57 & -1.08 & -5.78 & -0.276 \\ -1.45 & -1.94 & -5.52 & -14.4 & 5.73 \\ 0.384 & 0 & 0.556 & -0.481 & 0.505 \\ \hdashline 0 & 0 & 0 & 0 & -1.00 \end{array} \right]
$$

$$
\tilde{B} = \left[\begin{array}{cc} -3.74 & -1.07 \\ 0 & -0.926 \\ 0 & 0 \\ 0 & 0 \\ \hdashline 0 & 0 \end{array} \right]
$$

Thus $n_c = 4$ and the controllable subspace R is spanned by the first four columns of U. The fifth column of U spans $R^{\perp}$.

The unobservable subspace of (7.14) is computed in a similar way. Let $(V^T A V, CV)$ be the orthogonal canonical form of the pair (A, C), where $V \in \mathbb{R}^{n \times n}$ is an orthogonal matrix (see Section 4.4). Partition V as

$$
V = [V_1, V_2] \qquad V_2 \in \mathbb{R}^{n \times n_o} \tag{7.24}
$$

where $n_o = \dim(N)$. Then

$$
N = \text{Rg}(V_2) = \text{Ker}(V_1^T) \tag{7.25}
$$

and

$$N^\perp = \text{Rg}(V_1) = \text{Ker}(V_2^\text{T}) \tag{7.26}$$

Using the partitioning (7.21) and (7.24) of the orthogonal matrices U and V one gets orthonormal bases of the four subspaces L_i in (7.17) according to (7.18). In particular, using (7.22), (7.23), (7.25) and (7.26) we obtain

$$L_1 = \text{Ker}(M^\text{T}) = \text{Ker}(W_1^\text{T}) = \text{Rg}(W_2)$$

$$L_1^\perp = \text{Rg}(M) = \text{Rg}(W_1) = \text{Ker}(W_2^\text{T})$$

$$L_2 = L_1^\perp \cap R = \text{Ker}\begin{bmatrix} W_2^\text{T} \\ U_2^\text{T} \end{bmatrix}$$

$$= (L_1 + R^\perp)^\perp = (\text{Rg}\,[W_2, U_2])^\perp$$

$$L_3 = L_1^\perp \cap N = \text{Ker}\begin{bmatrix} W_2^\text{T} \\ V_1^\text{T} \end{bmatrix}$$

$$= (L_1 + N^\perp)^\perp = (\text{Rg}\,[W_2, V_1])^\perp$$

$$L_4 = (L_1 \oplus L_2 \oplus L_3)^\perp = L_1^\perp \cap L_2^\perp \cap L_3^\perp$$

where $M = [U_2, V_1]$ and W_1 and W_2 are orthonormal bases of $\text{Rg}(M)$ and $\text{Rg}(M)^\perp$, respectively.

In this way we have the following algorithm.

Algorithm 7.2 Computing the Kalman decomposition of a linear system

Compute orthonormal bases U_1 and U_2 of the controllable and uncontrollable subspaces, respectively, using Algorithm 4.8

Compute orthonormal bases V_1 and V_2 of the observable and unobservable subspaces, respectively, using Algorithm 4.8 for the pair (A^T, C^T)

Set $M = [U_2, V_1]$

Compute orthonormal bases W_1 and W_2 of $\text{Rg}(M)$ and $\text{Rg}(M)^\perp$, respectively, using QR decomposition with column pivoting or SVD

Take $S_1 = W_2$ as orthonormal basis of L_1

Compute S_2 as orthonormal basis for $L_2 = (\text{Rg}\,[W_2, U_2])^\perp$ using QR decomposition with column pivoting

Compute S_3 as orthonormal basis for $L_3 = (\text{Rg}\,[W_2, V_1])^\perp$

Compute S_4 as orthonormal basis for the orthogonal complement of

$$L_1 \oplus L_2 \oplus L_3, \ (L_1 \oplus L_2 \oplus L_3)^{\perp} = (\mathrm{Rg}\,[S_1, S_2, S_3])^{\perp}$$

Compute $\tilde{A} = S^{-1}AS$, $\tilde{B} = S^{-1}B$ and $\tilde{C} = CS$, where $S = [S_1, S_2, S_3, S_4]$

Since Algorithm 7.2 produces orthonormal bases S_i for the subspaces L_i, the resulting transformation matrix S is 'almost' an orthogonal matrix in the sense that

$$S_i^T S_i = I_{n_i} \qquad i = 1, 2, 3, 4$$
$$S_1^T S_j = 0 \qquad j = 2, 3, 4$$
$$S_4^T S_j = 0 \qquad j = 1, 2, 3$$

Only the product $Z = S_2^T S_3 \in \mathbb{R}^{n_2 \times n_3}$ may be nonzero, since L_2 and L_3 need not be orthogonal. It may be shown (see Boley (1984)) that the matrix S is best conditioned among all matrices that reduce (7.14) into the form (7.19) and (7.20).

Let $\sigma_1 \geqq \sigma_2 \geqq \cdots \geqq \sigma_\nu$ be the nonzero singular values of Z. Then the principal angles

$$\theta_i = \arccos(\sigma_i) \qquad i = 1, 2, \ldots, \nu$$

between L_2 and L_3 can be considered as internal characteristics of the system. The smallest angle between L_2 and L_3 is $\theta_1 = \arccos(\sigma_1) > 0$.
Having in mind that

$$S^T S = \begin{bmatrix} I_{n_1} & 0 & 0 & 0 \\ 0 & I_{n_2} & Z & 0 \\ 0 & Z^T & I_{n_3} & 0 \\ 0 & 0 & 0 & I_{n_4} \end{bmatrix}$$

we see that

$$\mathrm{cond}_2(S) = \mathrm{cond}_2 \begin{bmatrix} I_{n_2} & Z \\ Z^T & I_{n_3} \end{bmatrix} = \mathrm{cond}_2 \begin{bmatrix} I_\nu & \Sigma \\ \Sigma & I_\nu \end{bmatrix}$$

where $\Sigma = \mathrm{diag}(\sigma_1, \sigma_2, \ldots, \sigma_\nu)$. Since the eigenvalues of

$$\begin{bmatrix} I_\nu & \Sigma \\ \Sigma & I_\nu \end{bmatrix}$$

are $1 \pm \sigma_i$, $i = 1, 2, .., \nu$, we see that

$$\mathrm{cond}_2(S) = \sqrt{\left(\frac{1 + \sigma_1}{1 - \sigma_1}\right)} = \sqrt{\left(\frac{1 + \|Z\|_2}{1 - \|Z\|_2}\right)}$$

Thus if the subspaces L_2 and L_3 lie 'near' each other, then the matrix S, yielding the Kalman decomposition, is ill conditioned.

Example 7.11

Consider an eighth-order system with matrices

$$A = \begin{bmatrix} -10 & -2 & 12 & 14 & 34 & 22 & 38 & 16 \\ 4 & 1 & -4 & -6 & -22 & -10 & -20 & -8 \\ -6 & 2 & 3 & 10 & 14 & 14 & 12 & 16 \\ -1 & -6 & 4 & -1 & 24 & 6 & 25 & 0 \\ -4 & 1 & 1 & 8 & 5 & 8 & 2 & 8 \\ 1 & 4 & -5 & 0 & -21 & -3 & -22 & 3 \\ 2 & -1 & 1 & -4 & -6 & -6 & -1 & -8 \\ -4 & 1 & 7 & 6 & 14 & 6 & 16 & 3 \end{bmatrix}$$

$$B = \begin{bmatrix} 6 & 12 \\ -7 & -8 \\ 7 & 8 \\ 4 & 10 \\ -2 & 1 \\ -2 & -6 \\ -1 & -4 \\ 1 & 2 \end{bmatrix}$$

$$C = \begin{bmatrix} 2 & -2 & -3 & -4 & 1 & -1 & 0 & 0 \\ 3 & -11 & -6 & -6 & 4 & 0 & 9 & -5 \end{bmatrix}$$

Using Algorithm 4.8 we find that $\dim(R) = 5$, $\dim(N) = 4$. In the given case Algorithm 7.2 yields (up to three decimal digits)

$$S_1 = \begin{bmatrix} -0.550 & 0.705 \\ -0.394 & -0.307 \\ 0.394 & 0.307 \\ -0.275 & 0.353 \\ 0.0 & 0.0 \\ -0.394 & -0.307 \\ 0.0 & 0.0 \\ 0.394 & 0.307 \end{bmatrix}$$

$$S_2 = \begin{bmatrix} 0.212 & -0.0391 & 0.272 \\ -0.327 & 0.224 & 0.256 \\ 0.327 & -0.224 & -0.256 \\ -0.424 & 0.0783 & -0.544 \\ -0.404 & -0.519 & 0.363 \\ 0.592 & -0.273 & 0.0838 \\ 0.209 & 0.719 & 0.0632 \\ -0.0620 & 0.175 & 0.596 \end{bmatrix}$$

$$S_3 = \begin{bmatrix} -0.0432 & 0.251 \\ 0.266 & 0.154 \\ -0.375 & 0.474 \\ 0.0864 & -0.502 \\ -0.533 & -0.307 \\ -0.375 & 0.474 \\ 0.533 & 0.307 \\ 0.266 & 0.154 \end{bmatrix} \qquad S_4 = \begin{bmatrix} -0.277 \\ 0 \\ 0 \\ 0.555 \\ 0 \\ 0.555 \\ 0 \\ 0.555 \end{bmatrix}$$

As a result we obtain

$$\tilde{A} = S^{-1}AS = \begin{bmatrix} -0.569 & 0.106 & -2.87 & -0.617 & 1.14 & -0.363 & -1.59 & 0.696 \\ 6.37 & 0.569 & 2.37 & 5.88 & 19.6 & -8.97 & 12.5 & 38.1 \\ 0 & 0 & -0.587 & 0.839 & 0.0637 & 0 & 0 & -3.99 \\ 0 & 0 & 2.93 & 2.18 & -7.00 & 0 & 0 & 2.15 \\ 0 & 0 & -0.864 & 0.855 & -4.59 & 0 & 0 & 6.28 \\ 0 & 0 & 0 & 0 & 0 & 1.04 & -0.00476 & -27.8 \\ 0 & 0 & 0 & 0 & 0 & 3.75 & 0.958 & 10.2 \\ 0 & 0 & 0 & 0 & 0 & 0 & 0 & -2.0 \end{bmatrix}$$

$$\tilde{B} = S^{-1}B = \begin{bmatrix} 2.30 & 1.116 \\ 10.9 & 19.4 \\ 3.51 & -1.38 \\ -2.02 & -4.68 \\ -4.49 & -5.48 \\ 0 & 0 \\ 0 & 0 \\ 0 & 0 \end{bmatrix} \qquad \text{and}$$

$$\tilde{C} = CS = \begin{bmatrix} 0 & 0 & 0.797 & -0.414 & 3.26 & 0 & 0 & -3.33 \\ 0 & 0 & 5.39 & 1.81 & 1.84 & 0 & 0 & -6.94 \end{bmatrix}$$

In the given case

$$Z = \begin{bmatrix} -0.167 & 0.830 \\ 0.961 & 0.157 \\ 0.0732 & 0.299 \end{bmatrix}$$

$\sigma_1(Z) = 0.982$ and the smallest angle between L_2 and L_3 is $\theta_1 = 0.191$ rad. The matrix S is well conditioned, $\text{cond}_2(S) = 10.4$.

More detailed information about the structure of the spaces R and N can be obtained by using the concepts of ith controllable and ith unobservable subspaces. Specifically, the subspace

$$R_i = \sum_{j=1}^{i-1} A^j \text{Rg}(B) = \text{Rg}\,[B, AB, ..., A^{i-1}B]$$

is said to the *ith controllable subspace* of (7.14) while

$$N_i = \sum_{j=0}^{i-1} \text{Ker}(CA^j) = \text{Ker} \begin{bmatrix} C \\ CA \\ \vdots \\ CA^{i-1} \end{bmatrix}$$

is the *ith unobservable subspace.*
We have that

$$R_1 \subset R_2 \subset \cdots \subset R_p = R_{p+1} = \cdots = R_n$$
$$N_1 \supset N_2 \supset \cdots \supset N_q = N_{q+1} = \cdots = N_n$$

where p and q are the controllability and observability indices of the system, respectively.

Orthonormal bases of the ith controllable and unobservable subspaces can be obtained reliably by using the orthogonal canonical forms of the system. Let U and V be the orthogonal transformation matrices which reduce the pairs (A, B) and (A, C), respectively, in orthogonal canonical forms. Partitioning these matrices as

$$U = [\underbrace{U_{11}}_{m_1}, \underbrace{U_{12}}_{m_2}, ..., \underbrace{U_{1p}}_{m_p}, \underbrace{U_2}_{n - n_c}]$$

$$V = [\underbrace{V_{11}}_{r_1}, \underbrace{V_{12}}_{r_2}, ..., \underbrace{V_{1q}}_{r_q}, \underbrace{V_2}_{n_o}]$$

where $m_1, ..., m_p$ and $r_1, ..., r_q$ are the conjugate Kronecker indices of the pairs (A, B) and (A, C), respectively (see Section 4.4), we have that

$$
\begin{aligned}
R_1 &= \text{Rg}(B) = \text{Rg}(U_{11}) \\
R_2 &= \text{Rg}\,[B, AB] = \text{Rg}\,[U_{11}, U_{12}] \\
&\ \vdots \\
R_p &= \text{Rg}\,[B, AB, ..., A^{p-1}B] = \text{Rg}\,[U_{11}, U_{12}, ..., U_{1p}]
\end{aligned}
\tag{7.27}
$$

and

$$
N_1 = \text{Ker}(C) = \text{Rg}\,[V_{12}, V_{13}, ..., V_{1q}, V_2]
$$

$$
N_1 = \text{Ker} \begin{bmatrix} C \\ CA \end{bmatrix} = \text{Rg}\,[V_{13}, V_{14}, ..., V_{1q}, V_2]
$$

$$
\vdots
$$

$$
N_q = \text{Ker} \begin{bmatrix} C \\ CA \\ \vdots \\ CA^{q-1} \end{bmatrix} = \text{Rg}(V_2)
$$

$$\tag{7.28}$$

Consider finally a measure of the nearness between A-invariant and controllable, respectively, unobservable subspaces. If L is an A-invariant subspace, then it may coincide or may be 'near' to certain R_i or N_j. In the latter case we speak about 'nearly controllable' or 'nearly unobservable' A-invariant subspaces.

Let $\gamma(L, M)$ be the gap between the subspaces $L, M \subset \mathbb{R}^n$. The A-invariant subspace $L \subset \mathbb{R}^n$ is said to be *nearly controllable* (with respect to a small positive constant $\mu > 0$) if

$$\gamma(L, R_i) < \mu$$

for some i. Similarly, L is referred to as *nearly unobservable* if

$$\gamma(L, N_j) < \mu$$

for some j. In the above definitions we assume that $\dim(L) = \dim(R_i)$ or $\dim(L) = \dim(N_j)$, respectively. Note that if an A-invariant subspace is nearly controllable then the given system is near to an uncontrollable system.

Example 7.12

Given is a stable, controllable system with matrices

$$A = \begin{bmatrix} 13 & 20 & -36 & -30 & 40 \\ 0 & 3 & -4 & 0 & 8 \\ -6 & -8 & 13 & 12 & -16 \\ 11 & 14 & -26 & -24 & 28 \\ -4 & -8 & 12 & 8 & -17 \end{bmatrix} \text{ and } B = \begin{bmatrix} -0.02 & -1 \\ -2 & -1 \\ 0 & 1.01 \\ 0 & -1 \\ 1 & 1 \end{bmatrix}$$

The eigenvalues of A are

$$-1, -1, -2, -3, -5$$

and the matrix of corresponding eigenvectors is (to three decimal digits)

$$W = \begin{bmatrix} -1.64 & -9.50 & 2.36 & 2.40 & 0.768 \\ 3.04 & 4.21 & 0 & 1.20 & 0.128 \\ 1.64 & 9.50 & 0 & -0.601 & -0.256 \\ -1.64 & -9.50 & 1.18 & 1.20 & 0.512 \\ -0.702 & 2.64 & 0 & -1.20 & -0.256 \end{bmatrix}$$

Orthonormal bases of the controllable subspaces are found by taking the corresponding columns of the orthogonal matrix

$$
U = \begin{bmatrix}
-0.446 & 0.326 & -0.703 & 0.411 & 0.177 \\
-0.446 & -0.784 & -0.0555 & -0.134 & 0.407 \\
0.451 & -0.341 & -0.694 & -0.279 & -0.348 \\
-0.446 & 0.337 & -0.0676 & -0.818 & -0.118 \\
0.446 & 0.223 & -0.125 & -0.260 & 0.817
\end{bmatrix}
$$

which transforms the pair (A, B) to canonical form with $m_1 = 2$, $m_2 = 2$ and $m_3 = 1$.

Denoting by

$$
\begin{aligned}
L_1 &= \text{span}\{w_1, w_2\} & L_2 &= \text{span}\{w_1, w_3\} \\
L_3 &= \text{span}\{w_3, w_4\} & L_4 &= \text{span}\{w_4, w_5\} \\
L_5 &= \text{span}\{w_1, w_2, w_3, w_4\} & L_6 &= \text{span}\{w_1, w_3, w_4, w_5\}
\end{aligned}
$$

the corresponding invariant subspaces of A, and computing the gaps between the invariant and controllable subspaces by using orthogonal projections, we obtain that

$$
\begin{aligned}
\gamma(L_1, R_1) &= 0.0111 & \gamma(L_2, R_1) &= 0.776 \\
\gamma(L_3, R_1) &= 0.737 & \gamma(L_4, R_1) &= 0.413 \\
\gamma(L_5, R_2) &= 0.539 & \gamma(L_6, R_2) &= 0.260
\end{aligned}
$$

Thus the invariant subspace associated with eigenvalues $\{-1, -1\}$ is nearly controllable, since it lies near to $\text{Rg}(B)$.

EXERCISES

7.12 Show that

$$
N = \left(\sum_{j=0}^{n-1} \text{Rg}\,[C^{\mathrm{T}}(A^{\mathrm{T}})^j] \right)^{\perp}
$$

7.13 Discuss the influence of perturbations in A and B on $\dim(R)$ and $\dim(N)$.

7.14 Write a computer program performing the Kalman decomposition according to Algorithm 7.2.

7.15 Verify (7.27) and (7.28).

7.16 For particular small perturbations in the matrices of the system considered in Example 7.12, find the changes in the gaps between the invariant and controllable subspaces.

7.4 COMPUTING CONTROLLED INVARIANT AND CONTROLLABILITY SUBSPACES

Several important properties of a linear control system may be formulated in terms of its trajectories as sets of points in the state space. This approach gives rise to a number of new geometric objects, which, in particular, are subspaces of the state space.

Consider the system

$$\dot{x}(t) = Ax(t) + Bu(t) \qquad x(0) = x_0$$
$$y(t) = Cx(t)$$

(7.29)

where $x(t) \in \mathbb{R}^n$, $u(t) \in \mathbb{R}^m$, $y(t) \in \mathbb{R}^r$ and $A \in \mathbb{R}^{n \times n}, B \in \mathbb{R}^{n \times m}, C \in \mathbb{R}^{r \times m}$. We shall deal with controls determined by

$$u(t) = Kx(t) + Gv(t)$$

(7.30)

where $K \in \mathbb{R}^{m \times n}$ is the feedback matrix, $G \in \mathbb{R}^{m \times m}$ is the input gain matrix and v belongs to the set Ω of continuous vector functions $w : [0, \infty] \rightarrow \mathbb{R}^m$. The trajectory of the state vector $x_u(t)$, corresponding to the control $u \in \Omega$, is denoted by

$$X_u = \{x_u(t) : t \geqq 0\}$$

Extensions of the concepts of A-invariant and controllable subspaces associated with the system (7.29) and (7.30) are the controlled invariant and controllability subspaces.

A subspace $L \subset \mathbb{R}^n$ is said to be *controlled invariant* (or (A, B)-*invariant*) if it is $(A + BK)$-invariant for some matrix $K \in \mathbb{R}^{m \times n}$, i.e. if $AL \subset L + \mathrm{Rg}(B)$.

It may be shown that $L \subset \mathbb{R}^n$ is a controlled invariant subspace if and only if for each $x_0 \in L$ there exists $u \in \Omega$ such that $X_u \subset L$. Thus a controlled invariant subspace is simply an *invariant set* for the solutions of (7.29).

For every subspace $N \subset \mathbb{R}^n$ there is a unique largest controlled invariant subspace L_N^* contained in N. This follows from the fact that the set of all controlled invariant subspaces is closed under addition (i.e. the sum $L = L_1 + L_2$ of two controlled invariant subspaces L_1 and L_2 is again controlled invariant). Of particular interest here is the space

$$L^* = L_{\mathrm{Ker}(C)}^*$$

which is referred to as the *supremal controlled invariant subspace* for the system (7.29).

If $x_0 \in L$, where $L \subset \mathbb{R}^n$ is not a controlled invariant subspace, then the trajectory X_u does not lie in L. However, there exist subspaces L_a such that every trajectory starting from L_a remains arbitrary close to

L_a in the sense of the following definition. A subspace $L_a \subset \mathbb{R}^n$ is said to be *almost controlled invariant* if for each $x_0 \in L_a$ and $\eta > 0$ there exists a control $u \in \Omega$ such that $d(x(t), L_a) \leq \eta$, $t \geq 0$, where

$$d(x, Z) = \min\{\|x - z\| : z \in Z\}$$

is the distance from the point $x \in \mathbb{R}^n$ to the set $Z \subset \mathbb{R}^n$.

A subspace M is said to be a *controllability subspace* if it is the controllable subspace

$$\mathrm{Rg}[BG, (A + BK)BG, \ldots, (A + BK)^{n-1}BG]$$

of the pair $(A + BK, BG)$ for some matrices $K \in \mathbb{R}^{m \times n}$, $G \in \mathbb{R}^{m \times m}$ (note that G may be singular).

Obviously, the controllable subspace R of (A, B) is a controllability subspace (corresponding to an arbitrary K and a nonsingular G).

A trajectory characterization of a controllability subspace is as follows: $M \subset \mathbb{R}^n$ is a controllability subspace if for each x_0, $x_T \in L$, there exist $T > 0$ and $u \in \Omega$ such that $x(T) = x_T$ and $X_u \subset M$.

As for the controlled invariant subspaces, given a subspace $N \subset \mathbb{R}^n$ there exists a unique largest controllability subspace M_N^* contained in N. The subspace

$$M^* = M_{\mathrm{Ker}(C)}^*$$

is of special interest and it is known under the name *supremal controllability subspace*. Since every controllability subspace is also a controlled invariant one, it follows that $M^* \subset L^*$.

If a point $x_T \in \mathbb{R}^n$ may be reached from x_0, and x_0 and x_T lie in a subspace M_a which is not a controllability one, then no trajectory X_u, connecting x_0 and x_T, lies in M_a. Nevertheless, any two points from certain subspaces M_a (regarded as sets in $\mathbb{R}^n$) may be connected by trajectories which lie in an arbitrary small neighbourhood of M_a. More precisely, a subspace $M_a \subset \mathbb{R}^n$ is referred to as an *almost controllability subspace* if for each $x_0, x_T \subset M_a$ and $\eta > 0$ there exists a control $u \in \Omega$ such that $x(T) = x_T$ for some $T > 0$ and $d(x(t), M_a) \leq \eta$, $t \geq 0$.

A number of properties pertaining to controlled invariant and controllability subspaces are considered in the exercises at the end of this section.

Similarly to the existence of the supremal subspaces L_N^* and M_N^* it may be shown that there exist *supremal almost controlled invariant subspace* $L_{a,N}^*$ and *supremal almost controllability subspace* $M_{a,N}^*$ contained in a subspace $N \subset \mathbb{R}^n$. Special notations L_a^* and M_a^* are reserved for the subspaces $L_{a,\mathrm{Ker}(C)}^*$ and $M_{a,\mathrm{Ker}(C)}^*$.

The four subspaces L_N^*, M_N^*, $L_{a,N}^*$ and $M_{a,N}^*$ may be obtained from

$$L_N^* = L^{(\nu)} = L^{(\nu+1)} = \cdots$$
$$M_{a,N}^* = M^{(\nu)} = M^{(\nu+1)} = \cdots$$
$$L_{a,N}^* = L_N^* + M_{a,N}^* \tag{7.31}$$
$$M_N^* = L_N^* \cap M_{a,N}^*$$

where ν is at most $\dim(N)$ and $L^{(k)}$ and $M^{(k)}$ are determined by the recurrence relations

$$L^{(0)} = N$$
$$L^{(k+1)} = N \cap A^{-1}[L^{(k)} + \mathrm{Rg}(B)] \subset L^{(k)} \tag{7.32}$$

$$M^{(0)} = 0$$
$$M^{(k+1)} = N \cap [AM^{(k)} + \mathrm{Rg}(B)] \supset M^{(k)} \tag{7.33}$$

The computations in (7.31)–(7.33) involve intersection and summation of subspaces and may be done reliably using QR or singular value decomposition (see Section 7.1).

It should be pointed out that generically (i.e. with probability 1 if the elements of the matrices A, B and C are considered as random variables) one has

$$M^* = M_a^* = L^* = L_a^* = 0 \qquad\qquad \text{if } m < r$$
$$M^* = M_a^* = 0 \qquad L^* = L_a^* = \mathrm{Ker}(C) \qquad \text{if } m = r$$
$$M^* = M_a^* = L^* = L_a^* = \mathrm{Ker}(C) \qquad \text{if } m > r$$

Example 7.13

If

$$A = \begin{bmatrix} 1 & 0 & 0 & 0 & 0 & 0 & 0 & 0 \\ 1 & -6 & -7 & -6 & -2 & -4 & 0 & 4 \\ 1 & -15 & -14 & -14 & -6 & -8 & 0 & 8 \\ 5 & 10 & 6 & 7 & 4 & 4 & 0 & -4 \\ -7 & 0 & 4 & 2 & -1 & 2 & 0 & -2 \\ 0 & 0 & 0 & 0 & 0 & -1 & 0 & 0 \\ 0 & 0 & 0 & 0 & 0 & 0 & 1 & 0 \\ 7 & -20 & -24 & -22 & -8 & -14 & 0 & 13 \end{bmatrix}$$

$$B = \begin{bmatrix} -1 & 2 \\ -3 & 2 \\ -7 & 8 \\ 2 & -1 \\ 4 & -5 \\ -2 & 2 \\ 1 & -1 \\ -13 & 14 \end{bmatrix}$$

$$C = \begin{bmatrix} 0 & -2 & -2 & -2 & -1 & -1 & 0 & 1 \end{bmatrix}$$

then using (7.32) and (7.33) we obtain that (to three digits)

$$L^* = L^{(0)} = \mathrm{Rg}(U_1)$$

$$U_1 = \begin{bmatrix} 0.516 & 0.516 & 0.516 & 0.258 & 0.258 & 0 & -0.258 \\ 0.733 & -0.267 & -0.267 & -0.133 & -0.133 & 0 & 0.133 \\ -0.267 & 0.733 & -0.267 & -0.133 & -0.133 & 0 & 0.133 \\ -0.267 & -0.267 & 0.733 & -0.133 & -0.133 & 0 & 0.133 \\ -0.133 & -0.133 & -0.133 & 0.933 & -0.0667 & 0 & 0.0667 \\ -0.133 & -0.133 & -0.133 & -0.0667 & 0.933 & 0 & 0.0667 \\ 0 & 0 & 0 & 0 & 0 & 1 & 0 \\ 0.133 & 0.133 & 0.133 & 0.0667 & 0.0667 & 0 & 0.933 \end{bmatrix}$$

$$M_a^* = M^{(2)} = \mathrm{Rg}(U_2)$$

$$U_2 = \begin{bmatrix} -0.001\,77 & 0.415 \\ 0.449 & -0.496 \\ 0.446 & 0.334 \\ -0.449 & 0.496 \\ -0.446 & -0.334 \\ 0 & 0 \\ 0 & 0 \\ 0.446 & 0.334 \end{bmatrix}$$

Utilizing (7.31) we find that

$$L_a^* = L^*$$
$$M^* = M_a^*$$

EXERCISES

7.17 Prove the trajectory characterizations of controlled invariant and controllability subspaces.

7.18 Show that

$$\{L\} \quad \overset{\{L_a\}}{\underset{\{M\}}{}} \quad \{M_a\}$$

where $\{L\}$ is the set of all controlled invariant subspaces L and so on.

7.19 Show that $\tilde{M}_a \in \{M_a\}$ if and only if there exist a matrix K and a chain $Rg(B) \supset N_0 \supset N_1 \supset \cdots \supset N_{n-1}$ of subspaces $N_i \subset \mathbb{R}^n$ such that

$$\tilde{M}_a = \sum_{i=0}^{n-1} (A + BK)^i N_i$$

7.20 Prove that $\tilde{L} \in \{L_a\}$ and $\tilde{L} \cap Rg(B) = 0$ imply $\tilde{L} \in \{L\}$.

7.21 Verify (7.31)–(7.33).

7.22 Develop an algorithm and a computer program to determine supremal subspaces via (7.31)–(7.33).

NOTES AND REFERENCES

The definition and properties of basic geometric objects arising in linear spaces may be found in Halmos (1974), Strang (1976), Kato (1984) and Gohberg, Lancaster and Rodman (1986). The application of QR and singular value decomposition to the computation of subspaces, projections, distances and angles is considered in Golub and Van Loan (1983), Björck and Golub (1973) and Klema and Laub (1980). A perturbation analysis of projections and related objects is presented in Stewart (1977).

The geometric approach to linear systems theory is presented in depth in Wonham (1986). In this book one can find a detailed consideration of the basic subspaces of linear multivariable systems as well as their application to the solution of several analysis and design problems.

The method for computing the Kalman decomposition, presented in Section 7.3, is due to Boley (1984). The concept of nearly controllable and nearly unobservable subspaces is introduced in Lindner, Babendreier and Hamden (1989).

The concepts of controlled invariant and controllability subspaces were introduced by Basile and Marro (1969), while almost controlled invariant and controllability subspaces were studied in Willems (1980, 1981). The computation of supremal controlled invariant and controllability subspaces was considered by Moore and Laub (1978).

Supremal controlled invariant and controllability subspaces are used in solving problems of the analysis and design of linear control systems, such as disturbance decoupling (Wonham, 1986), disturbance decoupling with simultaneous pole assignment (Willems and Commault, 1981), Linnemann (1987), and several others.

Computation of several objects arising in the geometric theory of linear systems is considered in Laub (1978), Van Dooren (1981a) and Demmel and Kågström (1986).

APPENDIX

Sources of Information about Software for Analysis and Design of Linear Control Systems

The development of software for the analysis and design of control systems is done by many individuals and professional groups around the world. The achievements and prospects in the area of Computer-Aided Control System Design (CACSD) are discussed on the IFAC Symposia on CACSD, held in the years 1979–88 (Cuenod, 1980; Leininger, 1983; Larsen and Hansen, 1986), on the IEEE Control Systems Society Workshops and Symposia, held in the years 1981–88, as well as on many conferences and informal meetings. To this subject are devoted the Special Issue on CACSD of the *Control Systems Magazine* (December 1982) and the Special Section on CACSD in the *Proceedings of the IEEE* (December 1984). A thorough and up-to-date account of Computer-Aided Control System Engineering is done in Jamshidi and Herget (1985). This volume contains nineteen contributions from leading experts in the field as well as Software Summaries describing 37 packages for CACSD. A large number of software libraries and packages is described in the *Inventory of Basic Software for CACSD*, published by the Benelux Working Group on Software (WGS) in 1986. The most complete description of software for CACSD up to now is *The Extended List of Control Software (ELCS)* edited by D.K. Frederick, C.J. Herget, R. Kool and M. Rimvall. *ELCS* is a collection of software summaries, divided into the following five different sections:

1. Standard subroutine libraries, containing programs for solving standard mathematical problems.
2. Subroutine libraries, containing algorithms for solving control problems.
3. Software packages for CACSD.

4. MATLAB-family software packages, representing interactive CACSD packages which are extensions of MATLAB or use MATLAB's command mode of interaction.
5. Simulation packages, offering the user some kind of modelling/ simulation environment.

Currently, ELCS contains summaries of 90 libraries, packages and interactive systems for CACSD. A copy of ELCS can be obtained from the following individuals:

1. In the United States:

Professor Dean K. Frederik
ECSE Department Rensselaer
Polytechnic Institute
Troy, NY 12180, USA

Dr Charles J. Herget
Lawrence Livermore National
Laboratory
PO Box 808, L-156
Livermore, CA 94550, USA

2. Elsewhere:

R. Kool, Secretary of the WGS
Eindhoven University of
Technology
Department of Mathematics
and Computer Science
PO Box 513
NL-6600 MB Eindhoven
The Netherlands

References

ANDERSON, B.D.O. (1978) 'Second order convergent algorithms for the steady state Riccati equation', *Int. J. Control*, **28**, pp. 295–306.

ANDERSON, B.D.O. and J.B. MOORE (1971) *Linear Optimal Control*, Prentice Hall: Englewood Cliffs, NJ.

ANDERSON, B.D.O. and J.B. MOORE (1979) *Optimal Filtering*, Prentice Hall: Englewood Cliffs, NJ.

ARGOUN, M.B. (1986) 'On sufficient conditions for the stability of interval matrices', *Int. J. Control*, **44**, pp. 1245–50.

ARMSTRONG, E.S. (1975) 'An extension of Bass' algorithm for stabilizing linear continuous constant systems', *IEEE Trans. Autom. Control*, **AC-20**, pp. 153–4.

ARMSTRONG, E.S. and G.T. RUBLEIN (1976) 'A stabilization algorithm for linear discrete constant systems', *IEEE Trans. Autom. Control*, **AC-21**, pp. 629–31.

ARNOLD, W.F. (1983) 'On the numerical solution of algebraic matrix Riccati equations', *Ph.D. Thesis*, Dept of Electr. Eng. Systems, Univ. of South California.

ARNOLD, W.F. and A.J. LAUB (1984) 'Generalized eigenproblem algorithms and software for algebraic Riccati equations', *Proc. IEEE*, **72**, pp. 1746–54.

ÅSTRÖM, K.J. and B. WITTENMARK (1986) *Computer Controlled Systems*, Prentice Hall: Hemel Hempstead.

BAI, Z., J. DEMMEL and A. McKENNEY (1989) *On the Conditioning of the Nonsymmetric Eigenproblem: Theory and software*, LAPACK Working Note 13, Courant Institute: New York.

BARNETT, S. and R.G. CAMERON (1985) *Introduction to Mathematical Control Theory* (2nd edn), Clarendon Press: Oxford.

BARNETT, S. and C. STOREY (1970) *Matrix Methods in Stability Theory*, Nelson: London.

BARRAUD, A.Y. (1977) 'A numerical algorithm to solve $A^T X A - X = Q$', *IEEE Trans. Autom. Control*, **AC-22**, pp. 883–5

BARTELS, R.H. and G.W. STEWART (1972) 'Algorithm 432: solution of the matrix equation $AX + XB = C$', *Comm. ACM*, **15**, pp. 820–6.

BASILE, G. and G. MARRO (1969) 'Controlled and conditioned invariant subspaces in linear system theory', *J. Optim. Theory Appl.*, **3**, pp. 306–15.

BAVELY, C.A. and G.W. STEWART (1979) 'An algorithm for computing reducing subspaces by block diagonalization', *SIAM J. Numer. Anal.*, **16**, pp. 359–67.

BEAVERS, A. and E. DENMAN (1974) 'A new solution method for quadratic matrix equations', *Math. Biosci.*, **20**, pp. 135–43.

BENDER, D.J. and A.J. LAUB (1987a) 'The linear-quadratic optimal regulator for descriptor systems: discrete-time case', *Automatica*, **23**, pp. 71–85.

BENDER, D.J. and A.J. LAUB (1987b) 'The linear-quadratic optimal regulator for descriptor systems', *IEEE Trans. Autom. Control*, **AC-32**, pp. 672–88.

BIERMAN, G.J. (1984) 'Computational aspects of the matrix sign function solution to the ARE', in *Proc. 23rd IEEE CDC*, Las Vegas, pp. 514–19.

BJÖRCK, Å. and G.H. GOLUB (1973) 'Numerical methods for computing angles between linear subspaces', *Math. Comput.*, **27**, pp. 579–94.

BOLEY, D.L. (1981) 'Computing the controllability–observability decomposition of a linear time-invariant dynamic system, a numerical approach', *Ph.D. Thesis*, Dept of Comput. Sci., Stanford Univ.

BOLEY, D.L. (1984) 'Computing the Kalman decomposition: an optimal method', *IEEE Trans. Autom. Control*, **AC-29**, pp. 51–3.

BOLEY, D.L. (1985) 'A perturbation result for linear control problems', *SIAM J. Alg. Discr. Meth.*, **6**, pp. 66–72.

BOLEY, D.L. and W.S. LU (1986) 'Measuring how far a controllable system is from an uncontrollable one', *IEEE Trans. Autom. Control*, **AC-31**, pp. 249–51.

BROCKETT, R.W. (1970) *Finite Dimensional Linear Systems*, Wiley: New York.

BRUNOVSKY, P. (1970) 'A classification of linear controllable systems', *Kybernetika*, **6**, pp. 173–88.

BUNSE-GERSTNER, A. and V. MEHRMANN (1986) 'A symplectic QR like algorithm for the solution of the real algebraic Riccati equation', *IEEE Trans. Autom. Control*, **AC-31**, pp. 1104–13.

BYERS, R. (1983) 'Hamiltonian and symplectic algorithms for the algebraic Riccati equation', *Ph.D. Thesis*, Dept of Comput. Sci., Cornell Univ.

BYERS, R. (1984) 'A LINPACK-style condition estimator for the equation $AX - XB^T = C$', *IEEE Trans. Autom. Control*, **AC-29**, pp. 926–8.

BYERS, R. (1985) 'Numerical condition of the algebraic Riccati equation', *Contemporary Math.*, **47**, pp. 35–49.

BYERS, R. (1986a) 'A Hamiltonian QR algorithm', *SIAM J. Sci. Stat. Comput.*, **7**, pp. 212–29.

BYERS, R. (1986b) 'Numerical stability and instability in matrix sign function based algorithms', in *Computational and Combinatorial Methods in System Theory*, C.I. BYRNES and A. LINDQUIST (eds.), pp. 185–200, North-Holland: New York.

BYERS, R. (1987) 'Solving the algebraic Riccati equation with the matrix sign function', *Lin. Alg. & Appl.*, **85**, pp. 267–79.

BYERS, R. (1988) 'A bisection method for measuring the distance of a stable matrix to the unstable matrices', *SIAM. J. Sci. Stat. Comput.*, **9**, pp. 875–81.

BYERS, R. and S.G. NASH (1989) 'Approaches to robust pole assignment', *Int. J. Control*, **49**, pp. 97–117.

CADZOW, J.A. (1973) *Discrete-Time Systems*, Prentice Hall: Englewood Cliffs, NJ.

CHAN, S.P., R. FELDMAN and B.N. PARLETT (1977) 'Algorithm 517: A program for computing the condition numbers of matrix eigenvalues without computing eigenvectors', *ACM Trans. Math. Software*, **3**, pp. 186–203.

CHARLIER, J.-B. and P. VAN DOOREN (1989) 'A systolic algorithm for Riccati and Lyapunov equations', *Math. Control Signals Systems*, **2**, pp. 109–36.

CHEN, C.T. (1984) *Linear System Theory and Design*, Holt, Rinehart & Winston: New York.

CHU, K.-W.E. (1988) 'A controllability condensed form and a state feedback pole assignment algorithm for descriptor systems', *IEEE Trans. Autom. Control*, **AC-33**, pp. 366–70.

COBB, J.D. (1981) 'Feedback and pole-placement in descriptor variable systems', *Int. J. Control*, **33**, pp. 1135–46.

COBB, J.D. (1984) 'Controllability, observability and duality in singular systems', *IEEE Trans. Autom. Control*, **AC-29**, pp. 1076–82.

CODY, W.J. (1988) 'Algorithm 665: MACHAR: A subroutine to dynamically determine machine parameters', *ACM Trans. Math. Software*, **14**, pp. 303–11.

COX, C.L. and W.F. MOSS (1989) 'Backward error analysis for a pole assignment algorithm', *SIAM J. Matrix Anal. Appl.*, **10**, pp. 446–56.

CUENOD M.A. (ed.) (1980) *Proc. IFAC Symp. on Computer Aided Design of Control Systems*, Zurich, 29–31 August, Pergamon Press: Oxford.

DATTA, B.N. (1987) 'An algorithm to assign eigenvalues in a Hessenberg matrix: single input case', *IEEE Trans. Autom. Control*, **AC-32**, pp. 414–17.

DAVIS, G.J. and C.B. MOLER (1978) Sensitivity of matrix eigenvalues. *Int. J. Numer. Meth. Engr.*, **12**, pp. 1367–73.

DAVISON, E.J. and M.C. MAKI (1973) 'The numerical solution of the matrix Riccati differential equation', *IEEE Trans. Autom. Control*, **AC-18**, pp. 71–3.

DEMMEL, J.W. (1987a) 'A counter example for two conjectures about stability', *IEEE Trans. Autom. Control*, **AC-32**, pp. 340–2.

DEMMEL, J.W. (1987b) 'On condition numbers and the distance to the nearest ill-posed problem', *Numer. Math.*, **51**, pp. 251–89.

DEMMEL, J.W. and B. KÅGSTRÖM (1986) 'Accurate solution of ill-posed problems in control theory', in *Proc. 25th IEEE CDC*, Athens, 10–12 Dec., pp. 558–63.

DESOER, C.A. (1970) *Notes for a Second Course on Linear Systems*, Van Nostrand Reinhold: New York.

DONGARRA, J.J., J.R. BUNCH, C.B. MOLER AND G.W. STEWART (1979) *LINPACK Users' Guide*, SIAM: Philadelphia.

EISING, R. (1984a) 'A collection of numerically reliable algorithms for the deadbeat control problem', *System & Control Lett.*, **4**, pp. 189–93.

EISING, R. (1984b) 'Between controllable and uncontrollable', *Systems & Control Lett.*, **4**, pp. 263–4.

ENRIGHT, W. (1979) 'On the efficient and reliable numerical solution of large systems of ODE's', *IEEE Trans. Autom. Control*, **AC-24**, pp. 905–8.

FALLSIDE, F. (ed.) (1977) *Control System Design by Pole-Zero Assignment*, Academic Press: London.

FARIAS, M.C. and S.P. BINGULAC (1978) 'On the calculation of initial guess for iterative solution of Riccati equation', *Regelungstechnik*, Heft 1, pp. 28–30.

FARRAR, F.A. and R.C. DI PIETRO (1977) 'Comparative evaluation of numerical methods for solving the algebraic matrix Riccati equation', in *Proc. 1977 Joint ACC*, San Francisco, pp. 1543–8.

FERNANDO, K.Y.M. and N. NICHOLSON (1978) 'Modified Newton's algorithm for the Riccati equation', *Electronic Lett.*, **14**, pp. 576–8.

FLAMM, D.S. and R.A. WALKER (1982) 'Remark on Algorithm 506', *ACM Trans. Math. Software*, **8**, pp. 219–20.

FLETCHER, L.R., J. KAUTSKY, G.K.G. KOLKA and N.K. NICHOLS (1985) 'Some necessary and sufficient conditions for eigenstructure assignment', *Int. J. Control*, **42**, pp. 1457–68.

FORSYTHE, G.E. and MOLER C.B. (1967) *Computer Solution of Linear Algebraic Systems*, Prentice Hall: Englewood Cliffs, NJ.

FORSYTHE, G.E., M.A. MALCOLM AND C.B. MOLER (1977) *Computer Methods for Mathematical Computations*, Prentice Hall: Englewood Cliffs, NJ.

FRANKLIN, G.F. and J.D. POWELL (1980) *Digital Control of Dynamic Systems*, Addison-Wesley: Reading, MA.

GARBOW, B.S., J.M. BOYLE, J.J. DONGARRA and C.B. MOLER (1977) *Matrix Eigensystem Routines: EISPACK guide extension*, Springer-Verlag: Berlin.

GARDINER, J.D. and A.J. LAUB (1986) 'A generalization of the matrix-sign-function solution for algebraic Riccati equations', *Int. J. Control*, **44**, pp. 823–32.

GEAR. C.W. (1971) *Numerical Initial Value Problems in Ordinary Differential Equations*, Prentice Hall: Englewood Cliffs, NJ.

GENTLEMAN, V.M. and S.B. MAROVICH (1974) 'More on algorithms that reveal properties of floating-point arithmetic units', *Comm. ACM*, **17**, pp. 276–7.

GOHBERG, I., P. LANCASTER and L. RODMAN (1986) *Invariant Subspaces of Matrices with Applications*, John Wiley: New York.

GOLUB, G.H., S. NASH and C. VAN LOAN (1979) 'A Hessenberg–Schur method for the problem $AX + XB = C$', *IEEE Trans. Autom. Control*, **AC-24**, pp. 909–13.

GOLUB, G.H. and C.F. VAN LOAN (1983) *Matrix Computations*, The John Hopkins Univ. Press: Baltimore.

GOLUB, G.H. and J.H. WILKINSON (1976) 'Ill-conditioned eigensystems and the computation of the Jordan canonical form', *SIAM Rev.*, **18**, pp. 578–619.

HALMOS, P.R. (1974) *Finite-Dimensional Vector Spaces* (2nd edn), Springer-Verlag: New York.

HAMMARLING, S.J. (1982a) 'Numerical solution of the stable, nonnegative definite Lyapunov equation', *IMA J. Numer. Anal.*, **2**, pp. 303–23.

HAMMARLING, S.J. (1982b) *Newton's Method for Solving the Algebraic Riccati Equation*, Nat. Phys. Lab. Rpt DITC 12/82.

HAUTUS, M.L.J. (1969) 'Controllability and observability conditions of linear autonomous systems', *Proc. Kon. Nederl. Akad. Wetensch.*, ser. A, **72**, pp. 443–8.

HEALEY, M. (1973) 'Study of methods of computing transition matrices.' *Proc. IEE*, **120**, pp. 905–12.

HEATH, M.T., A.J. LAUB, C.C. PAIGE and R.C. WARD (1986) 'Computing the singular value decomposition of a product of two matrices', *SIAM J. Sci. Stat. Comput.*, **7**, pp. 1147–59.

HEWER, G.A. (1971) 'An iterative technique for the computation of the steady state gains for the discrete optimal regulator', *IEEE Trans. Autom. Control*, **AC-16**, pp. 382–3.

HEWER, G. and C. KENNEY (1988) 'The sensitivity of the stable Lyapunov equation', *SIAM J. Control & Optim.*, **26**, pp. 321–44.

HEWER, G. A. and G. NAZAROFF (1973) *A Survey of Numerical Methods for the Solution of Algebraic Matrix Riccati Equations*, Rpt. Naval Weapons Center, China Lake, CA.

HINRICHSEN, D. and A.J. PRITCHARD (1986) 'Stability radii of linear systems', *Systems & Control Lett.*, **7**, pp. 1–10.

HITZ, K. and B.D.O. ANDERSON (1972) 'Iterative method of computing the limiting solution of the matrix Riccati differential equation', *Proc. IEE*, **119**, pp. 1402–6.

HO, W.C. and L.R. FLETCHER (1988) 'Perturbation theory of output feedback pole assignment', *Int. J. Control*, **48**, pp. 1075–88.

HORN, R.A. and C.R. JOHNSON (1986) *Matrix Analysis*, Cambridge Univ. Press: Cambridge.

INCERTIS, F.C. (1983) 'An extension of a new formulation of the algebraic Riccati equation problem', *IEEE Trans. Autom. Control*, **AC-28**, pp. 235–8.

JAMSHIDI, M. (1980) 'An overview on the solutions of the algebraic matrix Riccati equation and related problems', *Large Scale Syst.*, **1**, pp. 167–92.

JAMSHIDI, M. and C.J. HERGET (eds.) (1985) *Computer-Aided Control Systems Engineering*, North-Holland: Amsterdam.

JOHNSON, J.C. and C.L. PHILIPS (1971) 'An algorithm for the computation of the integral of the state transition matrix', *IEEE Trans. Autom. Control*, **AC–18**, pp. 204–5.

JONES, E.L. (1976) 'A reformulation of the algebraic Riccati equation problem', *IEEE Trans. Autom. Control*, **AC-21**, pp. 113–14.

JURY, E.I. (1974) *Inners and Stability of Dynamic Systems*, Wiley-Interscience: New York.

KÅGSTRÖM, B. (1977) 'Bounds and perturbation bounds for the matrix exponential', *BIT*, **17**, pp. 39–57.

KÅGSTRÖM, B. (1986) 'RGSVD – An algorithm for computing the Kronecker structure and reducing subspaces of singular $A - \lambda B$ pencils', *SIAM J. Sci. Stat. Comput.*, **7**, pp. 185–211.

KÅGSTRÖM, B. and A. RUHE (1980) 'An algorithm for numerical computation of the Jordan normal form of a complex matrix', *ACM Trans. Math. Software*, **6**, pp. 398–419; 'Algorithm 560: JNF, an algorithm for numerical computation of the Jordan form of a complex matrix', *ibid.*, pp. 437–43.

KÅGSTRÖM, B. and L. WESTIN (1989) 'Generalized Schur methods with condition estimators for solving the generalized Sylvester equation', *IEEE Trans. Autom. Control*, **AC-34**, pp. 745–51.

KAILATH, T. (1980) *Linear Systems*, Prentice Hall: Englewood Cliffs, NJ.

KALMAN, R.E. (1963) 'Mathematical description of linear dynamical systems', *SIAM J. Control*, **1**, pp. 152–92.

KALMAN, R.E. and T.S. ENGLAR (1966) *A User's Manual for the Automatic Synthesis Program*, NASA Rpt CR-475, Washington.

KALMAN, R.E., P.L. FALB and M.A. ARBIB (1969) *Topics in Mathematical System Theory*, McGraw-Hill: New York.

KATO, T. (1984) *Perturbation Theory for Linear Operators* (2nd edn), Springer-Verlag: New York.

KAUTSKY, J. and N.K. NICHOLS (1984) *Robust Multiple Eigenvalue Assignment by State Feedback in Linear Systems*, Num. Anal. Rpt 7/84, Dept Math., Univ. of Reading.

KAUTSKY, J., N.K. NICHOLS and P. VAN DOOREN (1985) 'Robust pole assignment in linear state feedback', *Int. J. Control*, **41**, pp. 1129–55.

KENNEY, C. and G. HEWER (1987) 'Sensitivity of algebraic Riccati equations', in *Proc. 26th IEEE CDC*, Los Angeles, Dec., pp. 814–15.

KENNEY, C., A.J. LAUB and E.A. JONCKHEERE (1989) 'Positive and negative solutions of dual Riccati equations by matrix sign function iteration', *Systems & Control Lett.*, **13**, pp. 109–16.

KENNEY, C., A.J. LAUB and M. WETTE (1989) 'A stability-enhancing scaling procedure for Schur–Riccati solvers', *Systems & Control Lett.*, **12**, pp. 241–50.

KENNEY, C. S. and R.B. LEIPNIK (1985) 'Numerical integration of the differential matrix Riccati equation', *IEEE Trans. Autom. Control*, **AC-30**, pp. 962–70.

KHARITONOV, V.L. (1978) 'Asymptotic stability of an equilibrium position of a family of linear differential equations', *Differential Equations*, **14**, pp. 2086–8.

KLEIN, G. and B.C. MOORE (1977) 'Eigenvalue-generalized eigenvector assignment with state feedback', *IEEE Trans. Autom. Control*, **AC-22**, pp. 140–1.

KLEINMAN, D.L. (1968) 'On an iterative technique for Riccati equation computations', *IEEE Trans. Autom. Control*, **AC-13**, pp. 114–15.

KLEINMAN, D.L. (1970) 'An easy way to stabilize a linear constant system', *IEEE Trans. Autom. Control*, **AC-15**, p. 692.

KLEINMAN, D.L. (1974) 'Stabilizing a discrete, constant linear system with application to iterative methods for solving the Riccati equation', *IEEE Trans. Autom. Control*, **AC-19**, pp. 252–4.

KLEMA, V.C. and A.J. LAUB (1980) 'The singular value decomposition: its computation and some applications', *IEEE Trans. Autom. Control*, **AC-25**, pp. 164–76.

KONSTANTINOV, M.M., N.D. CHRISTOV and P.Hr. PETKOV (1987). 'Perturbation analysis of linear control problems', *Prepr. IFAC 10th Congress*, Munich, 27–31 July, vol. 9, pp. 16–21, Pergamon Press: Oxford.

KONSTANTINOV, M.M., P.Hr. PETKOV and N.D. CHRISTOV (1980) 'Synthesis of linear systems with desired equivalent form', *J. Comput. & Appl. Math.*, **6**, pp. 27–35.

KONSTANTINOV, M.M., P.Hr. PETKOV and N.D. CHRISTOV (1981) 'Invariants and canonical forms for linear multivariable systems under the action of orthogonal transformation groups', *Kybernetika*, **17**, pp. 413–24.

KONSTANTINOV, M.M., P.Hr. PETKOV and N.D. CHRISTOV (1982) 'Orthogonal invariants and canonical forms for linear controllable systems', in *Proc. 8th IFAC Congress*, Kyoto, August 1981, vol. 1, pp. 49–54, Pergamon Press: Oxford.

KONSTANTINOV, M.M., P.Hr. PETKOV and N.D. CHRISTOV (1986) 'Perturbation analysis of the continuous and discrete matrix Riccati equations', in *Proc. 1986 ACC*, Seattle, vol. 1, pp. 636–9.

KUČERA, V. (1972) 'A contribution to matrix quadratic equations', *IEEE Trans. Autom. Control*, **AC-17**, pp. 344–7.

KUNKEL, P. and V. MEHRMANN (1989) *Numerical Solution of Differential Algebraic Riccati Equations*, Techn. Rpt TR 89.03.006, Heidelberg Sci. Center: Heidelberg.

KUO, B.C. (1980) *Digital Control Systems*, Holt, Rinehart and Winston: New York.

KWAKERNAAK, H. and R. SIVAN (1972) *Linear Optimal Control Systems*, Wiley-Interscience: New York.

LANCASTER, P. (1969) *Theory of Matrices*, Academic Press: New York.

LARSEN, P.M. and N.E. HANSEN (eds.) (1986) *Proc. 3rd IFAC/IFIP Symp. on Computer Aided Design in Control and Engineering Systems*, Lyngby, 31 July–2 August 1985, Pergamon Press: Oxford.

LASTMAN, G.J. and N.K. SINHA (1985) 'An error analysis of the balanced matrix method of modal reduction, with application to the selection of reduced-order models', *Large Scale Systems*, **9**, pp. 63–71.

LAUB, A.J. (1978) *Linear Multivariable Control. Numerical Considerations*. Rpt ESL–P–833, Electronic Syst. Lab., MIT: Cambridge.

LAUB, A.J. (1979) 'A Schur method for solving algebraic Riccati equations', *IEEE Trans. Autom. Control*, **AC-24**, pp. 913–21.

LAUB, A.J. (1980) 'On computing "balancing" transformations', in *Proc. 1980 Joint ACC*, San Francisco, pp. FA8–E.

LAUB, A.J. (1982) 'Schur techniques for Riccati differential equations', in *Feedback Control of Linear and Nonlinear Systems*, D. HINRICHSEN and A. ISIDORI (eds.), pp. 165–74, Springer-Verlag: New York.

LAUB, A.J. (1983) 'Numerical aspects of solving algebraic Riccati equations', in *Proc. 22nd IEEE CDC*, San Antonio, Texas, Dec., pp. 184–6.

LAUB, A.J. (1985) Numerical linear algebra aspects of control design computations', *IEEE Trans. Autom. Control*, **AC-30**, pp. 97–108.

LAUB, A.J., M.T. HEATH, C.C. PAIGE and R.C. WARD (1987) 'Computation of system balancing transformations and other applications of simultaneous diagonalization algorithms', *IEEE Trans. Autom. Control*, **AC-32**, pp. 115–22.

LAUB, A.J. and A. LINNEMANN (1986) 'Hessenberg and Hessenberg/triangular forms in linear system theory', *Int. J. Control*, **44**, pp. 1523–47.

LAWSON, C.L. and R.J. HANSON (1974) *Solving Least Squares Problems*, Prentice-Hall: Englewood Cliffs, NJ.

LAYTON, J.M. (1976) *Multivaribale Control Theory*, Peter Peregrinus: Stevenage.

LEE, T.-T. and G.-T. LIAW (1986) 'Pole assignment in real Schur form', *Int. J. Systems Sci.*, **17**, pp. 337–43.

LEHNIGK, S.H. (1966) *Stability Theorems for Linear Motions with an Intro-duction to Lyapunov's Direct Method*, Prentice Hall: Englewood Cliffs, NJ.

LEIGH, J.R. (1985) *Applied Digital Control. Theory, Design and Implemen-tation*. Prentice Hall: Hemel Hempstead.

LEININGER, G.G. (ed.) (1983) *Proc. 2nd IFAC Symp. on Computer Aided Design of Multivariable Technol. Systems*, Purdue Univ., West Lafayette, 15–17 Sept. 1982, Pergamon Press: Oxford.

LEWKOWICZ, I. and R. SIVAN (1988) 'Maximal stability robustness for state equations', *IEEE Trans. Autom. Control*, **AC-33**, pp. 297–300.

LINDNER, D.K., J. BABENDREIER and A.M.A. HAMDEN (1989) 'Measures of control-lability and observability and residues', *IEEE Trans. Autom. Control*, **AC-34**, pp. 648–50.

LINNEMANN, A. (1987) 'A condensed form for disturbance decoupling with simultaneous pole placement using state feedback', *Prepr. IFAC 10th Congress*, Munich, 27–31 July, vol. 9, pp. 92–7.

LUENBERGER, D.G. (1966) 'Observers for multivariable systems', *IEEE Trans. Autom. Control,* **AC-11**, pp. 190–7.

LUENBERGER, D.G. (1967) 'Canonical forms for linear multivariable systems', *IEEE Trans. Autom. Control*, **AC-12**, pp. 290–3.

LUENBERGER, D.G. (1971) 'An introduction to observers', *IEEE Trans. Autom. Control*, **AC-16**, pp. 596–602.

LUENBERGER, D.G. (1978) 'Time-invariant descriptor systems', *Automatica*, **14**, pp. 473–80.

MacFARLANE, A.G.J. (1963) 'An eigenvector solution of the optimal linear regulator problem', *J. Electron. Control*, **14**, pp. 643–54.

MAHMOUD, M.S. and M.G. SINGH (1981) *Large Scale Systems Modelling*, Pergamon Press: Oxford.

MARCUS, M. and H. MINC (1964) *A Survey of Matrix Theory and Matrix Inequalities*, Allyn & Bacon: Boston.

MAROULAS, J. and St. BARNETT (1979) 'Canonical forms for time-invariant linear control systems: a survey with extensions. Part I. Single-input systems', *Int. J. Systems Sci.*, **9**, pp. 497–514; 'Part II. Multivariable case', *ibid.*, **10**, pp. 33–50.

MÅRTENSSON, K. (1971) 'On the matrix Riccati equation', *Inform. Sci.*, **3**, pp. 17–49.

MARTIN, J.M. and G.A. HEWER (1987) 'Smallest destabilizing perturbations for linear systems', *Int. J. Control*, **45**, pp. 1495–1504.

MERRIAM, C.W. (1974) *Automated Design of Control Systems*, Gordon & Breach: New York.

MIMINIS, G.S. and C.C. PAIGE (1982) 'An algorithm for pole assignment of time invariant linear systems', *Int. J. Control*, **35**, pp. 341–54.

MIMINIS, G.S. and C.C. PAIGE (1988) 'A direct algorithm for pole assignment of time-invariant multi-input linear systems using state feedback', *Automatica*, **24**, pp. 343–56.

MISRA, P. and R.J. PATEL (1989) 'Numerical algorithms for eigenvalue assign-ment by constant and dynamic output feedback', *IEEE Trans. Autom. Control*, **AC-34**, pp. 579–88.

MOLER, C.B. (1982) *MATLAB User's Guide*, Dept of Comput Sci., Univ. of New Mexico.

MOLER, C.B., J. LITTLE and S. BANGERT (1987) *PC-MATLAB User's Guide*, The MathWorks, Inc.: Sherborn, MA.

MOLER, C.B. and G.W. STEWART (1973) 'An algorithm for generalized matrix eigenvalue problem', *SIAM J. Numer. Anal.*, **10**, pp. 241–56.

MOLER, C.B. and C.F. VAN LOAN (1978) 'Nineteen dubious ways to compute the exponential of a matrix', *SIAM Rev.*, **20**, pp. 801–36.

MOORE, B.C. (1976) 'On the flexibility offered by state feedback in multivariable systems beyond closed-loop eigenvalue assignment', *IEEE Trans. Autom. Control*, **AC-21**, pp. 689–92.

MOORE, B.C. (1981) 'Principal component analysis in linear systems: controllability, observability and model reduction', *IEEE Trans. Autom. Control*, **AC-26**, pp. 17–32.

MOORE, B.C. and A.J. LAUB (1978) 'Computation of supremal (A, B)-invariant and controllability subspaces', *IEEE Trans. Autom. Control*, **AC-23**, pp. 783–92.

NICHOLS, N.K. (1986) 'On computational algorithms for pole assignment', *IEEE Trans. Autom. Control*, **AC-31**, pp. 643–5.

OGATA, K. (1987) *Modern Control Engineering*, Prentice Hall: Hemel Hempstead.

O'REILLY, J. (1983) *Observers for Linear Systems*, Academic Press: London.

ORTEGA, J.M. (1988) *Introduction to Parallel and Vector Solution of Linear Systems*, Plenum Press: New York.

ORTEGA, J.M. and W.G. POOLE, Jr. (1981) *Numerical Methods for Differential Equations*, Pitman Publishing: London.

OSHMAN, Y. and I.Y. BAR-ITZHACK (1985) 'Eigenfactor solution of the matrix Riccati equation – a continuous square root algorithm', *IEEE Trans. Autom. Control*, **AC-30**, pp. 971–8.

OWENS, D.H. (1981) *Multivariable and Optimal Systems*, Academic Press: London.

PAIGE, C.C. (1981) 'Properties of numerical algorithms related to computing controllability', *IEEE Trans. Autom. Control*, **AC-26**, pp. 130–8.

PAIGE, C.C. and C.F. VAN LOAN (1981) 'A Schur decomposition for Hamiltonian matrices', *J. Linear Algebra & Appl.*, **41**, pp. 11–32.

PAPPAS, T., A.J. LAUB and N.R. SANDELL, Jr. (1980) 'On the numerical solution of the discrete time algebraic Riccati equation', *IEEE Trans. Autom. Control*, **AC-25**, pp. 631–41.

PARLETT, B.N. (1976) 'A recurrence among the elements of functions of triangular matrices', *Linear Algebra & Appl.*, **29**, pp. 323–46.

PATEL, R.V. (1985) 'Algorithms for eigenvalue assignment in multivariable systems', in *Computer Aided Control Systems Engineering*, M. Jamshidi and C.J. Herget (eds.), pp. 315–45, North-Holland: Amsterdam.

PATEL, R.V. and N. MUNRO (1982) *Multivariable System Theory and Design*, Pergamon Press: Oxford.

PEARCE, C.E.M. (1986) 'On the solution of a class of algebraic matrix Riccati equations', *IEEE Trans. Autom. Control*, **AC-31**, pp. 252–5.

PERNEBO, L. and L.M. SILVERMAN (1982) 'Modal reduction via balanced state space representations', *IEEE Trans. Autom. Control*, **AC-27**, pp. 382–7.

PETKOV, P.Hr., N.D. CHRISTOV and M.M. KONSTANTINOV (1984a) 'A computational algorithm for pole assignment of linear single-input systems', *IEEE Trans. Autom. Control*, **AC-29**, pp. 1045–8.

PETKOV, P.Hr., N.D. CHRISTOV and M.M. KONSTANTINOV (1984b) 'Remark on Algorithm 590', *ACM Trans. Math. Software*, **10**, p. 207.

PETKOV, P.Hr., N.D. CHRISTOV and M.M. KONSTANTINOV (1984c) 'Contributions to the synthesis of state observers', in *Proc. 6th Congress on Cybern. & Systems*, 10–14 Sept., Paris, vol. 1, pp. 389–94.

PETKOV, P.Hr., N.D. CHRISTOV and M.M. KONSTANTINOV (1986a) 'Numerical analysis of the reduction of linear systems into orthogonal canonical form', *Systems & Control Lett.*, **7**, pp. 361–4.

PETKOV, P.Hr., N.D. CHRISTOV and M.M. KONSTANTINOV (1986b) 'A computational algorithm for pole assignment of linear multiinput systems', *IEEE Trans. Autom. Control*, **AC-31**, pp. 1044–7.

PETKOV, P.Hr., N.D. CHRISTOV and M.M. KONSTANTINOV (1987) 'On the numerical properties of the Schur approach for solving the matrix Riccati equation', *Systems & Control Lett.*, **9**, pp. 197–201.

PETKOV, P.Hr., N.D. CHRISTOV and M.M. KONSTANTINOV (1988a) 'Comments on "On computational algorithms for pole assignment"', *IEEE Trans. Autom. Control*, **AC-33**, pp. 892–3.

PETKOV, P.Hr., N.D. CHRISTOV and M.M. KONSTANTINOV (1988b) 'Algorithms to estimate the distance between a stable matrix to the nearest unstable one', in *Proc. 1988 ACC*, Atlanta, 15–17 June, pp. 1508–9.

POPOV V.M. (1972) 'Invariant description of linear time invariant controllable systems', *SIAM J. Control*, **10**, pp. 252–84.

PORTER, B. and R. CROSSLEY (1972) *Modal Control Theory and Applications*, Taylor & Francis: London.

POTTER, J.E. (1966) 'Matrix quadratic solutions', *SIAM J. Appl. Math.*, **14**, pp. 496–501.

PRUSSING, J.E. (1972) 'A simplified method for solving the matrix Riccati equation', *Int. J. Control*, **15**, pp. 995–1000.

REPPERGER, D.W. (1976) 'A square root of a matrix approach to obtain the solution to a steady state matrix Riccati equation', *IEEE Trans. Autom. Control*, **AC-21**, pp. 786–7

ROBERTS, J.D. (1980) 'Linear model reduction and solution of the algebraic Riccati equation by use of the sign function', *Int. J. Control*, **32**, pp. 677–87.

ROSENBROCK, H.H. (1970) *State–Space and Multivariable Theory*, Nelson: London.

RUHE, A. (1970) 'An algorithm for numerical determination of the structure of a general matrix', *BIT*, **10**, pp. 196–216.

SAAD, Y. (1988) 'Projection and deflation methods for partial pole assignment in linear state feedback', *IEEE Trans. Autom. Control*, **AC-33**, pp. 290–7.

SAFONOV, M.G. and R.Y. CHIANG (1989) 'A Schur method for balanced-truncation model reduction', *IEEE Trans. Autom. Control*, **AC-34**, pp. 729–33.

SAGE, A.P. and C.C. WHITE (1977) *Optimum Systems Control* (2nd edn), Prentice Hall: Englewood Cliffs, NJ.

SAHA, R.K. and J.R. BORIOTTI (1981) 'Truncation and round-off errors in computation of matrix exponentials', *Int. J. Control*, **33**, pp. 137–47.

SANDELL, N.R., Jr. (1974) 'On Newton's method for Riccati equation solution', *IEEE Trans. Autom. Control*, **AC-19**, pp. 254–5.

SHAFAI, B. and S.P. BHATTACHARYYA (1988) 'An algorithm for pole assignment in high order multivariable systems', *IEEE Trans. Autom. Control*, **AC-33**, pp. 870–6.

SHAMPINE, L.F. and C.W. GEAR (1979) 'A user's views of solving stiff ordinary differential equations', *SIAM Rev.*, **21**, pp. 1–17.

SHAMPINE, L.F. and M.K. GORDON (1975) *Computer Solution of Ordinary Differential Equations: The initial value problem*, W.H. Freeman: San Francisco.

SHAMPINE, L.F. and H.A. WATTS (1977) 'The art of writing a Runge–Kutta code. Part I', in *Mathematical Software III*, J.R. Rice (ed.), pp. 257–75, Academic Press: New York; 'Part II', *Appl. Math. & Comput.*, **5**, pp. 93–121.

SHAMPINE, L.F. and H.A. WATTS (1979) *DEPAC – Design of a User Oriented Package Solvers*, Sandia Lab. Rpt SAND 79-2374, Albuquerque, NM.

SHAMPINE, L.F., H.A. WATTS and S.M. DAVENPORT (1976) 'Solving nonstiff ordinary differential equations – the state of the art', *SIAM Rev.*, **18**, pp. 376–411.

SHOKOOHI, S., L.M. SILVERMAN and P. VAN DOOREN (1983) 'Linear time-variable systems: balancing and model reduction', *IEEE Trans. Autom. Control*, **AC-28**, pp. 810–22.

SHUBERT, M.A. (1974) 'An analytic solution for an algebraic Riccati equation', *IEEE Trans. Autom. Control*, **AC-19**, pp. 255–6.

SIMA, V. (1981) 'An efficient Schur method to solve the stabilizing problem', *IEEE Trans. Autom. Control*, **AC-26**, pp. 724–5.

SINCOVEC, R.F., A.M. ERISMAN, E.L. YIP and M.A. EPTON (1981) 'Analysis of descriptor systems using numerical algorithms', *IEEE Trans. Autom. Control*, **AC-26**, pp. 139–47.

SINGER, M.A. and S.J. HAMMARLING (1983) *The Algebraic Riccati Equation: A summary review of some available results*, National Physical Laboratory Rpt DITC 23/83.

SINHA, P.K. (1984) *Multivariable Control: An introduction*, Marcel Dekker: New York.

SMITH, B.T., J.M. BOYLE, J.J. DONGARRA, B.S. GARBOW, Y. IKEBE, V.C. KLEMA and C.B. MOLER (1976) *Matrix Eigensystem Routines: EISPACK guide*, Springer-Verlag: Berlin.

STEWART, G.W. (1972) 'On the sensitivity of the eigenvalue problem $Ax = \lambda Bx$', *SIAM J. Numer. Anal.*, **9**, pp. 669–86.

STEWART, G.W. (1973a) 'Error and perturbation bounds for subspaces associated with certain eigenvalue problem', *SIAM Rev.*, **15**, pp. 727–64.

STEWART, G.W. (1973b) *Introduction to Matrix Computations*, Academic Press: New York.

STEWART, G.W. (1976) 'Algorithm 506: HQR3 and EXCHNG: Fortran subroutines for calculating and ordering the eigenvalues of a real upper Hessenberg matrix', *ACM Trans. Math. Software*, **2**, pp. 275–80.

STEWART, G.W. (1977) 'On the perturbation of pseudo-inverses, projections and linear least squares problems', *SIAM Rev.*, **19**, pp. 634–62.

STRANG, G. (1976) *Linear Algebra and its Applications*, Academic Press: New York.

SUN, J.-G. (1987) 'Perturbation of angles between linear subspaces', *J. Comput. Math.*, **5**, pp. 58–61.

THERAPOS, C.P. (1989) 'Balancing transformations for unstable nonminimal linear systems', *IEEE Trans. Autom. Control*, **AC-34**, pp. 455–7.

TSUI, C.-C. (1985) 'A new algorithm for the design of multifunctional observers', *IEEE Trans. Autom. Control*, **AC-30**, pp. 89–93.

TSUI, C.-C. (1986) 'An algorithm for computing state feedback in multiinput linear systems', *IEEE Trans. Autom. Control*, **AC-31**, pp. 243–6.

TSUI, C.-C. (1987) 'A complete analytical solution to the equation TA – FT = LC and its application', *IEEE Trans. Autom. Control*, **AC-32**, pp. 742–4.

VANDERGRAFT, J.S. (1978) *Introduction to Numerical Computations*, Academic Press: New York.

VAN DOOREN, P. (1981a) 'The generalized eigenstructure problem in linear system theory', *IEEE Trans. Autom. Control*, **AC-26**, pp. 111–29.

VAN DOOREN, P. (1981b) 'A generalized eigenvalue approach for solving Riccati equations', *SIAM J. Sci. Stat. Comput.*, **2**, pp. 121–35.

VAN DOOREN, P. (1982) 'Algorithm 590: DSUBSP and EXCHQZ: FORTRAN subroutines for computing deflating subspaces with specified spectrum', *ACM Trans. Math. Software*, **8**, pp. 377–82.

VAN DOOREN, P. (1984a) 'Deadbeat control: a special inverse eigenvalue problem', *BIT*, **24**, pp. 681–99.

VAN DOOREN, P. (1984b) 'Reduced order observers: a new algorithm and proof', *Systems & Control Lett.*, **4**, pp. 243–51.

VAN DOOREN, P., A. EMAMI-NAEINI and L. SILVERMAN (1979) 'Stable extraction of the Kronecker structure of pencils', *Proc. 17th IEEE CDC*, pp. 521–4.

VAN LOAN, C.F. (1977) 'The sensitivity of the matrix exponential', *SIAM J. Numer. Anal.*, **14**, pp. 971–81.

VAN LOAN, C.F. (1978) 'Computing integrals involving the matrix exponential', *IEEE Trans. Autom. Control*, **AC-23**, pp. 395–404.

VAN LOAN, C.F. (1982) *A Symplectic Method for Approximating all the Eigenvalues of a Hamiltonian Matrix*, Tech. Rpt TR 82–494, Dept of Comput. Sci., Cornell Univ.

VAN LOAN, C.F. (1985a) 'How near is a stable matrix to an unstable matrix', *Contemporary Math.*, **47**, pp. 465–78.

VAN LOAN, C.F. (1985b) 'Computing the CS and the generalized singular value decomposition', *Numer. Math.*, **46**, pp. 479–91.

VAN LOAN, C.F. (1987) 'On estimating the condition of eigenvalues and eigenvectors', *Linear Alg. Appl.*, **88/89**, pp. 715–32.

VARGA, A. (1979) 'Numerically reliable algorithm to test controllability', *Electronic Lett.*, **15**, pp. 452–3.

VARGA, A. (1981) 'A Schur method for pole assignment', *IEEE Trans. Autom. Control*, **AC-26**, pp. 517–19.

VAUGHAN, D.R. (1970) 'A nonrecursive algebraic solution for the discrete Riccati equation', *IEEE Trans. Autom. Control*, **AC-15**, pp. 597–9.

VIT, K. (1972) 'Iterative solution of the Riccati equation', *IEEE Trans. Autom. Control*, **AC-17**, pp. 258–9.

WALKER, R.A., A. EMAMI-NAEINI and P. VAN DOOREN (1982) 'A general algorithm for solving the algebraic Riccati equation', in *Proc. 21st IEEE CDC*, Orlando, FL, Dec., pp. 68–72.

WARD, R.C. (1977) 'Numerical computation of the matrix exponential with accuracy estimate', *SIAM J . Numer. Anal.*, **14**, pp. 600–10.

WARD, R.C. (1981) 'Balancing the generalized eigenvalue problem', *SIAM J. Sci. Stat. Comput.*, **2**, pp. 141–52.

WICKS, M.A. and R.A. DECARLO (1987) 'On the distance from an uncontrollable pair: a summary,' in *Proc. 25th Ann. Allert Conf. Commun., Control & Comput.*, Manticello, IL, 30 Sept.–20 Oct., pp. 556–7.

WILKINSON, J.H. (1963) *Rounding Errors in Algebraic Processes*, Prentice Hall: Englewood Cliffs, NJ.

WILKINSON, J.H. (1965) *The Algebraic Eigenvalue Problem*, Clarendon Press: Oxford.

WILKINSON, J.H. (1984) 'Sensitivity of eigenvalues', *Utilitas Math.*, **25**, pp. 5–76.

WILLEMS, J.L. (1970) *Stability Theory of Dynamical Systems*, Nelson: London.

WILLEMS, J.C. (1980) 'Almost A(modB)-invariant subspaces', *Asterisque*, **75/76**, pp. 239–48.

WILLEMS, J.C. (1981) 'Almost invariant subspaces: an approach to high gain feedback design – Part I: almost controlled invariant subspaces', *IEEE Trans. Autom. Control*, **AC-26**, pp. 235–52.

WILLEMS, J.C. and C. COMMAULT (1981) 'Disturbance decoupling by measurement feedback with stability or pole placement', *SIAM J. Control & Optim.*, **19**, pp. 490–504.

WOLOVICH, W.A. (1974) *Linear Multivariable Systems*, Springer-Verlag: New York.

WONHAM, W.M. (1967) 'On pole assignment in multi-input controllable linear systems', *IEEE Trans. Autom. Control*, **AC-12**, pp. 660–5.

WONHAM, W.H. (1986) *Linear Multivariable Control: A geometric approach* (3rd edn), Springer-Verlag: Berlin.

YEDAVALLI, R.K. (1985) 'Improved measures of stability robustness for linear state space models', *IEEE Trans. Autom. Control*, **AC-30**, pp. 577–9.

YEDAVALLI, R.K. (1986) 'Stability analysis of interval matrices: another sufficient condition', *Int. J. Control*, **43**, pp. 767–72.

YOUNG, N.J. (1985) 'Balanced realizations via model operators', *Int. J. Control*, **42**, pp. 369–89.

ZADEH, L.A. and C.A. DESOER (1963) *Linear System Theory: A state–space approach*, McGraw-Hill: New York.

ZHOU, K. and P.P. KHARGONEKAR (1987) 'Stability robustness for linear state-space models with structured uncertainty', *IEEE Trans. Autom. Control*, **AC-32**, pp. 621–3.

Index